ECOMETABOLOMICS

METABOLIC FLUXES VERSUS ENVIRONMENTAL STOICHIOMETRY

SUMIRA JAN

PARVAIZ AHMAD

ELSEVIER

ACADEMIC PRESS

An imprint of Elsevier

Academic Press is an imprint of Elsevier
125 London Wall, London EC2Y 5AS, United Kingdom
525 B Street, Suite 1650, San Diego, CA 92101, United States
50 Hampshire Street, 5th Floor, Cambridge, MA 02139, United States
The Boulevard, Langford Lane, Kidlington, Oxford OX5 1GB, United Kingdom

Library of Congress Cataloging-in-Publication Data
A catalog record for this book is available from the Library of Congress

British Library Cataloguing-in-Publication Data
A catalogue record for this book is available from the British Library

ISBN 978-0-12-814872-3

For information on all Academic Press publications visit our website at https://www.elsevier.com/books-and-journals

Publisher: Charlotte Cockle
Acquisition Editor: Nancy Maragioglio
Editorial Project Manager: Ruby Smith
Production Project Manager: Maria Bernard
Cover Designer: Christian Bilbow

Typeset by SPi Global, India

ECOMETABOLOMICS

DEDICATION

Dedicated to our respected and beloved Teacher Muhammad Iqbal.

The goal of life is to make your heartbeat match the beat of the universe, to match your nature with Nature.

Joseph Campbell

CONTENTS

PREFACE

Ecometabolomics is an emerging new discipline that concentrates on applications of metabolomics, wherein we can integrate metabolomic approaches with ecology to decipher the biochemical interactions of organisms across the diverse spatial and temporal scales. Metabolomics is an untargeted biochemical approach to estimate the numerous thousands of metabolites in different plant and animal species. Any variation in metabolite concentration can contribute to our understanding of mechanistic evidence for biochemical processes that are pertinent at ecological scales. This variation is comprised of physiological, phenotypic, and morphological responses of plants and plant communities to ecological variations, and also their interactions with other organisms. Conventionally, research in biochemistry and ecology represents totally discreet paradigms emerging from two different directions and moving at diverse spatiotemporal scales. Typically, biochemistry adopts the bottom-up approach, commencing from cellular processes and advancing from spatiotemporally refined to expansive scales. On the contrary, ecology is based on extrinsic processes that occur at the population and community levels, and generally investigates the top-down and bottom-up processes together. Ecometabolomics is a trans-disciplinary subject that integrates biochemistry and ecology and inter-relates the discreet spatiotemporal levels. In this book, we have summarized developments and analyzed the progress made in ecometabolomics over the past decade from varied angles. Furthermore, we have tried to cover metabolomics topics such as (1) targeted metabolite profiling and metabolite fingerprinting; (2) an integrated approach to plant biology via multi-analogous methods; (3) metabolomics studies of stress in plants; (4) climatic change and metabolome fluxes; and (5) statistical analyses and bioinformatics software tools. The presented solutions to challenges in these areas of study are likely to advance our understanding by connecting the distinct spatiotemporal scales and bridging the concepts of biochemistry and ecology. We have tried our best to cast a glance over the comprehensive development that has taken place in metabolomic studies, but there is always scope for further improvement. We will be pleased if students or readers/naturalists add to our efforts through valuable comments and constructive criticism.

Dr.Sumira Jan
Dr.Parvaiz Ahmed

ACKNOWLEDGEMENTS

We are thankful to our institutes, ICAR—Central Institute of Temperate Horticulture, and Jamia Hamdard, wherein we have been working on metabolite analysis of medicinal herbs, and have investigated the ecophysiology of diverse plant species. I am grateful to my father and my teachers for inculcating in me the deep reverence for nature and natural laws. I was introduced to the rhapsody of nature using subtle equations of statistics, as instructed by Dr. Gowher Hussain Bhat. Thank you so much for paving solid ground under my wandering and fluid imagination. We are thankful to the reviewers and all the members of the production team at Elsevier.

ABOUT THE BOOK

Ecometabolomics offers an exclusive window into metabolomic investigations concerning organisms working against climatic change under diverse stress conditions. With the advancement in analytical techniques, together with large datasets from varied metabolic fluxes, we now have a snapshot of the metabolome. Hence, we have a holistic view of the metabolism of organisms at the species, population, and ecosystem levels, which paves the way toward fascinating insights about ecology. We can inter-relate metabolomics datasets with omics technologies to enhance our extensive analysis, from phenomes to genomes. Metabolomics thus enables intricate investigations of ecological systems, and provides us with a picture of ecosystem health. This book summarizes the latest research, including an elucidative summary of the techniques and applications of metabolomics with respect to the environment. This book also provides a concise assemblage of the latest data on metabolomics and ecology for researchers, as well as postgraduates, to advance the understanding of metabolomics in relation to biogeocenosis.

CHAPTER 1

Introducing Metabolomics

1.1 INTRODUCTION

The disparity in metabolites varies immensely in response to physical and chemical characteristics that can be analyzed via a metabolomic approach. There are different methods other than the metabolomic approach for metabolite variation, such as genome sequencing using DNA, RNA, and proteins (Lindon, Nicholson, & Holmes, 2011; Okazaki & Saito, 2012). However, metabolomics is entirely analytical, and is comprised of varied analytical strategies. In the past decade we have seen the evolution of numerous analytical approaches, such as chromatography-coupled mass spectrometry-based metabolomic and nuclear magnetic resonance approaches to ascertain non-targeted analysis procedures (Patti, Yanes, & Siuzdak, 2012; Roux, Lison, Junot, & Heilier, 2011). But the core purpose of metabolomics can be achieved via integration of data procured from diverse analytic methods (Allard et al., 2016; Farag, Porzel, Mahrous, Mo'men, & Wessjohann, 2015). Non-targeted analysis of metabolites via a metabolomic approach identifies an extensive range of metabolites in biological samples, and identifies new metabolites that could function as biomarkers as well (Floegel et al., 2013; Naz, Vallejo, García, & Barbas, 2014). However, it is cumbersome to integrate all of the data procured from diverse instruments in varied formats using different instruments and ascertain the identification of unknown metabolites (Koek, Jellema, van der Greef, Tas, & Hankemeier, 2011; Yuan, Breitkopf, Yang, & Asara, 2012). Therefore, it can be conjectured that non-targeted metabolomics of bulk metabolites is not practically feasible. Hence, there is an immediate need for advancing or upgrading the existing methodology for targeted and high throughput metabolomics. During recent years there has been an increased proclivity toward drug discovery, which was made possible through identification, isolation, and characterization of new metabolites (Gika, Theodoridis, Plumb, & Wilson, 2014; Monteiro, Carvalho, Bastos, & Guedes de Pinho, 2013). This has led to the discovery of the global analysis of all metabolites (metabolomics) in a sample, and analysis of all metabolic responses to particular diseases or drugs (metabonomics) (Jorge et al.,

Ecometabolomics
https://doi.org/10.1016/B978-0-12-814872-3.00001-1

2016; Zhang, Lei, Huhman, Sumner, & Zhao, 2015). Metabolomics is actually an extension of existing methods that digs deep into target metabolites and their biochemical transformations. New advancements in mass detectors and approaches, such as liquid chromatography (LC)-mass spectroscopy (MS), gas chromatography (GC)-MS, and capillary electrophoresis (CE)-MS maintain the strategies of estimating the whole metabolome instead of only a target metabolite (Bingol et al., 2015, 2016). The prime significant difference from preexisting methodologies is the concept of displaying metabolite dynamics in an unbiased way, as per the idea of the omics approach (Aw & Fukuda, 2015; Turroni et al., 2016). The metabolomics approach provides a spontaneous, specified, and exact identification of metabolites, followed by pattern recognition of biological samples, characterization of biomarkers, and direct evaluation of biochemical transformations as a result of mutation or drug treatment (Baig, Pechlaner, & Mayr, 2016; Patel & Ahmed, 2015). On the contrary to post-genomic approaches such as proteomics, transcriptomics metabolomics has been established as regular method for procurement of metabolite data (Feussner & Polle, 2015; Keeling et al., 2016). Chromatographic techniques, including LC-MS, GC-MS, and nuclear resonance spectroscopy (NMR) yield hundreds of individual chemical structures that encompass everything from metabolite fingerprinting to metabolite profiling (Rabinowitz et al., 2017; Satou et al., 2014). These methods further evaluate unidentified chemical structures, and can also pave the way toward identification of potential candidates allied with diseases for biological interpretation (reference). Following, basic metabolic characterization metabolomics encompass new advancements in structural elucidation, biostatistics, and metabolite profiling so as to procure a snapshot of metabolic networks (Cavill, Jennen, Kleinjans, & Briedé, 2015; Gioria et al., 2016). The integration of metabolomic data with protein and transcript data can elucidate and simplify the intricacy of most complex regulatory networks within a cell. The essential components of metabolomes reveal the cellular processes that regulate the biochemical phenotype of cells, tissues, or whole organisms. The estimation of intracellular metabolites via qualitative or quantitative methods reflects the metabolic status of an organism, which can subsequently be used to monitor and assess gene function (Cambiaghi, Ferrario, & Masseroli, 2016; Krumsiek, Bartel, & Theis, 2016). Metabolomics emerges as a significant strategy in functional genomics in estimating gene activity, more so than presently used transcriptomics and proteomics approaches. In most research studies using yeast, transcriptomic and proteomic data with mRNA and protein levels utilized cannot be correlated with gene function due to allosteric modulations occurring during transcription, or at the protein level (Fahrmann et al., 2017; Sumner, Lei, Nikolau, & Saito, 2015). The coherent assembly of extensive data sets with metabolomic approaches from carefully designed experiments allows the expansion of propositions by inductive reasoning (Peters, 2017; Villar et al., 2015). The quality of inductive strategy is enhanced by expanding the size of the data set. It is quite difficult to scan an entire metabolome within an organism, because not a single extraction method or analytical instrument can isolate and detect it wholly (Peng, Li, & Peng, 2015). These issues are further amplified by experimental errors that happen during sample preparation, including extraction, due to instrumental maltreatment/poor method development (Kuhlisch & Pohnert, 2015). In order to tackle this issue, a metabolomic standard initiative has been imitated to elaborate the standardization of metabolomic methods from growth of materials and their collection to analyze metabolite levels and data processing (Considine, Thomas, Boulesteix, Khashan, & Kenny, 2018; Fiehn et al., 2007).

Metabolomics is further convoluted by an extensive assortment of metabolites in any plant species containing 100,000–200,000 metabolites. This comes at a considerable price, whether we opt for targeted analysis, or complete metabolic profiling, including primary and secondary metabolism (Haug et al., 2012; Sud et al., 2015). Comprehensive metabolic profiling encompasses a minor fraction of carbohydrates, amino acids, organic acids, vitamins, and diverse classes, for example, phenylpropaniods, terpeniods, alkaloids, and other metabolites from varied organisms (Ma et al., 2009; Schauer et al., 2006). Different metabolomic areas have been utilized, with differing terminology for each metabolic strategy (Table 1). Metabolomic research mainly encompasses metabolite fingerprinting and profiling strategies. Fingerprinting encompasses the detection of metabolites within a sample, excluding their identification. Different techniques such as NMR, Fourier transform infrared (FT-IR) spectroscopy, Raman spectroscopy, and electrospray ionization (ESI)-MS are mostly utilized to screen biological systems to determine whether differential metabolites are present among control and test materials, thereby facilitating the preliminary method preceding expensive metabolite profiling (Boughton, Thinagaran, Sarabia, Bacic, & Roessner, 2016; Cubero-Leon, Peñalver, & Maquet, 2014; Dunn & Ellis, 2005). Metabolic profiling is comprised of detection and estimation, followed by identification via GC or LC, coupled with MS approaches. MS entails detection of metabolites prior to the selection and identification of

Table 1 Summary of the different techniques employed for metabolomic analysis

Technique	Throughput	Advantages	Disadvantages	Cost
High-resolution NMR spectroscopy	~10 min	Instantaneous identification of various metabolites like carbohydrates, amino acids, organic and fatty acids, amines, and lipids without any initial sample pre-treatment Technique is non-destructive Good libraries of spectra Convenient to process	Poor sensitivity Co-resonances for 1D-NMR spectroscopy 2D-NMR spectroscopy is time consuming	Low
In vivo NMR spectroscopy	~30 min with preparation time ~15 min	Employed for recording the working of living tissues Can provide map of metabolite distribution.	Low sensitivity which can be improved by hyper-polarization or higher field intensity	High
High-resolution magic-angle spinning NMR	~15 min	Assessment of cellular environment like compartmentation in living tissue	Tissue cannot be perfused, so their viability is limited	Low

GC-MS	20–30 min for fatty acids 30–45 min for aqueous metabolites	Metabolite characterization is facilitated by acquiring generalized ionization parameters in electron impact and extensive accessibility of spectral libraries	Non-volatile metabolites need to derivatized which is not possible for all metabolite classes	Low-medium
LC-MS	~15–30 min	Reduced ion suppression and segregate isobaric species Appropriate for estimation of intact lipids, di-peptides, tri-peptides, and other macromolecules	Difficult metabolite characterization due to chromatographic drift	Medium
Triple-quadrupole (targeted) MS	15 min per chromatography run ~60 min for more comprehensive screens	Highly sensitive Highly quantitative targeted Results readily transferable because concentrations can be measured	Targeted, so discovery of novel biomarkers unlikely Time consuming to set up quantitative assay	Medium to high
Capillary-electrophoresis (CE-MS)	CE-MS analysis time was less than 20 min per sample and allowed the simultaneous and reproducible analysis of more than	High-analyte resolution, providing information mainly on polar or ionic compounds in biological fluid	CE-MS is a valid approach for targeted profiling of metabolites and not targeted metabolite profiling	High

Continued

Table 1 Summary of the different techniques employed for metabolomic analysis—cont'd

Technique	Throughput	Advantages	Disadvantages	Cost
	80 metabolites in a single run with a minimum consumption of sample and reagents			
UPLC-MS	Separations on UHPLC columns often produce narrow peaks (few seconds or even less) Analytical separations using conventional size particles take on an average 10–30 min or more	Higher peak capacity, enhanced specificity and high-throughput capabilities compared to conventional HPLC columns Rapid analysis without loss of resolution UHPLC systems are capable of handling high pressures (15,000–18,000 psi) in comparison to normal analytical range High performance liquid chromatography (5000–6000 psi) Pumps are able to withstand such conditions with minimum wear and tear of components	Life of columns, during analysis high pressure developed because the particle size Increased pressure reduces the life of the columns Due to increased pressure requires more maintenance and reduces the life of the columns of these types	High

Matrix-assisted laser desorption/ionization Fourier-transform ion cyclotron resonance mass spectrometry imaging (MALDI-FT-ICR MSI)	~15–30 min	MALDI/TOF is a diagnostic tool with much potential because it allows for the rapid identification of proteins and changes to proteins without the cost or computing power of sequencing nor the skill or time needed to solve a crystal structure in X-ray crystallography. High resolution imaging of metabolites	Competitive ionization effects, which are especially common in complex mixtures Ion suppression in MALDI-based methods that cannot be rectified only by optimized sample processing procedure	High

differential metabolites between control and test organisms, usually by computer aided inductive strategies (Aretz & Meierhofer, 2016; Gika et al., 2014). This is subsequently followed by target metabolite identification, generally by MS-MS or FT-MS-based instruments (Schuhmann et al., 2017; Xiao, Zhou, & Ressom, 2012). The then identified metabolites are estimated and associated with one another, and also with their probable metabolite pathways. The modified metabolic pathways are then associated back with the system's biological significance manifested in the phenotype, or to allocate the gene function (reference). In plants, this is usually commenced by a correlation analysis approach termed 'clique metabolite matrices,' which was established by researchers working at the Max Planck Institute (Putri et al., 2013; Zhang, Sun, Wang, Han, & Wang, 2012). It is an imperative step to identify modified metabolite levels and correlate them with variations. However, flux analysis of particular pathways via heavy and stable isotope labels, called 'isotopomers analysis,' has also been utilized to trace metabolite uptake and utilization. This chapter will summarize the identification and quantification of targeted and non- targeted metabolites, and methods such as metabolite fingerprinting and footprinting, and their potential in advancing metabolomics.

1.2 NON-TARGETED IDENTIFICATION AND QUANTIFICATION OF ALL METABOLITES

Non-targeted metabolomics is a technique for generation of unbiased analysis of all the low molecular weight metabolites that exist within an organism in a particular environmental condition (Christians, Klawitter, Hornberger, & Klawitter, 2011; Naz et al., 2014). Non-targeted analysis includes different steps, such as collection of biological samples, sample preparation, method development, and data procurement, including chemometrics database searches and biological interpretation (Klepacki et al., 2016; Tsugawa et al., 2011). Non-targeted metabolomics provide a holistic strategy in biomedical research to advance disease diagnosis and comprehend its pathological mechanisms (Menni et al., 2013; Sévin, Kuehne, Zamboni, & Sauer, 2015). Several analytical procedures have been advanced based on NMR and MS, coupled with diverse separation techniques (Lubes & Goodarzi, 2017; Vernocchi et al., 2012). The central theme of any analytical procedure is the authentication of each step to procure consistent and reproducible data sets, even though analytical challenges are entirely novel and discreet to target methods (Beisken, Eiden, & Salek, 2015; Schrimpe-Rutledge,

Codreanu, Sherrod, & McLean, 2016). In this section we will describe accessible authenticated methods, and we will recommend some procedures to be considered for non-targeted metabolomics method development. Non-targeted metabolomics has been effectively appropriate for some biological samples, such as urine and serum/plasma, which can provide identification of biomarkers for various diseases (Fuhrer & Zamboni, 2015; Khamis, Adamko, & El-Aneed, 2017). In addition to metabolomics of biological samples of urine and serum, tissue analysis is significant, because not all variations are affected in the same way at the tissue level, and vice versa, which has been widely noted in the research, and is significant primarily for mechanism research, and in animal models (Mizuno et al., 2017). Metabolic variations in particular disease states are primarily at the tissue level. Hence, tissue metabolomics can advance non-targeted research to saliva, bile extract, cell lines, and microbial and food extracts, even in ecological toxicology (García-Villalba et al., 2015; Cheng, Lan, Zheng, & Gao, 2018). Non-targeted analysis of the amalgamation of diverse biological samples such as urine, plasma, and tissue paves the way toward deciphering molecular mechanisms (Athersuch, 2016; Sajic, Liu, & Aebersold, 2015). There are numerous research papers that have been published on plants using non-targeted strategies comprised of flowering plants, herbal medicines, the food industry, and so forth (Kibble et al., 2015); however, analytical characteristics are quite dissimilar. In addition to the selection of biological samples, it is important to note that no single analytical method is appropriate for identification, detection, and estimation of all metabolites with diverse physiochemical characteristics in non-targeted analysis (Aceña, Stampachiacchiere, Pérez, & Barceló, 2015; Phillips, Wambaugh, Grulke, Dionisio, & Isaacs, 2017). To tackle this, a combination of diverse techniques should be used, such as NMR, LC, GC, CE, and a combination of LC and GC (Fall et al., 2016; Mizuno et al., 2017). Authenticated sample preparation is a prerequisite for any analytical technique, so as to cover an extensive range of metabolites, together with convenient preparation steps limiting sample volume (Mullard et al., 2015; Ortmayr, Causon, Hann, & Koellensperger, 2016). Following sample preparation and instrumental validation, the next step is to upgrade the data processing algorithms to get actual numbers. There are no directives for authenticating the analytical techniques in non-targeted methods. The objective of non-targeted metabolomics is to explore statistically significant biomarkers via unbiased differential analysis of several signals, as much as possible, emanating from biological samples that will compare diverse circumstances being analyzed.

Non-targeted analysis can be utilized for comparing samples under similar ecological conditions, and variation in signals should be linked to the concentration of the components (Fraser et al., 2014; Raffler et al., 2015). All steps commencing from sample preparation to data processing should be in sync with instrumental techniques, and yield an articulate and extensive range of metabolites to prove the utility of the method for a non-targeted approach (Carvalho et al., 2014). Consequently, researchers are utilizing several different strategies to authenticate the analytical methodology. Current validation approaches include the following steps.

1.2.1 Quality Control Samples

For non-targeted metabolomics, high-throughput techniques encounter major limitations, such as repeatability and low reproducibility, which could arise in both chromatographic and MS development. These limitations need to be identified and executed to eradicate impartiality that arises due to steady variations in method development (Jamari, Dohmann, Raab, Krupp, & Feldmann, 2017; Tumanov et al., 2016). To assess the performance of chromatographic and MS instrumental systems, we have randomized the order of the quality control samples (Dunn et al., 2017; Kirwan, Weber, Broadhurst, & Viant, 2014). Quality control (QC) is collection of samples made from biological test samples to be evaluated, or a representative bulk control, and it should be measured against predefined norms to facilitate an analytical run (Liu, Liang, & Liu, 2016; Godzien, Alonso-Herranz, Barbas, & Armitage, 2015). It is immensely worthwhile to execute numerous injections at the commencement of the analytical run to assuage the system, and the analytical run should be constantly evaluated at regular intervals during the sequential run (Haggarty & Burgess, 2017; Soule, Longnecker, Johnson, & Kujawinski, 2015). This would consequently explicate whether the system is inequitable in a complete sequence run, and will explain the quality and integrity of the analyzed data sets (Reference). The FDA has been recommended that researchers include a relative standard deviation (RSD) for calculation of natural compounds present in QCs (Oliveira et al., 2015). Non-targeted metabolomics by LC-MS values near 30% RSD should be allowed; however, RSD greater than 30% would not be listed as potential bio-markers (He et al., 2017; Li et al., 2017). RSD in QC is being scrutinized by researchers who are trying to authenticate the theme by evaluating the linearity of the diluted pooled QCs (Jiao et al., 2018). Along with RSD, matrix effects and recovery were evaluated as well in the QCs (Bauer, Luetjohann, Rohn, Kuballa, & Jantzen, 2018). Evaluating the linearity in diluted QCs facilitates data processing based on the criteria that only those metabolites pass linear trends, which will require further data analysis (Styczynski et al., 2007). However, metabolites that are present in extremely low concentrations will not exhibit a linear trend, and some diluted samples are under the LOD (limit of detection), which leads to the loss of such samples. Moreover, dilution causes further deviation from linearity because ionic inhibition will be diverse to actual samples.

Quality Control Procedures

The utilization of quality control in non-targeted metabolomics is measured as equivalent to the use of standards in generalized target analysis, as compared with authenticating methods. Several research studies described the significance of QC samples in non-targeted MS metabolomics (Bouhifd et al., 2015; Fiehn et al., 2008). Using spiked standards is the most employed method in the pharmaceutical industry to facilitate consistency of the methodology (Naz et al., 2014; Wolfender, Marti, & Ferreira Queiroz, 2010). Numerous research studies followed this principle in non-targeted analysis. Some researchers employed this principle along with quantitative method validation in nutritional plasma metabolomics by calculating the matrix effect and recovery following spiking six discreet xenobiotic standards (Bailey, Oven, Holmes, Zenk, & Nicholson, 2004; Beyoğlu & Idle, 2013). For massive batch analysis, the method was authenticated in terms of the reproducibility of retention times and accurate masses in QCs. However, in different methods, evaluation of analytical reproducibility within a run was done by assessing sample-specific identified metabolites with diverse physico-chemical characteristics comprised of identified endogenous, along with xenobiotics (Barouki, 2010; Ramel, Sulmon, Serra, Gouesbet, & Couée, 2012). However, spiking of particular metabolites could produce better results, as this method cannot be applied to all metabolites of samples (Wang et al., 2003). Utilization of internal standards can be employed to verify instrumental reproducibility. Non-targeted metabolomics provides data pertaining to enormous metabolites present in particular samples. It is impossible to utilize all of them to check the reproducibility of the method. However, it would be better to target some of them from non-targeted metabolites (considering diverse class/category, retention/migration time, extra profuse or lesser copious, numerous physicochemical properties), and authenticate the method (Mohler et al., 2013; Petrakis, Cagliani, Polissiou, & Consonni, 2015). However, this authenticated method cannot substantiate final biomarkers for clinical purposes that require a targeted method with appropriate and conventional method authentication, but results are more authentic in comparison with non-validated methods, at least for further data processing. There are few reports documenting these principles for validating reproducibility, but to effectively authenticated and applied on biofluids and tissues (Bujak, Daghir-Wojtkowiak, Kaliszan, & Markuszewski, 2016; Dixit & Reddy, 2017). In addition to utilization of spiked standards, statistical model validation can facilitate disease diagnosis in which non-targeted metabolomics results were compiled with calculation of the class using a multivariate statistical algorithm, generally first employing non-supervised principal component analysis (PCA). PCA describes a summary of all variables entered in statistical analysis by finding a correlation among the variables. Following correlation, it sorts the variables into a smaller number of principal components that are accountable for the probable variance in the observed variables. PCA compiles the entire data in a global and qualitative visual pattern, signifying similarities and disparity among and within the sample (Nyamundanda, Brennan, & Gormley, 2010; Smilde et al., 2005; Zwanenburg, Hoefsloot, Westerhuis, Jansen, & Smilde, 2011). The quality of both targeted and non-targeted in metabolomics can be authenticated by close clustering of QCs in the PCA model. While assessing inconspicuous or wider clustering of QCs, a primary idea of the data could be acquired. Prior to drafting the PCA model, exclusion of QCs is done, and then later predicted on the same model. The clustering of QCs following prediction exhibits the repeatability of the instrumental reproducibility, along with illustration of separation among groups due to sample (Cajka & Fiehn, 2015; Hirayama, Wakayama, & Soga, 2014).

In addition to the preceding principle, some research reports document the different methods to validate the analytical method by investigating the instrumental efficiency, such as signal to noise ratio, plate number, and the retention factor, either in pooled or diluted samples (Rozet et al., 2007; Shrivastava & Gupta, 2011). However, employing these approaches only allows relative quantification of metabolites (Xiao et al., 2012; Wishart, 2011). The most popular authentication for non-targeted metabolomics employs all probable compounds that are varying in particular conditions, and is comprised of the evaluation of biological samples in an analytical platform. The process also includes the production of a predictive model from an assortment of identified multiple compounds that were utilized to categorize new samples into specific groups with moderate sensitivity and specificity (Kenar et al., 2014; Lei, Huhman, & Sumner, 2011). Finally, the important metabolites should be associated with biological pathways to identify potential markers and their biological significance. Therefore, methodically regulated, authenticated metabolites should be considered while developing a non-targeted strategy, and authentication should be based on the principal purpose of the research.

1.2.2 Steps Involved in Non-Targeted Metabolomics

The general steps implicated in non-targeted metabolomics are comprised of the following steps, from sample selection to analytical method development.

Selection of Sample

For any research analysis in non-targeted metabolomics, variation within population is to be validated within the population for which both meager and massive data sets are to be assessed meticulously. The primary requisite for the non-targeted approach is to reduce non-controlled variation among the samples, then arbitrarily selecting a sufficient set of samples (Sawada et al., 2008; Christians et al., 2011). It is important to ensure that groups have equal scores as far as dependent variables go, at commencement of research analysis. The variables should also be randomly allocated; irrelevant variables are regulated so that no group is affected by them during analysis, and each group receives identical evaluation.

Sample Treatment and Analytical Conditions

The prerequisite of method development originates from the requirement for method authentication and appropriate sample treatment. The most indispensable step in the non-targeted strategy is the sample preparation, and it is essential that the results be precise (Fuhrer & Zamboni, 2015; Ogura, Bamba, & Fukusaki, 2013). Sample treatment is determined by the analytical tools and metabolites of interest. Convenient sample treatment has to be implemented to cover extensive analyte signals attuned with techniques with moderate sensitivity and reproducibility in terms of retention/migration time, abundance, pressure (LC), current (CE), and reducing deferred effects (LC) (Kim & Verpoorte, 2010; Römisch-Margl et al., 2012). For biofluids, the sample preparation is comparatively simple; while for cell or tissues or extracts, particular attention is required to enfold an extensive range of metabolites (Cesbron et al., 2017; Skov, Hadrup, Smedsgaard, & Frandsen, 2015). Sample preparation on a large scale can be carried out by analyzing all samples in a single day, on different days, and segregating them into numerous batches, or preparing a sample every day. The protocol of sample pretreatment will be entirely determined on the primary aim of the study. Regardless of this fact, special care should be taken during the analytical method development. However, more attention should be diverted toward procurement of consistent and reproducible data acquisition, together with a wider signal range (Deda, Gika, Wilson, & Theodoridis, 2015; Filla, Sanders, Filla, & Edwards, 2016). Numerous protocols are available in biological samples such as urine, plasma/serum, and tissue metabolomics analysis (reference).

Analytical Method Validation

Non-targeted metabolomics help illustrate the changes in all the possible metabolites identified, and interpret the specific conditions. So a reliable analytical method could give us a primary reliable assumption from the identified distinctive metabolites. The fundamental of any analytical technique is to obtain dependable, consistent, and accurate data. As described in the previous section, an analytical non-targeted method can be validated by the following steps: first, following the published guidelines for targeted metabolomics, some known compounds that have already been identified in the specific sample should be selected (considering different physiochemical properties, and intensities along the entire chromatogram). If the method is validated in terms of linearity, accuracy, and precision for the selected compounds, the non-targeted approach will generate reliable and acceptable data sets for further evaluation. Moreover, sample treatment and method development steps should be optimized. For routine work, the use of QCs is recommended. These are, theoretically, identical biological samples, with a metabolic and sample matrix composition similar to those of the biological samples under study. Two types of QCs are used in metabolomics: (i) pooled QCs, where small aliquots (same volume) of each biological sample to be studied are aliquoted and thoroughly mixed, and (ii) commercially available biological samples. Pooled QCs are useful, as they are very close to the composition of the biological sample to be studied. The signal variations of any metabolite could be reflected in the QCs. However, pooled QCs can only be applicable in small-scale studies; whereas, for large batch analysis, it is not possible to aliquote in order to make a pool for QCs. Therefore, a replacement for QCs is required. As the replacement QCs do not completely resemble the original population, there are some metabolites that will not be seen. During data processing, RSDs for these metabolites will be unknown, and they could be removed during data filtering, leading to a loss of metabolites that could be important biomarkers. QCs help measure the repeatability within an analytical run, and allow filtering data before analysis, considering the drift in signal variations, retention time, and accurate mass. QCs are used in metabolomics for three reasons. The first is to 'condition' or equilibrate the analytical system with injections of a matrix prior to the main analytical run, in order to ensure that reproducible data are acquired. Second, it provides data to calculate analytical and technical precision. And third, the data from QCs are used for signal corrections within or between analytical run/batch analyses. An analytical method can be validated by diluting the pooled QCs and checking the linearity. By doing so, the metabolites that are showing a linear trend can be considered for further data processing. There are some limitations in this methodology that have been previously discussed. To get a snapshot of the instrumental repeatability before starting any data treatment, the total signal from each chromatogram could be plotted against each sample (as per the sequence/run order). The dispersion of QCs in the plot could reflect the instrumental repeatability. Moreover, for validation, the use of blanks during method development is recommended. Blanks should contain the solvent resembling the study samples, and should be processed following the same procedure, along with the real samples. The goal of including blanks is to detect and minimize the carryover during method development, and consider the identified features in blanks as background, mainly for GC-MS. Another indirect way of analytical method validation is to check the signal variation in all identified metabolites in terms of RSDs in QCs, and filtering data by frequency present in any study group. Thus, the repeatability and sensitivity of the method can be measured. In chemometric models validation, a non-targeted approach generates huge data sets. Using multivariate algorithms, these data sets can often be presented as a prediction of class separation, and it is very important that this prediction is carried out in a relevant way. There are several strategies that are already being used in validating statistical models, and they are out of the scope of this work. At the very least, to avoid the over-fitting of a multivariate statistical model, a cross-validation scheme should be employed. In biomarker validation, at the starting point of metabolomics, a non-targeted approach was thought to be a diagnostic tool, but scientists and clinicians determined that the methodology was better suited for a discovery phase. Therefore, the end result of every non-targeted approach comes with one or more statistically significant compounds related to specific pathways. That can provide insight into the progress, or pathogenesis, of specific conditions. However, determining one, or several, biomarkers from the identified differentiating significant compounds is beyond the scope of biomarker validation, and involves several other validation steps. Biomarker authentication needs targeted, specific analytical methods, respective method validation, internal standards, and competence evaluation. Non-targeted metabolomics pave the way toward the discovery phase in diagnosis, treatment, and disease prognosis. Non-targeted metabolomics can be utilized for unraveling the independent pathways implicated in chemical diagnostics biomarkers in biological samples, allowing them to be characterized. The analytical approaches implicated in metabolomics require validation so that authenticated principles for non-targeted study can be established. Numerous efforts are underway by research teams to ascertain authentication principles with approved ranges following recognized methodologies, despite current validation approaches for non-targeted strategies that are still unexplained. First, the optimization of the method for an extensive range of physicochemically diverse classes of metabolites is the indication for authentication of the analytical method in non-targeted metabolomics. This method includes acquiring purposeful substantiation regarding numerous possibilities following defined principles of accuracy and linearity by spiking or diluting the quality control. Metabolomics method authentication should now be recognized as an open discussion among researchers.

1.3 QUANTITATIVE DETECTION OF METABOLITES

Analytical chemistry has advanced the accuracy and sensitivity of evaluating predefined metabolites in biological tissues, which has led to the identification of single molecules in single living cells, with enormous prospects for demonstrating biological responses to cellular events in vivo (Fernandez-Peralbo, Santiago, Priego-Capote, & de Castro, 2015; Vargas et al., 2016). Numerous regular methods have been advanced to selectively identify few metabolite classes, while abounding others (Cajka & Fiehn, 2015). Polyamines are known to be implicated in numerous metabolic processes for plant systems, for example, drought stress; and varied analytical methods are available for quantification of plant material (Creek et al., 2014; Liu et al., 2017). Vitamins are prime targets of current analytical research, particularly when the concurrent evaluation of diverse isoforms is required (Chen, Li, Cui, Yu, & Zhai, 2015; Fanali, D'Orazio, Fanali, & Gentili, 2017). However, most elaborately demonstrated compounds are still tedious to analyze, due to high instability, which results in the flip of oxidized forms to reduced forms; for example, glutathione (Pandey, Patel, Mishra, & Jha, 2015; Bruno et al., 2018), or in diverse stereoisomers, for example, zeaxanthins (Arathi, Sowmya, Vijay, Baskaran, & Lakshminarayana, 2015; Arathi, Sowmya, Vijay, Baskaran, & Lakshminarayana, 2016; Moradi, Ford-Lloyd, & Pritchard, 2017). However, the most sophisticated analysis involves the identification of trace compounds in intricate matrices; for example, phytohormones in plants. Varied protocols have been established for identification of indole-3-acetic acid (Kim, Choi, & Verpoorte, 2010), for abscisic acid (Urano et al., 2009), and for indole-3-pyruvate (Novák, Pěnčík, & Ljung, 2014; Kumavath, Barh, Azevedo, & Kumar, 2017). Methods for phytohormones' evaluation will certainly have to become more advanced compared with preceding years, in order to achieve advanced detection limits and convenient sample clean-ups (Rawlinson, Kamphuis, Gummer, Singh, & Trengove, 2015; Van Meulebroek, Bussche, De Clercq, Steppe, & Vanhaecke, 2015). Target analysis will remain the most widely used technique, with prospects in all areas of biological research. Though for comparative analysis in functional genomic approaches, target analysis is only of restricted application, because the level of target analytes might be modified by sudden effects that cannot be comprehended without more inclusive strategies (Thomas et al., 2003; Tohge et al., 2005). Consequently, an extensive evaluation of metabolite modification is required to limit expansive data illustration. Chromatographic methods began to evolve after the 1960s, and made peak identification possible, depending on chromatography alone (Bujak, Struck-Lewicka, Markuszewski, & Kaliszan, 2015; Forcisi et al., 2013). With the advent of sensitive detectors, analytical methods were upgraded further, and utilized in biological samples spanning from human urine to plant tissues, so as to screen significant metabolite classes, for example, amino acids (Alonso, Marsal, & Julià, 2015; Gika, Zisi, Theodoridis, & Wilson, 2016). Using standard compounds with known retention times, shifts in retention times could be monitored, which facilitates the detection of massive metabolites, such as organic acids in human plasma for clinical diagnosis (Engskog, Haglöf, Arvidsson, & Pettersson, 2016; Roullier-Gall, Witting, Gougeon, & Schmitt-Kopplin, 2014).

Mass spectrometry presents as an auxiliary, and entirely independent, method for compound identification. By coupling chromatography with mass spectrometry, diverse human diseases could be diagnosed simultaneously (Nicholson & Wilson, 2003; Zhang et al., 2012). Currently, bioinformatics tools and computational contrivances facilitate more advanced classification of human metabolic disorders (Qin, Yalamanchili, Qin, Yan, & Wang, 2015; Reali, Priami, & Marchetti, 2017) and tumor-associated tissues (Ding, Wendl, McMichael, & Raphael, 2014; Kan et al., 2010). For automatic metabolite identification, authenticated data pertaining to retention time and mass spectra is required (Smith et al., 2005; Xia, Psychogios, Young, & Wishart, 2009). However, mass spectra of metabolites can be subjugated by coeluting metabolites in intricate chromatograms, and may be covered at trace levels via chemical noise (Barbu, 2008; Yan, Zhao, Feng, Wei, & Lin, 2009). For high throughput of mass spectra-quality regular identifications, mass spectra consequently needs to be purified (Colinge, Masselot, Giron, Dessingy, & Magnin, 2003; van Veen, Claas, & Kuijper, 2010). Through utilization of mass spectral deconvolution software, peak identification was performed for 68 target metabolites for instantaneous detection of inborn errors, when compared with diseased and normal children (Tebani, Abily-Donval, Afonso, Marret, & Bekri, 2016; Thiboonboon et al., 2015). Relatively less research has been done on the comparative analysis of profiling of plant metabolites (Hall, 2006). The instantaneous analysis of macromolecules such as carbohydrates, sterols, amino acids, and sugar alcohols by GC-MS was primarily carried out by Sauter, Li, and Aubrun (2009) for comparing the effects of pesticide applications on plants. Improper or absent sample prefractionation results in cluttered chromatograms and less abundant metabolites; for example, lysine is not screened. Derivatization of protocols can be further advanced by limiting analysis to polar compounds (Garcia et al., 2007; Halket et al., 2004). For example, profiles of polar metabolites in apricots can be easily procured, and more specifically, Roessner-Tunali et al. (2004) demonstrated the application of GC-MS quantification for analysis of polar metabolites in potato tubers. However, metabolic pathways are differently fashioned, not only in different organisms, but inside different tissues of the same organism, and hence, metabolic pathways need to be re-evaluated via analytical contrivances. Consequently, the basic metabolic pathway needs to be re-investigated via utilization of modern analytical tools. For example, GC-MS can be utilized for evaluating metabolic networks comprised of few metabolites using stable isotopes and evaluating the limited metabolome (Cuadros-Inostroza et al., 2016; Styczynski et al., 2007). For some metabolite classes, for example, bis- and trisphosphates or lipids, LC is the preferred separation method. Using absorption spectra of ultraviolet (UV) light, profiles of aromatic and desaturated organics can be procured, for example, carotenes, xanthophylls, ubiquinones, tocopherols, and plastoquinones. LC coupled with UV has been extensively used to identify transgenic and mutant tomato genotypes, and for identifying Arabidopsis mutants (Zhang, Li, Tsang, & Chetwyn, 2013). Though mass spectrometers are more lucid and accurate than UV spectrometers, they were able to analyze not only isoprenoids and aromatics, but also compounds without UV-absorbing moieties (such as oligosaccharides).

The advent of ESI has immense potential, and can provide inclusive information as we align LC with mass spectrometry (Seeram, Lee, Scheuller, & Heber, 2006; Smith, Loo, Edmonds, Barinaga, & Udseth, 1990). For the validated quantification of complex sugars, such as polyols, using internal standards, electro-spray ionization provides an articulate and precise analysis (El-Hawiet, Shoemaker, Daneshfar, Kitova, & Klassen, 2011; Wan & Yu, 2007). Similarly, LC-MS analysis of ceramides exhibits detection limits in the range of femtomolar, in the case of cultured T-cells (van Smeden et al., 2011). An advanced level of sensitivity in metabolite profiling can be acquired by linking capillary electrophoresis to laser-induced fluorescence identification (Couderc, Ong-Meang, & Poinsot, 2017; Hodáková, Preisler, Foret, & Kubáň, 2015). This technique even measures steroids in the atto-molar range, in comparison with femto-molar sensitivity, when coupled with mass spectrometry. Generally, metabolic profiling refers to the catabolic degradation of a certain compound in an organism. Numerous analytical strategies are required to analyze degradation pathways, for example, degradation oflinoleic acid in apples was analyzed via GC-MS, LC-MS/MS, and LC linked with radiodetection (Beuerle & Schwab, 1999; Szolar, 2007). Pharmaceutical research determines the metabolic fate of administered drugs, for example, the demonstration of propanolol degradation in rats via LC-MS/MS (Johnson & Lewis, 2005; Johnson & Lewis, 2006). LC-MS/MS can be coupled with bioassay-directed fractionation; for example, binding affinity of catabolites to particular receptors so as to explore inclusive information pertaining to the toxicological significance of catabolites (Christian & Luckas, 2008; Qu et al., 2011).

1.4 IDENTIFICATION AND QUANTIFICATION OF METABOLITES

Metabolites are chemical compounds that can be analyzed by the standard contrivances of chemical evaluation; for example, molecular spectroscopy and mass spectrometry (Tautenhahn, Patti, Rinehart, & Siuzdak, 2012; Wehrens, Weingart, & Mattivi, 2014). Molecular and mass spectroscopy offer high resolution, enhanced sensitivity, and improved specificity when they are coupled with GC or LC steps. These hyphenated technologies are utilized for diverse metabolomic studies. The technologies are employed as per type of samples to be analyzed. Some samples, such as biofluids, are the most convenient to analyze via NMR, as they can be used with a minimum level of preparation (Cambiaghi et al., 2016; Wanichthanarak, Fahrmann, & Genomic, 2015). On the contrary, animal and plant tissues and microbial systems require exhaustive sample preparation (Dunn et al., 2017; Kim & Verpoorte, 2010). Target analysis and metabolite profiling require precise and specific evaluation, along with established calibration methods (Salek et al., 2015; Shah, Brock, & Rai, 2015). These methodologies can be authenticated by efficient separation via GC or HPLC coupled with sensitive detectors (Erxleben, Gewssler, Vervliet-Scheebaum, & Reski, 2012; Sangster et al., 2006; Viant, Pincetich, Hinton, & Tjeerdema, 2006). Remarkable improvement has been seen in the utilization of so-called hydrophobic interaction chromatography for the precise identification of polar plant compounds in an MS-friendly manner (Schlichtherle-Cerny, Affolter, & Cerny, 2003; Wang & He, 2011). NMR spectroscopy provides inclusive information on metabolites, generally in samples from biofluids with the least sample preparation (Nicholson, Lindon, & Holmes, 1999; Dieterle, Ross, Schlotterbeck, & Senn, 2006). NMR reveals all the structural information, as it is based on the nuclear spin of ^{1}H, ^{13}C, ^{31}P, which coexists in diverse energy levels in a strong magnetic field, because it possesses nuclear spin (Kikuchi, Shinozaki, & Hirayama, 2004; Leiss, Choi, Abdel-Farid, Verpoorte, & Klinkhamer, 2009). Magical angle spinning NMR is a highly advanced offering for examining the intricate details of living tissues (Leiss, Choi, Verpoorte, & Klinkhamer, 2011; Lewis et al., 2007). For exclusive analysis of the metabolome, it is necessary to use methodologies that cover an extensive range of metabolites. The exhaustive sample preparation is implicated in several steps into aliquots, prior to specific fortification and evaluation of diverse metabolites in each aliquot. These sample prefractionation steps and the succeeding analogous evaluation are necessary to optimize analysis, and to assist in the identification of even inconsequential amendments in a structurally assorted metabolome data set (Lindon & Nicholson, 2008a, 2008b; Moco, Bino, De Vos, & Vervoort, 2007). The integrated approach comprised of numerous analytical methodologies might be utilized for such studies; for example, parallel LC separations can be coupled with MS- or NMR-based identification methods (Oksman-Caldentey & Saito, 2005; Sumner, Mendes, & Dixon, 2003).

A majority of extraction methodologies reported in the literature so far have been less inclusive, and thus biased, because they bounce some metabolites, leading to modeling of only a portion of the metabolome (Tucker et al., 2010; van der Meer, 2006). Presently, the most utilized methodology for metabolite analysis includes GC-MS. Although this method is restricted to targeted molecular mass, it can estimate why temperature sensitive metabolites are not screened (Michaud & Denlinger, 2007; Trenkamp, Eckes, Busch, & Fernie, 2009). Non-volatile polar metabolites are generally required for derivatization for conversion to less polar, volatile, and thermally stable derivatives prior to the separation on the GC column. Even though improved derivatization approaches are available, low sample throughput can be a limiting factor in massive metabolite profiling, because characteristic metabolite acquisition times are on the order of 10–30 min. Linearization is then required to estimate metabolites that are not quantified by GC, and this is facilitated by MS. Appropriate linearization is established via algorithms. Advanced linearization of algorithms and instant spectral acquisition via time of flight (TOF) measurements (Rutherford, 2004) have led to identification of thousands of metabolites from leaf tissues at a throughput of more than 1000 samples per month (Spicer, Salek, & Steinbeck, 2017; Steinbeck et al., 2012). Another development is the discovery of the MSFACTS program developed by Sumner and colleagues (Sumner et al., 2007). The ultimate upshot of this process is an inventory of metabolites (either identified metabolites found in databases, for example, KEGG or exclusive MS profiles linked with a particular GC retention time), and a catalogue of their comparative concentrations (Kanehisa, Goto, Sato, Furumichi, & Tanabe, 2012). In recent years there has been an increased proclivity toward enormous metabolites that can be separated in a single analysis run (Allwood, Clarke, Goodacre, & Mur, 2010; Allwood, Ellis, & Goodacre, 2008). The methods implicating flow injection evaluation via direct infusion into ESI coupled with TOF or Fourier transform ion cyclotron resonance (FT-ICR) MS has also been popularized (Bae, Na, Chung, Kim, & Kim, 2010; Sampson, Hawkridge, & Muddiman, 2006). This technique is highly sensitive, and provides high mass resolution coupled with inclusive data on isotope patterns that could yield empirical formulae for metabolites directly (Nikolaev, Kostyukevich, & Vladimirov, 2016). Metabolic fingerprinting is instant, and would be most appropriate for quick identification if important variations in the metabolome were detained in a reproducible manner (Chatterjee et al., 2010; Ellis, Dunn, Griffin, Allwood, & Goodacre, 2007; Gidman, Goodacre, Emmett, Smith, & Gwynn-Jones, 2003). These techniques can process massive samples with negligible sample preparation, and at same time, generate significant information that is suitable for generating instant fingerprints. In this context, crude extracts or whole cells can be analyzed via MS (Domon & Aebersold, 2006; Nesvizhskii, Keller, Kolker, & Aebersold, 2003), NMR (Fotakis, Kokkotou, Zoumpoulakis, & Zervou, 2013; Krishnan, Kruger, & Ratcliffe, 2004), or FT-IR or Raman spectroscopies (Movasaghi, Rehman, & ur Rehman, 2008; Rohman & Man, 2010). At present, NMR fingerprinting is most suitable for metabonomics, but numerous researchers link this technique with MS technologies, offering sensitive and articulate processing (Schripsema, 2010; Kim et al., 2010). NMR is relatively less sensitive, however, which provides consistent data for all proton-containing molecules, in comparison with MS, which is sensitive to matrix effects referred to as 'ion suppression,' and discards few classes of analyte (Kim, Choi, & Verpoorte, 2011; Lindon & Nicholson, 2008a, 2008b). In recent years, metabolomics has emerged as a new field of omics, dealing with identification of the metabolome, the low molecular weight metabolite complement of the biological system via investigation (Lindon et al., 2011; Patti et al., 2012). The field of metabolomics offers an extensive range of biological systems, from simple microbes, to plants and intricate multi-cellular organisms, such as mammals (Fukushima, Kusano, Redestig, Arita, & Saito, 2009; Ryan & Robards, 2006). The basic objective of most metabolomics studies is to produce an instant of the metabolic state of biological sample, and to identify the variations in abundance of the estimated metabolites originating from natural fluctuations or external, experimental biotic or abiotic alterations (Cambiaghi et al., 2016; Fuhrer & Zamboni, 2015). Metabolomics offers a multifaceted platform for identification of diversity of metabolites of living systems (Kusano et al., 2015; Lee et al., 2015). The integration of analytical approaches generates data on a common platform. For example, in the field of food safety, untargeted GC-MS-based evaluation is performed via two diverse methods for evaluation of volatile and polar small metabolites for the identification of meat spoilage (Luo, 2015; Vinaixa et al., 2016). There are two complementary methods available for use in current metabolomics; targeted and untargeted analysis. In targeted analysis, known metabolites are monitored, which usually offers absolute quantification and explicit recognition when restrained to metabolites available as internal standards (Johnson, Ivanisevic, & Siuzdak, 2016; Schubert et al., 2015). On the contrary, untargeted profiling methods aim to investigate analytical characteristics of all identifiable compounds. Consequently, they are able to investigate the whole metabolome-comprising substance, which is unidentified at the time of estimation (Naz et al., 2014; Palermo, Botrè, de la Torre, & Zamboni, 2017). Untargeted analysis is appropriate for identification variation in the unknown terrain of the metabolome, which often leads to novel scientific hypotheses. In untargeted analysis, effectual, automatic, proficient, computerized data processing, precise quantification, and consistent annotation/detection of metabolites still creates major confrontations in metabolomics (Stultz et al., 2018; Wallace et al., 2017). In this context, stable isotope-assisted profiling approaches offer great potential to tackle these challenges. Furthermore, the advancement and application of the appropriate quality measures, comprised of quality control, data processing algorithms, method authentication, and assessment of measurement improbability, are more and more being identifiable as central requirements for assessing analogous output for prolonged time periods across diverse samples and biological experiments, or between diverse laboratories (DeFelice et al., 2017; Guida et al., 2016; Sud et al., 2015).

1.5 TARGETED METABOLITE ANALYSIS

This refers to the inclusive estimation and detection of a single metabolite, or a set of linked metabolites, following extensive sample preparation and extraction before analysis by one or more investigative platform (Alonso et al., 2015; Roberts, Souza, Gerszten, & Clish, 2012). In addition to diverse analytic approaches, there is also an extensive range of analytical strategies used in metabolomics; the prime analytical juncture is comprised of GC mass spectrometry (GC-MS) (Mondello, Tranchida, Dugo, & Dugo, 2008), LC mass spectrometry (LC/MS), FT-IR, Raman spectroscopy, and NMR spectroscopy (Al-Saffar, Stephanson, & Beck, 2013; Pitt, 2009). These analytical approaches have their limitations as far as applications in the identification of diverse classes of metabolites. Of all analytical approaches, ultra-high-performance LC mass spectrometry (UPLC-MS) has e great potential for inclusive profiling (Want et al., 2010; Want et al., 2013). Due to an intricate and varied assortment of metabolome sizes, the inclusive estimation of metabolites is highly difficult, and presently inaccessible (Bino et al., 2004; Tanaka et al., 2009). There is no single, precise, analytical approach for estimating and identifying every metabolite. To overcome this problem, multifaceted targeted extraction, integrated with efficient sample preparation on diverse analytical platforms is used in an effort to screen biological systems at an inclusive level, with the least presumption. This multifaceted strategy reduces the effects of exclusive extraction and instrumental bias toward certain classes of metabolites, and allows a more inclusive range of possible metabolites to be identified in order to cover maximum metabolites.

Recent technological innovations in MS-based approaches have led to an exponential rise in performance. The preamble of Orbi-traps (OT) and Fourier transform ion cyclotron resonance (FT-ICR) mass analyzers provide an enhanced level of sensitivity, selectivity, resolution, and hence, accuracy (Kussmann & Blum, 2007; Satake et al., 2007). Instant and lower-size electric devices have provided MS analyzers to acquire near-total 3D ion confinement, and ultrafine ion control. This integrated approach facilitates association, and, along with modern analyzers, increases sensitivity to the order of sub-milliseconds, which ensures the screening of thousands of metabolites instantly at higher speeds and magnitude (Luzzatto-Knaan, Melnik, & Dorrestein, 2015; Sidebottom, Johnson, Karty, Trader, & Carlson, 2013). The elevated scan rate ensures better peal resolution to be accessed, even in ultra-low-level metabolites, leading to precise peak linearization, followed by their detection. The utilization of collision-induced dissociation and chemical fragmentation has allowed an extensive range of structural data to be obtained and linked to unknown metabolites, assisting in detection (Dongre, Somogyi, & Wysocki, 1996; Laskin, Denisov, & Futrell, 2001). The utilization of robotic automation associated with condensed sample evaluation ensures significantly high throughput to the point where an enormous amount of samples can be evaluated precisely and reproducibly in a comparatively short span of time with the least amount of operator regulation. Presently, MS-based instruments provide explicit and exact methods of detection and estimating of metabolites present in low concentrations in intricate sample matrices (reference).

1.5.1 Metabolic Flux Analysis

Metabolic flux analysis (MFA) is another significant field of metabolomics in which the substrate flow via multifaceted biochemical pathways at a steady state is assessed and evaluated. This is acquired by the use of isotopic labeling experiments, and estimating these allocations via either NMR or MS. This technique provides a time-wise assessment of changes in the substrate, and product concentration, by evaluating the flow and redistribution of labeled precursors into other metabolites (Antoniewicz, 2015; Templeton, 2016; Wiechert, Niedenführ, & Nöh, 2015). This approach can be successfully implemented in plants, human cells, and bacteria (Crown, Long, & Antoniewicz, 2015; Gopalakrishnan & Maranas, 2015). MFA is imperative for comprehending the control and regulatory procedures among metabolic networks (Cortassa et al., 2015; Imam, Schäuble, Brooks, Baliga, & Price, 2015). Metabolic fluxes are executed by both gene expression, and post-transcriptional and post-translational modulations, although such procedures may lead to meager variations in enzyme concentration. These changes have imperative effects on the concentration of different metabolites. With the advent of the post-genomic era, together with emerging fields of transcriptomics, proteomics, and metabolomics, there is still a need to comprehend the metabolic network within the biological system, to the point where rational metabolic engineering is a practical alternative. Conversely, MFA has been recognized as a significant contrivance for the demonstration of metabolic pathways, improving our competence to explicitly detect the effect of gene knockouts on mutant flux networks, and assisting the advancement of exact mechanistic pathway models (Long & Antoniewicz, 2014; Kogadeeva & Zamboni, 2016). Therefore, MFA proves to be a valuable contrivance to metabolomics for the inclusive identification of metabolic networks (Gerstl, Ruckerbauer, Mattanovich, Jungreuthmayer, & Zanghellini, 2015; Peng et al., 2015).

The conventional description of volatile organic compounds (VOC) is exceptionally wide and ambiguous due to overlap between discreet classes of metabolites. Any low-molecular-weight metabolite with a molecular mass of less than 400 amu with a carbon backbone, vapor pressure higher than 0.27 kPa, and a boiling point between 50–260°C that freely exists as a gas at standard temperatures and pressures can be termed VOCs (WHO, 1989; EPA, 1999). This description provides no disparity of natural and synthetic forms of VOCs encompassing everything from organic solvents, such as acetone and dichloromethane, to polyaromatic hydrocarbons, and environmental pollutants such as BTEX (benzene, toluene, ethylbenzene, and xylene). Biogenic VOCs (BVOCs) are comparatively new terms that define any VOCs that are synthesized by living beings. However, their production is determined by varied biotic and abiotic factors, such as light, temperature, and nutrient availability. In addition, BVOCs are produced against diverse pathogenic attacks by insects and herbivores (Kigathi et al., 2009; Clavijo Mccormick, Gershenzon, & Unsicker, 2014). This can also be manifested as a defense response of plants against climatic fluctuations. In addition to their functional significance in abiotic and biotic responses, BVOCs are also implicated in ecological behaviors of insects by facilitating their communications (Atkinson & Arey, 2003; Pinto, Blande, Souza, Nerg, & Holopainen, 2010). Flavor and fragrance metabolites comprised of aromatic compounds such as isoprene, terpenes, alkanes, alcohols, esters, carbonyls, and acids are also categorized in BVOC. BVOC profiling can be utilized for determining food quality, as they release VOCs. BVOCs profiling of fruits and vegetables are generally evaluated at different phases of transport for quality control purposes (Dudareva, Klempien, Muhlemann, & Kaplan, 2013; Muhlemann, Klempien, & Dudareva, 2014). BVOC profiling has been found to be a potential diagnostic tool, used on human skin and breath for personalized health care and disease screening (Amann et al., 2014; Shirasu & Touhara, 2011). GC, coupled with headspace sampling, provides evaluation of VOCs. Headspace analysis is one the most significant techniques for determination of VOCs at minimal sample preparation, and an enhanced level of instrumental sensitivity, making this technique a preferable choice for sampling an extensive range of analytes and concentrations. Zhang et al. (2010) and Zhang and Li (2010) have illustrated diverse methodologies comprised of liquid-liquid extraction (LLE), stream distillation (SD), simultaneous distillation extraction (SDE), and supercritical fluid extraction (SFE) techniques (Sardjono, Athiana, Gumilar, & Rachmawati, 2017; Wang et al., 2014). Headspace sampling can be divided into two prime types, static and dynamic, which describes the two diverse sampling methods, and the different surfaces employed for the separation of samples. Solid phase microextraction (SPME) is generally referred to as the prime standard for static headspace (HS) sampling. The extraction of samples via SPME exhibits an elevated level of toughness. SPME prepares samples in sealed vials, wherein sample is bound to HS in a 1:1 ratio, and low air flow is passed through a magnetic bead. This is significant for a representative sampling of a high volume of HS onto a meager surface area of fiber, which assists in launching enough VOCs into the HS (Risticevic, Vuckovic, & Pawliszyn, 2010; Souza-Silva, Jiang, Rodríguez-Lafuente, Gionfriddo, & Pawliszyn, 2015). Solid samples are placed directly into vials and sealed to build up VOCs in approximately 45–60 min. The SPME fiber was incorporated into HS in the preceding sample, and exposed for around 30 min, which is sufficient for the collection of VOCs. Following the complete sampling procedure, SPME fiber is withdrawn and detached; and it is then inculcated into a GC injector, where it is thermally treated to desorb the VOCs that are transferred into the GC to undergo chromatographic separation, following recognition (reference). SPME fiber offers an exceptional surface for holding the VOCs with diverse activated surfaces, making them accessible for sampling diverse classes of VOCs. SPME exhibits high sample flexibility, and each fiber can evaluate 200 injections prior to degradation (reference).

SPME fiber throughput can be restricted, as it needs to be cleaned separately prior to, and following, each injection; otherwise, this could limit its utilization in extensive experiments (Asfaram, Ghaedi, & Goudarzi, 2016; Haggerty, Bowyer, Jiranek, & Taylor, 2016). Moreover, the fiber is exceedingly prone to contamination, and should be kept clean and employed instantly so as to curtail a backdrop signature during sampling (Boyacı et al., 2015; Xu & Ouyang, 2017). The estimation of BVOC in human skin via polydimethylsilioxane (PDMS) elastomer patches was initially designed by Gallagher et al., 2008; Vuckovic, Zhang, Cudjoe, & Pawliszyn, 2010. These patches are modified to function as passive VOC sampling tools. PDMS is analogous to SPME, as a patch is suspended over the HS of the sample. A collection of patches is followed by thermal desorption on a silico-coated tube for GC-MS analysis (Xia et al., 2009; Kokosa, 2013). PDMS provides the appropriate sensitivity and reproducibility for trapping of polar compounds, as the massive batches are resistant to degradation through cleaning. These batches can be stored within clean thermal desorption tubes, and remain stable for prolonged time periods.

Dynamic sampling refers to the volume of the headspace within a sample vial that is evacuated many times, generally via an inert carrier gas such as helium, nitrogen, or argon, which can flow through an absorbent material such as Tenax-TA or PDMS for the VOCs to be adsorbed on it. Dynamic headspace ensures the detection of low level VOCs due to significant sample enhancement, which is acquired via the sampling of massive air flow over a comparatively low surface area (Boczkaj, Przyjazny, & Kamiński, 2014). However, such sampling is immensely intricate, as it requires the utilization of specific sampling rigs. The intricacy of sampling increases if numerous comparable samples are used. Analogue needle valve flow controllers are usually used because of their versatility and low cost, but digital flow controllers are commercially accessible, and more accurate; although they are considerably more exclusive, and their lack of resilience is a general issue (Manzini et al., 2011; Saraji, Khayamian, Siahpoosh, & Farajmand, 2012). Suitable flow regulators and controllers are necessary to achieve reproducible results. Precautions should be taken to ensure uniform flow rate and sampling time, as lower variations in either parameter will be augmented into considerable disparity in the total volume of HS extracted. Tenax-TA is a porous polymer resin of 2,6-diphenylene oxide. It is an extremely malleable absorbent material employed in DHS. Tenax-TA is granulated, and offers an enormous surface area for the VOCs adsorption on the trapping material, which is common for diverse VOCs (polar or non-polar). Additionally, it offers superior resilience for recurring exploitation (Arnts, 2010; Hogg et al., 2011). Moreover, Tenax-TA has low backdrop noise, and if adequate precautions are taken during the cleaning process, each Tenax-TA packed thermal tube can be employed for up to 250 injections prior to any observed degradation. However, a change of packing material following 250 injections is highly recommended for maintaining the appropriate trapping performance (Lai, Batra, & Cohen, 2005). Tenax-TA exhibits numerous disadvantages; for example, it is liable to sample degradation, and remains stable for a relatively short time following cleaning. Thermal desorption offers robust sample introductions for transferring VOCs from absorbent material with an extensive surface area, for example, Tenax-TA or PDMS, into a GC in an exceedingly reproducible and mechanized way. The Tenax-TA-packed thermal desorption tube is the primary intervention for thermal desorption of VOCs, and concentrating the compact volume into a cold trap (Jo, Kim, & Kwon, 2017; Wong et al., 2015). This cold trap is heated, and VOCs are then launched into GC as a diminutive, distinct quantity of gas. This method of sample injection will reduce peak expansion, and lower coelution effects, considerably enhancing the general resolution and sensitivity of the analysis (Wolfender, Marti, Thomas, & Bertrand, 2015; van Stee, 2017).

1.6 METABOLITE FINGERPRINTING

Metabolite fingerprinting is an instantaneous and high-throughput evaluation of unprocessed metabolites with the least amount of sample preparation. Detection and estimation is limited in this approach, as higher significance is given to the differentiation of samples, depending on origin and physiological status (Scholz, Gatzek, Sterling, Fiehn, & Selbig, 2004; Scott et al., 2010). Fingerprinting surpasses the assignment problem presented by the multitude of signals in a high-resolution ^{1}H NMR spectrum, and instead utilizes multivariate analysis to compare sets of spectra, and the samples from where spectra are derived. The signals in these spectra have their origin in the metabolites present in biological samples employed for NMR analysis, but the identity of the metabolites is secondary to the task of confirming whether the spectra is from a set of different or similar samples. NMR spectra offers pattern recognition of compounds in the biomedical field. This method has also been used for the evaluation of extracts and materials of plants (Griffin, 2003; Smolinska, Blanchet, Buydens, & Wijmenga, 2012). NMR fingerprinting in plant samples was initially utilized for evaluation of food products, particularly fruit juices, with the objective of discerning between juices originated from diverse varieties of the same fruit. In the study of apple juices, it was shown that ^1H NMR spectra could be used to differentiate between three types of apples, with a success rate of up to 100% under favorable conditions (Belton et al., 1998). This research is primarily based on principal components analysis, and investigation of loadings detected sucrose and malate as the prime, but not exclusive, basis for the classification. Actually, the NMR fingerprinting principle demonstrated by these research reports can be extended to numerous types of comparisons, and significant determination has been demonstrated in numerous areas, including the assessment of ecological factors, for example, changing growth conditions (reference) and chemical contamination (He & Liu, 2008; Holmes, Tang, Wang, & Seger, 2006), along with comparisons between ecotypes of *Arabidopsis thaliana* (Ward, Harris, Lewis, & Beale, 2003) and between wild-type and transgenic genotypes of tomatoes (Le Gall, Colquhoun, Davis, Collins, & Verhoeyen, 2003). Fingerprinting provides the foundation for inclusive data on the classification of samples in each application, and it can be used to produce tangible concluding points. NMR can be utilized for determining the considerable variability in the metabolic composition of plants thriving under normal conditions. Metabolite profiling in several horticultural crops is evaluated via NMR analysis (Hohmann, Christoph, Wachter, & Holzgrabe, 2014; Vandendriessche et al., 2013). NMR was used to identify change in the fingerprints of pesticide-stressed maize seedlings that displayed stunted growth (Blondel, Khelalfa, Reynaud, Fauvelle, & Raveton, 2016). Mass spectrometry is also used for obtaining such results, and it is obvious that the prevailing conditions should be regulated if the precision and sensitivity of the analysis is not to be filled by factors that are irrelevant to the analysis. Further, proper regulated growth conditions are indispensable to pool samples to decrease biological variation (Sumner et al., 2003). Moreover, higher control of sample preparation, NMR data collection, and spectrum processing are significant if the advantages of the fingerprinting strategy are to be amplified (Johnson, 2004; Lewis et al., 2007). The prime objectives of fingerprinting are to increase efficiency of NMR data collection, and to decrease disorganization of NMR signals prior to the statistical evaluation. The two similar biological samples should generate identical NMR fingerprints, ideally, but actually there will be disparity in the line shape and chemical shift, either of which results in unclear spectra (Ghosh, Sengupta, Sharma, & Sonawat, 2012; Yilmaz, Nyberg, Mølgaard, Asili, & Jaroszewski, 2010). Variation in line shape can be further reduced by employing exactly the same sample volume, and by optimizing the magnetic field homogeneity prior to data acquisition. This should be enough to produce reproducible line shapes, but if not, then the line shapes can be modulated during processing, for example, by changing the line expansion parameter to certify that the processed line width for an exact signal is the same in every spectrum. Variation in chemical shift can also be challenging, and the negative effect of such variation on principal components analysis has been illustrated (Cavalli, Salvatella, Dobson, & Vendruscolo, 2007; Alonso-Salces et al., 2010). Controlled sample preparation with exact pH regulation and specific temperature regulation during the spectrum acquisition should certify that most signals occur at exactly the same chemical shift (Aliev & Harris, 1994; Beckonert et al., 2010). Disparity in ionic affinity and the exact association between components in a sample may result in disparity of peak positions. These can be only be rectified by aligning spectral regions prior to the statistical analysis (Claridge, 2016). Moreover, numerous research reports have provided probable explanations for NMR fingerprinting by producing fascinating biological examinations of systems that have been evaluated. For example, evaluation of transgenic plant material has demonstrated the impending price of NMR fingerprinting as a contrivance for evaluating the significant equivalence of traditionally and genetically engineered crops (Manetti et al., 2004; Ren, Wang, Peng, Xia, & Qu, 2009). Comparative studies on transgenic and wild-type tomatoes and peas reveal significant variations in metabolites, as induced by external factors. These research findings further illustrate the comprehensive investigation of even meager variations in metabolites, so that they could assist in differentiating transgenics from wild-type plants. Metabolite fingerprinting can be used for analyzing abiotic stresses. ^{1}H NMR followed by PCA demonstrates the obvious discrepancy between control and Cd-stressed plants, and plots are drafted, because the primary principal component was subjugated by variations in meager signals that could be easily assigned to specific metabolites (Liu et al., 2011). Following this ^{1}H NMR, the data was utilized to test a hypothesis pertaining to the metabolic response to Cd stress, which was beyond the capacity of the fingerprinting study. NMR fingerprints of herbicide-stressed maize seedlings were used as the input for the normal network evaluation (Blondel et al., 2016). This resulted in discerning among 19 modes of action, and it was debated whether it was adequately consistent to be a functional contrivance in the discovery of bioactive compounds with new modes of action.

1.6.1 Metabolite Profiling of Plant Tissues Using NMR

Multivariate analysis of ^{1}H NMR spectra procured from plant samples is a comparatively modern expansion; NMR spectroscopy has been utilized as an analytical contrivance to identify the metabolite in such samples for many years (Longobardi et al., 2012; Rasmussen, Cloarec, Tang, Stærk, & Jaroszewski, 2006). Using multivariate analysis as a method of detecting the interesting characteristics of a spectrum before embarking on an assignment is an effective method of unravelling the plant metabolome, and it is expected to become a standard practice in examinations that progress beyond the discrimination step in a fingerprinting exercise (Le Gall, Colquhoun, & Defernez, 2004). However, profiling applications of NMR in plant tissues have usually focused first on the identification of particular metabolites, and so the techniques that have been developed to extract useful analytical information from the spectra of complex mixtures have been developed from a purely analytical perspective (Krishnan et al., 2004). ^{1}H NMR spectroscopy is a widespread type of NMR employed for metabolite detection, and it has been used for metabolite profiling on plant tissues in vivo (Mahrous & Farag, 2015; Zhao et al., 2016). These studies have further advanced to record the metabolic response of the tissues to anoxia, so that they can be examined directly. It was also possible to reduce data pertaining to the subcellular distribution of limited metabolites. The limited resolution of in vivo spectra was made difficult both by increasing overlap, and by covering the quality splitting of spin-coupled resonance; and it was conjectured that the comprehensive metabolic profile of tissue can be acquired with tissue extracts. The method of two-dimensional NMR techniques was easily implemented by narrowing the lines in extract spectra so as to correlate signals recorded on different regions of the spectrum (Frydman, Lupulescu, & Scherf, 2003; Schanda & Brutscher, 2005). This further amplified the metabolite magnitude that could be detected in the extract, and it amplified and enhanced the quantum of experiments that are retrospectively in vivo. Thus, it can be conjectured that quantitative data on 20 metabolites could be readily acquired from unprocessed plant extracts, and, to maximize this quantity, it is necessary to utilize two-dimensional spectroscopy (Brixner et al., 2005).

The method of ^{1}H NMR spectroscopy utilized for metabolite profiling has numerous advantages:

(a) This method determines the comprehensive metabolite profile procured from a minimally crude extract with complete structural information pertaining to chemical and physical properties such as volatility, polarity, or the existence of explicit chromophores.

(b) This method is constructive, allowing successive evaluation via other methods, and this method can be used with minimal information on the metabolite profile of the sample.

(c) Though ^{1}H NMR spectroscopy has a wide scope in metabolite profiling of animal and human samples, the method is significantly less relevant in plant samples. However, this method is employed for the analysis of root exudates and fruit juices. There are numerous research reports that demonstrate the metabolite profile explaining various organic acids, amino acids, and derivatives of mugineic acid in crude root exudates. One- and two-dimensional NMR was often complemented by GC-MS to detect metabolites, even at very low concentrations, and by high resonance mass spectrometry for estimating molecular mass determination. Fruit juices are prime targets for metabolite analysis of crude plant samples. NMR analysis of transgenic and wild-type tomatoes was prepared from freeze dried fruits, which resulted in identification of more than 40 compounds (Mattoo et al., 2006).

In these profiling evaluations, the focus has been on the estimation of composition and the associated classification of the samples into diverse groups, instead of testing the hypothesis pertaining to control and regulation of metabolic processes. Thus, compared with metabolite profiling by GC-MS, where one of the objectives is to utilize the data for comprehending the intermediary metabolism and its modulation (Sekiyama, Chikayama, & Kikuchi, 2010), the data from NMR has been entirely used for analytical and screening purposes. NMR has numerous advantages as a method for metabolite profiling, and it ensures inline separation and a detailed spectral record of crude extracts to surpass technical intricacies (Leiss et al., 2011; Tang, Xiao, & Wang, 2009). This method has a significant effect on the use of metabolic analysis in the biomedical field (Kim, Kim, Kim, Hwang, & Yoo, 2016), and LC-NMR is now established as a chief technique for phytochemical characterization (Akira et al., 2010; Cloarec et al., 2007). This strategy does not require explicit sample preparation prior to analysis, and NMR spectra can be recorded in a continuous or stopped flow mode.

This technique basically bifurcates output from the HPLC column into two pathways; one leads to NMR analysis, and the other to MS. This technique has numerous applications in xenobiotics metabolism (Halabalaki, Vougogiannopoulou, Mikros, & Skaltsounis, 2014; Hammerl, Frank, & Hofmann, 2017), and in detecting natural compounds in plant extracts (Amberg et al., 2017). The sensitivity of this strategy can be further advanced by a solid phase extraction (SPE) step between the HPLC column and the NMR spectrometer, and a cryogenic probe to detect the NMR signals (Lima et al., 2016). This SPE cartridge concentrates the samples, and reduces the need for costly deuterated solvents that are required for the elution of samples from cartridges (Mazumder, Gutch, & Dubey, 2015). These hyphenated NMR techniques can potentially enhance the magnitude of metabolites that can be detected in ^{1}H NMR fingerprints of unprocessed extracts. Methods for instances LC-UV-SPE-NMR-MS are used in the isolation of natural products, and in xenobiotic metabolism, and they can integrate with the fingerprinting method, as described in the previous section (Cakova et al., 2015). Hence, it can be conjectured that NMR is a prime technique for revealing plant metabolomes when fractionation of tissue extracts becomes a regular contrivance for procuring spectra employed in fingerprinting analysis.

1.7 METABOLITE FOOTPRINTING

Metabolic footprinting is new approach for the functional evaluation and characterization of cells via employing metabolomes. This method entails the measurement of metabolites generated from the intracellular complement of an organism into the external surrounding growth medium (Behrends, Williams, & Bundy, 2014; Kell et al., 2005). This technique is not only dependent on the estimation of intracellular metabolites, but also on assessment of metabolites utilized from, and released into, the growth media by a group of microbes via direct injection into an ESI mass spectrometer. Metabolic footprinting has been established as a suitable contrivance for pattern identification of metabolites in single-gene knockout strains in functional genomics evaluation (Kell & Oliver, 2016; Samota, Bhatt, Yadav, Garg, & Bajiya, 2017), together with the mode of action reports, in which the metabolite pattern is released when microbial strains are stimulated with a sublethal dosage of growth inhibitors, which enables it to discern the side or mode of action of these inhibitors. A genetic programming box facilitates the advancement of easy rules that could both differentiate the mode of action, and explain how this differentiation works (Kjeldahl et al., 2014). Metabolic footprinting is a valuable contrivance in chemical genetics and chemogenomics, in which xenobiotic metabolites impact cells at indefinite interaction sites (Patil et al., 2016). It is a well-established fact that microbes secrete enormous metabolites into external surroundings, particularly under severe environmental conditions. Metabolite footprinting reveals metabolite profiles of fermented beverages, a wide range of intracellular signaling molecules, and numerous other signaling metabolites, and biologically active compounds resulting from a primary and secondary metabolism are prime targets (Chaijaroenkul, Mubaraki, Ward, & Na-Bangchang, 2014; Haggerty et al., 2016). Metabolic secretions are products of cellular metabolic activity. Along with point of transcription and translation of relevant genes, it led to the comprehensive estimation of secreted metabolites that may be manipulated to achieve analytical benefits (DeBerardinis, Lum, Hatzivassiliou, & Thompson, 2008). The instant quenching that is needed to analyze intracellular metabolites should be averted so that the metabolite yield can be diluted into larger extracellular space, which declines corresponding to the comparative ratios of extracellular and intracellular volumes. Even though identification of metabolites in a microbial culture has previously been conceded via identification approaches (Bunge et al., 2008; Dorrestein, Mazmanian, & Knight, 2014), and via electronic noses, such evaluations are restricted to explicit metabolites, and are not carried out on an inclusive level (Garcia-Alcega et al., 2017; Schmidberger, Gutmann, Bayer, Kronthaler, & Huber, 2014). The metabolic footprinting approach commences with sample analysis, and is followed by data processing until identification can be completed, so a conclusion can be drawn. Metabolic footprinting has emerged as a suitable, reproducible, and high-throughput technique for the extensive genomic and physiological detection of microbes (Chumnanpuen, Hansen, Smedsgaard, & Nielsen, 2014). Extracellular metabolic flux is determined by metabolic pathways, and a majority of metabolic footprints of even single gene knockouts differ distinctly from other isogenic wild-type microbes (Schripsema, Lemos, Dianin, Vianna, & Dagnino, 2014). Metabolic footprinting can be integrated with knockout genes with known functions to recommend the function of other genes with unknown functions. Consistently, expansive and reproducible variations in the exometabolome occur when cells are exposed to sublethal concentrations of inhibitors with explicit modes of action (Jacoby, Martyn, & Kopriva, 2018; Swenson, Karaoz, Swenson, Bowen, & Northen, 2018). Increased technical advancements, for example, CE-MS24, high-throughput LC-MS, and multidimensional chromatography, along with GC × GC-TOF MS will ensure extensive coverage of the metabolome, and make these estimations more concise and inequitable (Niell et al., 2014; Rager et al., 2016). Metabolic footprinting has wider significance in the field of metabolic engineering (Kell & Oliver, 2016; Patejko, Jacyna, & Markuszewski, 2017). Metabolic engineering intends to ascertain metabolic products that augment the productivity of biological processes, and modulates their concentration or functions (Georgiev & Pavlov, 2017; Kiran, Ali, & Abdin, 2017). Metabolite footprinting provides the identification of undesired metabolites in extracellular space that are secreted as a result of non-productive pathways that should be eradicated (Sharma, Verma, Mathur, & Mathur, 2018). Metabolic footprinting will enable the comprehensive drafting of mathematical models of cell behavior that will explain the intricate relationships within cellular metabolism that are characteristic of systems biology (Pinu, Villas-Boas, & Aggio, 2017; García-Barrera, Rodríguez-Moro, Callejón-Leblic, Arias-Borrego, & Gómez-Ariza, 2017; Kostidis, 2018). Other metabolic models of microbes cannot generate metabolic secretions. The quantitative models with fewer possibilities for unraveling the steady state also require parameter optimization. Consequently, estimation of metabolic footprints of exometabolomes, as well as endometabolomes, can supply significant data about the limitations of metabolic models.

1.8 CONCLUSION

Metabolite footprinting and metabolite profiling are effective strategies for high-throughput systems of metabolic detection in plant samples. These analyses have the potential to complement high-throughput, system-wide evaluations via MS. Further, the application of coupled techniques that provide parallel MS and NMR analysis on similar plant samples would seem to be a suitable option to augment the metabolomic fraction that can be revealed by routine evaluation. Remarkably, the two factors that have induced the current progress in metabolomics; the requirement for more detailed evaluation to emphasize our perception of metabolic pathways, and the prospect of employing metabolic phenotypes based on structure to reveal gene function; have been advanced immensely by NMR and MS strategies for system-wide metabolite analysis. It is noteworthy that the immensely proficient strategy of employing metabolome data to display the phenotype of still mutation-manipulated metabolite fingerprints is based on ^{1}H NMR analysis. Metabolite fingerprinting implicates the arrangement into classes, so that conclusions can be drawn pertaining to the sorting of independent samples. Metabolite fingerprinting has immense potential for averting the frequently prolonged process of signal assignment. Prior to signal assignment, interest is focused on the spectra that are most suitable to the questions being addressed. Metabolic profiling and metabolite footprinting are instantaneous, and are expansively impartial in metabolite identification in samples, making metabolic fingerprinting an economic analytical approach for metabolic phenotypes.

REFERENCES

Aceña, J., Stampachiacchiere, S., Pérez, S., & Barceló, D. (2015). Advances in liquid chromatography–high-resolution mass spectrometry for quantitative and qualitative environmental analysis. *Analytical and Bioanalytical Chemistry*, *407*(21), 6289–6299.

Akira, K., Mitome, H., Imachi, M., Shida, Y., Miyaoka, H., & Hashimoto, T. (2010). LC-NMR identification of a novel taurine-related metabolite observed in 1H NMR-based metabonomics of genetically hypertensive rats. *Journal of Pharmaceutical and Biomedical Analysis*, *51*(5), 1091–1096.

Aliev, A. E., & Harris, K. D. (1994). Simple technique for temperature calibration of a MAS probe for solid-state NMR spectroscopy. *Magnetic Resonance in Chemistry*, *32*(6), 366–369.

Allard, P. M., Péresse, T., Bisson, J., Gindro, K., Marcourt, L., Pham, V. C., et al. (2016). Integration of molecular networking and in-silico MS/MS fragmentation for natural products dereplication. *Analytical Chemistry*, *88*(6), 3317–3323.

Allwood, J. W., Clarke, A., Goodacre, R., & Mur, L. A. J. (2010). Dual metabolomics: a novel approach to understanding plant-pathogen interactions. *Phytochemistry*, *71*, 590–597.

Allwood, J. W., Ellis, D. I., & Goodacre, R. (2008). Metabolomic technologies and their application to the study of plants and plant-host interactions. *Physiologia Plantarum*, *132*, 117–135.

Alonso, A., Marsal, S., & Julià, A. (2015). Analytical methods in untargeted metabolomics: state of the art in 2015. *Frontiers in Bioengineering and Biotechnology*, *3*, 23.

Alonso-Salces, R. M., Héberger, K., Holland, M. V., Moreno-Rojas, J. M., Mariani, C., Bellan, G., et al. (2010). Multivariate analysis of NMR fingerprint of the unsaponifiable fraction of virgin olive oils for authentication purposes. *Food Chemistry*, *118*(4), 956–965.

Al-Saffar, Y., Stephanson, N. N., & Beck, O. (2013). Multicomponent LC–MS/MS screening method for detection of new psychoactive drugs, legal highs, in urine—experience from the Swedish population. *Journal of Chromatography B*, *930*, 112–120.

Amann, A., Lacy Costello, B., de, Miekisch, W., Schubert, J., Buszewski, B., Pleil, J., et al. (2014). The human volatilome: volatile organic compounds (VOCs) in exhaled breath, skin emanations, urine, feces and saliva. *Journal of Breath Research*, *8*(3)https://doi.org/10.1088/1752-7155/8/3/03400.

Amberg, A., Riefke, B., Schlotterbeck, G., Ross, A., Senn, H., Dieterle, F., et al. (2017). NMR and MS methods for metabolomics. In *Drug Safety Evaluation* (pp. 229–258). New York, NY: Humana Press.

Antoniewicz, M. R. (2015). Methods and advances in metabolic flux analysis: a mini-review. *Journal of Industrial Microbiology & Biotechnology*, *42*(3), 317–325.

Arathi, B. P., Sowmya, P. R. R., Vijay, K., Baskaran, V., & Lakshminarayana, R. (2015). Metabolomics of carotenoids: the challenges and prospects–A review. *Trends in Food Science & Technology*, *45*(1), 105–117.

Arathi, B. P., Sowmya, P. R. R., Vijay, K., Baskaran, V., & Lakshminarayana, R. (2016). Biofunctionality of carotenoid metabolites: an insight into qualitative and quantitative analysis. In *Metabolomics-Fundamentals and Applications*: InTech.

Aretz, I., & Meierhofer, D. (2016). Advantages and pitfalls of mass spectrometry based metabolome profiling in systems biology. *International Journal of Molecular Sciences*, *17*(5), 632.

Arnts, R. R. (2010). Evaluation of adsorbent sampling tube materials and Tenax-TA for analysis of volatile biogenic organic compounds. *Atmospheric Environment*, *44*(12), 1579–1584.

Asfaram, A., Ghaedi, M., & Goudarzi, A. (2016). Optimization of ultrasound-assisted dispersive solid-phase microextraction based on nanoparticles followed by spectrophotometry for the simultaneous determination of dyes using experimental design. *Ultrasonics Sonochemistry*, *32*, 407–417.

Athersuch, T. (2016). Metabolome analyses in exposome studies: profiling methods for a vast chemical space. *Archives of Biochemistry and Biophysics*, *589*, 177–186.

Atkinson, R., & Arey, J. (2003). Atmospheric degradation of volatile organic compounds. *Chemical Reviews*, *103*(12), 4605–4638.

Aw, W., & Fukuda, S. (2015). Toward the comprehensive understanding of the gut ecosystem via metabolomics-based integrated omics approach. *Seminars in Immunopathology*, Vol. 37(1), 5–16.

Bae, E., Na, J. G., Chung, S. H., Kim, H. S., & Kim, S. (2010). Identification of about 30 000 chemical components in shale oils by electrospray ionization (ESI) and atmospheric pressure photoionization (APPI) coupled with 15 T Fourier transform ion cyclotron resonance mass spectrometry (FT-ICR MS) and a comparison to conventional oil. *Energy & Fuels*, *24*(4), 2563–2569.

Baig, F., Pechlaner, R., & Mayr, M. (2016). Caveats of untargeted metabolomics for biomarker discovery. *Journal of the American College of Cardiology*, *68*(12).

Bailey, N. J., Oven, M., Holmes, E., Zenk, M. H., & Nicholson, J. K. (2004). An NMR-based metabolomic approach to the analysis of the effects of xenobiotics on endogenous metabolite levels in plants. *Journal of Spectroscopy*, *18*(2), 279–287.

Barbu, I. M. (2008). *Fourier transform ion cyclotron resonance mass spectrometry. Development of instrumentation, data acquisition software and processing methods.* [Doctoral dissertation] Utrecht University.

Barouki, R. (2010). Linking long-term toxicity of xeno-chemicals with short-term biological adaptation. *Biochimie*, *92*(9), 1222–1226.

Bauer, A., Luetjohann, J., Rohn, S., Kuballa, J., & Jantzen, E. (2018). Ion chromatography tandem mass spectrometry (IC-MS/MS) multimethod for the determination of highly polar pesticides in plant-derived commodities. *Food Control*, *86*, 71–76.

Beckonert, O., Coen, M., Keun, H. C., Wang, Y., Ebbels, T. M., Holmes, E., et al. (2010). High-resolution magic-angle-spinning NMR spectroscopy for metabolic profiling of intact tissues. *Nature Protocols*, *5*(6), 1019.

Behrends, V., Williams, H. D., & Bundy, J. G. (2014). Metabolic footprinting: extracellular metabolomic analysis. In *Pseudomonas methods and protocols* (pp. 281–292). New York, NY: Humana Press.

Beisken, S., Eiden, M., & Salek, R. M. (2015). Getting the right answers: understanding metabolomics challenges. *Expert Review of Molecular Diagnostics*, *15*(1), 97–109.

Belton, P. S., Colquhoun, I. J., Kemsley, E. K., Delgadillo, I., Roma, P., Dennis, M. J., et al. (1998). Application of chemometrics to the 1H NMR spectra of apple juices: discrimination between apple varieties. *Food Chemistry*, *61*(1-2), 207–213.

Beuerle, T., & Schwab, W. (1999). Metabolic profile of linoleic acid in stored apples: formation of 13 (*R*)-hydroxy-9 (*Z*), 11 (*E*)-octadecadienoic acid. *Lipids*, *34*(4), 375.

Beyoğlu, D., & Idle, J. R. (2013). Metabolomics and its potential in drug development. *Biochemical Pharmacology*, *85*(1), 12–20.

Bingol, K., Bruschweiler-Li, L., Li, D., Zhang, B., Xie, M., & Brüschweiler, R. (2016). Emerging new strategies for successful metabolite identification in metabolomics. *Bioanalysis*, *8*(6), 557–573.

Bingol, K., Bruschweiler-Li, L., Yu, C., Somogyi, A., Zhang, F., & Brüschweiler, R. (2015). Metabolomics beyond spectroscopic databases: a combined MS/NMR strategy for the rapid identification of new metabolites in complex mixtures. *Analytical Chemistry*, *87*(7), 3864–3870.

Bino, R. J., Hall, R. D., Fiehn, O., Kopka, J., Saito, K., Draper, J., et al. (2004). Potential of metabolomics as a functional genomics tool. *Trends in Plant Science*, *9*(9), 418–425.

Blondel, C., Khelalfa, F., Reynaud, S., Fauvelle, F., & Raveton, M. (2016). Effect of organochlorine pesticides exposure on the maize root metabolome assessed using high-resolution magic-angle spinning 1H NMR spectroscopy. *Environmental Pollution*, *214*, 539–548.

Boczkaj, G., Przyjazny, A., & Kamiński, M. (2014). Characteristics of volatile organic compounds emission profiles from hot road bitumens. *Chemosphere*, *107*, 23–30.

Boughton, B. A., Thinagaran, D., Sarabia, D., Bacic, A., & Roessner, U. (2016). Mass spectrometry imaging for plant biology: a review. *Phytochemistry Reviews*, *15*(3), 445–488.

Bouhifd, M., Beger, R., Flynn, T., Guo, L., Harris, G., Hogberg, H., et al. (2015). Quality assurance of metabolomics. *ALTEX*, *32*(4), 319–326.

Boyacı, E., Rodriguez-Lafuente, A., Gorynski, K., Mirnaghi, F., Souza-Silva, E. A., Hein, D., et al. (2015). Sample preparation with solid phase microextraction and exhaustive extraction approaches: Comparison for challenging cases. *Analytica Chimica Acta*, *873*, 14–30.

Brixner, T., Stenger, J., Vaswani, H. M., Cho, M., Blankenship, R. E., & Fleming, G. R. (2005). Two-dimensional spectroscopy of electronic couplings in photosynthesis. *Nature*, *434*(7033), 625.

Bruno, C., Patin, F., Bocca, C., Nadal-Desbarats, L., Bonnier, F., Reynier, P., et al. (2018). The combination of four analytical methods to explore skeletal muscle metabolomics: better coverage of metabolic pathways or a marketing argument? *Journal of Pharmaceutical and Biomedical Analysis*, *148*, 273–279.

Bujak, R., Daghir-Wojtkowiak, E., Kaliszan, R., & Markuszewski, M. J. (2016). PLS-based and regularization-based methods for the selection of relevant variables in non-targeted metabolomics data. *Frontiers in Molecular Biosciences*, *3*, 35.

Bujak, R., Struck-Lewicka, W., Markuszewski, M. J., & Kaliszan, R. (2015). Metabolomics for laboratory diagnostics. *Journal of Pharmaceutical and Biomedical Analysis*, *113*, 108–120.

Bunge, M., Araghipour, N., Mikoviny, T., Dunkl, J., Schnitzhofer, R., Hansel, A., et al. (2008). On-line monitoring of microbial volatile metabolites by proton transfer reaction-mass spectrometry. *Applied and Environmental Microbiology*, *74*(7), 2179–2186.

Cajka, T., & Fiehn, O. (2015). Toward merging untargeted and targeted methods in mass spectrometry-based metabolomics and lipidomics. *Analytical Chemistry*, *88*(1), 524–545.

Cakova, V., Urbain, A., Antheaume, C., Rimlinger, N., Wehrung, P., Bonté, F., et al. (2015). Identification of phenanthrene derivatives in *Aerides rosea* (Orchidaceae) using the combined systems HPLC–ESI–HRMS/MS and HPLC–DAD–MS–SPE–UV–NMR. *Phytochemical Analysis*, *26*(1), 34–39.

Cambiaghi, A., Ferrario, M., & Masseroli, M. (2016). Analysis of metabolomic data: tools, current strategies and future challenges for omics data integration. *Briefings in Bioinformatics*, *18*(3), 498–510. Epub 2016/04/15.

Carvalho, L. M., Carvalho, F., de Lourdes Bastos, M., Baptista, P., Moreira, N., Monforte, A. R., et al. (2014). Non-targeted and targeted analysis of wild toxic and edible mushrooms using gas chromatography–ion trap mass spectrometry. *Talanta, 118*, 292–303.
Cavalli, A., Salvatella, X., Dobson, C. M., & Vendruscolo, M. (2007). Protein structure determination from NMR chemical shifts. *Proceedings of the National Academy of Sciences, 104*(23), 9615–9620.
Cavill, R., Jennen, D., Kleinjans, J., & Briedé, J. J. (2015). Transcriptomic and metabolomic data integration. *Briefings in Bioinformatics, 17*(5), 891–901.
Cesbron, N., Royer, A. L., Guitton, Y., Sydor, A., Le Bizec, B., & Dervilly-Pinel, G. (2017). Optimization of fecal sample preparation for untargeted LC-HRMS based metabolomics. *Metabolomics, 13*(8), 99.
Chaijaroenkul, W., Mubaraki, M. A., Ward, S. A., & Na-Bangchang, K. (2014). Metabolite footprinting of *Plasmodium falciparum* following exposure to *Garcinia mangostana* Linn. crude extract. *Experimental Parasitology, 145*, 80–86.
Chatterjee, S., Srivastava, S., Khalid, A., Singh, N., Sangwan, R. S., Sidhu, O. P., et al. (2010). Comprehensive metabolic fingerprinting of *Withania somnifera* leaf and root extracts. *Phytochemistry, 71*(10), 1085–1094.
Chen, J., Li, B. Q., Cui, Y. Q., Yu, E., & Zhai, H. L. (2015). A fast and effective method of quantitative analysis of VB1, VB2 and VB6 in B-vitamins complex tablets based on three-dimensional fluorescence spectra. *Journal of Food Composition and Analysis, 41*, 122–128.
Cheng, J., Lan, W., Zheng, G., & Gao, X. (2018). Metabolomics: a high-throughput platform for metabolite profile exploration. In *Computational systems biology* (pp. 265–292). New York, NY: Humana Press.
Christian, B., & Luckas, B. (2008). Determination of marine biotoxins relevant for regulations: from the mouse bioassay to coupled LC-MS methods. *Analytical and Bioanalytical Chemistry, 391*(1), 117–134.
Christians, U., Klawitter, J., Hornberger, A., & Klawitter, J. (2011). How unbiased is non-targeted metabolomics and is targeted pathway screening the solution? *Current Pharmaceutical Biotechnology, 12*(7), 1053–1066.
Chumnanpuen, P., Hansen, M. A. E., Smedsgaard, J., & Nielsen, J. (2014). Dynamic metabolic footprinting reveals the key components of metabolic network in yeast *Saccharomyces cerevisiae*. *International Journal of Genomics, 2014*.
Claridge, T. D. (2016). *High-resolution NMR techniques in organic chemistry*. Vol. 27. Elsevier.
Clavijo Mccormick, A., Gershenzon, J., & Unsicker, S. B. (2014). Little peaks with big effects: establishing the role of minor plant volatiles in plant–insect interactions. *Plant, Cell & Environment, 37*(8), 1836–1844.
Cloarec, O., Campbell, A., Tseng, L. H., Braumann, U., Spraul, M., Scarfe, G., et al. (2007). Virtual chromatographic resolution enhancement in cryoflow LC – NMR experiments via statistical total correlation spectroscopy. *Analytical Chemistry, 79*(9), 3304–3311.
Colinge, J., Masselot, A., Giron, M., Dessingy, T., & Magnin, J. (2003). OLAV: towards high-throughput tandem mass spectrometry data identification. *Proteomics, 3*(8), 1454–1463.
Considine, E. C., Thomas, G., Boulesteix, A. L., Khashan, A. S., & Kenny, L. C. (2018). Critical review of reporting of the data analysis step in metabolomics. *Metabolomics, 14* (1), 7.
Cortassa, S., Caceres, V., Bell, L. N., O'Rourke, B., Paolocci, N., & Aon, M. A. (2015). From metabolomics to fluxomics: a computational procedure to translate metabolite profiles into metabolic fluxes. *Biophysical Journal, 108*(1), 163–172.
Couderc, F., Ong-Meang, V., & Poinsot, V. (2017). Capillary electrophoresis hyphenated with UV-native-laser induced fluorescence detection (CE/UV-native-LIF). *Electrophoresis, 38*(1), 135–149.

Creek, D. J., Dunn, W. B., Fiehn, O., Griffin, J. L., Hall, R. D., Lei, Z., et al. (2014). Metabolite identification: are you sure? And how do your peers gauge your confidence? *Metabolomics*, *10*(3), 350–353.
Crown, S. B., Long, C. P., & Antoniewicz, M. R. (2015). Integrated 13 C-metabolic flux analysis of 14 parallel labeling experiments in *Escherichia coli*. *Metabolic Engineering*, *28*, 151–158.
Cuadros-Inostroza, A., Ruíz-Lara, S., González, E., Eckardt, A., Willmitzer, L., & Peña-Cortés, H. (2016). GC–MS metabolic profiling of Cabernet Sauvignon and Merlot cultivars during grapevine berry development and network analysis reveals a stage-and cultivar-dependent connectivity of primary metabolites. *Metabolomics*, *12*(2), 39.
Cubero-Leon, E., Peñalver, R., & Maquet, A. (2014). Review on metabolomics for food authentication. *Food Research International*, *60*, 95–107.
DeBerardinis, R. J., Lum, J. J., Hatzivassiliou, G., & Thompson, C. B. (2008). The biology of cancer: metabolic reprogramming fuels cell growth and proliferation. *Cell Metabolism*, *7* (1), 11–20.
Deda, O., Gika, H. G., Wilson, I. D., & Theodoridis, G. A. (2015). An overview of fecal sample preparation for global metabolic profiling. *Journal of Pharmaceutical and Biomedical Analysis*, *113*, 137–150.
DeFelice, B. C., Mehta, S. S., Samra, S., Čajka, T., Wancewicz, B., Fahrmann, J. F., et al. (2017). Mass spectral feature list optimizer (MS-FLO): a tool to minimize false positive peak reports in untargeted liquid chromatography–mass spectroscopy (LC-MS) data processing. *Analytical Chemistry*, *89*(6), 3250–3255.
Di Guida, R., Engel, J., Allwood, J. W., Weber, R. J., Jones, M. R., Sommer, U., et al. (2016). Non-targeted UHPLC-MS metabolomic data processing methods: a comparative investigation of normalisation, missing value imputation, transformation and scaling. *Metabolomics*, *12*(5), 93.
Dieterle, F., Ross, A., Schlotterbeck, G., & Senn, H. (2006). Probabilistic quotient normalization as robust method to account for dilution of complex biological mixtures. Application in 1H NMR metabonomics. *Analytical Chemistry*, *78*, 4281–4290.
Ding, L., Wendl, M. C., McMichael, J. F., & Raphael, B. J. (2014). Expanding the computational toolbox for mining cancer genomes. *Nature Reviews Genetics*, *15*(8), 556.
Dixit, D., & Reddy, C. R. K. (2017). Non-targeted secondary metabolite profile study for deciphering the cosmeceutical potential of red marine macro alga *Jania rubens*—an LCMS-based approach. *Cosmetics*, *4*(4), 45.
Domon, B., & Aebersold, R. (2006). Mass spectrometry and protein analysis. *Science*, *312* (5771), 212–217.
Dongre, A. R., Somogyi, Á., & Wysocki, V. H. (1996). Surface-induced dissociation: an effective tool to probe structure, energetics and fragmentation mechanisms of protonated peptides. *Journal of Mass Spectrometry*, *31*(4), 339–350.
Dorrestein, P. C., Mazmanian, S. K., & Knight, R. (2014). Finding the missing links among metabolites, microbes, and the host. *Immunity*, *40*(6), 824–832.
Dudareva, N., Klempien, A., Muhlemann, J. K., & Kaplan, I. (2013). Biosynthesis, function and metabolic engineering of plant volatile organic compounds. *New Phytologist*, *198*(1), 16–32.
Dunn, W. B., Broadhurst, D. I., Edison, A., Guillou, C., Viant, M. R., Bearden, D. W., et al. (2017). Quality assurance and quality control processes: summary of a metabolomics community questionnaire. *Metabolomics*, *13*(5), 50.
Dunn, W. B., & Ellis, D. I. (2005). Metabolomics: current analytical platforms and methodologies. *TrAC Trends in Analytical Chemistry*, *24*(4), 285–294.
El-Hawiet, A., Shoemaker, G. K., Daneshfar, R., Kitova, E. N., & Klassen, J. S. (2011). Applications of a catch and release electrospray ionization mass spectrometry assay for carbohydrate library screening. *Analytical Chemistry*, *84*(1), 50–58.

Ellis, D. I., Dunn, W. B., Griffin, J. L., Allwood, J. W., & Goodacre, R. (2007). Metabolic fingerprinting as a diagnostic tool. *Pharmacogenomics*, *8*(9), 1243–1266.

Engskog, M. K., Haglöf, J., Arvidsson, T., & Pettersson, C. (2016). LC–MS based global metabolite profiling: the necessity of high data quality. *Metabolomics*, *12*(7), 114.

EPA (1999). *TO-11 Compendium of Methods for the Determination of Toxic Organic Compounds in Ambient Air: Determination of Formaldehyde in Ambient Air Using Adsorbent Cartridge Followed by High Performance Liquid Chromatography (HPLC)*. US Environmental Protection Agency. Available from:*http://www.epa.gov/ttnamti1/files/ambient/airtox/to-11ar.pdf* (Accessed January 2009).

Erxleben, A., Gewssler, A., Vervliet-Scheebaum, M., & Reski, R. (2012). Metabolite profiling of the moss *Physcomitrella patens* reveals evolutionary conservation of osmoprotective substances. *Plant Cell Reports*, *31*, 427–436.

Fahrmann, J. F., Grapov, D., Wanichthanarak, K., DeFelice, B. C., Salemi, M. R., Rom, W. N., et al. (2017). Integrated metabolomics and proteomics highlight altered nicotinamide and polyamine pathways in lung adenocarcinoma. *Carcinogenesis*, *38*(3), 271–280.

Fall, T., Salihovic, S., Brandmaier, S., Nowak, C., Ganna, A., Gustafsson, S., et al. (2016). Non-targeted metabolomics combined with genetic analyses identifies bile acid synthesis and phospholipid metabolism as being associated with incident type 2 diabetes. *Diabetologia*, *59*(10), 2114–2124.

Fanali, C., D'Orazio, G., Fanali, S., & Gentili, A. (2017). Advanced analytical techniques for fat-soluble vitamin analysis. *TrAC Trends in Analytical Chemistry*, *87*, 82–97.

Farag, M. A., Porzel, A., Mahrous, E. A., Mo'men, M., & Wessjohann, L. A. (2015). Integrated comparative metabolite profiling via MS and NMR techniques for Senna drug quality control analysis. *Analytical and Bioanalytical Chemistry*, *407*(7), 1937–1949.

Fernandez-Peralbo, M. A., Santiago, M. C., Priego-Capote, F., & de Castro, M. L. (2015). Study of exhaled breath condensate sample preparation for metabolomics analysis by LC–MS/MS in high resolution mode. *Talanta*, *144*, 1360–1369.

Feussner, I., & Polle, A. (2015). What the transcriptome does not tell—proteomics and metabolomics are closer to the plants' patho-phenotype. *Current Opinion in Plant Biology*, *26*, 26–31.

Fiehn, O., Robertson, D., Griffin, J., van der Werf, M., Nikolau, B., Morrison, N., et al. (2007). The metabolomics standards initiative (MSI). *Metabolomics*, *3*(3), 175–178.

Fiehn, O., Wohlgemuth, G., Scholz, M., Kind, T., Lee, D. Y., Lu, Y., et al. (2008). Quality control for plant metabolomics: reporting MSI-compliant studies. *The Plant Journal*, *53* (4), 691–704.

Filla, L. A., Sanders, K. L., Filla, R. T., & Edwards, J. L. (2016). Automated sample preparation in a microfluidic culture device for cellular metabolomics. *Analyst*, *141*(12), 3858–3865.

Floegel, A., Stefan, N., Yu, Z., Mühlenbruch, K., Drogan, D., Joost, H. G., et al. (2013). Identification of serum metabolites associated with risk of type 2 diabetes using a targeted metabolomic approach. *Diabetes*, *62*(2), 639–648.

Forcisi, S., Moritz, F., Kanawati, B., Tziotis, D., Lehmann, R., & Schmitt-Kopplin, P. (2013). Liquid chromatography–mass spectrometry in metabolomics research: Mass analyzers in ultra high pressure liquid chromatography coupling. *Journal of Chromatography A*, *1292*, 51–65.

Fotakis, C., Kokkotou, K., Zoumpoulakis, P., & Zervou, M. (2013). NMR metabolite fingerprinting in grape derived products: An overview. *Food Research International*, *54*(1), 1184–1194.

Fraser, K., Lane, G. A., Otter, D. E., Harrison, S. J., Quek, S. Y., Hemar, Y., et al. (2014). Non-targeted analysis by LC–MS of major metabolite changes during the oolong tea manufacturing in New Zealand. *Food Chemistry*, *151*, 394–403.

Frydman, L., Lupulescu, A., & Scherf, T. (2003). Principles and features of single-scan two-dimensional NMR spectroscopy. *Journal of the American Chemical Society*, *125*(30), 9204–9217.
Fuhrer, T., & Zamboni, N. (2015). High-throughput discovery metabolomics. *Current Opinion in Biotechnology*, *31*, 73–78.
Fukushima, A., Kusano, M., Redestig, H., Arita, M., & Saito, K. (2009). Integrated omics approaches in plant systems biology. *Current Opinion in Chemical Biology*, *13*(5-6), 532–538.
Gallagher, M., Wysocki, C. J., Leyden, J. J., Spielman, A. I., Sun, X., & Preti, G. (2008). Analyses of volatile organic compounds from human skin. *British Journal of Dermatology*, *159*(4), 780–791.
Garcia, B. A., Mollah, S., Ueberheide, B. M., Busby, S. A., Muratore, T. L., Shabanowitz, J., et al. (2007). Chemical derivatization of histones for facilitated analysis by mass spectrometry. *Nature Protocols*, *2*(4), 933.
Garcia-Alcega, S., Nasir, Z. A., Ferguson, R., Whitby, C., Dumbrell, A. J., Colbeck, I., et al. (2017). Fingerprinting outdoor air environment using microbial volatile organic compounds (MVOCs)–a review. *TrAC Trends in Analytical Chemistry*, *86*, 75–83.
García-Barrera, T., Rodríguez-Moro, G., Callejón-Leblic, B., Arias-Borrego, A., & Gómez-Ariza, J. L. (2017). Mass spectrometry based analytical approaches and pitfalls for toxicometabolomics of arsenic in mammals: a tutorial review. *Analytica Chimica Acta*,.
García-Villalba, R., Tomás-Barberán, F. A., Fança-Berthon, P., Roller, M., Zafrilla, P., Issaly, N., et al. (2015). Targeted and untargeted metabolomics to explore the bioavailability of the secoiridoids from a seed/fruit extract (*Fraxinus angustifolia* Vahl) in human healthy volunteers: a preliminary study. *Molecules*, *20*(12), 22202–22219.
Georgiev, V., & Pavlov, A. (2017). Genetic engineering and manipulation of metabolite pathways in *Salvia* spp. In *Salvia biotechnology* (pp. 399–414). Cham: Springer.
Gerstl, M. P., Ruckerbauer, D. E., Mattanovich, D., Jungreuthmayer, C., & Zanghellini, J. (2015). Metabolomics integrated elementary flux mode analysis in large metabolic networks. *Scientific Reports*, *5*, 8930.
Ghosh, S., Sengupta, A., Sharma, S., & Sonawat, H. M. (2012). Metabolic fingerprints of serum, brain, and liver are distinct for mice with cerebral and noncerebral malaria: a 1H NMR spectroscopy-based metabonomic study. *Journal of Proteome Research*, *11* (10), 4992–5004.
Gidman, E., Goodacre, R., Emmett, B., Smith, A. R., & Gwynn-Jones, D. (2003). Investigating plant–plant interference by metabolic fingerprinting. *Phytochemistry*, *63*(6), 705–710.
Gika, H. G., Theodoridis, G. A., Plumb, R. S., & Wilson, I. D. (2014). Current practice of liquid chromatography–mass spectrometry in metabolomics and metabonomics. *Journal of Pharmaceutical and Biomedical Analysis*, *87*, 12–25.
Gika, H. G., Zisi, C., Theodoridis, G., & Wilson, I. D. (2016). Protocol for quality control in metabolic profiling of biological fluids by U (H) PLC-MS. *Journal of Chromatography B*, *1008*, 15–25.
Gioria, S., Lobo Vicente, J., Barboro, P., La Spina, R., Tomasi, G., Urbán, P., et al. (2016). A combined proteomics and metabolomics approach to assess the effects of gold nanoparticles in vitro. *Nanotoxicology*, *10*(6), 736–748.
Godzien, J., Alonso-Herranz, V., Barbas, C., & Armitage, E. G. (2015). Controlling the quality of metabolomics data: new strategies to get the best out of the QC sample. *Metabolomics*, *11*(3), 518–528.
Gopalakrishnan, S., & Maranas, C. D. (2015). 13C metabolic flux analysis at a genome-scale. *Metabolic Engineering*, *32*, 12–22.
Griffin, J. L. (2003). Metabonomics: NMR spectroscopy and pattern recognition analysis of body fluids and tissues for characterisation of xenobiotic toxicity and disease diagnosis. *Current Opinion in Chemical Biology*, 7(5), 648–654.

Haggarty, J., & Burgess, K. E. (2017). Recent advances in liquid and gas chromatography methodology for extending coverage of the metabolome. *Current Opinion in Biotechnology, 43*, 77–85.

Haggerty, J., Bowyer, P. K., Jiranek, V., & Taylor, D. K. (2016). Optimisation and validation of a high-throughput semi-quantitative solid-phase microextraction method for analysis of fermentation aroma compounds in metabolomic screening studies of wines. *Australian Journal of Grape and Wine Research, 22*(1), 3–10.

Halabalaki, M., Vougogiannopoulou, K., Mikros, E., & Skaltsounis, A. L. (2014). Recent advances and new strategies in the NMR-based identification of natural products. *Current Opinion in Biotechnology, 25*, 1–7.

Halket, J. M., Waterman, D., Przyborowska, A. M., Patel, R. K., Fraser, P. D., & Bramley, P. M. (2004). Chemical derivatization and mass spectral libraries in metabolic profiling by GC/MS and LC/MS/MS. *Journal of Experimental Botany, 56*(410), 219–243.

Hall, R. D. (2006). Plant metabolomics: from holistic hope, to hype, to hot topic. *New Phytologist, 169*(3), 453–468.

Hammerl, R., Frank, O., & Hofmann, T. (2017). Differential off-line LC-NMR (DOLC-NMR) metabolomics to monitor tyrosine-induced metabolome alterations in *Saccharomyces cerevisiae*. *Journal of Agricultural and Food Chemistry, 65*(15), 3230–3241.

Haug, K., Salek, R. M., Conesa, P., Hastings, J., de Matos, P., Rijnbeek, M., et al. (2012). MetaboLights—an open-access general-purpose repository for metabolomics studies and associated meta-data. *Nucleic Acids Research, 41*(D1), D781–D786.

He, L., Li, C., Liu, X., Yang, Q., Zhang, H., Xu, W., et al. (2017). Comparative study on the interaction between 3 CYP2C9 allelic isoforms and benzbromarone by using LC–MS/MS method. *Journal of Chromatography B, 1070*, 97–103.

He, X., & Liu, R. H. (2008). Phytochemicals of apple peels: isolation, structure elucidation, and their antiproliferative and antioxidant activities. *Journal of Agricultural and Food Chemistry, 56*(21), 9905–9910.

Hirayama, A., Wakayama, M., & Soga, T. (2014). Metabolome analysis based on capillary electrophoresis-mass spectrometry. *TrAC Trends in Analytical Chemistry, 61*, 215–222.

Hodáková, J., Preisler, J., Foret, F., & Kubáň, P. (2015). Sensitive determination of glutathione in biological samples by capillary electrophoresis with green (515 nm) laser-induced fluorescence detection. *Journal of Chromatography A, 1391*, 102–108.

Hogg, D., Alfeeli, B., Hoover, A., Zhang, T., Rice, G., & Agah, M. (2011). Trace organics monitoring in drinking water using Tenax-coated fibers. In *Sensors, 2011 IEEE* (pp. 1093–1096): IEEE.

Hohmann, M., Christoph, N., Wachter, H., & Holzgrabe, U. (2014). 1H NMR profiling as an approach to differentiate conventionally and organically grown tomatoes. *Journal of Agricultural and Food Chemistry, 62*(33), 8530–8540.

Holmes, E., Tang, H., Wang, Y., & Seger, C. (2006). The assessment of plant metabolite profiles by NMR-based methodologies. *Planta Medica, 72*(09), 771–785.

Imam, S., Schäuble, S., Brooks, A. N., Baliga, N. S., & Price, N. D. (2015). Data-driven integration of genome-scale regulatory and metabolic network models. *Frontiers in Microbiology, 6*, 409.

Jacoby, R. P., Martyn, A., & Kopriva, S. (2018). Exometabolomic profiling of bacterial strains cultivated using Arabidopsis root extract as the sole carbon source. *Molecular Plant-Microbe Interactions*.

Jamari, N. L. A., Dohmann, J. F., Raab, A., Krupp, E. M., & Feldmann, J. (2017). Novel non-target analysis of fluorine compounds using ICPMS/MS and HPLC-ICPMS/MS. *Journal of Analytical Atomic Spectrometry, 32*(5), 942–950.

Jiao, C., Song, C., Zheng, S., Zhu, Y., Jin, Q., Cai, Y., et al. (2018). Metabolic profiling of *Dendrobium officinale* in response to precursors and methyl jasmonate. *International Journal of Molecular Sciences, 19*(3), 728.

Jo, S. H., Kim, K. H., & Kwon, K. (2017). The combined effects of sampling parameters on the sorbent tube sampling of phthalates in air. *Scientific Reports*, 7, 45677.
Johnson, B. A. (2004). Using NMRView to visualize and analyze the NMR spectra of macromolecules. In *Protein NMR Techniques* (pp. 313–352): Humana Press.
Johnson, C. H., Ivanisevic, J., & Siuzdak, G. (2016). Metabolomics: beyond biomarkers and towards mechanisms. *Nature Reviews Molecular Cell Biology*, *17*(7), 451.
Johnson, R. D., & Lewis, R. J. (2005). *Simultaneous Quantitation of Atenolol, Metoprolol, and Propranolol in Biological Matrices Via LC/MS*. [No. DOT/FAA/AM-05/10]. Washington, DC: Office of Aviation Medicine, Federal Avaition Administration
Johnson, R. D., & Lewis, R. J. (2006). Quantitation of atenolol, metoprolol, and propranolol in postmortem human fluid and tissue specimens via LC/APCI-MS. *Forensic Science International*, *156*(2-3), 106–117.
Jorge, T. F., Rodrigues, J. A., Caldana, C., Schmidt, R., van Dongen, J. T., Thomas-Oates, J., et al. (2016). Mass spectrometry-based plant metabolomics: metabolite responses to abiotic stress. *Mass Spectrometry Reviews*, *35*(5), 620–649.
Kan, Z., Jaiswal, B. S., Stinson, J., Janakiraman, V., Bhatt, D., Stern, H. M., et al. (2010). Diverse somatic mutation patterns and pathway alterations in human cancers. *Nature*, *466*(7308), 869.
Kanehisa, M., Goto, S., Sato, Y., Furumichi, M., & Tanabe, M. (2012). KEGG for integration and interpretation of largescale molecular data sets. *Nucleic Acids Research*, *40*(Database issue). D109–14. Epub 2011/11/15.
Keeling, C. I., Li, M., Sandhu, H. K., Henderson, H., Saint Yuen, M. M., & Bohlmann, J. (2016). Quantitative metabolome, proteome and transcriptome analysis of midgut and fat body tissues in the mountain pine beetle, *Dendroctonus ponderosae* Hopkins, and insights into pheromone biosynthesis. *Insect Biochemistry and Molecular Biology*, *70*, 170–183.
Kell, D. B., Brown, M., Davey, H. M., Dunn, W. B., Spasic, I., & Oliver, S. G. (2005). Metabolic footprinting and systems biology: the medium is the message. *Nature Reviews Microbiology*, *3*(7), 557.
Kell, D. B., & Oliver, S. G. (2016). The metabolome 18 years on: a concept comes of age. *Metabolomics*, *12*(9), 148.
Kenar, E., Franken, H., Forcisi, S., Wörmann, K., Häring, H. U., Lehmann, R., et al. (2014). Automated label-free quantification of metabolites from liquid chromatography–mass spectrometry data. *Molecular & Cellular Proteomics*, *13*(1), 348–359.
Khamis, M. M., Adamko, D. J., & El-Aneed, A. (2017). Mass spectrometric based approaches in urine metabolomics and biomarker discovery. *Mass Spectrometry Reviews*, *36*(2), 115–134.
Kibble, M., Saarinen, N., Tang, J., Wennerberg, K., Mäkelä, S., & Aittokallio, T. (2015). Network pharmacology applications to map the unexplored target space and therapeutic potential of natural products. *Natural Product Reports*, *32*(8), 1249–1266.
Kigathi, R. N., Unsicker, S. B., Reichelt, M., Kesselmeier, J., Gershenzon, J., & Weisser, W. W. (2009). Emission of volatile organic compounds after herbivory from *Trifolium pratense* (L.) under laboratory and field conditions. *Journal of Chemical Ecology*, *35*(11), 1335.
Kikuchi, J., Shinozaki, K., & Hirayama, T. (2004). Stable isotope labeling of *Arabidopsis thaliana* for an NMR-based metabolomics approach. *Plant & Cell Physiology*, *45*, 1099–1104.
Kim, H. K., Choi, Y. H., & Verpoorte, R. (2010). NMR-based metabolomic analysis of plants. *Nature Protocols*, *5*(3), 536.
Kim, H. K., Choi, Y. H., & Verpoorte, R. (2011). NMR-based plant metabolomics: where do we stand, where do we go? *Trends in Biotechnology*, *29*(6), 267–275.
Kim, H. K., & Verpoorte, R. (2010). Sample preparation for plant metabolomics. *Phytochemical Analysis*, *21*, 4–13.
Kim, S. J., Kim, S. H., Kim, J. H., Hwang, S., & Yoo, H. J. (2016). Understanding metabolomics in biomedical research. *Endocrinology and Metabolism*, *31*(1), 7–16.

Kiran, U., Ali, A., & Abdin, M. Z. (2017). Metabolic engineering of secondary plant metabolism. In *Plant biotechnology: principles and applications* (pp. 173–190). Singapore: Springer.
Kirwan, J. A., Weber, R. J., Broadhurst, D. I., & Viant, M. R. (2014). Direct infusion mass spectrometry metabolomics dataset: a benchmark for data processing and quality control. *Scientific Data, 1*, 140012.
Kjeldahl, K., Rasmussen, M. A., Hasselbalch, A. L., Kyvik, K. O., Christiansen, L., Rezzi, S., et al. (2014). No genetic footprints of the fat mass and obesity associated (FTO) gene in human plasma 1H CPMG NMR metabolic profiles. *Metabolomics, 10*(1), 132–140.
Klepacki, J., Klawitter, J., Klawitter, J., Karimpour-fard, A., Thurman, J., Ingle, G., et al. (2016). Amino acids in a targeted versus a non-targeted metabolomics LC-MS/MS assay. Are the results consistent? *Clinical Biochemistry, 49*(13-14), 955–961.
Koek, M. M., Jellema, R. H., van der Greef, J., Tas, A. C., & Hankemeier, T. (2011). Quantitative metabolomics based on gas chromatography mass spectrometry: status and perspectives. *Metabolomics*, 7(3), 307–328.
Kogadeeva, M., & Zamboni, N. (2016). SUMOFLUX: a generalized method for targeted 13C metabolic flux ratio analysis. *PLoS Computational Biology, 12*(9)e1005109.
Kokosa, J. M. (2013). Advances in solvent-microextraction techniques. *TrAC Trends in Analytical Chemistry, 43*, 2–13.
Kostidis, S. (2018). Quantitative analysis of central energy metabolism in cell culture samples. In *Clinical metabolomics* (pp. 329–342). New York, NY: Humana Press.
Krishnan, P., Kruger, N. J., & Ratcliffe, R. G. (2004). Metabolite fingerprinting and profiling in plants using NMR. *Journal of Experimental Botany, 56*(410), 255–265.
Krumsiek, J., Bartel, J., & Theis, F. J. (2016). Computational approaches for systems metabolomics. *Current Opinion in Biotechnology, 39*, 198–206.
Kuhlisch, C., & Pohnert, G. (2015). Metabolomics in chemical ecology. *Natural Product Reports, 32*(7), 937–955.
Kumavath, R. N., Barh, D., Azevedo, V., & Kumar, A. P. (2017). Potential pharmacological applications of enzymes associated with bacterial metabolism of aromatic compounds. *Journal of Microbiology and Antimicrobials, 9*(1), 1–13.
Kusano, M., Yang, Z., Okazaki, Y., Nakabayashi, R., Fukushima, A., & Saito, K. (2015). Using metabolomic approaches to explore chemical diversity in rice. *Molecular Plant, 8*(1), 58–67.
Kussmann, M., & Blum, S. (2007). OMICS-derived targets for inflammatory gut disorders: opportunities for the development of nutrition related biomarkers. *Endocrine, Metabolic & Immune Disorders Drug Targets*, 7(4), 271–287.
Lai, S. K., Batra, A., & Cohen, C. (2005). Characterization of polydimethylsiloxane elastomer degradation via cross-linker hydrolysis. *Polymer, 46*(12), 4204–4211.
Laskin, J., Denisov, E., & Futrell, J. (2001). Comparative study of collision-induced and surface-induced dissociation. 2. Fragmentation of small alanine-containing peptides in FT-ICR MS. *The Journal of Physical Chemistry B, 105*(9), 1895–1900.
Le Gall, G., Colquhoun, I. J., Davis, A. L., Collins, G. J., & Verhoeyen, M. E. (2003). Metabolite profiling of tomato (*Lycopersicon esculentum*) using 1H NMR spectroscopy as a tool to detect potential unintended effects following a genetic modification. *Journal of Agricultural and Food Chemistry, 51*(9), 2447–2456.
Le Gall, G., Colquhoun, I. J., & Defernez, M. (2004). Metabolite profiling using 1H NMR spectroscopy for quality assessment of green tea, *Camellia sinensis* (L.). *Journal of Agricultural and Food Chemistry, 52*(4), 692–700.
Lee, J. E., Lee, B. J., Chung, J. O., Kim, H. N., Kim, E. H., Jung, S., et al. (2015). Metabolomic unveiling of a diverse range of green tea (*Camellia sinensis*) metabolites dependent on geography. *Food Chemistry, 174*, 452–459.
Lei, Z., Huhman, D. V., & Sumner, L. W. (2011). Mass spectrometry strategies in metabolomics. *Journal of Biological Chemistry, 286*(29), 25435–25442.

Leiss, K. A., Choi, Y. H., Abdel-Farid, I. B., Verpoorte, R., & Klinkhamer, P. G. L. (2009). NMR metabolomics of thrips (*Frankliniella occidentalis*) resistance in Senecio hybrids. *Journal of Chemical Ecology*, *35*, 219–229.

Leiss, K. A., Choi, Y. H., Verpoorte, R., & Klinkhamer, P. G. L. (2011). An overview of NMR metabolomic to identify secondary plant compounds involved in host plant resistance. *Phytochemistry Reviews*, *10*, 205–216.

Lewis, I. A., Schommer, S. C., Hodis, B., Robb, K. A., Tonelli, M., Westler, W. M., et al. (2007). Method for determining molar concentrations of metabolites in complex solutions from two-dimensional 1H–13C NMR spectra. *Analytical Chemistry*, *79*, 9385–9390.

Li, B., Tang, J., Yang, Q., Li, S., Cui, X., Li, Y., et al. (2017). NOREVA: Normalization and evaluation of MS-based metabolomics data. *Nucleic Acids Research*, *45*(W1), W162–W170.

Lima, R. D. C. L., Gramsbergen, S. M., Van Staden, J., Jäger, A. K., Kongstad, K. T., & Stærk, D. (2016). HPLC-HRMS-SPE-NMR for accelerated identification of compounds in complex plant extracts: new coumarine derivatives from *Coleonema album* (Thunb.) Bartl. & Wendl. *Planta Medica*, *82*(S01), P121.

Lindon, J. C., & Nicholson, J. K. (2008a). Spectroscopic and statistical techniques for information recovery in metabonomics and metabolomics. *Annual Review of Analytical Chemistry*, *1*, 45–69.

Lindon, J. C., & Nicholson, J. K. (2008b). Analytical technologies for metabonomics and metabolomics, and multi-omic information recovery. *TrAC Trends in Analytical Chemistry*, *27*(3), 194–204.

Lindon, J. C., Nicholson, J. K., & Holmes, E. (Eds.), (2011). *The handbook of metabonomics and metabolomics*. Elsevier.

Liu, S., Liang, Y. Z., & Liu, H. T. (2016). Chemometrics applied to quality control and metabolomics for traditional Chinese medicines. *Journal of Chromatography B*, *1015*, 82–91.

Liu, X., Yang, C., Zhang, L., Li, L., Liu, S., Yu, J., et al. (2011). Metabolic profiling of cadmium-induced effects in one pioneer intertidal halophyte *Suaeda salsa* by NMR-based metabolomics. *Ecotoxicology*, *20*(6), 1422–1431.

Liu, X. G., Lu, X., Wang, J. X., Wu, B., Lin, L., Wang, H. Y., et al. (2017). Combining paired analytical metabolomics and common garden trial to study the metabolism and gene variation of *Ginkgo biloba* L. cultivated varieties. *RSC Advances*, 7(87), 55309–55317.

Long, C. P., & Antoniewicz, M. R. (2014). Metabolic flux analysis of *Escherichia coli* knockouts: lessons from the Keio collection and future outlook. *Current Opinion in Biotechnology*, *28*, 127–133.

Longobardi, F., Ventrella, A., Napoli, C., Humpfer, E., Schütz, B., Schäfer, H., et al. (2012). Classification of olive oils according to geographical origin by using 1H NMR fingerprinting combined with multivariate analysis. *Food Chemistry*, *130*(1), 177–183.

Lubes, G., & Goodarzi, M. (2017). Analysis of volatile compounds by advanced analytical techniques and multivariate chemometrics. *Chemical Reviews*, *117*(9), 6399–6422.

Luo, J. (2015). Metabolite-based genome-wide association studies in plants. *Current Opinion in Plant Biology*, *24*, 31–38.

Luzzatto-Knaan, T., Melnik, A. V., & Dorrestein, P. C. (2015). Mass spectrometry tools and workflows for revealing microbial chemistry. *Analyst*, *140*(15), 4949–4966.

Ma, C., Wang, H., Lu, X., Wang, H., Xu, G., & Liu, B. (2009). Terpenoid metabolic profiling analysis of transgenic *Artemisia annua* L. by comprehensive two-dimensional gas chromatography time-of-flight mass spectrometry. *Metabolomics*, *5*(4), 497–506.

Mahrous, E. A., & Farag, M. A. (2015). Two dimensional NMR spectroscopic approaches for exploring plant metabolome: a review. *Journal of Advanced Research*, *6*(1), 3–15.

Manetti, C., Bianchetti, C., Bizzarri, M., Casciani, L., Castro, C., D'Ascenzo, G., et al. (2004). NMR-based metabonomic study of transgenic maize. *Phytochemistry, 65*(24), 3187–3198.
Manzini, S., Durante, C., Baschieri, C., Cocchi, M., Sighinolfi, S., Totaro, S., et al. (2011). Optimization of a dynamic headspace–thermal desorption–gas chromatography/mass spectrometry procedure for the determination of furfurals in vinegars. *Talanta, 85*(2), 863–869.
Mattoo, A. K., Sobolev, A. P., Neelam, A., Goyal, R. K., Handa, A. K., & Segre, A. L. (2006). NMR spectroscopy-based metabolite profiles of transgenic tomato fruit engineered to accumulate polyamines spermidine and spermine reveal enhanced anabolic and nitrogen: carbon interactions. *Plant Physiology, 142*(4), 1759–1770.
Mazumder, A., Gutch, P. K., & Dubey, D. K. (2015). A derivatization strategy for the detection and identification of volatile trialkylphosphites using liquid chromatography-online solid phase extraction and offline nuclear magnetic resonance spectroscopy. *Journal of Chromatography A, 1393*, 26–36.
Menni, C., Fauman, E., Erte, I., Perry, J. R., Kastenmüller, G., Shin, S. Y., et al. (2013). Biomarkers for type 2 diabetes and impaired fasting glucose using a non-targeted metabolomics approach. *Diabetes*. DB_130570.
Michaud, M. R., & Denlinger, D. L. (2007). Shifts in the carbohydrate, polyol, and amino acid pools during rapid cold-hardening and diapause-associated cold-hardening in flesh flies (*Sarcophaga crassipalpis*): a metabolomic comparison. *Journal of Comparative Physiology B, 177*, 753–763.
Mizuno, H., Ueda, K., Kobayashi, Y., Tsuyama, N., Todoroki, K., Min, J. Z., et al. (2017). The great importance of normalization of LC–MS data for highly-accurate non-targeted metabolomics. *Biomedical Chromatography, 31*(1).
Moco, S., Bino, R. J., De Vos, R. C. H., & Vervoort, J. (2007). Metabolomics technologies and metabolite identification. *Trends in Analytical Chemistry, 26*, 855–866.
Mohler, R. E., O'Reilly, K. T., Zemo, D. A., Tiwary, A. K., Magaw, R. I., & Synowiec, K. A. (2013). Non-targeted analysis of petroleum metabolites in groundwater using GC × GC–TOFMS. *Environmental Science & Technology, 47*(18), 10471–10476.
Mondello, L., Tranchida, P. Q., Dugo, P., & Dugo, G. (2008). Comprehensive two-dimensional gas chromatography-mass spectrometry: a review. *Mass Spectrometry Reviews, 27*(2), 101–124.
Monteiro, M. S., Carvalho, M., Bastos, M. L., & Guedes de Pinho, P. (2013). Metabolomics analysis for biomarker discovery: advances and challenges. *Current Medicinal Chemistry, 20*(2), 257–271.
Moradi, P., Ford-Lloyd, B., & Pritchard, J. (2017). Metabolomic approach reveals the biochemical mechanisms underlying drought stress tolerance in thyme. *Analytical Biochemistry, 527*, 49–62.
Movasaghi, Z., Rehman, S., & ur Rehman, D. I. (2008). Fourier transform infrared (FTIR) spectroscopy of biological tissues. *Applied Spectroscopy Reviews, 43*(2), 134–179.
Muhlemann, J. K., Klempien, A., & Dudareva, N. (2014). Floral volatiles: from biosynthesis to function. *Plant, Cell & Environment, 37*(8), 1936–1949.
Mullard, G., Allwood, J. W., Weber, R., Brown, M., Begley, P., Hollywood, K. A., et al. (2015). A new strategy for MS/MS data acquisition applying multiple data dependent experiments on Orbitrap mass spectrometers in non-targeted metabolomic applications. *Metabolomics, 11*(5), 1068–1080.
Naz, S., Vallejo, M., García, A., & Barbas, C. (2014). Method validation strategies involved in non-targeted metabolomics. *Journal of Chromatography A, 1353*, 99–105.
Nesvizhskii, A. I., Keller, A., Kolker, E., & Aebersold, R. (2003). A statistical model for identifying proteins by tandem mass spectrometry. *Analytical Chemistry, 75*(17), 4646–4658.

Nicholson, J., Lindon, J., & Holmes, E. (1999). 'Metabonomics': understanding the metabolic responses of living systems to pathophysiological stimuli via multivariate statistical analysis of biological NMR spectroscopic data. *Xenobiotica*, *29*, 1181–1189.

Nicholson, J. K., & Wilson, I. D. (2003). Understanding 'global' systems biology: metabonomics and the continuum of metabolism. *Nature Reviews Drug Discovery*, *2*(8), 668.

Niell, S., Cesio, V., Hepperle, J., Doerk, D., Kirsch, L., Kolberg, D., et al. (2014). QuEChERS-based method for the multiresidue analysis of pesticides in beeswax by LC-MS/MS and GC× GC-TOF. *Journal of Agricultural and Food Chemistry*, *62*(17), 3675–3683.

Nikolaev, E. N., Kostyukevich, Y. I., & Vladimirov, G. N. (2016). Fourier transform ion cyclotron resonance (FT ICR) mass spectrometry: Theory and simulations. *Mass Spectrometry Reviews*, *35*(2), 219–258.

Novák, O., Pěnčík, A., & Ljung, K. (2014). Identification and profiling of auxin and auxin metabolites. In *Auxin and its role in plant development* (pp. 39–60). Vienna: Springer.

Nyamundanda, G., Brennan, L., & Gormley, I. C. (2010). Probabilistic principal component analysis for metabolomic data. *BMC Bioinformatics*, *11*(1), 571.

Ogura, T., Bamba, T., & Fukusaki, E. (2013). Development of a practical metabolite identification technique for non-targeted metabolomics. *Journal of Chromatography A*, *1301*, 73–79.

Okazaki, Y., & Saito, K. (2012). Recent advances of metabolomics in plant biotechnology. *Plant Biotechnology Reports*, *6*(1), 1–15.

Oksman-Caldentey, K. M., & Saito, K. (2005). Integrating genomics and metabolomics for engineering plant metabolic pathways. *Current Opinion in Biotechnology*, *16*, 174–179.

Oliveira, T. S., Murphy, M., Mendola, N., Wong, V., Carlson, D., & Waring, L. (2015). Characterization of pharmaceuticals and personal care products in hospital effluent and waste water influent/effluent by direct-injection LC-MS-MS. *Science of the Total Environment*, *518*, 459–478.

Ortmayr, K., Causon, T. J., Hann, S., & Koellensperger, G. (2016). Increasing selectivity and coverage in LC-MS based metabolome analysis. *TrAC Trends in Analytical Chemistry*, *82*, 358–366.

Palermo, A., Botrè, F., de la Torre, X., & Zamboni, N. (2017). Non-targeted LC-MS based metabolomics analysis of the urinary steroidal profile. *Analytica Chimica Acta*, *964*, 112–122.

Pandey, S., Patel, M. K., Mishra, A., & Jha, B. (2015). Physio-biochemical composition and untargeted metabolomics of cumin (*Cuminum cyminum* L.) make it promising functional food and help in mitigating salinity stress. *PLoS One*, *10*(12)e0144469.

Patejko, M., Jacyna, J., & Markuszewski, M. J. (2017). Sample preparation procedures utilized in microbial metabolomics: an overview. *Journal of Chromatography B*, *1043*, 150–157.

Patel, S., & Ahmed, S. (2015). Emerging field of metabolomics: big promise for cancer biomarker identification and drug discovery. *Journal of Pharmaceutical and Biomedical Analysis*, *107*, 63–74.

Patil, C., Calvayrac, C., Zhou, Y., Romdhane, S., Salvia, M. V., Cooper, J. F., et al. (2016). Environmental metabolic footprinting: a novel application to study the impact of a natural and a synthetic β-triketone herbicide in soil. *Science of the Total Environment*, *566*, 552–558.

Patti, G. J., Yanes, O., & Siuzdak, G. (2012). Innovation: metabolomics: the apogee of the omics trilogy. *Nature Reviews Molecular Cell Biology*, *13*(4), 263–269.

Peng, B., Li, H., & Peng, X. X. (2015). Functional metabolomics: from biomarker discovery to metabolome reprogramming. *Protein & Cell*, *6*(9), 628–637.

Peters, S. (2017). *Measurements of transcripts, proteome and metabolite profiles*. Fuel4Me.

Petrakis, E. A., Cagliani, L. R., Polissiou, M. G., & Consonni, R. (2015). Evaluation of saffron (*Crocus sativus* L.) adulteration with plant adulterants by 1H NMR metabolite fingerprinting. *Food Chemistry*, *173*, 890–896.

Phillips, K. A., Wambaugh, J. F., Grulke, C. M., Dionisio, K. L., & Isaacs, K. K. (2017). High-throughput screening of chemicals as functional substitutes using structure-based classification models. *Green Chemistry*, *19*(4), 1063–1074.

Pinto, D. M., Blande, J. D., Souza, S. R., Nerg, A. M., & Holopainen, J. K. (2010). Plant volatile organic compounds (VOCs) in ozone (O3) polluted atmospheres: the ecological effects. *Journal of Chemical Ecology*, *36*(1), 22–34.

Pinu, F. R., Villas-Boas, S. G., & Aggio, R. (2017). Analysis of intracellular metabolites from microorganisms: quenching and extraction protocols. *Metabolites*, 7(4), 53.

Pitt, J. J. (2009). Principles and applications of liquid chromatography-mass spectrometry in clinical biochemistry. *Clinical Biochemist Reviews*, *30*(1), 19.

Putri, S. P., Nakayama, Y., Matsuda, F., Uchikata, T., Kobayashi, S., Matsubara, A., et al. (2013). Current metabolomics: practical applications. *Journal of Bioscience and Bioengineering*, *115*(6), 579–589.

Qin, Y., Yalamanchili, H. K., Qin, J., Yan, B., & Wang, J. (2015). The current status and challenges in computational analysis of genomic big data. *Big Data Research*, *2*(1), 12–18.

Qu, G., Shi, J., Wang, T., Fu, J., Li, Z., Wang, P., et al. (2011). Identification of tetrabromobisphenol A diallyl ether as an emerging neurotoxicant in environmental samples by bioassay-directed fractionation and HPLC-APCI-MS/MS. *Environmental Science & Technology*, *45*(11), 5009–5016.

Rabinowitz, J. S., Robitaille, A. M., Wang, Y., Ray, C. A., Thummel, R., Gu, H., et al. (2017). Transcriptomic, proteomic, and metabolomic landscape of positional memory in the caudal fin of zebrafish. *Proceedings of the National Academy of Sciences*, *114*(5), E717–E726.

Raffler, J., Friedrich, N., Arnold, M., Kacprowski, T., Rueedi, R., Altmaier, E., et al. (2015). Genome-wide association study with targeted and non-targeted NMR metabolomics identifies 15 novel loci of urinary human metabolic individuality. *PLoS Genetics*, *11* (9)e1005487.

Rager, J. E., Strynar, M. J., Liang, S., McMahen, R. L., Richard, A. M., Grulke, C. M., et al. (2016). Linking high resolution mass spectrometry data with exposure and toxicity forecasts to advance high-throughput environmental monitoring. *Environment International*, *88*, 269–280.

Ramel, F., Sulmon, C., Serra, A. A., Gouesbet, G., & Couée, I. (2012). Xenobiotic sensing and signalling in higher plants. *Journal of Experimental Botany*, *63*(11), 3999–4014.

Rasmussen, B., Cloarec, O., Tang, H., Stærk, D., & Jaroszewski, J. W. (2006). Multivariate analysis of integrated and full-resolution 1H-NMR spectral data from complex pharmaceutical preparations: St. John's wort. *Planta Medica*, *72*(06), 556–563.

Rawlinson, C., Kamphuis, L. G., Gummer, J. P., Singh, K. B., & Trengove, R. D. (2015). A rapid method for profiling of volatile and semi-volatile phytohormones using methyl chloroformate derivatisation and GC–MS. *Metabolomics*, *11*(6), 1922–1933.

Reali, F., Priami, C., & Marchetti, L. (2017). Optimization algorithms for computational systems biology. *Frontiers in Applied Mathematics and Statistics*, *3*, 6.

Ren, Y., Wang, T., Peng, Y., Xia, B., & Qu, L. J. (2009). Distinguishing transgenic from non-transgenic Arabidopsis plants by 1 H NMR-based metabolic fingerprinting. *Journal of Genetics and Genomics*, *36*(10), 621–628.

Risticevic, S., Vuckovic, D., & Pawliszyn, J. (2010). Solid-phase microextraction. In *Handbook of sample preparation* (pp. 81–101). .

Roberts, L. D., Souza, A. L., Gerszten, R. E., & Clish, C. B. (2012). Targeted metabolomics. *Current Protocols in Molecular Biology*, 30–32.

Roessner-Tunali, U., Liu, J., Leisse, A., Balbo, I., Perez-Melis, A., Willmitzer, L., et al. (2004). Kinetics of labelling of organic and amino acids in potato tubers by gas chromatography-mass spectrometry following incubation in 13C labelled isotopes. *The Plant Journal, 39*(4), 668–679.
Rohman, A., & Man, Y. C. (2010). Fourier transform infrared (FTIR) spectroscopy for analysis of extra virgin olive oil adulterated with palm oil. *Food Research International, 43*(3), 886–892.
Römisch-Margl, W., Prehn, C., Bogumil, R., Röhring, C., Suhre, K., & Adamski, J. (2012). Procedure for tissue sample preparation and metabolite extraction for high-throughput targeted metabolomics. *Metabolomics, 8*(1), 133–142.
Roullier-Gall, C., Witting, M., Gougeon, R. D., & Schmitt-Kopplin, P. (2014). High precision mass measurements for wine metabolomics. *Frontiers in Chemistry, 2*, 102.
Roux, A., Lison, D., Junot, C., & Heilier, J. F. (2011). Applications of liquid chromatography coupled to mass spectrometry-based metabolomics in clinical chemistry and toxicology: A review. *Clinical Biochemistry, 44*(1), 119–135.
Rozet, E., Ceccato, A., Hubert, C., Ziemons, E., Oprean, R., Rudaz, S., et al. (2007). Analysis of recent pharmaceutical regulatory documents on analytical method validation. *Journal of Chromatography A, 1158*(1-2), 111–125.
Rutherford, A. (2004). Designing experiments and analyzing data: a model comparison perspective. *The British Journal of Mathematical and Statistical Psychology, 57*, 185–187.
Ryan, D., & Robards, K. (2006). Metabolomics: the greatest omics of them all? *Analytical Chemistry, 78*(23), 7954–7958.
Sajic, T., Liu, Y., & Aebersold, R. (2015). Using data-independent, high-resolution mass spectrometry in protein biomarker research: perspectives and clinical applications. *PROTEOMICS-Clinical Applications, 9*(3-4), 307–321.
Salek, R. M., et al. (2015). COordination of Standards in MetabO- lomicS (COSMOS): facilitating integrated metabolomics data access. *Metabolomics, 11*(6), 1587–1597.
Samota, M. K., Bhatt, L., Yadav, D. K., Garg, N., & Bajiya, R. (2017). Metabolomics for functional genomics. *International Journal of Current Microbiology and Applied Sciences, 6*(5), 2531–2537.
Sampson, J. S., Hawkridge, A. M., & Muddiman, D. C. (2006). Generation and detection of multiply-charged peptides and proteins by matrix-assisted laser desorption electrospray ionization (MALDESI) Fourier transform ion cyclotron resonance mass spectrometry. *Journal of the American Society for Mass Spectrometry, 17*(12), 1712–1716.
Sangster, T., et al. (2006). A pragmatic and readily implemented quality control strategy for HPLC-MS and GC-MS-based metabonomic analysis. *Analyst, 131*(10), 1075–1078.
Saraji, M., Khayamian, T., Siahpoosh, Z. H., & Farajmand, B. (2012). Determination of volatile residual solvents in pharmaceutical products by static and dynamic headspace liquid-phase microextraction combined with gas chromatography-flame ionization detection. *Analytical Methods, 4*(6), 1552–1559.
Sardjono, R. E., Athiana, A. F., Gumilar, G. G., & Rachmawati, R. (2017). Extraction of essential aroma compounds from several malodorous Indonesian plants using simultaneous steam distillation-extraction. *Asian Journal of Chemistry, 29*(3), 679.
Satake, H., Hasegawa, H., Hirabayashi, A., Hashimoto, Y., Baba, T., & Masuda, K. (2007). Fast multiple electron capture dissociation in a linear radio frequency quadrupole ion trap. *Analytical Chemistry, 79*(22), 8755–8761.
Satou, M., Enoki, H., Oikawa, A., Ohta, D., Saito, K., Hachiya, T., et al. (2014). Integrated analysis of transcriptome and metabolome of *Arabidopsis albino* or pale green mutants with disrupted nuclear-encoded chloroplast proteins. *Plant Molecular Biology, 85*(4-5), 411–428.
Sauter, D., Li, S., & Aubrun, C. (2009). Robust fault diagnosis of networked control systems. *International Journal of Adaptive Control and Signal Processing, 23*(8), 722–736.

Sawada, Y., Akiyama, K., Sakata, A., Kuwahara, A., Otsuki, H., Sakurai, T., et al. (2008). Widely targeted metabolomics based on large-scale MS/MS data for elucidating metabolite accumulation patterns in plants. *Plant and Cell Physiology*, *50*(1), 37–47.

Schanda, P., & Brutscher, B. (2005). Very fast two-dimensional NMR spectroscopy for real-time investigation of dynamic events in proteins on the time scale of seconds. *Journal of the American Chemical Society*, *127*(22), 8014–8015.

Schauer, N., Semel, Y., Roessner, U., Gur, A., Balbo, I., Carrari, F., et al. (2006). Comprehensive metabolic profiling and phenotyping of interspecific introgression lines for tomato improvement. *Nature Biotechnology*, *24*(4), 447.

Schlichtherle-Cerny, H., Affolter, M., & Cerny, C. (2003). Hydrophilic interaction liquid chromatography coupled to electrospray mass spectrometry of small polar compounds in food analysis. *Analytical Chemistry*, *75*(10), 2349–2354.

Schmidberger, T., Gutmann, R., Bayer, K., Kronthaler, J., & Huber, R. (2014). Advanced online monitoring of cell culture off-gas using proton transfer reaction mass spectrometry. *Biotechnology Progress*, *30*(2), 496–504.

Scholz, M., Gatzek, S., Sterling, A., Fiehn, O., & Selbig, J. (2004). Metabolite fingerprinting: detecting biological features by independent component analysis. *Bioinformatics*, *20*(15), 2447–2454.

Schrimpe-Rutledge, A. C., Codreanu, S. G., Sherrod, S. D., & McLean, J. A. (2016). Untargeted metabolomics strategies—challenges and emerging directions. *Journal of the American Society for Mass Spectrometry*, *27*(12), 1897–1905.

Schripsema, J. (2010). Application of NMR in plant metabolomics: techniques, problems and prospects. *Phytochemical Analysis*, *21*(1), 14–21.

Schripsema, J., Lemos, M. A., Dianin, M. F., Vianna, D. S., & Dagnino, D. S. (2014). New approaches in metabolic fingerprinting: improved extraction method and automatic reduction of NMR spectra to essential data. In *Instrumental methods for the analysis and identification of bioactive molecules* (pp. 361–373). American Chemical Society.

Schubert, O. T., Gillet, L. C., Collins, B. C., Navarro, P., Rosenberger, G., Wolski, W. E., et al. (2015). Building high-quality assay libraries for targeted analysis of SWATH MS data. *Nature Protocols*, *10*(3), 426.

Schuhmann, K., Thomas, H., Ackerman, J. M., Nagornov, K. O., Tsybin, Y. O., & Shevchenko, A. (2017). Intensity-independent noise filtering in FT MS and FT MS/MS spectra for shotgun lipidomics. *Analytical Chemistry*, *89*(13), 7046–7052.

Scott, I. M., Vermeer, C. P., Liakata, M., Corol, D. I., Ward, J. L., Lin, W., et al. (2010). Enhancement of plant metabolite fingerprinting by machine learning. *Plant Physiology*, *153*(4), 1506–1520.

Seeram, N. P., Lee, R., Scheuller, H. S., & Heber, D. (2006). Identification of phenolic compounds in strawberries by liquid chromatography electrospray ionization mass spectroscopy. *Food Chemistry*, *97*(1), 1–11.

Sekiyama, Y., Chikayama, E., & Kikuchi, J. (2010). Profiling polar and semipolar plant metabolites throughout extraction processes using a combined solution-state and high-resolution magic angle spinning NMR approach. *Analytical Chemistry*, *82*, 1643–1652.

Sévin, D. C., Kuehne, A., Zamboni, N., & Sauer, U. (2015). Biological insights through nontargeted metabolomics. *Current Opinion in Biotechnology*, *34*, 1–8.

Shah, J. S., Brock, G. N., & Rai, S. N. (2015). Metabolomics data analysis and missing value issues with application to infarcted mouse hearts. *BMC Bioinformatics*, *16*, P16. Suppl. 15.

Sharma, A., Verma, P., Mathur, A., & Mathur, A. K. (2018). Genetic engineering approach using early Vinca alkaloid biosynthesis genes led to increased tryptamine and terpenoid indole alkaloids biosynthesis in differentiating cultures of *Catharanthus roseus*. *Protoplasma*, *255*(1), 425–435.

Shirasu, M., & Touhara, K. (2011). The scent of disease: volatile organic compounds of the human body related to disease and disorder. *The Journal of Biochemistry*, *150*(3), 257–266.
Shrivastava, A., & Gupta, V. (2011). Methods for the determination of limit of detection and limit of quantitation of the analytical methods. *Chronicles of Young Scientists*, *2*(1), 21.
Sidebottom, A. M., Johnson, A. R., Karty, J. A., Trader, D. J., & Carlson, E. E. (2013). Integrated metabolomics approach facilitates discovery of an unpredicted natural product suite from *Streptomyces coelicolor* M145. *ACS Chemical Biology*, *8*(9), 2009–2016.
Skov, K., Hadrup, N., Smedsgaard, J., & Frandsen, H. (2015). LC–MS analysis of the plasma metabolome—a novel sample preparation strategy. *Journal of Chromatography B*, *978*, 83–88.
Smilde, A. K., Jansen, J. J., Hoefsloot, H. C., Lamers, R. J. A., Van Der Greef, J., & Timmerman, M. E. (2005). ANOVA-simultaneous component analysis (ASCA): a new tool for analyzing designed metabolomics data. *Bioinformatics*, *21*(13), 3043–3048.
Smith, C. A., O'Maille, G., Want, E. J., Qin, C., Trauger, S. A., Brandon, T. R., et al. (2005). METLIN: a metabolite mass spectral database. *Therapeutic Drug Monitoring*, *27* (6), 747–751.
Smith, R. D., Loo, J. A., Edmonds, C. G., Barinaga, C. J., & Udseth, H. R. (1990). New developments in biochemical mass spectrometry: electrospray ionization. *Analytical Chemistry*, *62*(9), 882–899.
Smolinska, A., Blanchet, L., Buydens, L. M., & Wijmenga, S. S. (2012). NMR and pattern recognition methods in metabolomics: from data acquisition to biomarker discovery: a review. *Analytica Chimica Acta*, *750*, 82–97.
Soule, M. C. K., Longnecker, K., Johnson, W. M., & Kujawinski, E. B. (2015). Environmental metabolomics: analytical strategies. *Marine Chemistry*, *177*, 374–387.
Souza-Silva, É. A., Jiang, R., Rodríguez-Lafuente, A., Gionfriddo, E., & Pawliszyn, J. (2015). A critical review of the state of the art of solid-phase microextraction of complex matrices I. Environmental analysis. *TrAC Trends in Analytical Chemistry*, *71*, 224–235.
Spicer, R., Salek, R., & Steinbeck, C. (2017). Compliance with minimum information guidelines in public metabolomics repositories. *Scientific Data*, *4*, 17137.
Steinbeck, C., et al. (2012). MetaboLights: towards a new COSMOS of metabolomics data management. *Metabolomics*, *8*(5), 757–760.
Stultz, C., Jobst, K. J., Haimovici, L., Jones, R., Besevic, S., Byer, J., et al. (2018). Evaluation of multiple alternative instrument platforms for targeted and non-targeted dioxin and furan analysis. *Journal of Mass Spectrometry*.
Styczynski, M. P., Moxley, J. F., Tong, L. V., Walther, J. L., Jensen, K. L., & Stephanopoulos, G. N. (2007). Systematic identification of conserved metabolites in GC/MS data for metabolomics and biomarker discovery. *Analytical Chemistry*, *79*(3), 966–973.
Sud, M., Fahy, E., Cotter, D., Azam, K., Vadivelu, I., Burant, C., et al. (2015). Metabolomics workbench: an international repository for metabolomics data and metadata, metabolite standards, protocols, tutorials and training, and analysis tools. *Nucleic Acids Research*, *44* (D1), D463–D470.
Sumner, L. W., Lei, Z., Nikolau, B. J., & Saito, K. (2015). Modern plant metabolomics: advanced natural product gene discoveries, improved technologies, and future prospects. *Natural Product Reports*, *32*(2), 212–229.
Sumner, L. W., Mendes, P., & Dixon, R. A. (2003). Plant metabolomics: largescale phytochemistry in the functional genomics era. *Phytochemistry*, *62*, 817–836.
Sumner, L. W., et al. (2007). Proposed minimum reporting standards for chemical analysis. *Metabolomics*, *3*(3), 211–221.
Swenson, T. L., Karaoz, U., Swenson, J. M., Bowen, B. P., & Northen, T. R. (2018). Linking soil biology and chemistry in biological soil crust using isolate exometabolomics. *Nature Communications*, *9*(1), 19.

Szolar, O. H. J. (2007). Environmental and pharmaceutical analysis of dithiocarbamates. *Analytica Chimica Acta*, *582*(2), 191–200.
Tanaka, K., Nakamura, K., Saito, T., Osada, H., Hirai, A., Takahashi, H., et al. (2009). Metabolic pathway prediction based on inclusive relation between cyclic substructures. *Plant Biotechnology*, *26*(5), 459–468.
Tang, H. R., Xiao, C. N., & Wang, Y. L. (2009). Important roles of the hyphenated HPLC-DAD-MS-SPE-NMR technique in metabonomics. *Magnetic Resonance in Chemistry*, *47*, S157–S162.
Tautenhahn, R., Patti, G. J., Rinehart, D., & Siuzdak, G. (2012). XCMS online: a web-based platform to process untargeted metabolomic data. *Analytical Chemistry*, *84*, 50355039.
Tebani, A., Abily-Donval, L., Afonso, C., Marret, S., & Bekri, S. (2016). Clinical metabolomics: the new metabolic window for inborn errors of metabolism investigations in the post-genomic era. *International Journal of Molecular Sciences*, *17*(7), 1167.
Templeton, N. (2016). Fluxomics: The integration of metabolic flux analysis (MFA) with multivariate data analysis (MVDA) to identify key process parameters for CHO cell culture. In R. Kiss, G. S. Harcum, Clemson University, J. Chalmers, Ohio State University (Eds.), *Cell Culture Engineering XV, ECI Symposium Series*. Available from: http://dc.engconfintl.org/cellculture_xv/45
Thiboonboon, K., Leelahavarong, P., Wattanasirichaigoon, D., Vatanavicharn, N., Wasant, P., Shotelersuk, V., et al. (2015). An economic evaluation of neonatal screening for inborn errors of metabolism using tandem mass spectrometry in Thailand. *PLoS One*, *10*(8)e0134782.
Thomas, J. W., Touchman, J. W., Blakesley, R. W., Bouffard, G. G., Beckstrom-Sternberg, S. M., Margulies, E. H., et al. (2003). Comparative analyses of multi-species sequences from targeted genomic regions. *Nature*, *424*(6950), 788.
Tohge, T., Nishiyama, Y., Hirai, M. Y., Yano, M., Nakajima, J. I., Awazuhara, M., et al. (2005). Functional genomics by integrated analysis of metabolome and transcriptome of Arabidopsis plants over-expressing an MYB transcription factor. *The Plant Journal*, *42*(2), 218–235.
Trenkamp, S., Eckes, P., Busch, M., & Fernie, A. R. (2009). Temporally resolved GC-MS-based metabolic profiling of herbicide treated plants treated reveals that changes in polar primary metabolites alone can distinguish herbicides of differing mode of action. *Metabolomics*, *5*, 277–291.
Tsugawa, H., Bamba, T., Shinohara, M., Nishiumi, S., Yoshida, M., & Fukusaki, E. (2011). Practical non-targeted gas chromatography/mass spectrometry-based metabolomics platform for metabolic phenotype analysis. *Journal of Bioscience and Bioengineering*, *112*(3), 292–298.
Tucker, D. J., Wallis, I. R., Bolton, J. M., Marsh, K. J., Rosser, A. A., Brereton, I. M., et al. (2010). A metabolomic approach to identifying chemical mediators of mammal-plant interactions. *Journal of Chemical Ecology*, *36*, 727–735.
Tumanov, S., Zubenko, Y., Obolonkin, V., Greenwood, D. R., Shmanai, V., & Villas-Bôas, S. G. (2016). Calibration curve-free GC–MS method for quantitation of amino and non-amino organic acids in biological samples. *Metabolomics*, *12*(4), 64.
Turroni, F., Milani, C., Duranti, S., Mancabelli, L., Mangifesta, M., Viappiani, A., et al. (2016). Deciphering bifidobacterial-mediated metabolic interactions and their impact on gut microbiota by a multi-omics approach. *The ISME Journal*, *10*(7), 1656–1668.
Urano, K., Maruyama, K., Ogata, Y., Morishita, Y., Takeda, M., Sakurai, N., et al. (2009). Characterization of the ABA-regulated global responses to dehydration in Arabidopsis by metabolomics. *The Plant Journal*, *57*(6), 1065–1078.
van der Meer, J. (2006). Metabolic theories in ecology. *Trends in Ecology & Evolution*, *21*, 136–140.

Van Meulebroek, L., Bussche, J. V., De Clercq, N., Steppe, K., & Vanhaecke, L. (2015). A metabolomics approach to unravel the regulating role of phytohormones towards carotenoid metabolism in tomato fruit. *Metabolomics*, *11*(3), 667–683.
van Smeden, J., Hoppel, L., van der Heijden, R., Hankemeier, T., Vreeken, R. J., & Bouwstra, J. A. (2011). LC/MS analysis of stratum corneum lipids: ceramide profiling and discovery. *Journal of Lipid Research*, *52*(6), 1211–1221.
van Stee, L. L. P. (2017). *Analysis of complex samples using multidimensional gas chromatography and selective detection*. [Academisch Proefschrift] Vrije Universiteit.
van Veen, S. Q., Claas, E. C. J., & Kuijper, E. J. (2010). High-throughput identification of bacteria and yeast by matrix-assisted laser desorption ionization-time of flight mass spectrometry in conventional medical microbiology laboratories. *Journal of Clinical Microbiology*, *48*(3), 900–907.
Vandendriessche, T., Schäfer, H., Verlinden, B. E., Humpfer, E., Hertog, M. L., & Nicolaï, B. M. (2013). High-throughput NMR based metabolic profiling of Braeburn apple in relation to internal browning. *Postharvest Biology and Technology*, *80*, 18–24.
Vargas, L. H. G., Neto, J. C. R., de Aquino Ribeiro, J. A., Ricci-Silva, M. E., Souza, M. T., Rodrigues, C. M., et al. (2016). Metabolomics analysis of oil palm (*Elaeis guineensis*) leaf: evaluation of sample preparation steps using UHPLC–MS/MS. *Metabolomics*, *12*(10), 153.
Vernocchi, P., Vannini, L., Gottardi, D., Del Chierico, F., Serrazanetti, D. I., Ndagijimana, M., et al. (2012). Integration of datasets from different analytical techniques to assess the impact of nutrition on human metabolome. *Frontiers in Cellular and Infection Microbiology*, *2*, 156.
Viant, M. R., Pincetich, C. A., Hinton, D. E., & Tjeerdema, R. S. (2006). Toxic actions of dinoseb in medaka (*Oryzias latipes*) embryos as determined by in vivo P-31 NMR, HPLC-UV and H-1 NMR metabolomics. *Aquatic Toxicology*, *76*, 329–342.
Villar, M., Ayllón, N., Alberdi, P., Moreno, A., Moreno, M., Tobes, R., et al. (2015). Integrated metabolomics, transcriptomics and proteomics identifies metabolic pathways affected by *Anaplasma phagocytophilum* infection in tick cells. *Molecular & Cellular Proteomics*, *14*(12), 3154–3172.
Vinaixa, M., Schymanski, E. L., Neumann, S., Navarro, M., Salek, R. M., & Yanes, O. (2016). Mass spectral databases for LC/MS-and GC/MS-based metabolomics: state of the field and future prospects. *TrAC Trends in Analytical Chemistry*, *78*, 23–35.
Vuckovic, D., Zhang, X., Cudjoe, E., & Pawliszyn, J. (2010). Solid-phase microextraction in bioanalysis: new devices and directions. *Journal of Chromatography A*, *1217*(25), 4041–4060.
Wallace, M. A. G., Pleil, J. D., Mentese, S., Oliver, K. D., Whitaker, D. A., & Fent, K. W. (2017). Calibration and performance of synchronous SIM/scan mode for simultaneous targeted and discovery (non-targeted) analysis of exhaled breath samples from firefighters. *Journal of Chromatography A*, *1516*, 114–124.
Wan, E. C., & Yu, J. Z. (2007). Analysis of sugars and sugar polyols in atmospheric aerosols by chloride attachment in liquid chromatography/negative ion electrospray mass spectrometry. *Environmental Science & Technology*, *41*(7), 2459–2466.
Wang, H., Li, P., Sun, S. H., Zhang, Q. D., Su, Y., Zong, Y. L., et al. (2014). Comparison of liquid–liquid extraction, simultaneous distillation extraction, ultrasound-assisted solvent extraction, and headspace solid-phase microextraction for the determination of volatile compounds in jujube extract by gas chromatography/mass spectrometry. *Analytical Letters*, *47*(4), 654–674.
Wang, P. G., & He, W. (Eds.), (2011). *Hydrophilic interaction liquid chromatography (HILIC) and advanced applications*: CRC Press.
Wang, W., Zhou, H., Lin, H., Roy, S., Shaler, T. A., Hill, L. R., et al. (2003). Quantification of proteins and metabolites by mass spectrometry without isotopic labeling or spiked standards. *Analytical Chemistry*, *75*(18), 4818–4826.

Wanichthanarak, K., Fahrmann, J. F., & Genomic, G. D. (2015). Proteomic, and metabolomic data integration strategies. *Biomarker Insights*, *10*(Suppl 4), 1–6.

Want, E. J., Masson, P., Michopoulos, F., Wilson, I. D., Theodoridis, G., Plumb, R. S., et al. (2013). Global metabolic profiling of animal and human tissues via UPLC-MS. *Nature Protocols*, *8*(1), 17.

Want, E. J., Wilson, I. D., Gika, H., Theodoridis, G., Plumb, R. S., Shockcor, J., et al. (2010). Global metabolic profiling procedures for urine using UPLC–MS. *Nature Protocols*, *5*(6), 1005.

Ward, J. L., Harris, C., Lewis, J., & Beale, M. H. (2003). Assessment of 1H NMR spectroscopy and multivariate analysis as a technique for metabolite fingerprinting of *Arabidopsis thaliana*. *Phytochemistry*, *62*(6), 949–957.

Wehrens, R., Weingart, G., & Mattivi, F. (2014). metaMS: an open-source pipeline for GC-MS based untargeted metabolomics. *Journal of Chromatography B*, *966*, 109–116.

WHO (1989). *Formaldehyde. Environmental Health Criteria, No. 89, International Agency for Research on Cancer IARC, Formaldehyde. Wood dust and formaldehyde.* Geneva: World Health Organization.

Wiechert, W., Niedenführ, S., & Nöh, K. (2015). A primer to 13C metabolic flux analysis. *Fundamental Bioengineering*, 97–142.

Wishart, D. S. (2011). Advances in metabolite identification. *Bioanalysis*, *3*(15), 1769–1782.

Wolfender, J. L., Marti, G., & Ferreira Queiroz, E. (2010). Advances in techniques for profiling crude extracts and for the rapid identification of natural products: dereplication, quality control and metabolomics. *Current Organic Chemistry*, *14*(16), 1808–1832.

Wolfender, J. L., Marti, G., Thomas, A., & Bertrand, S. (2015). Current approaches and challenges for the metabolite profiling of complex natural extracts. *Journal of Chromatography A*, *1382*, 136–164.

Wong, G. K., Lim, L. Z., Lim, M. J. W., Ong, L. L., Khezri, B., Pumera, M., et al. (2015). Evaluation of the sorbent properties of single- and multiwalled carbon nanotubes for volatile organic compounds through thermal desorption–gas chromatography/mass spectrometry. *ChemPlusChem*, *80*(8), 1279–1287.

Xia, J., Psychogios, N., Young, N., & Wishart, D. S. (2009). MetaboAnalyst: a web server for metabolomic data analysis and interpretation. *Nucleic Acids Research*, *37*(suppl_2), W652–W660.

Xiao, J. F., Zhou, B., & Ressom, H. W. (2012). Metabolite identification and quantitation in LC-MS/MS-based metabolomics. *TrAC Trends in Analytical Chemistry*, *32*, 1–14.

Xu, J., & Ouyang, G. (2017). Development of novel solid-phase microextraction fibers. In *Solid phase microextraction* (pp. 17–61). Berlin, Heidelberg: Springer.

Yan, W., Zhao, L., Feng, Q., Wei, Y., & Lin, J. M. (2009). Simultaneous determination of ten estrogens and their metabolites in waters by improved two-step SPE followed by LC–MS. *Chromatographia*, *69*(7-8), 621–628.

Yilmaz, A., Nyberg, N. T., Mølgaard, P., Asili, J., & Jaroszewski, J. W. (2010). 1H NMR metabolic fingerprinting of saffron extracts. *Metabolomics*, *6*(4), 511–517.

Yuan, M., Breitkopf, S. B., Yang, X., & Asara, J. M. (2012). A positive/negative ion-switching, targeted mass spectrometry-based metabolomics platform for bodily fluids, cells, and fresh and fixed tissue. *Nature Protocols*, 7(5), 872–881.

Zhang, S., Zu, Y. G., Fu, Y. J., Luo, M., Li, J., & Efferth, T. (2010). Supercritical carbon dioxide extraction of seed oil from yellow horn (Xanthoceras sorbifolia Bunge.) and its anti-oxidant activity. *Bioresour Technol*, *101*, 2537–2544.

Zhang, Z., & Li, G. (2010). A review of advances and new developments in the analysis of biological volatile organic compounds. *Microchemical journal*, *95*(2), 127–139.

Zhang, A., Sun, H., Wang, P., Han, Y., & Wang, X. (2012). Modern analytical techniques in metabolomics analysis. *Analyst*, *137*(2), 293–300.

Zhang, K., Li, Y., Tsang, M., & Chetwyn, N. P. (2013). Analysis of pharmaceutical impurities using multi-heartcutting 2D LC coupled with UV-charged aerosol MS detection. *Journal of Separation Science*, *36*(18), 2986–2992.

Zhang, W., Lei, Z., Huhman, D., Sumner, L. W., & Zhao, P. X. (2015). MET-XAlign: a metabolite cross-alignment tool for LC/MS-based comparative metabolomics. *Analytical Chemistry*, *87*(18), 9114–9119.
Zhao, L., Huang, Y., Hu, J., Zhou, H., Adeleye, A. S., & Keller, A. A. (2016). 1H NMR and GC-MS based metabolomics reveal defense and detoxification mechanism of cucumber plant under nano-Cu stress. *Environmental Science & Technology*, *50*(4), 2000–2010.
Zwanenburg, G., Hoefsloot, H. C., Westerhuis, J. A., Jansen, J. J., & Smilde, A. K. (2011). ANOVA–principal component analysis and ANOVA–simultaneous component analysis: a comparison. *Journal of Chemometrics*, *25*(10), 561–567.

FURTHER READING

Gurumoorthy, S., Muppalaneni, N. B., & Gao, X. Z. (2018). *Computational intelligence techniques in diagnosis of brain diseases*. Springer.
Schmelz, E. A., Engelberth, J., Alborn, H. T., O'donnell, P., Sammons, M., Toshima, H., et al. (2003). Simultaneous analysis of phytohormones, phytotoxins, and volatile organic compounds in plants. *Proceedings of the National Academy of Sciences*, *100*(18), 10552–10557.
Snart, C. J., Hardy, I. C., & Barrett, D. A. (2015). Entometabolomics: applications of modern analytical techniques to insect studies. *Entomologia Experimentalis et Applicata*, *155*(1), 1–17.
Tang, Z. L., Wang, Q. L., & Peng, L. M. (2016). Study of the effect of porous particles on vapor-liquid equilibrium using the headspace gas chromatography. *AIChE Journal*, *62*(1), 256–263.
Xia, X. R., & Leidy, R. B. (2001). Preparation and characterization of porous silica-coated multifibers for solid-phase microextraction. *Analytical Chemistry*, *73*(9), 2041–2047.
Zamboni, N., Fendt, S. M., Rühl, M., & Sauer, U. (2009). 13C-based metabolic flux analysis. *Nature Protocols*, *4*(6), 878.

CHAPTER 2

An Integrated Approach to Plant Biology via Multi-Analogous Methods

2.1 INTRODUCTION

Due to escalating progress in the field of molecular genetics and physiology, there has been a paradigm shift from "vertical" analysis of the functions of one or a few genes, to "horizontal" holistic approaches, thereby facilitating the function of numerous genes of organisms concurrently (Fiehn, 2001; Jones et al., 2013). This variation is advancing us beyond genomes to transcriptomes, proteomes, and metabolomes, and has led to a perspective on life at a completely novel level (Horgan & Kenny, 2011). Profiling methods are accelerating this alteration, via the massive data generation that necessitates the present bioinformatics strategies, enabling them to adapt and grow in order to make the most of these data. An immense investment in genome sequencing and gene analysis was initially made with the concept that, even though they were precious for only a short span, they would be highly commercial in the end. Now, gene discovery via utilization of heterologous probes has been expansively substituted by using sequence data generated via large-scale sequencing programs (Bino et al., 2004; Zhou et al., 2015). The considerable value of partial cDNA sequence (expressed sequence tag, EST) collections was initially outlined in *Arabidopsis thaliana* (Rudd, 2003), and has demonstrated the parallel relevance of this method to the varied plant species, including crop species (Seki et al., 2002). The expansive relevance of EST sequencing unraveled not only the potential, but also the constraints of this method. These obstructions are imposed principally by the fact that the majority of genes in organisms are not widely represented in any given library (Legendre, Pochet, Pak, & Verstrepen, 2007). The identification of this constraint, and the need to acquire extra information regarding genome structure (e.g., about regulatory elements implicated the regulation of gene expression) stimulated the commencement of genomic sequencing programs for *Arabidopsis* and rice (Hu et al.,

https://doi.org/10.1016/B978-0-12-814872-3.00002-3

2011; Yu et al., 2002). Even though it was initially recognized as a gene discovery tool, sequencing provided numerous new resources, and incited the advancement of varied analytical methods, together with the paradigm shift in plant biology research (Nathan et al., 2008). The accessibility of the complete genome sequence of an organism, and the resultant gene catalogue, enables holistic analytic methods at the molecular level. For the first time, an organism can be considered a whole genetic system. Numerous multifaceted analytical strategies have been progressed that contribute to the study of gene/genome expression at the transcript (mRNA), protein, and metabolite (the substrates and the products of reactions catalyzed and regulated by the gene products) levels (Kaplan et al., 2007; Yang, Hua, & Shimizu, 2002). It is necessary to assess all constituents of the system at all of these levels. The intricacy of the data sets produced via these analyses will require knowledge beyond our present bioinformatics competence.

With the advent of full-scale DNA sequencing projects, biologists started to look at organisms in a holistic way. Contrary to last century's paradigm, genes are not the be all and end all of the analysis of the mechanisms of life (Krebs & Davies, 2009). Sequencing is an initial step in comprehending organisms at a molecular level, and a majority of the queries remain unraveled. In order to completely comprehend the gene function, the biological system must be evaluated at multiple regulatory levels (Alon, 2007; Kitano, 2002). Transcript profiling via DNA microarrays provides an expansive set of physiologically pertinent data about plants and other organisms. However, metabolic regulation is imposed not only at the mRNA level, but significantly at the metabolite and protein/enzyme levels (Hewitt et al., 2007). Certainly, modified metabolite levels, proteins, and post-translational protein alterations are of prime importance for evident phenotypic plasticity. Therefore, the investigation of molecular phenotypes requires the relevance of post-genomic strategies, for example, metabolomics and proteomics. Most of these methods are MS-based, while metabolomics and proteomics developed along with the advancement of new mass analyzers and strategies such as gas chromatography, coupled with time of flight mass spectrometry (GC-TOF-MS), and multidimensional liquid chromatography coupled with mass spectrometry (LC-MS). Even though illustrating the entire metabolome/proteome, these strategies contribute to an expansive richness, along with sample throughput, that is significant for a systematic examination of dynamic systems via many replicate evaluations. Due to recent advancements, data integration, as the significant objective of system biology, uses many replicate evaluations. In the following sections, the methods for high through-put metabolite and protein profiling are illustrated. The chapter concludes with a final word on data integration, including references to systems complexity.

2.2 TRANSCRIPT PROFILING

There are numerous analytical methods for transcript profiling, such as microarray based methods (cDNAs or oligonucleotides) (Oksman-Caldentey & Saito, 2005; Schauer & Fernie, 2006); sequencing-based approaches such as serial analysis of gene expression (SAGE) (Wang, Gerstein, & Snyder, 2009), or massively parallel signature sequencing (MPSS) (Crawford et al., 2006). Additionally, there are differential-display-based approaches, such as arbitrarily primed polymerase chain reaction (AP-PCR) (Micheli & Bova, 2013) and cDNA-amplified fragment length polymorphism (AFLP) (Parisi et al., 2010). Among all the approaches, microarray technology is a universal method that is applicable to all fields of biology (Mestdagh et al., 2009). The approaches have been extensively employed in yeast and human tissue samples, and in procuring the thousands of gene expression profiles (Metzker, 2010). Although the earnest revelation of the feasibility of the microarray was acquired with a set of *Arabidopsis* cDNAs (Gaeta et al., 2009; Leviatan, Alkan, Leshkowitz, & Fluhr, 2013), only a few microarray based research reports on plants have been published. With the progress in genome sequencing in *Arabidopsis*, there is upgradation of relevance of this technology in diverse organisms such as yeast (Afzal, Gul, & Kazi, 2016). Desprez, Amselem, Caboche, & Höfte, 1998 first used cDNA fragments containing 864 genes to describe gene expression in light-grown and dark-grown seedlings. Later, Ruan, Gilmore, and Conner (1998) generated expression data for 1443 genes in leaves, roots, and floral tissues of *A. thaliana*. Microarray-based differential analysis of gene expression symbolized by 1701 randomly selected strawberry fruit cDNA led to the identification of sets of genes differentially expressed during fruit ripening (Bombarely et al., 2010). In this way, new gene SAAT functions were discovered to have a significant role in flavor biogenesis. An assortment of 150 *Arabidopsis* cDNAs was employed to investigate and evaluate wound, dehydration, and insect feeding-induced gene expression in wild-type, ethylene-insensitive and coronatine/jasmonate insensitive mutants (Aharoni & O'Connell, 2002; Bombarely et al., 2010). In response to pathogen infection or elicitation with salicylic acid, methyl jasmonate or ethylene displays coordinated alterations in gene expression using microarrays of 5524 genes (Salzman et al., 2005). Standard Microarray Database reports offer an extensive range of entries that signify the huge expression profiling data sets procured for microarrays for up to 11,000 cDNA fragments available for *Arabidopsis* (Wisman & Ohlrogge, 2000). Furthermore, primary investigations have been carried out using *Arabidopsis* oligonucleotide arrays signifying 8775 genes (Zhu et al., 2001; Zimmermann, Hirsch-Hoffmann, Hennig, & Gruissem, 2004). Sequencing technology such as mRNA-Seq is emerging as the preferred technology for transcript profiling due its high reproducibility, dynamic range, and richness of data. Whereas microarray expression profiling exhibits three major limitations: profiling is restricted to only known genes/ splice variants, it hybridizes artifacts, and results are difficult to reproduce.

Conventional digital gene expression techniques such as SAGE produce data such as mRNA sequences, but are restricted by comparatively high costs imposed by Sanger sequencing, per base and requirements for bacterial cloning steps. Conversely, they offer numerous advantages, for example:

- mRNA sequences exhibit a high dynamic range up to 5 orders of magnitude, which is considerably greater than microarray technologies (Wang et al., 2009).
- mRNA sequencing does not require technical replicates, as this technology displays concordance between identical samples (Marioni, Mason, Mane, Stephens, & Gilad, 2008).
- Characterization in relative abundance of splice variants is likely via paired-end reads known as RNA splice patterns (Trapnell, Pachter, & Salzberg, 2009).
- Highly abundant transcripts can be employed to catalog patient haplotypes directly from RNA, which is known as SNP (single nucleotide substitutions or polymorphisms) typing and allelic expression (Verlaan et al., 2009). This approach can be further expanded to detect allelic inequity in expression.

Identification of unusual transcripts is restricted only by sequencing intensity. Previous studies of almost 8m reads introduced a new perspective (Sultan et al., 2008), but sequencing technology has progressed considerably. An especially profound (10 gigabase) mouse transcript sequencing report displayed that 80 M reads is adequate to identify exclusive sequence tags, in which a majority of data is originated from the first 10–40 M reads (Mortazavi, Williams, McCue, Schaeffer, & Wold, 2008).

2.2.1 Library Preparation from RNA Samples

This method selects polyadenylated mRNA transcripts from total RNA, fractionates them, and then converts them to dsDNA for sequencing. This process advances as follows:

- Take 1 μg of high quality RNA and incubate with oligo(dT) magnetic beads to augment mRNA with poly(A) tails.
- The RNA thus eluted is immersed in a tris buffer containing potassium and magnesium acetate and incubated at 94°C. This results in fractionated RNA ranging from 200 to 500 nt.
- The RNA is then precipitated with sodium acetate and ethanol, and thereafter washed in water.
- This is followed by reverse transcription using random oligonucleotide hexamers, which is performed to produce cDNA.
- Then, the RNA is degraded by adding RNase and DBA polymerase to produce a second strand. The DNA is then prepared for standard illumina adaptor ligation for sequencing (Kozarewa et al., 2009).

The platform illumina uses an in-situ amplification strategy followed by dye terminator sequencing (Bentley et al., 2008). The short oligonucleotides are covalently bound to the sequencing flow cell, which is employed to fix the DNA strand being sequenced. Each molecule in DNA library encloses two specific sequences at its ends. These are introduced by DNA ligase and the method is controlled as follows:

Following the second strand synthesis, the dsDNA is then comprised of ends with 5′ and 3′ overhangs, which are filled in via T4 polymerase and T4 polynucleotide kinase, leading to a blunt-ended DNA molecule. A deoxyadenosine (dA) 5′ tail is added to DNA strands employing the Klenow fragment (exo-). Double-stranded DNA adapters with 3′ thymidine overhangs are ligated to the dA tailed library via T4 ligase. The adapters comprise the sequences required for binding to the flow cell and sequencing primer binding sites.

The mRNA-Seq protocol requires about two days to complete. The initial step requires RNA preparation to validate RNA quality with the Bioanalyzer, then it is necessary to choose the poly(A) RNA with oligo(dT) beads, and perform the first and second synthesis with reverse transcriptase and DNA clearout. Following DNA clean up, the second day is intended for library preparation comprised of end-polishing of dsDNA, ligating adaptors, size collection by agarose gel/SPRI beads, enrichment of the library by PCR, and substantiation of the library with the BioAnalyzer (Zhong et al., 2011).

RNA sample quality is important. The mRNA-Seq employs oligo(dT) beads to choose polyadenylated RNA from a total RNA population. This leads to a strong bias toward sequencing reads due to fragmented RNA molecules. This is alleviated partly by selecting high-quality RNA prior to sequencing (strong 18S/28S peaks on BioAnalyzer). Downstream bioinformatics evaluation for authenticating the bias reads are counted, and per-gene expression is computed. Reverse transcriptase might have varied efficacies based on templates. This might also be a source of bias in read counts, but that is cumbersome to eliminate. Further, reverse transcriptase template switching can be a source of errors. This is generally dealt with via software that detects chameric reads with low counts and eliminates them.

2.2.2 MicroRNA Profiling

MicroRNA is small (about 22 nt long). It post-transcriptionally regulates the gene expression in an extensive range of organisms, in both normal physiologic and disease contexts (Lakshmipathy, Love, Adams, Thyagarajan, & Chesnut, 2007). MicroRNA expression profiling is emerging as a popular technique because miRNA are functional regulators in gene expression networks, which can affect numerous biological processes. They have also exhibited potential as biomarkers for disease detection (Tang, Galili, & Zhuang, 2007; Zhong et al., 2011). Technological progress has facilitated the advancement of varied platforms for miRNA profiling, and a sufficient understanding of the advantages and limitations of diverse strategies can facilitate the efficient utilization of miRNA profiling in varied applications (Pritchard, Cheng, & Tewari, 2012). There are many recommendations for execution, and for interpretation of results of miRNA profiling studies; in addition to novel methods for, and the increasing relevance of miRNA profiling (Baker, 2010). MicroRNA has emerged as the critical regulator of gene expression in plants and animals. miRNA are short (~22 nucleotides), noncoding RNAs that control gene expression by binding to particular messenger RNA (mRNA) targets, advancing their degradation and translational inhibition (Gu & Kay, 2010; Shruti, Shrey, & Vibha, 2011). In the majority of organisms, there are comparatively numerous miRNAs in contrast to the number of mRNAs and proteins. However, one miRNA may control hundreds of mRNA, and lead to significant effects on gene expression networks. Therefore, miRNA expression patterns can be particularly loaded with biological data, because alterations in the expression of hundreds of mRNAs may, to an extent, be reflected in the expression patterns of one or a few miRNA that control them. miRNA expression profiling has proved to be valuable for detection of miRNA that are significant in the control of diverse processes, including organism growth, and organization and maintenance of tissue differentiation (Kim, 2005). Consequently, miRNA expression biomarkers are finding relevance in the diagnosis of tissue differentiation states in cancers of novel origins, and miRNA are being evaluated as reagents for reprogramming the cell fate in stem cell relevance (Martinez & Gregory, 2010). The estimation of miRNA expression can update system level models of gene regulation, particularly when miRNA data is blended with mRNA profiling and other genome-scale data. The most significant relevance of miRNA is their potential as biomarkers for different molecular diagnostics applications in cancer, and cardiovascular and autoimmune diseases, together to advance miRNA-based approaches for forensic investigation where extremely sensitive estimations of these molecules may contribute data.

2.3 MICROARRAY-BASED APPROACHES

Microarrays were originally developed to investigate gene expression. The technology has evolved, and has since been tailored to diverse applications, among which is the evaluation of genomes to explore DNA dosage and sequence content (Bloom et al., 2004; Okou et al., 2007). An increase in the development of microarray fabrication and the execution of expansive projects on genome sequencing have facilitated the instant progress of array-based approaches for genome-wide monitoring of DNA modifications (Barbulovic-Nad et al., 2006; Suo et al., 2010). There are diverse types of DNA modifications that can occur; for example, DNA amplification or deletion of single base pairs to thousands of base pairs, SNPs, and depletion of heterozygosity (Tanaka et al., 2003). An extensive range of human diseases is related with genomic variations. Some of these diseases have an intricate genetic component in which numerous transcripts or DNA modifications contribute to disease (Cookson, Liang, Abecasis, Moffatt, & Lathrop, 2009). Detection of genomic regions will advance our perception pertaining to biology linked with these diseases, and is significant in determining disease susceptibility (Green & Guyer, 2011).

2.3.1 Comparative Genomic Hybridization to Microarrays

This is the most efficient method for illustrating DNA modification, and it has been utilized to acquire more data pertaining to DNA copy number gains and losses in diverse diseases since the early 1990s (Barrett et al., 2004; Zhao, Fung-Leung, Bittner, Ngo, & Liu, 2014). This method utilizes labeled-test and normal-reference DNA, which are hybridized to normal chromosome metaphase spreads. The addition or deletion of some regions of chromosomes caused by duplications, amplifications, or deletions are identified as rations between the labeled test and normal reference DNA, together with the length of the chromosome. Even though innovative, conventional CGH is now recognized as labor intensive, with poor resolution (~20 Mb) incapable of identifying explicit transcripts inside areas of gains and losses across the genome that contribute to disease (Donaghue et al., 2017). During the late 1990s, array CGH was developed that blends conventional CGH with microarrays and advanced the declaration of copy number identification (Autio et al., 2003). During array CGH, DNA from test and reference populations are labeled and hybridized to microarrays that have been printed with either genomic fragments, for example, bacterial artificial chromosome (BAC) clones from significant genomic regions (Fiegler et al., 2003), and microarrays comprised of cDNA clones, which are employed for gene expression investigations (Wang et al., 1999). Following hybridization, the microarrays are scanned, and the ration of test to reference DNA is utilized to estimate copy number disparity between the samples. DNA microarrays were then generated to span the human genome, and were comprised of 2400 BAC clones, which symbolized the sequenced human genome at an average resolution of ~1.4 Mb (Snijders et al., 2001). These are trailed by tiling resolution BAC microarrays comprised of >30,000 probes (Barrett & Caren, 2011). However, these BAC arrays still had limitations related to their incompetency to identify SNPs or loss of heterozygosities (LOHs). The majority of array-based contrivances of genomic modifications are oligo-array-based techniques for genotyping thousands of SNPs concurrently (Elingarami, Li, & He, 2013). The data from these oligo SNP arrays can then be modified to perform high-resolution evaluation of DNA copy number variation and LOH (Curnutte, 2012; Pfeifer et al., 2007). Current progress in array technology and genome data from projects such as HapMap has facilitated the advancement of whole genome SNP arrays, which are discussed as follows.

2.3.2 The International HapMap Project

SNPs are DNA alterations that present in single nucleotides and illustrate considerable heterozygosity in human populations (Thorisson, Smith, Krishnan, & Stein, 2005). SNPs are present within the coding region of genes, thereby altering the amino-acid sequence of the protein or noncoding or intergenic regions (Brookes, 1999). SNPs are the main form of genetic diversity in the human genome, and there are thought to about to be about 10 million SNPs (IHC, 2007). Alleles of SNPs are closely packed, and tend to be inherited together in haplotype blocks. Therefore, assaying a single SNP, considered as a tag SNP, will contribute data on the neighboring SNPs, and will advance the power of association studies. The International HapMap Project (www.hapmap.org) has facilitated the development of whole-genome SNP genotyping by illustrating these common genetic variation patterns to facilitate groups of tag SNPs (Montpetit et al., 2006).

2.3.3 SNP Genotyping Arrays

The microarrays presently accessible to researchers can be used for expansive genome SNP genotyping, high density copy numbering, and LOH evaluation (Gunnarsson et al., 2008; Wang et al., 2007). The resolution of SNP arrays is rising exponentially, and the main collections that are currently accessible facilitate genotyping of more than 1 million SNPs in a single experiment (Alkan, Coe, & Eichler, 2011). These arrays are the Affymetrix™ Genome-Wide Human SNP Array (www.affymetrix.com) and the Illumina™ Human1M Bead Chip (www.illumina.com) (Choy, Setlur, Lee, & Lau, 2010; Duan, Zhang, Bleibel, Cox, & Dolan, 2008). Both of these arrays are contrived in a similar fashion to their gene expression equivalents, but possess oligo probes for genotyping (Ronald et al., 2005). The assays are easy to run for the Affymetrix arrays, and genomic DNA is then digested, ligated, and amplified by PCR (McCarroll et al., 2008). The DNA is then labeled and hybridized to the arrays, and the fluorescence intensities are estimated for each allele of each SNP. On the contrary to Illumina BeadChips, the DNA is amplified, fragmented, and hybridized to the BeadChips, which is comprised of allele-specific oligo probes (Waddell, 2008). Once hybridized, a single-base extension of the oligos on the BeadChip, via the hybridized DNA as a template, integrates labels on the BeadChip, which are employed to establish the genotype after scanning. Approximately half of the SNPs are targeted by these array-based approaches, and are known as tag SNP markers that have originated from the human HapMap Project. The rest of the SNPs on the Affymetrix array primarily consist of an unbiased selection of SNPs; those on the Illumina arrays are intended for the SNPs that are near, or in, genes. Both chips are also comprised of an assortment of copy number variant probes that are spaced along the genome (Bibikova et al., 2011; Hester et al., 2009).

2.3.4 SNP Genotyping Array Studies

SNP arrays are employed for genome-wide association research studies to genotype genetic markers across case and control populations (Ganal et al., 2011; Jones et al., 2012). The genomic regions that exhibit statistical significance bear a candidate disease locus. This kind of study has facilitated the localization and detection of genes associated with diseases (Howie, Fuchsberger, Stephens, Marchini, & Abecasis, 2012; Keating et al., 2008). Numerous research reports have documented the significance of genomic SNP arrays that advances our perception pertaining to various diseases comprised of lymphoblastic leukemia (Mullighan et al., 2008), celiac disease (Trynka et al., 2011), improper mental development (Matukumalli et al., 2009), diabetes (Unoki et al., 2008), and breast cancer (Hunter et al., 2007). A genomic-wide association research study has detected breast cancer susceptibility loci (Easton et al., 2007), and further genomic-wide association studies described disparity in vulnerability to HIV-1 infection (Rotger et al., 2010).

Microarray-based methods to investigate genomic aberrations have considerably progressed in recent years. It is now easier to resolve copy number variation, SNP genotyping, and LOH at high resolutions. The progress of rapid, expansive genome strategies employing SNP genotyping microarrays will enable researchers to acquire a new perspective on the biology of human diseases, and will detect several loci of disease susceptibility.

2.4 SEQUENCING-BASED APPROACHES

The transcriptome is a whole set of transcripts in a cell, and their estimation for a particular developmental stage or physiological status (An, Hsiao, & Chan, 2011; Hirst & Marra, 2010). The complete elucidation of transcriptome is indispensable for decoding the functional elements of the genome and unraveling the molecular components of cells and tissues, and for comprehending the growth and advancement of diseases (Barabási, Gulbahce, & Loscalzo, 2011; Gerling, Solomon, & Bryer-Ash, 2003). The significant functions of transcriptomics are to catalogue all species of transcripts comprising mRNAs, noncoding RNAs, and small RNAs to resolve the transcriptional modifications as known in their start sites' 5′ and 3′ ends, splicing patterns, and other post-transcriptional modifications; and to estimate the varying expression levels of each transcript during development and under varied conditions (Helm & Motorin, 2017; Nejat & Mantri, 2018). Different technologies have been progressed to presume and estimate the transcriptome comprising hybridization or sequence-based methods (Kogenaru, Yan, Guo, & Wang, 2012). Hybridization-based methods generally implicate incubating fluorescently labeled cDNA with customized microarrays, or marketable high-density oligo microarrays. Specific microarrays have also been premeditated; for example, arrays with probes spanning exon junctions can be employed to identify and estimate different spliced isoforms (Finotello & Di Camillo, 2015; Wang et al., 2009). Genomic tiling microarrays that symbolize the genome at high density have been manufactured and facilitate the mapping of transcribed regions to enhanced resolution from numerous base pairs to ~100 bp 2–5 (Lippman, Gendrel, Colot, & Martienssen, 2005). Hybridization-based approaches are high throughput and comparatively economical, except for high-resolution tiling arrays that query expansive genomes (Ahmadloo et al., 2017; Reuter, Spacek, & Snyder, 2015; Sasidharan, Agarwal, Rozowsky, & Gerstein, 2009). However, these approaches have numerous limitations, which include dependence on present data pertaining to the genome sequence, high background levels due to cross hybridization (Goodwin, McPherson, & McCombie, 2016; Rimmer et al., 2014), and an inadequate dynamic range of identification due to surroundings and the infiltration of signals. Furthermore, evaluating expression levels across diverse experiments is frequently intricate and can require complex normalization approaches. On the contrary to microarray approaches, sequence-based methods openly resolve the cDNA sequence. At first, Sanger sequencing of cDNA or EST libraries was employed, but this method is comparatively low throughput, exclusive, and typically not quantitative (Lagarde et al., 2016). Tag-based approaches were advanced to further prevail over restrictions, and included SAGE (Tuteja & Tuteja, 2004; Velculescu, Zhang, Vogelstein, & Kinzler, 1995), cap analysis of gene expression (Kanamori-Katayama et al., 2011; Kodzius et al., 2006), and MPSS (Brenner et al., 2000; Habib et al., 2017). These tag-based sequencing methods are high-throughput, and can contribute to correct digital gene expression levels. However, the majority of methods are based on exclusive Sanger sequencing technology, and a considerable fraction of the short tags cannot be exclusively mapped to the reference genome (Hirst & Marra, 2010; Reuter et al., 2015). Furthermore, only the fraction of the transcript is evaluated, and isoforms are typically similar to each other. These limitations restrict the utilization of traditional sequencing technology in annotating the structure of transcriptomes (Ekblom & Wolf, 2014; Wang et al., 2009). In recent times, the progress of new, high-throughput DNA sequencing approaches has contributed to novel approaches for both mapping and estimating transcriptomes (Garber, Grabherr, Guttman, & Trapnell, 2011; Grabherr et al., 2011). This technique, known RNA-Seq (RNA sequencing), has obvious applications over prevailing methods, and is supposed to transform the way in which eukaryotic transcriptomes are investigated (Mortazavi et al., 2008; Trapnell et al., 2010). This approach has been applied to *Saccharomyces cerevisiae*, *Schizosaccharomyces pombe*, *A. thaliana*, and mouse and human cells (Marguerat & Bähler, 2010; Nookaew et al., 2012). Here, we illustrate how RNA-Seq works, enumerate its constraints, and give away an outline of research reports that have employed this strategy, which have already commenced to diverge our outlook of eukaryotic transcriptomes (Li & Dewey, 2011; O'Rourke et al., 2012).

2.4.1 Advantages of RNA-Seq Technology

RNA-Seq is still under-developed technology that contributes to numerous applications in many prevailing technologies, which are summarized as follows:

1. In contrast to hybridization-based methods, RNA-Seq is not restricted to identification of transcripts that correspond to an accessible genomic sequence. RNA-Seq is applicable to non-model organisms with genomic sequences that are yet to be determined. For example, 454-based RNA-Seq has been employed to sequence the transcriptome of the *Glanville fritillary* butterfly (Agarwal et al., 2010; Somervuo et al., 2014).
2. RNA-Seq can exhibit the accurate site of transcription boundaries to a single-base resolution. Additionally, 30-bp short reads from RNA-Seq contributes data about how two exons are associated; while longer reads, or pair-end short reads should unravel connectivity between multiple exons. This factor contributes RNA-Seq that is valuable for illustrating intricate transcriptomes. Further, RNA-Seq also unravels sequence disparity; for example, SNPs in the transcribed regions (Hensman, Papastamoulis, Glaus, Honkela, & Rattray, 2015; Su et al., 2014).
3. In contrast to DNA microarrays, RNA-Seq has very low background signal because DNA sequences can be seen explicitly mapped to a unique genomic region (Fu et al., 2009; Zhao et al., 2014). RNA-Seq does not have an upper limit of estimation that associates with the number of sequences procured. Therefore, it has an expansive dynamic range of expression levels over which transcripts can be identified. A greater than 90,000-fold range was quantified in a study that evaluated 16 million

mapped reads in *Saccharomyces cerevisiae* (McCarthy, Chen, & Smyth, 2012; Zhao et al., 2014), and a range spanning five orders of magnitude was quantified for 40 million mouse sequence reads (Anders & Huber, 2010; Langmead, Trapnell, Pop, & Salzberg, 2009).

4. In contrast, DNA microarrays are required for genes expressed either at low or very high levels, and consequently have a much smaller dynamic range. RNA-Seq has also been shown to exhibit extremely precise results for estimating expression levels, which is achieved by employing quantitative PCR (qPCR) (Anders & Huber, 2010) and spike-in RNa controls of known concentrations (Jiang et al., 2011). The findings of RNA-Seq also exhibit high levels of reproducibility, for both technical and biological replicates (Marioni et al., 2008). Ultimately, because there are no clone steps, and with Helicos technology, there is no amplification step, RNA-Seq needs very little RNA sample matter.
5. Taking all these applications into account, RNA-Seq is the primary sequencing-based approach that permits the complete transcriptome to be investigated in a very high-throughput and quantitative manner. This approach contributes both single-base resolution for annotation, and digital gene expression levels at the genomic scale, and frequently at a lesser value than either tiling arrays or expansive Sanger EST sequencing.

2.4.2 Challenges for RNA-Seq Library Construction

The ultimate approach for transcriptomics should be able to easily detect and estimate all the RNAs, small or large. Even though there are few steps in RNA-Seq, it does require numerous exploitation stages during the making of cDNA libraries, which can create hurdles for its utilization in profiling all kinds of transcripts. In contrast to small RNAs (microRNAs (miRNA), piwi-interacting (piRNAs), short interfering RNAs (siRNAs), and numerous others), which can be openly sequenced following adaptor ligation, larger RNA molecules must be split into smaller pieces (200–500 bp) to be compatible with most deep-sequencing technologies. Common fragmentation approaches include RNA fragmentation (RNA hydrolysis or nebulization) and cDNA fragmentation (DNase I treatment or sonication). Each of these approaches generates a diverse bias in the result. For example, RNA fragmentation has a slight predisposition over the transcript body (Shimkets et al., 1999), but is lost for transcript ends in comparison to other approaches. On the contrary, cDNA fragmentation is generally immensely biased toward the detection of sequences from the 3′ ends of transcripts, and thus contributes expensive data pertaining to the exact identity of these ends. Some modifications during library construction also confound the investigation of RNA-Seq results. For example, numerous short reads that are similar to each other can be procured from cDNA libraries that have been amplified. These might be authentic evidence of rich RNA species; they should be PCR artifacts. One of the major ways to differentiate between potential is to resolve whether the same sequences are illustrated in diverse biological replicates. Another significant decision regarding library construction is whether or not to prepare strand-specific libraries, as done in previous studies (Gleason & Burton, 2015; Head et al., 2014). These libraries have the benefits data regarding the direction of transcripts, which is expensive for transcriptome annotation, particularly for regions with overlapping transcription from opposite directions (Costa, Angelini, De Feis, & Ciccodicola, 2010). However, strand-specific libraries are presently difficult to generate because they require numerous steps, or direct RNA-RNA ligation, which is ineffective. Furthermore, it is necessary to ensure that the antisense transcripts are not artifacts of reverse transcription. Due to these intricacies, the majority of research studies investigated so far have investigated cDNAs without strand data. Similar to other high-throughput sequencing technologies, RNA-Seq is subjected to numerous informatics constraints, including the advancement of competent methods to preserve, recover, and process expansive data, which must be surmounted to reduce errors in image investigation and base calling, and eliminate low-quality reads. Once high-quality reads have been procured, the primary objective of data analysis is to map the short reads from RNA-Seq to the reference genome, or to collect them into contigs prior to aligning them to the genome sequence to unravel the transcription structure (Lister, Gregory, & Ecker, 2009). There are numerous programs for mapping reads to the genomes, including ELAND, SOAP31, MAQ32, and RMAP33 (data regarding these can be found at the illumine forum and at SEQ answers) (Pareek, Smoczynski, & Tretyn, 2011). However, short transcriptomic reads also possess reads that span exon junctions, or that contain poly(A) ends, and these cannot be investigated in the similar way. There are some explicit examples, such as *Saccharomyces cerevisiae*, wherein genomic splicing is uncommon; hence, special focus should be given to poly(A) tails and to a small number of exon-exon junctions (Cloonan et al., 2008; Oshlack, Robinson, & Young, 2010)). Poly(A) tails can be detected easily by the occurrence of mani-fold As or Ts at the end of some reads. Exon-exon junctions can be detected by the incidence of the particular sequence milieu (the GT-AG di-nucleotides that flank splice sites), and established by the low expression of intronic sequences that are eliminated during splicing (Li et al., 2015; Trapnell et al., 2009). For *Saccharomyces cerevisiae*, transcriptome maps have been produced in this way. It is more complicated to map reads in complex transcriptomes due to the splice junctions; specifically, because of the occurrence of wide substitutive splicing and trans-splicing (Jhuang, Lee, & Leu, 2017). The fractional solution is to assemble a junction library that includes all identified and predicted junction sequences, and map reads to this library (Trapnell et al., 2010). The constraint for the future is to advance computationally easy procedures to detect new splicing that takes place between two distant sequences, or between exons from two diverse genes (Zhang et al., 2014). For an expansive range of transcriptomes, configuration is also convoluted by the detail that a considerable fraction of sequence reads are equivalent to manifold sites in the genome (Dominissini, Moshitch-Moshkovitz, Salmon-Divon, Amariglio, & Rechavi, 2013; Mercer et al., 2012). For expansive transcriptomes, alignment is convoluted by the fact that a considerable portion of sequence reads match many genomic sites. The elucidation is to allocate these multifaceted reads by proportionally conveying them based on the numerous reads mapped to their adjacent distinct sequences (Grabherr et al., 2011). This procedure has been thriving for low copy repetitive sequences. Short reads that have high copy numbers (>100) and extended stretches of repetitive regions provide an intricate issue (Ozsolak & Milos, 2011). Multi-matching issues can be lessened with the aid of 454 technologies so as to procure longer sequence reads (Saliba, Westermann, Gorski, & Vogel, 2014). On the other hand, a paired end sequencing technology in which short sequences are established from both ends of a DNA fragment expand the mapped fragment length to 200–500 bp, and is anticipated to be valuable in the coming years (Head et al., 2014).

Sequencing errors and polymorphism can harbor more mapping problems, not only for repetitive DNA, but for all genomes (Conesa et al., 2016; Kumar et al., 2012). Usually, single-base disparities are not challenging, because the majority of mapping algorithms contain one or two base differences. However, determination of expansive disparities will need reference genome annotation for polymorphism and in-depth sequencing documentation. Another significant subject is sequence coverage, on the percentage of transcripts surveyed, which has implications for cost. To identify different transcripts or variants, significant elucidation is required. In simple transcriptomes, for example, yeast (both *Schizosaccharomyces pombe* and

Saccharomyces cerevisiae), for which there is no proof of substitutive splicing, 30 million 35-nucleotide reads from poly(A) mRNA libraries are adequate to scrutinize transcription from a majority of genes (>90%) for cells grown under single-nutrient-rich conditions (Wittenberg & Reed, 2005). This process is likely more than adequate for the majority of purposes, as the numerous expressed genes identified by RNA-Seq attain 80% coverage at 4 million distinctly mapped reads, following which doubling the depth only amplifies the coverage by 10% (Honaas et al., 2016; Renaut, Grassa, Moyers, Kane, & Rieseberg, 2012). Investigating numerous diverse conditions can further amplify the coverage in *Saccharimyces pombe* 122 million reads from six diverse growth conditions identified in transcription from >99% of annotated genes (Nowrousian, 2010). Generally, the larger the genome, the more intricate the transcriptome, and the more sequencing intensity is required for enough coverage (Mortazavi et al., 2008). In contrast to genome sequencing coverage, it is relatively simple to quantify the reporting of the transcriptome, as the precise number and extent of diverse transcript isoforms is not generally recognized, and because transcription activity changes immensely across the genome (Bonasio & Shiekhattar, 2014). Our study employed numerous distinct transcription start sites as an estimate of coverage in mouse embryonic cells, and illustrated that at 80 million reads, the number of start sites reached a plateau (Landt et al., 2012). However, this method does not tackle transcriptome intricacy in substitutive splicing and transcription termination sites, where apparently, advance sequencing can display extra variants (Lagarde et al., 2017). If it is determined that the transcription level of RNA-Seq is quantitative, it can be employed to resolve RNA expression levels more precisely than microarrays (Łabaj et al., 2011; Trapnell et al., 2013). It has the potential to resolve the total amount of each molecule in a cell population, and easily evaluate results between experiments.

One of the basic advantages of RNA-Seq is that it can capture transcriptome dynamics across diverse tissues or conditions without the complicated normalization of data sets (Poelchau, Reynolds, Denlinger, Elsik, & Armbruster, 2011; Zheng, Kawane, Bottomly, & Wilmot, 2014). RNA-Seq has been employed to precisely assess gene expression during yeast vegetative growth (Marguerat & Bähler, 2010), yeast meiosis (Nagalakshmi et al., 2008), and mouse embryonic stem-cell differentiation (Yan et al., 2013), and to track gene expression changes during development, and to give a "digital measurement" of gene expression disparity between diverse tissues (Macosko et al., 2015). Because of these applications, RNA-Seq will certainly be valuable for comprehending trancriptomic dynamics during development and normal physiological changes, and in the analysis of biomedical samples, where it will permit intensive assessment between diseased and normal tissues along with determination of disease states. Although RNA-Seq is still in the initial stages of utilization, it has obvious applications over formerly developed trancriptomic approaches. The subsequent immense confrontation for RNA-Seq is to target more intricate transcriptomes to detect and track the expression variation of uncommon RNA isoforms from all the genes. The technological advancements that acquire pair-end sequencing, strand-specific sequencing, and utilization of longer reads to amplify coverage and depth should be implemented. With the drop in sequencing cost, RNA-Seq is anticipated to substitute microarrays for numerous applications that implicate establishing the structure and dynamics of the transcriptome.

2.5 DIFFERENTIAL-DISPLAY-BASED APPROACHES

The PCR-based DNA fingerprinting in which nucleotide sequences are randomly selected is classified as a differential display-based method. Among AP-PCR is PCR-based DNA fingerprinting, employing primers with randomly chosen nucleotide sequences (Gellatly, Ash, & Taylor, 2001; Welsh & McClelland, 1990; Williams, Kubelik, Livak, Rafalski, & Tingey, 1990). This approach is known as random amplified polymorphic DNA (RAPD) (Lu, Gosslau, Liu, & Chen, 2008; Xiao et al., 2016). The first steps (cycles) of the reactions are investigated under comparatively low temperatures during the annealing step, with a high magnesium concentration in the reaction buffer. These arbitrary primers anneal to the complimentary template under the aforementioned conditions (Li et al., 2018; Ziegenhain et al., 2017). The competition between these annealing events leads to reproducible and quantitative amplification of numerous distinct bands. Moreover, amplification of these sequences under exceedingly severe conditions generates a fingerprint as determined by gel electrophoresis (Zhao et al., 2018). The band pattern procured in an ordinary experiment is typical, and characteristic of the genome employed as a template. Even though aspects controlling the definite priming and amplification are somewhat intricate and are incomprehensible, the experimental substantiation reveals that for an already-established set of experimental conditions, the amplified pattern of bands is distinct and reproducible. Further, in contrast to other DNA fingerprinting methods, the AP-PCR approach allows instant cloning of the in vitro amplified DNA sequences (Lehr, Kilpatrick, Wilkerson, & Conn, 2014). Two features of this approach are particularly fascinating. One is that amplified bands generally derive from single-copy sequences. Second, the amplification is quantitative; that is, the intensities of the amplified bands are approximately comparative to the concentration of the resultant sequences in the DNA preparation. It can be determined that data pertaining to the general allelic components of the genome can be produced by a convenient experiment that is done in a single microtube. The expansive range of bands amplified with a single random primer produces an intricate fingerprint that can be employed to identify comparative disparity in the randomly amplified DNA sequences from two diverse, but closely linked genomes (Chaisson, Wilson, & Eichler, 2015). Such disparity corresponds to polymorphism, or somatic genetic modifications. The priming stages during the first low-stringency cycles are random because they were determined on the nucleotide sequence of the PCR primer, which is randomly selected. Therefore, the amplified sequences are, in principle, a representative, random small sample of the donor DNA cell genome. Consequently, this kind of investigation presents a great deal of relevant data, including the genetic map of polymorphisms, and the taxonomy of somatic mutations in cancer cells, which can lead to identification. Its utilization has been illustrated in varied prokaryotes and eukaryotes systems comprised of plants, rodents, and humans (Baird et al., 2008; Poland, Brown, Sorrells, & Jannink, 2012). The extent of association between individual organisms, or the inheritance in progeny populations can be estimated by the disparity in length or sequence of DNA segments, though the detection of these molecular markers needs previously collected data from the DNA sequence, cloned to identify probes, and to be used for substantial exploitation (Lister et al., 2009). This restriction can be prevented with manifold random amplicon profiling (MAAP) (Avelar and Linardi, 2010) approaches, such as RAPD evaluation (Nielsen et al., 2014), AP-PCR, and DNA amplification fingerprinting (DAF) (Houzet et al., 2015). These autonomously advanced approaches employ one or more random oligonucleotide primers to target explicit, but unidentified sites in the genome, most of which are polymorphic. Identified AFLP can be employed simply as markers for genetic typing and mapping relevance (John, Joseph, & Gbolagade, 2016; Winkler et al., 2017). MAAP strategies target manifold and random amplicons, but diverge in primer length, amplification stringency, and methods employed to determine DNA patterns. For example, AP-PCR (2s) employs primers of lengths analogous with those of PCR primers, RAPD (Gresshoff & Caetano-Anollés, 2018) evaluation employs shorter primers (generally 10 nucleotides in length

0, and DAF utilizes primers as short as 5 nucleotides, but usually 7 or 8 nucleotides in length. These, and other differences, lead to DNA profiles that are extremely discrete. While as RAPD determines some amplification products via agarose gel electrophoresis and ethidium bromide staining, DAF generates typical and intricate data-affluent DNA profiles via polyacrylamide gel electrophoresis and a silver stain that identifies DNA at the picogram level (Muzzalupo, Pisani, Greco, & Chiappetta, 2015). Fingerprints are confirmed by primer annealing to incompletely or completely complementary sites on each DNA strand. Consequently, primer design, amplification reaction conditions, and sufficient separation and identification of amplified products incite the reproducibility, intricacy, and polymorphic content of DNA profiles (Nybom, Weising, & Rotter, 2014; Passari et al., 2015). For example, RAPD investigation via decamer primer-generated fingerprints were, on average, 3–10 times easier compared with those produced via DAF with decamer and octamer primer. Comprehensive assessment of amplification products might be incompletely responsible for disparities in profile intricacy. However, a primer, magnesium, and deoxynucleoside triphosphate concentration, in addition to temperature cycling parameters and other amplification reaction components, can modify reaction stringency and considerably affect both the number and the nature of amplification products.

2.5.1 Amplified Fragment Length Polymorphism

During the past decade, different approaches have been developed for the detection and summary of prokaryotic and eukaryotic organisms at the DNA level (He, Liu, Cai, Pan, & Zhang, 2014; Rasmussen, 2012). The strategies diverge in their taxonomic range, differential ability, reproducibility, and direct elucidation and validation (Gerber, Mariette, Streiff, Bodenes, & Kremer, 2000; Savelkoul et al., 1999). The best genotyping approaches generate results that are inconsistent from laboratory to laboratory, and permit definite proportional investigation and the establishment of reliable databases. One of the latest and most significant approaches is AFLP investigations (Godschalk et al., 2006), developed by Keygene BV, Wageningen, The Netherlands. This procedure blends general relevance with efficient differentiation and reproducibility (Spooner, McLean, Ramsay, Waugh, & Bryan, 2005). Numerous research reports illustrate the application of AFLP evaluation for genetic mapping of plants and animals, disease diagnostics, phylogenetic investigations, and microbial typing (Meudt & Clarke, 2007).

The method of AFLP is classified as a selective restriction fragment amplification strategy that is primarily determined by the ligation of adapters (i.e., linkers and indexers) to genomic restriction fragments trailed by a PCR-based amplification with adapter specific primers. The major advantage with AFLP analysis is the requirement for a small amount of genomic DNA, which is to be digested with two restriction enzymes, such as *Eco*RI and *Mse*I or *Taq*I, with average and high cutting frequency, respectively (Vuylsteke, Peleman, & Van Eijk, 2007). In this method, double stranded oligonucleotide adapters are manufactured in a particular way so that the first restriction site is reinstated following ligation, which permits instant restriction and ligation, while relegated fragments are cleaved again. An aliquot is then exposed to two successive PCR amplifications under extremely strict conditions with adapter-specific primers that have in their 39 ends an addition of one to three nucleotides heading into unidentified chromosomal restriction fragments (Fry, Savelkoul, & Visca, 2009). An addition of one chosen nucleotide amplifies 1 of 4 of the ligated fragments, while as three selected nucleotides in both primers amplify 1 of 4096 of the fragments. The primers with average frequency restriction site s should be labeled. Following polyacrylamide gel electrophoresis, an extremely instructive pattern of 40–200 bands is acquired. The patterns procured from diverse strains are polymorphic because of mutations in the restriction site, mutation in the sequences adjoining to the restriction sites, and a complementary relationship to the specific primer extension, along with insertion or deletion among amplified fragments. Several enzyme combinations have been employed, for example, *Eco*RI, *Pst*I, *Hind*III, or *Apa*I combined with *Mse*I or *Taq*I (Gzyl et al., 2005). The restriction enzymes such as *Eco*RI and *Taq*I are most suitable for animal genomic studies. Another AFLP typing procedure is based on one enzyme with a single adapter, and analysis by agarose gel electrophoresis (Melles et al., 2007). The main advancement has been acquired by switching from radioactive to fluorescently labeled primers for identification of fragments in a usual sequence apparatus (Rademaker et al., 2000). Further, it has been illustrated that for small bacterial and fungal genomes, a single PCR amplification with one and two specific nucleotides, respectively, on both primers is adequate (Buermans & Den Dunnen, 2014).

2.5.2 Advantages of AFLP Analysis

- This analysis is the most precise, highly reproducible, and has discriminatory power.

- AFLP has significant applications in the phylogenetics of plants and animals. In plants, AFLP is a multilocus PCR technology that produces as many as 150 locus-specific bands, with a high percentage of polymorphic bands.
- Disparity in AFLP patterns depict genetic distances that are instructive pertaining to genetic diversity (Andrade et al., 2016), phylogeny (Semerikova & Semerikov, 2016), and the geographic origin of genotypes and the gene pool of plants (Busconi et al., 2015).
- AFLP has a wider scope in breeding, as it covers all the chromosomes, and is inherited in a Mendelian manner, which is a main requirement for molecular marker applications in breeding analysis. AFLP offers some applications in marker-assisted breeding, for variety in identification.
- AFLP can also be used for the detection of infected plant varieties, and can facilitate germplasm management.
- AFLP can also assist in indirect selection of agronomically significant characteristics. For example, for monogenic traits, AFLP analysis requires an expansive assortment of primer pairs to capitulate markers closely associated to these traits for diagnostic use. This technique facilitates the localization of genes that contribute in varied pathological diseases (reference), and positional cloning of the significant genes (Tsehaye et al., 2016).
- AFLP analysis permits the instant and competent manufacture of dense genetic maps (Zhou et al., 2016). Genotyping of expansive segregating populations can result in the localization of one or more quantitative trait loci.
- AFLP can be employed to map blood pressure's quantitative trait locus after incorporating AFLP markers in the microsatellite map (Li et al., 2014; Zou et al., 2016). The comprehensible AFLP genetic map of a rabbit, for which no microsatellite map was accessible, has been created in a comparatively short time period (El-Sabrout & El-Raffa, 2015).
- AFLP analysis is particularly valuable for quantifying the discrepancy of nuclear genomes of related species, as illustrated for wild and domestic cattle species. AFLP is more sensitive for this application than comparative genomics, and is dissimilar by assessment of mitochondrial sequences not altered by irregular inheritance of the maternal lineage. In contrast to elaborate microsatellite allele frequencies, AFLP yields direct assessment of genetic diversity among and within species/breeds (Muys et al., 2014).

- The most exciting application of AFLP analysis of genomes that have been sequenced completely includes direct localization of genomics, or expressed AFLP fragments (Arias et al., 2016). The conversion of polymorphism to a SNP is extremely instructive in genotyping in a microarray format.
- AFLP analysis is now recognized as an extensively relevant genotyping approach with a high degree of reproducibility and discriminatory power. Numerous applications in taxonomy, diagnostics, and epidemiology have already been established. Its reproducibility may permit assemblage of genotype database and data exchange between laboratories.

2.6 PROTEIN ANALYSIS

Following the transcript profiling, to procure effective separation, we need to switch to protein profiling. In protein profiling, we employ two-dimensional electrophoresis (2DE) using orthogonal physical features of proteins, for example, isoelectric points and obvious molecular masses (Chang & Zhang, 2017; Feist & Hummon, 2015). The process of 2DE is a primitive biochemical process that is currently being modified for plant proteomic research (Wu, Gong, & Wang, 2014), and is particularly relevant to the evaluation of massive alterations in exceedingly cytosolic proteins (Komatsu & Hossain, 2018). However, 2DE methods display intrinsic impediments if proteome detection or precise estimation is required (Aslam, Basit, Nisar, Khurshid, & Rasool, 2017). It has been demonstrated that conventional 2DE methods identify only extremely expressed proteins, even when exceedingly sensitive staining methods are used (Freeman & Hemby, 2004). Even when spots are picked up by sensitive stains, quantitative determination is restricted by low dynamic ranges. Capillary isoelectric focusing (IEF) is a substantive alternative to 2DE that can be directly coupled to ion cyclotron resonance mass spectrometers (Fourier-transform mass spectrometry, FT-MS) for the investigation of both basic protein mixtures (Breadmore, 2012) and protein complexes (Marshall, Hendrickson, & Jackson, 2006). There are other approaches wherein we can investigate complexes of nondigested proteins with contemptible time-of-flight mass spectrometers (Fagerquist et al., 2010). Coupling of non-porous liquid chromatographic (LC) with mass spectrometric instruments can carry out effective separation and peptide detection and amplify sample loading limits (D'Orazio, Rocchi, & Fanali, 2012; González-Ruiz, Olives, & Martín, 2015). Peptide sequences can be effectively interpreted by employing reverse phase LC coupled to conventional tandem MS methods (Fortin et al., 2009; Lu et al., 2010). This method can be useful when investigating the proteomes of plants in which sequencing has not been carried out so far. The hyphenated approaches permit more reliable and instant separation of intricate peptide mixtures when performing cation exchange coupled with a reverse-phase LC column (Köcher, Swart, & Mechtler, 2011). The hyphenated chromatographic techniques permit more reliable and instant separation of intricate peptide complexes when cation exchange is coupled with reversed-phase LC columns. This method has been demonstrated to be superior in speed and protein coverage in contrast to 2DE investigation (Köcher et al., 2011). The main advance, though, has been acquired by methods that connect chemical linkers to particular protein moieties prior to mass spectrometric determination (Donato, Cacciola, Tranchida, Dugo, & Mondello, 2012; Hennrich, Groenewold, Kops, Heck, & Mohammed, 2011). For example, an isotope allocation encoded tag has been employed to identify and characterize low copy number proteins (Lange, Picotti, Domon, & Aebersold, 2008), and isotope-coded tags have been employed to estimate protein expression when comparing two diverse cell states (Gygi et al., 1999; Thompson et al., 2003). By integrating these and forthcoming approaches, proteomic methods will be suitable in more biological laboratories offering more accuracy, repeatability, high throughput, and greater cost competence than 2DE methods (Tao & Aebersold, 2003; Wu, Wang, Baek, & Shen, 2006).

2.6.1 Two-Dimensional Gel Electrophoresis

This is the most efficient and widely used approach for the investigation of intricate protein mixtures extracted from cells, tissues, or other biological samples (Adams & Gallagher, 2004; Dunn, 2014; Rabilloud, Chevallet, Luche, & Lelong, 2010). This method extracts proteins in two steps, as per two independent characteristics; the first dimension is IEF, which separates proteins as per their isoelectric points (p*I*); the second dimension is sodium dodecyl sulfate polyacrylamide gel electrophoresis (SDS-PAGE), which separates proteins as per their molecular weights (MWs) (Sharma & Bisht, 2016; Terova et al., 2014). In such approaches, intricate mixtures comprised of thousands of diverse proteins can be determined, and comparative protein content can be determined. The method implicates incorporating sample in a gel with a pH gradient and employing significant disparity across it. The proteins migrate as per pH gradient in an electric field until there is no general charge (Pomastowski & Buszewski, 2014; Zargar, Kurata, Rakwal, & Fukao, 2015). The site of the protein in the gel constitutes the obvious p*I* of the protein (Hu et al., 2005; Westermeier, 2014). There are substantive methods to generate the pH gradient and carrier ampholites, and to immobilize pH gradient (IPG) gels. The commercial-range IPGs are employed for procuring reproducible results as documented by the Maiman Institute for Proteome Research (Mohanty, Yadav, & Choudhary, 2017). The IEF is highly significant for the 2DE method, and proteins must be solubilized without charged detergents, generally in a highly concentrated urea solution, with reducing agents and chaotrophs (Lai, Yan, Caruso, & Nice, 2012; Syahir, Usui, Tomizaki, Kajikawa, & Mihara, 2015). In order to procure high quality data, it is necessary to ensure low-ionic intensity conditions prior the IEF itself. Because diverse kinds of samples differ in their ion content, it is important to modify the IEF buffer and the electrical profile of each sample type. The separation by molecular size in the second dimension is investigated in the slab SDS-PAGE. In order to reduce variation in separations between individual gels, twelve parallel gels can be separated in a fixed temperature (Gallagher, 2012; Westermeier, 2014). 2D gel electrophoresis is an efficient method for purifying proteins from complex mixtures, depending on their isoelectric points and MWs. There is a plethora of protocols and article displays; 2DE demands persistence thorough optimization (Lu et al., 2012). The 2DE method employs four fundamental steps: sample solubilization, IEF, SDS-PAGE, and analysis.

Step 1: Sample Solubilization

Similar to the generalized SDS-PAGE, the tissues or blood samples for 2DE require processing and solubilizing prior to loading into the IEF gel. Preferably, all proteins would solubilize instantly without qualitative or quantitative variations (Rogowska-Wrzesinska, Le Bihan, Thaysen-Andersen, & Roepstorff, 2013; Wu et al., 2014). All proteins are not equal. The first step instantaneously tilts 2DE toward identifying highly soluble and profuse proteins. However, utilization of tween and triton should be limited, although solubilization needs IEF-compatible lysis reagents, for example, electrically neutral detergents such as CHAPS, and chaotropes such as urea (Battle, 2013; Wright et al., 2014). Urea can carbamylate proteins efficiently, while flaming detergents can impede downstream steps. Consequently, this step requires that immense work be done on sample-specific optimization to avoid predisposition and contamination.

Step 2: Isoelectric Focusing

Once the sample is solubilized, it has to be loaded onto the IEF gel SDS PAGE; then the proteins will be passed through the acrylamide gel by an electric field. When the IEF is excited, the gel incorporates a pH gradient, and each protein advances, but only until it accomplishes its isoelectric point (p*I*). The isoelectric point is the pH at which the proteins have no net charge, which demonstrates that the field has no effect, and it remains focused on the band within 0.01 pH units of its p*I*. IEF is often problematic when performing scientific tasks; for example, proteins can become less soluble, and can precipitate proteins near their p*I*, particularly in low salts and IEF compatible buffers (Augustsson, Karlsen, Su, Bruus, & Voldman, 2016; Catsimpoolas, 2012). Second, IEF gels and buffers interfere with sample preparation for mass spectrometry (MS), and can be difficult to stain for investigation. Therefore, IEF needs to be carried out prior to SDS-PAGE to make sample MS compatible. To carry out this step, we need to be more cautious, as samples are more liable to contamination. In order to tackle this issue, we require gloves, assiduous depilation, and a sneeze shield.

Step 3: SDS-PAGE

Once proteins are arrayed, they need to be solubilized in SDS prior to separation due to their MWs on an orthogonal second axis. In addition to equilibration, it is important to use and extra caution with timing and voltages in this step which employs MW to separate proteins via the already established SDS PAGE method. Finally, the gel will have proteins aligned across two axis isoelectric points, versus MW (Rabilloud, 2013).

Step 4: Analysis

The last step requires our particular focus on experimental endpoints, such as whether our objective is comparing protein expression and protein detection. If we are comparing protein expression across varied experimental samples, the gels generally need to be stained with silver or Coomassie blue for total protein. Diverse image analysis platforms can be employed to scan and evaluate the site and intensities of segregated proteins (Magdeldin et al., 2014). Actually, analysis is not as clean and easy as slapping gel into the scanner. In order to advance reproducibility between the samples, gels should be run in parallel, which results in logistic and discreet confrontations. Moreover, in order to interpret Morse-like blueprints of dots and splotches, it is necessary to perform protein detection and annotation earlier via internal tasks or through simple annotated databases of comparable samples. There are two broad alternatives for those in the protein identification phase. The primary step is to run a western blot, transferring proteins from the SDS-PAGE gel. The second strategy, which is more efficient, painstaking, and economical, is to cut desired proteins in the gel followed by their digestion and their subsequent detection via MS (reference). This could result in enormously high-resolution data; for example, post-translational alterations or disease-associated adducts to proteins. However, there are rigorous requirements for the sample to be compatible with MS. The process of clean up requires the general elimination of detergents, and proteins stains, particularly silver stains, which generally need development with a formaldehyde solution, such as the kind used to impede MS sample preparation (Kinoshita, Kinoshita-Kikuta, & Koike, 2012). Moreover, MS needs numerous hundreds of protein nanograms per sample, as very little sample is required for visualization of particular proteins by the less-sensitive Coomassie blue stain. There are numerous other alternatives for purifying target proteins for MS analysis, such as an antibody-based bead separation system, for example, Dynabeads. Research has been conducted in which researchers have used zymograms instead of the IEF step. The technique of 2DE is complex; therefore, we need to review some of the research studies focused on the annotated databases of explicit samples run via 2DE.

2.6.2 Capillary IEF

The process of separation of biomolecules, especially proteins and peptides under the influence of an electric field, is subjected to varied methodologies, which reduces the intricacy of samples used to investigate the physicochemical characteristics of such biomolecules (Chen et al., 2013; Grossman & Colburn, 2012; Ramautar et al., 2012). Generally, proteins and peptides are investigated by electrophoretic approaches. IEF is one such method that provides separation of ampholytic metabolites that function as weak acids and bases, as per isoelectric points (Gidhi, 2013). In IEF, metabolites migrate, as per their charge, under the influence of an electric field in the presence of a pH gradient, until the overall charge of the molecule is zero (Funk, 2015). The separation of peptides and proteins was carried out as per the composition of amino acids and exposed charged residues, which function as weak acids and bases (Cann, 2012). The ampholytic biomolecules migrate as per the principles of electrophoresis, although the mobility will vary in the presence of the pH gradient by slowing down migration at values close to the p*I* value (Kuhn & Hoffstetter-Kuhn, 2013). Amino acids are the simplest ampholytes that can generate pH gradients and function as an isoelectric buffer. The discovery of IEF dates back to primitive works by A.J.P. Martin, who assisted with numerous advances in the electrophoresis field (Consden, Gordon, & Martin, 1946). Martin was awarded with a Nobel Prize for his contributions to the field of chromatography. The contribution by P.G. Righetti has been pivotal in advancing the process of segregating biomolecules electrophoretically, especially as per isoelectric points (Righetti, Sebastiano, & Citterio, 2013).

Principle of IEF

Electrophoresis separation is based on the differential rate of the migration of charged species along the direction of the electric field (Janson, 2012; Righetti et al., 2013). While crossing through the separation channel, the analyte ions form zones as per their mass/charge ratio (Catsimpoolas, 2012). These zones are extended because of molecular diffusion and deliberate mass migration when convection or interaction with capillary surface occurs (Hühner, Lämmerhofer, & Neusüß, 2015). In IEF, a counteracting force is present, which, upon lengthening, originates in the amphoteric nature of both the analyte molecules and the constituents of the separation matrix (Herzog et al., 2016).

Proteins and peptides are amphoteric molecules containing carboxylic acid and amino acid groups, and their p*I* determines their particular amino-acid composition. Moreover, their three-dimensional structure effects the molecular configuration, and hence the ionization constants of each amino-acid composition. Additionally, on a flat surface, IEF in a gel solution has also been executed in a tube. Comparatively wide-diameter PTFE or glass capillaries (0.5-2mm id) have been employed (Qiao, Tobolkina, Liu, & Girault, 2013; Shintani, 2012). Nevertheless, executing IEF in a tube has been as laborious as slab gel IEF in relation to the preparation of the separation matrix. Only low voltages can be employed, so that thermal gradients in the gel are reduced. In spite of the low voltage, general cooling is necessary, and separation takes several hours. Hjertén (1985) was the first to discover that very narrow-diameter capillaries can be employed in electrophoresis, known as capillary electrophoresis, which is much better suited for dissipation of heat produced by the current. This would permit a much higher electric field to be employed, and thus accelerate the separation (Westermeier, 2016). Even though IEF can be done in a capillary or flat conduit without a supporting matrix, it has become a general way to utilize gel to facilitate in decreasing the iso-electro-osmotic flow and decline diffusion (Chang et al., 2012).

Instrumentation of Capillary Iso-electro Focusing

Capillary iso-electro focusing can be carried out on standard equipment for capillary electrophoresis; for example, a capillary electrophoresis system. Capillary iso-electrofocussing is one mode of operation in the arsenal of electrophoretic separation methods available on commercial CE instruments (Thomassen, van Eikenhorst, van der Pol, & Bakker, 2013). This instrument assists in the utilization of fused silica capillaries that carry low pH and high pH electrolytes, and conduct the electric field, the mobilization of the pH gradient, UV-Visible detection, and automation. There are numerous inferences that must be made in working with capillaries for capillary iso-electrofocussing. In contrast, slab-gel IEF or IPG gel electrophoresis, CIEF sample proteins, standards, carrier ampholytes, a polymer, and an optional additive are premixed and filled together in the fused silica capillary. This refers to two-step capillary IEF, where the electric field is applied in one end of the filled capillary with low pH electrolytes (anolytes), and in the other end, with high pH electrolytes (catholytes). The anolyte vial releases hydronium ion, which moves toward the cathode; while hydroxide ions move instantly from the catholyte vial toward the anode. During this process, the proteins and ampholytes become sorted in the capillary from low to high p*I*.

In single-step capillary IEF, the focusing of proteins and ampholytes takes place in the presence of residual EDF within fused silica capillaries (reference). Here, protein zones become focused, while being driven toward the point of detection by the EDF. If the EOF velocity is lower than the velocity of the focusing zones, the whole protein train can reach equilibrium prior to its emergence at the detector end (Bianga et al., 2014).

Applications of Capillary IEF

- Protein detection and Post-Translational Modifications:

  Detection of proteins from complex samples (e.g., cells, tissue lysates, plasma, etc.) has been enabled by IEF. Following the orthogonal separation, in any format IEF is employed, it is likely to determine highly complex samples, allowing for deeper investigation of the proteome. For effective separation by capillaries, IEF compounds need to be de-protonated, such as via amino groups, or conversely, via carboxylic acid groups. Therefore, significant candidates for separation by CIEF are proteins and peptides. However, small molecules may also be amphoteric, for which CIEF can be employed for isoelectric points (p*I*) and pK_a measurements. The utilization of narrow-range IPG-IEF strips has been coupled with nanoLC-MS to characterize the variation in peptide p*I* and retention time because of varied post-translational modifications.

- The main purpose of capillary IEF is the determination of the isoelectric point of the purified recombinant protein, or monitoring of the charge heterogeneity of the protein.
- In clinical investigations, CIEF has been immensely significant for the determination of hemoglobin charge variants as an indicator of the severity of diabetes.
- CIEF has wider applications in the separation of monoclonal antibodies. Mobilization was acquired by pressure voltage, as it exhibits its significance while working under high voltage.

2.7 METABOLITE ANALYSIS

In contrast to transcripts and proteins, metabolites, which are actually final gene products, have been consistently overlooked to date (Brotman et al., 2012; Van Emon, 2015). However, metabolite profiling can facilitate the discrepancy among healthy versus diseased human tissues (Luo et al., 2012). MS has been used for decades to determine metabolic target metabolites, and now, the modern metabolomic approach has explored the list of targets to screen specified primary metabolites such as apricot fruits (Saini & Keum, 2018), plant leaves (Silvente, Sobolev, & Lara, 2012), and potato tubers (Payyavula, Navarre, Kuhl, Pantoja, & Pillai, 2012). Further, there is more complexity in deciphering plant metabolomes due to massive plant metabolome ranges from 90,000 to 200,000 (Patti, Yanes, & Siuzdak, 2012). So in order to explore plant metabolomes, we must cope with absolute intricacy of mass spectra found in plant chromatograms and employ novel approaches for efficient de novo metabolite identification (Brown et al., 2016; Xia, Mandal, Sinelnikov, Broadhurst, & Wishart, 2012). The advancement and production of automated mass spectral deconvolution and identification software (AMDIS) has been undertaken to cope with the complexity issue with expansive metabolite output (Law & Lim, 2013; Niu, Knight, Xia, & McGarvey, 2014). For clarity in the mass spectra, complete de novo identification of relatively lower MW polar metabolites can be demonstrated by employing mass spectrometric approaches coupled with databases, as illustrated in *Arabidopsis* leaf extracts (Wolfender, Marti, Thomas, & Bertrand, 2015). Recent years have seen the advent of high-throughput profiling of metabolic snapshots, for the first time, in the context of plant functional genomics (Mitra, Carvunis, Ramesh, & Ideker, 2013). Metabolite profiling in two *Arabidopsis* ecotypes with two single gene mutants released about 326 metabolites that differentiated two discreet metabolic phenotypes via cluster analysis (Risticevic, Souza-Silva, DeEll, Cochran, & Pawliszyn, 2015). This research can be further advanced by combining the observed metabolic variations into hypotheses pertaining to variation in biochemical pathways; and gene expression levels, in the future, will be able to decipher plant metabolome-genome interfaces (Goulitquer, Potin, & Tonon, 2012).

2.7.1 Metabolic Fingerprinting

Metabolic fingerprinting is defined as the utilization of the software output as a significantly identifiable chemical pattern, particular to an individual sample (Arbona, Manzi, Ollas, & Gómez-Cadenas, 2013; Obata & Fernie, 2012). Nuclear magnetic resonance (NMR) provides an instant, competent contrivance for deciphering between groups of associated samples, and it detects the most significant regions of the spectra for further evaluation (Petrakis, Cagliani, Polissiou, & Consonni, 2015). Hence, it can be conferred that spectra processing is the transitional process between raw spectra and data evaluation (Fotakis, Kokkotou, Zoumpoulakis, & Zervou, 2013). It encompasses the conservation of appreciable variability comparative with the chemical metabolites contained in the NMR spectra, while dropping other kinds of variances induced by diverse sources of bias, for example, baseline and misalignment (Porzel, Farag, Mülbradt, & Wessjohann, 2014).

Therefore, the metabolite distinctiveness, which is immensely significant, is determined following statistical data evaluation of metabolic fingerprints, and this allows it:

- To emphasize that spectral regions having a discrepancy between the groups are statistically significant.
- To ensure that each of these regions implicates only a single metabolite, that is, that there is an exclusive association connecting a bucket and spectral signature of a metabolite.

The standard technique in NMR-based metabolomics entails the division of spectra into equally sized bins, thus simplifying the following data analysis.

However, there are limitations, including the loss of data, and the incidence of the artifacts caused by peak shifts (Farag, Porzel, Schmidt, & Wessjohann, 2012; Farag, Porzel, & Wessjohann, 2012). Consequently, we involved the Adaptive Intelligent Binning (AI-Binning) algorithm, which expansively averts these issues. This contrivance detects bin edges in accessible bins, needing only negligible user input, and averts the utilization of the random parameters or reference spectra (Mladenov et al., 2015). Following the data bucketing, you have to export the data matrix, and the exported matrix is formatted so that we can consequently execute statistical analysis via a Bio-Stat Flow web application. Hence, data file management is reduced. Metabolic fingerprinting cannot determine the identification or accurately estimate all the metabolites in the sample (Dunn et al., 2013). Instead, it is known as complete profiling, or fingerprinting as a discreet pattern identifying a snapshot of the metabolism in a specific cell line or tissue.

2.7.2 Metabolite Profiling

This refers to the estimation of metabolite levels in a sample. Metabolite profiling is now established as a main contrivance for diverse activities in the biomedical sciences, employed usually on biological fluids as a symbol of the diagnosis that facilitates the identification of the health status of the patient (Jain et al., 2012). These profiles are generally comprised of a limited number of metabolites; however, metabolomes have been expanded to a more expansive level (Theodoridis, Gika, Want, & Wilson, 2012). During recent years, researchers have used metabolite profiling as an extension of functional genomics (Fernie, Trethewey, Krotzky, & Willmitzer, 2004; Fiehn et al., 2000; Nielsen & Oliver, 2005). It has been hypothesized that metabolite profiles of intrinsic cell states could assist in the detection of the function of genes, particularly when mutants in that gene have no clear phenotype (Schauer et al., 2006; Sumner, Mendes, & Dixon, 2003). This is because when mutations have led to an effect, the regulatory mechanism of the cell functioned to neutralize the effect, leading in no obvious macroscopic discernible way (Weckwerth, Loureiro, Wenzel, & Fiehn, 2004). However, those regulatory mechanisms would have altered the metabolite concentration in the metabolic network, and if one could estimate the metabolite level, it could result in the function of the mutated gene (Daran-Lapujade et al., 2007; Goodacre, Vaidyanathan, Dunn, Harrigan, & Kell, 2004). The first proposition is that not knowing a priori which metabolites are expected to be varied, the profiles will then need to be as easily understandable as possible. The second inference is that evaluating the metabolite profiles of numerous mutants might result in the breakthrough of the fundamental metabolic network (DeBerardinis & Thompson, 2012). This relevance of metabolite profiling in functional genomics is then comparable to transcript and protein profiling, and like these, it will be employed to resolve the entire composition of the cell in terms of metabolites-metabolomes (Putri et al., 2013; Sanford, Soucaille, Whited, & Chotani, 2002). The most favorable utilization of genome sequences and gene expression resources needs potential phenotyping platforms comprising those for regular evaluation of metabolite composition (Kanehisa, Sato, Kawashima, Furumichi, & Tanabe, 2015; Kopka, Fernie, Weckwerth, Gibon, & Stitt, 2004). The most employed approaches for metabolite profiling, include mass spectral, NMR, and enzyme-based techniques, have varied applications and limitations. These techniques are useful in the explanation of the massive datasets generated. These approaches will be valuable for answering significant biological queries in the future. It is useful to differentiate between metabolite fingerprinting, metabolite profiling, and metabolomics (Dunn, Bailey, & Johnson, 2005; Fiehn, 2002). Metabolic fingerprinting is the relevance of an extensive analytic technology to unravel massive disparities between two samples; for example, two different genotypes. Metabolite profiling encompasses the estimation of hundreds, or even thousands of metabolites. It needs a streamlined pipeline for extraction, separation, and evaluation, so that the expansive number of metabolites can be estimated in a dynamic and quantitative manner, while in the presence of the extremely multifaceted mixture of chemicals ("matrix") that is found in cellular extracts (Dettmer, Aronov, & Hammock, 2007; Harrigan & Goodacre, 2012).

The estimation of metabolites contributes fundamental data pertaining to biological responses to physiological or ecological variations. Metabolite profiling permits a shift from hypothesis-induced research to the evaluation of system-wide responses, particularly when it is combined with other profiling technologies (Theodoridis, Gika, & Wilson, 2011). Following the identification of the response, the subsequent goal was to elaborate the regulatory mechanisms. The organized evaluation of all metabolites within a part or segment of a metabolic network contributes potent and impartial approaches for characterizing the site or sites at which functional processes function to modify fluxes. The regulated enzyme is displayed because the substrate level varies equally to the flux via the pathway (Cazzonelli & Pogson, 2010). Metabolite profiling datasets can function to be chemometrically investigated to reveal associations between individual metabolites and explicit gene expression (Metallo & Vander Heiden, 2013) or proteins (El-Mansi, Cozzone, Shiloach, & Eikmanns, 2006). Following the advent of the post-genomic era, metabolite profiling will be exceedingly utilized to phenotype mutants and transgenic organisms to describe the function of the gene. The major constraint in functional genomics is to allocate functions to the numerous inadequately or unannotated genes (Halligan & Keightley, 2006). Metabolite profiling will contribute a function for those that encode proteins implicated in metabolism (Kleffmannet al., 2004). Now the thematic issues that are associated with molecular composition and its involvement for nutrition and health are emerging. Upgraded technologies are escalating the advent of novel interventions in breeding programs, either via transgenic technology, or by employing molecular markers in integration with wide crosses (Motto, 2017). Metabolite profiling can be employed for identifying this diversity phenotypically, with reference to its metabolite composition contributing a potent tool to direct breeding programs, and to indicate valuable or undesirable traits in the emerging stages (Riedelsheimer et al., 2012). The competence of this method will be immensely increased if it can be integrated with an organized review of the metabolite composition of plant produce that is released to the market (Arbona, Iglesias, Talon, & Gomez-Cadenas, 2009; t'Kindt et al., 2010), along with contributing a baseline that gives a logical structure for risk assessment through 'substantial equivalence.' Metabolite profiling can be employed for the determination of metabolite composition of new and emerging products to compare them with the metabolite range already available (Patti et al., 2012; Wishart, 2011). Moreover, metabolite profiling will contribute significant input into nutritional research, and into public debate pertaining to the acceptability of alterations in food production chains (Neeha & Kinth, 2013; Van Emon, 2015). One of the most important constraints for the advancement of the life sciences is to progress beyond elaborate information pertaining to organisms and their responses in the controlled or laboratory environment to comprehend interactions in natural ecosystems and during evolution (Norgaard, 2006). The integration of high throughput expansion will allow metabolite profiling to be combined with ecological studies that need expansive sampling approaches, but have been frequently subjected to the drawbacks of preconception-driven research (Kuhlisch & Pohnert, 2015; Macel, van Dam, & Keurentjes, 2010). Metabolite profiling can also be employed to monitor genotypic diversity without requiring the prior development of molecular contrivances for explicit species (Jones et al., 2013; van Dam & van der Meijden, 2011). We can anticipate that metabolite profiling will emerge as a potent functional tool in diverse fields.

2.8 ANALYTICAL TECHNIQUES

There are numerous approaches that can be employed for metabolite profiling, and each technique has associated advantages and limitations. Thus, integration of diverse analytical techniques must be employed to acquire an extensive viewpoint of the tissue metabolome (Mehl et al., 2014; Zhang, Creek, Barrett, Blackburn, & Watson, 2012; Zhang, Sun, Wang, Han, & Wang, 2012). Analytical approaches that are mainly utilized for metabolite profiling are NMR (Leiss, Choi, Abdel-Farid, Verpoorte, & Klinkhamer, 2009; Weljie, Newton, Mercier, Carlson, & Slupsky, 2006), gas chromatography-mass spectrometry (GC-MS) (Kind et al., 2009; Koek, Jellema, van der Greef, Tas, & Hankemeier, 2011), liquid chromatography-mass spectrometry (LC-MS) (Vuckovic, 2012), and capillary electrophoresis-mass spectrometry (CE-MS) (Ibáñez et al., 2012; Sugimoto, Wong, Hirayama, Soga, & Tomita, 2010). As it is evident that no single analytical methodology is suitable for expansive metabolome profiling within a cell instead, the integration of techniques is required for the evaluation of a majority of metabolites in diverse polarity and MW ranges (Barbas, Moraes, & Villasenor, 2011).

2.8.1 Gas Chromatography Coupled With Mass Spectrometry

In this technique, gas chromatography is coupled with mass spectrometry; metabolites are separated by GC, and then transferred online to the mass spectrometer for further separation and identification. This integrates two strongly complementary approaches (Dunn, Broadhurst, et al., 2011a, b; Rocha et al., 2012). GC can separate metabolites that have nearly identical mass spectra; for example, isomers while MS fetches fragmentation that discriminates between coeluting chemically discreet metabolites (Allwood & Goodacre, 2010). GC-MS contributes quantitative data, and is extensively employed for clinical diagnostics and profiling of complex biological samples (Lei, Huhman, & Sumner, 2011; Smart, Aggio, Van Houtte, & Villas-Bôas, 2010). This technique has already been illustrated in Chapter 5. At present, the most established technology for instant metabolite profiling is gas chromatography coupled with electron impact (EI) quadrupole, or time of flight (TOF) MS (GC-MS) (Allwood et al., 2009; Zhu et al., 2013). The idea of automated GC-MS metabolite profiling was developed 25 years ago, and subsequently emerged as the main technology for metabolomics (Dunn et al., 2013; Hernández, Portolés, Pitarch, & López, 2011). The GC-MS technique can be utilized for the profiling of an expansive variety of metabolites, ranging from amino acids, sugars, sugar alcohols, aromatic amines, and fatty acids (Villas-Bôas, Smart, Sivakumaran, & Lane, 2011). GC-MS is the preferred method for the separation of the volatile metabolites; but nonvolatile metabolites need to be derivatized prior to being injected into GC-MS (Barderas et al., 2011; Molnár-Perl, 2005). Even though chemical derivatization contributes to considerable advancement in GC separation, it also commences artifacts because of the derivatization method (reference). The most significant advantage of GC-MS coupled with EI ionization is the accessibility of numerous exportable mass spectral libraries. The most expansive commercial EI libraries are 2005 NIST/EPA/NIH Mass Spectral Library (http://www.nist.gov/srd/nist1.htm), comprised of 190,825 spectra, and 7th edition of the Wiley Registry of Mass Spectral Data (http://www.wiley.com), which contains 338,000 EI mass spectra (or more than 460,000 EI mass spectra integrated with the NIST library). Further, commercially available EI libraries and several public libraries are also available. However, the majority of GC-MS libraries are tailored to the chemical industry, pharmaceutical sector, and natural product discovery, which results in limiting GC-MS relevance in metabolomic research studies (Chauhan, Goyal, & Chauhan, 2014; Safaei-Ghomi, Ebrahimabadi, Djafari-Bidgoli, & Batooli, 2009).

2.8.2 Nuclear Magnetic Resonance

NMR is the most suitable technique employed for metabolomic studies, as it could discreetly detect and instantly estimate an extensive range of organic metabolites in the micromolar range (Capitani, Sobolev, & Mannina, 2017; Gu et al., 2011). In contrast to MS, NMR is noncombustible, so samples can continue for further evaluation. Further, sample preparation for NMR is direct, and expansively automated (Motta et al., 2012). However, the evaluation of NMR spectra in the case of complex mixtures has conventionally been cumbersome; but NMR is widely employed for metabolite fingerprinting, profiling, and metabolic flux evaluation (Dunn, Broadhurst, et al., 2011a, b; Simmler, Napolitano, McAlpine, Chen, & Pauli, 2014). The main disadvantage of NMR for elaborate metabolite profiling is its comparatively low efficiency, thereby rendering it less efficient for the evaluation of expansive ranges of low abundance metabolites (Dieterle et al., 2011; Schripsema, 2010).

2.8.3 Mass Spectrometry

Due to the extensive range of sensitivity and broad range of metabolites, MS has emerged as the preferred technique in the field of metabolomics (Katajamaa & Orešič, 2007; Villas-Bôas, Mas, Åkesson, Smedsgaard, & Nielsen, 2005). MS can be employed for the evaluation of biological samples either directly, or via direct injection. With the advent of an extensive range of novel mass analyzers, together with advancements in mass accuracy there has been dramatic expansion in the metabolite range that can be evaluated by MS, and higher accuracy in metabolite detection (Scalbert et al., 2009; Yuan, Breitkopf, Yang, & Asara, 2012). A high-resolution mass spectrometer coupled with a direct injector provides an instant and robust technique for analyzing an expansive range of metabolites, and therefore, is widely used for metabolic fingerprinting and metabolite profiling (Zeng, Deng, & Wu, 2003). However, direct injection also has limitations, such as co-suppression, where the signals of numerous metabolites exhibiting low ionization efficiencies to be evaluated can be mislaid on the mass spectrometer frontier (Dettmer et al., 2007). In order to avoid these hindrances, and to reduce the intricacy of the sample, MS is generally used as a hyphenated technique, as it is coupled with GC, LC, and CE (Ballesteros-Gómez, de Boer, & Leonards, 2013). In these cases, the complex mixtures are separated by the chromatography or electrophoresis, and individual metabolites can be eluted from columns or capillaries, and then analyzed by MS.

2.8.4 Liquid Chromatography-Mass Spectrometry

LC-MS is the most widely used technique in metabolomics because of its high sensitivity and wider range of polarity and molecular mass than GC-MS. LC, particularly high performance liquid chromatography (HPLC), is an established technique that integrates high resolution and analytical elasticity (Creek et al., 2011; Gika, Theodoridis, Plumb, & Wilson, 2014). LC-MS can be modified for the analysis of explicit metabolites, or class of metabolites that belong to a broader range of metabolite classes (Wohlfarth, Scheidweiler, Chen, Liu, & Huestis, 2013). Several applications for an extensive range of metabolites are already accessible, or can be converted for employing LC in integration with MS (Brown et al., 2011; Paglia et al., 2014). In contrast to GC-MS, LC-MS does not require metabolite derivatization, which is usually utilized for analysis of nonvolatile metabolites by GC-MS (Farag, Porzel, Schmidt, & Wessjohann, 2012; Farag, Porzel, & Wessjohann, 2012; Vinaixa et al., 2016). With the discovery of soft ionization methods, such as electrospray (ESI), LC-MS have advanced further, thereby rendering mass spectrometers more sensitive and rapid for daily use (Cajka & Fiehn, 2015; Forcisi et al., 2013). In recent years, advancements in separation science employing monolithic capillary columns (Tolstikov, Lommen, Nakanishi, Tanaka, & Fiehn, 2003) and ultra-performance liquid chromatography (UPLC) (Wang, Sun, Zhang, Wang, & Han, 2011) have contributed to more efficient solutions for separation of complex biological mixtures than those that were previously attainable (Zhang, Creek, et al., 2012; Zhang, Sun, et al., 2012). The more distinct limitation for the LC-MS in its relevance for being a non-targeted profiling tool is the lack of convertible mass spectral libraries (Forcisi et al., 2013). On the contrary, LC-MS can be effectively employed to illustrate the structure of unidentifiable metabolites. Now ion-trap mass spectrometers can perform the multiple sequential steps of MS, contributing considerable help in structure explication (Roux, Lison, Junot, & Heilier, 2011).

2.8.5 Capillary Electrophoresis-Mass Spectrometry

This is a newer technique, compared with the GC-MS and LC-MS, but it is becoming a more popular analytical approach for deciphering metabolomic studies (Lapainis, Rubakhin, & Sweedler, 2009; Monton & Soga, 2007). CE-MS contributes numerous applications over other analytical approaches. It has comparatively high resolving power, with plate numbers of 100,000–1,000,000, small sample requirements, with an average injection range from 1 to 20 nL, and shorter analysis time (Hirayama, Wakayama, & Soga, 2014; Ramautar, Busnel, Deelder, & Mayboroda, 2011). CE can be employed for both targeted and non-targeted analysis of metabolites (reference) including the evaluation of inorganic ions (Ramautar, Somsen, & de Jong, 2013), organic acids (reference), amino acids (Soga & Heiger, 2000), nucleotides and nucleosides (Warnke et al., 2004), vitamins, carbohydrates, and peptides in foods (Ibáñez, Simó, García-Cañas, Cifuentes, & Castro-Puyana, 2013). The most marked application of CE-MS is its higher efficiency for the separation of the charged metabolites in the single analytical run; hence, CE can be used for the instantaneous profiling of numerous metabolite classes (García, Godzien, López-Gonzálvez, & Barbas, 2017; Neusüb, Römer, Höcker, & Jooß, 2018; Sugimoto et al., 2010). This characteristic renders CE-MS as a potential analytical technique for high throughput non-targeted metabolomics (Das et al., 2017; Wang, Li, Yao, Wang, & Van Schepdael, 2017).

2.8.6 Other Technologies

Even though NMR and MS are widely employed for extensive metabolite analysis, metabolomics is not limited to these techniques. Other options include thin-layer chromatography (Liu et al., 2018), HPLC with UV/visible absorbance, photodiode array (PDA) (da Silveira Petruci, Liebetanz, Cardoso, & Hauser, 2017; Kissoudi, Sarakatsianos, & Samanidou, 2018) or electrochemical detectors (Boeye, Gismera, Sevilla, & Procopio, 2017; Zhang et al., 2018), Fourier transform infrared spectroscopy (FT-IR) (Keymanesh, Darvishi, & Sardari, 2009), phenotype microarrays (Atanasova & Druzhinina, 2010), and diverse enzymatic assays (Tohge & Fernie, 2010). An integrated approach of multiple techniques or numerous detectors in online or parallel evaluation can considerably augment the metabolite coverage, amplify estimation limits, and advance metabolite detection from a single biological sample (Zeng et al., 2014). By integrating mass spectrometric methods and electrochemical array detection, we can acquire extensive metabolite ranges and advance the concentration range of endogenous metabolites (Zhao & Suo, 2008).

2.9 PATTERN RECOGNITION

Pattern recognition contrivances are used to categorize fingerprints and detect the particular characteristics of the profiles that are typical for each pattern. Metabolic fingerprinting is most suitable in biomarker discovery and diagnostics (Weckwerth & Morgenthal, 2005). Fingerprinting is generally carried out with spectroscopic approaches; for example, NMR (Defernez & Colquhoun, 2003), FT-IR (Ellis, Harrigan, & Goodacre, 2003), Fourier transform ion cyclotron resonance mass spectroscopy (FTICR-MS), or MS (Han et al., 2008), by directly procuring physical spectra exclusive of preceding separation methods, such as chromatography or electrophoresis (Huang, McLean, & Zheng, 2005). The main limitation with metabolic fingerprinting is that it does not permit a thorough assessment of the metabolic pathways, because metabolites in the samples are seldom identifiable. Metabolic fingerprinting has been extensively employed for unravelling exclusive metabolic patterns of diseases (Pechlivanis et al., 2010; Shulaev, 2006). Metabolic fingerprinting is specifically functional in integration with pattern recognition and discriminant analysis techniques (Smolinska, Blanchet, Buydens, & Wijmenga, 2012; van der Kooy, Maltese, Choi, Kim, & Verpoorte, 2009). No matter what kind of technique has been employed, metabolomic analyses result in the assemblage of expansive raw data, and with the rise in metabolic changes, there is a concomitant increase in metabolic profiles (Scherling, Roscher, Giavalisco, Schulze, & Weckwerth, 2010). However, if there are subtle metabolic changes, there is a probability of acquiring more lucid clusters in metabolic profiles of comparative experimental designs, such as mutant/wild types, healthy/diseased, young/old, and so forth. In case these comparisons are not verified by clustering applications, the data might appear too profuse or noisy, thereby rendering it difficult for further analysis (Xia, Psychogios, Young, & Wishart, 2009). The inaccessibility of intrinsic data, such as minor variations in metabolite levels, errors that arise during sample preparation, or data acquisition can result in some intricacies in deciphering metabolomes (Hagel & Facchini, 2008; Prosser, Larrouy-Maumus, & de Carvalho, 2014).

2.10 MULTIVARIATE STATISTICS TECHNIQUES

The multivariate statistics technique is a simple statistical analysis in which the determination of more than one statistical variable is carried out instantly, as multivariate statistics is an expansive discipline, with numerous usual and explicit procedures (Lindon & Nicholson, 2008; Lindon, Nicholson, & Holmes, 2011). Numerous research articles have illustrated the application of multivariate statistics for deciphering metabolomic data (Liland, 2011; Sun & Weckwerth, 2012). The main objective of metabolomics is to present a comprehensive snapshot of small MW metabolites in cells and biological fluids during unbiased observations for further elaborating on the research studies conducted on metabolism (Fumagalli et al., 2009; Kemsley et al., 2007). The expansive metabolomic data induces enormous constraints to resolving any metabolomic data with a conclusive biological relevance, which demonstrates a requirement for explicit contrivances in data analysis (Craig, Cloarec, Holmes, Nicholson, & Lindon, 2006; Worley & Powers, 2013). The main strategy for comprehending metabolomic datasets is the implication of multivariate analysis (MVA) methods; for example, principal component analysis (PCA) and partial least squares projection to latent structures (PLS), where spectral features, mainly disparity or separation, are detected for further investigation.

2.10.1 Principal Component Analysis

This is the most generalized MVA employed for metabolic fingerprinting, and in fact, chemometrics in general. The aim of PCA is to arrive at a linear transformation that retains as much of a variance in the original data as possible in the lower dimensionality output data (Nyamundanda, Brennan, & Gormley, 2010; OuYang, Xu, Huang, & Chen, 2011). It can be described that transformation A that acquires this goal is a matrix whose columns are the first $\boldsymbol{P}$ eigenvectors of the nonsingular fraction of the sample covariance matrix $\boldsymbol{S}$:

$$\boldsymbol{S} = \frac{1}{N-1}\boldsymbol{X}^T\boldsymbol{H}\boldsymbol{X} = \boldsymbol{Q}\boldsymbol{\Lambda}\boldsymbol{Q}^{-1}$$

Here, $\boldsymbol{H}$ is the N-by-N centering matrix employed to center each variable about its sample mean. The second equality describes the form of the eigen decomposition of $\boldsymbol{S}$, where $\boldsymbol{Q}$ is the matrix of eigenvectors of $\boldsymbol{S}$, and $\boldsymbol{\Lambda}$ is a diagonal matrix of the corresponding eigen values. When $\boldsymbol{X}$ is left unscaled, the eigenvalues in $\boldsymbol{\Lambda}$ equal the variances of the newly transformed data in T, giving means to estimate the ratio of variance retained during the transformation, compared with the original variance:

$$R_i^2 = \frac{\Lambda_{ii}}{\sum_{j=1}^{N} S_{jj}}$$

where R_i^2 is the total of variance in X conserved in the ith principal component.

2.10.2 Partial Least Squares

Although the unsubstantiated method of the PCA algorithm contributes a means to acquire impartial dimensionality reduction, its relevance only displays group structures, when within-group variation is adequately less than between-group variation (Gromski et al., 2015; Gu et al., 2011). Consequently, supervised forms of discriminant analysis, for example, partial least squares (PLS-DA; alternatively partial least squares projections to latent structures) (Mehmood, Liland, Snipen, & Sæbø, 2012) that depend on the class membership of each observation are also generally applied in metabolic fingerprinting experiments (Brereton & Lloyd, 2014). When class memberships are coded in matrix form into $\boldsymbol{Y}$ (Szymańska, Saccenti, Smilde, & Westerhuis, 2012) and the PLS components are controlled to be orthogonal, the dimensionality dropping transformation $\boldsymbol{A}$ is a matrix whose columns are the first $\boldsymbol{P}$ eigenvectors of a matrix formed by the covariances between $\boldsymbol{X}$ and $\boldsymbol{Y}$

$$\boldsymbol{S} = \boldsymbol{S}_{xy}\boldsymbol{S}_{yx} = \frac{1}{(N-1)^2}\boldsymbol{X}^T\boldsymbol{H}\boldsymbol{Y}\boldsymbol{Y}^T\boldsymbol{H}\boldsymbol{X}$$

where $\boldsymbol{H}$ is yet again the centring matrix, and the eigen decomposition takes a similar form to that shown for PCA. Therefore, the novel "latent variables" produced by this transformation are linear combinations of original variables that conserve as much covariance between $\boldsymbol{X}$ and $\boldsymbol{Y}$ as possible in the first transformed dimensions; simply put, the low dimensional scores' space is produced mostly by the predictive components of $\boldsymbol{X}$. This casting of PLS-DA as an eigen decomposition predicament (Brereton, 2009; Brückner & Heethoff, 2017) is certainly balancing out the common explanation of the algorithm as an iterative regression problem that more strongly resembles the roots of PLS regression (Szymańska et al., 2012). Conversely, it offers resources to compare the results of PLS-DA with those of PCA. Actually, placing every observation into its own class, effectively setting $\boldsymbol{Y}$ to the identity, yields an identical eigenvector problem to that of PCA. It is far more common, though, to find descriptions of PLS-DA that do not inflict orthogonality of the PLS components; these methods require, as a substitute, that the $\boldsymbol{X}$ scores be uncorrelated, closer to traditional PLS regression (Westerhuis et al., 2008). The established form of PLS-DA under this condition is then

$$\boldsymbol{X}=\hat{\boldsymbol{X}}+\boldsymbol{E}=\boldsymbol{T}\boldsymbol{P}^{T}+\boldsymbol{E}$$
$$\boldsymbol{Y}=\hat{\boldsymbol{Y}}+\boldsymbol{G}=\boldsymbol{U}\boldsymbol{C}^{T}+\boldsymbol{G}$$

where $\boldsymbol{T}$ and $\boldsymbol{P}$ are the scores and loadings for $\boldsymbol{X}$; $\boldsymbol{U}$ and $\boldsymbol{C}$ are the scores and loadings for $\boldsymbol{Y}$; and $\boldsymbol{E}$ and $\boldsymbol{G}$ are the residual errors of $\boldsymbol{X}$ and $\boldsymbol{Y}$ that are left unaccounted for in the model. Decomposition is evaluated such that $\boldsymbol{T}$ and $\boldsymbol{U}$ distribute the most covariance, in effect allocating $\boldsymbol{T}$ to function as an estimator of $\boldsymbol{U}$. These equations highlight the information that PLS reveals a small set of scores and loadings—the latent structures—which mainly précis $\boldsymbol{X}$ and $\boldsymbol{Y}$ in addition to illustrating their association. This conventional PLS regression method was referred to as PLS due to its utilization of nonlinear iterative partial least squares (NIPALS) for calculating model parameters. With the preamble of PLS to chemometrics, an optional significance of "projection to latent structures" has also been employed for the methods of partial least squares regression (Tran, Afanador, Buydens, & Blanchet, 2014).

2.10.3 Orthogonal Projection to Latent Structures

The exploitation of class memberships in PLS-DA permits the algorithm to competently unravel separations between classes in scores space. However, disparity is not directly associated with $\boldsymbol{Y}$, which is still present in the scores.

This complicates interpretation of PLS-DA scores and loadings plots, particularly as the class number amplifies (Galindo-Prieto, Eriksson, & Trygg, 2014; Kang et al., 2008). Orthogonal projections to latent structures (OPLS) tackle this interpretability issue by inculcating an orthogonal signal correction (OSC) filter (Duggan et al., 2011) into a PLS model, efficiently separating the ***Y***-predictive variation from the ***Y***-uncorrelated variation in ***X***:

$$\boldsymbol{X} = \hat{\boldsymbol{X}} + \hat{\boldsymbol{X}}_o + \boldsymbol{E} = \boldsymbol{T}\boldsymbol{P}^T + \boldsymbol{T}_o\boldsymbol{P}_o^T + \boldsymbol{E}$$

where $\boldsymbol{T_o}$ and $\boldsymbol{P_o}$ are the scores and loadings, correspondingly, for the ***Y***-uncorrelated variation detected by the OSC filter. The predictive OPLS-DA scores and loadings employed to calculate ***Y*** are then comprised of diversity directly associated with ***Y***, and free of interfering structured variation, yielding enhanced interpretability when compared with PLS-DA (Trygg & Wold, 2002). Ultimately, it is significant to write that OPLS-DA contributes no predictive benefits over PLS-DA (Bylesjö et al., 2006). When no ***Y*** uncorrelated variation exists in ***X***, OPLS-DA will harbor an indistinguishable model to PLS-DA.

2.11 BIOINFORMATICS FOR METABOLOMICS

In contrast to transcriptomics, proteomics and metabolomics do not produce as much data, but they will likely play a significant role in comprehending phenotypes in the future. The main focus of bioinformatics in the metabolomic field is, at present, metabolic pathway simulation and model construction (Karnovsky et al., 2011; Sugimoto, Kawakami, Robert, Soga, & Tomita, 2012; Wishart, 2009). Numerous research studies have been conducted to detect diverse metabolic pathways resulting in a given substance (Rzhetsky et al., 2004). The integration of simulated pathway data with pathway databases, for example KEGG (Demir et al., 2010), will function as a model in the endeavors to integrate other such fields (Luo & Brouwer, 2013). Like other scientific fields, metabolomics produces expansive data. The biggest challenge before researchers was handling, processing, and analyzing metabolomic data; hence, it required sophisticated mathematical, statistical, and bioinformatical contrivances (Hegeman, 2010; Kusari & Spiteller, 2012). Functional genomics data sets are widespread and multidimensional, providing their association in suitably designed databases a requirement rather than an alternative (Quanbeck et al., 2012; Saito &

Matsuda, 2010). The investigation of these data sets is likewise difficult, and in no way is it a recognized method for ascertaining functional algorithm selections (reference). Metabolomics has discreet bioinformatics requirements for the generalization of microarray or proteomics data (van den Berg, Hoefsloot, Westerhuis, Smilde, & van der Werf, 2006). The main sectors where progress in bioinformatics are central to the further advancement of metabolomics standards and ontology are statistical evaluation and data mining, and data integration and mathematical modelling of metabolic networks inside the framework of systems biology (Joyce & Palsson, 2006). Metabolomics raw data processing is possibly the most difficult and time consuming step in data analysis (Xia & Wishart, 2011). Usually, processing a set of raw chromatograms implicates noise deduction, spectrum deconvolution, peak identification and integration, chromatograms alignment, and metabolite detection and estimation (Wei et al., 2012). Metabolomics requires automated data processing solutions. The software utilized for data processing includes AMDIS (automated mass spectral deconvolution and identification system; http://chemdata.nist.gov/mass-spc/amdis/). This software uses well-established algorithms that are exceptionally valuable in processing the GC-MS data (Maštovská & Lehotay, 2003), but its relevance to LC-MS or CE-MS is rather inadequate (Dakna, He, Yu, Mischak, & Kolch, 2009). ESI-LC-MS data can be processed via a component detection algorithm (CODA) (Christin et al., 2008), or windowed mass selection method (WMSM) (reference). Even though a chain of commercial and public tools exists (Arnhard, Gottschall, Pitterl, & Oberacher, 2015), none of them contributes an illustrative solution to meet the challenges of metabolomics. The prospective advancement in this area is crucial for development. Accessibility of a vendor-independent data processing software pipeline that is modular and wide enough to inculcate new algorithms could considerably enhance the progress in metabolomics. Recent advancements of the MZmine (http://mzmine.sourceforge.net/index.shtml), a platform-independent software for processing the LC-MS data in metabolomics and proteomics relevance contributes a significant step in this direction (Katajamaa, Miettinen, & Orešič, 2006; Smita, Singh, Akhoon, Gupta, & Gupta, 2013). The software uses a modular infrastructure with the competence to combine novel algorithms and relevance. Another characteristic of this software is that it is flexible to other kinds of mass spectral data like GC-MS and CE-MS. Metabolomics data have been investigated with an extensive range of statistical and machine learning algorithms (Broadhurst &

Kell, 2006; Katajamaa & Orešič, 2007). These can be classified into two main groups, including unsupervised algorithms and supervised algorithms. The unsupervised methods that have been regularly utilized in evaluating metabolomics data are hierarchical clustering (Catchpole et al., 2005), PCA (Smilde et al., 2005), and self-organizing maps (Patterson et al., 2008). Supervised methods comprise ANOVA (Xu & Goodacre, 2012), PLS (Chun & Keleş, 2010), and discriminant function analysis (DFA). In order to carry out statistical analysis of metabolomic data, it is imperative to decrease the number of variables so as to acquire uncorrelated characteristics in the data. This can be acquired either via significance methods in ANOVA and *t*-tests via the linear combination of variables in PCA, or by employing evolutionary algorithms; for example, genetic algorithms or genetic programming (Shulaev, 2006). Evolutionary algorithms are generally performed in integration with second evaluation algorithms; for example, PLS or DFA, that explore integrations of variables most competent in the secondary algorithm, and are directed principles of evolution and species selection. Numerous databases, data management, analysis, and visualization contrivances are presently publicly accessible. These are comprised of, among others, metabolic pathway databases and pathway viewers KEGG (http://www.genome.ad.jp/kegg/), MetaCyc (http://metacyc.org/), AraCyc (http://www.*Arabidopsis*.org/tools/aracyc/), MapMan (http://gabi.rzpd.de/projects/MapMan/) and KaPPA-View (http://kpv.kazusa.or.jp/kappa-view/), the data model for plant metabolomics experiments ArMet (http://www.armet.org/), functional genomics databases MetNet (http://metnet.vrac.iastate.edu/) and DOME (http://medicago.vbi.vt.edu/). DOME, a database advanced by the Mendes group at the Virginia Bioinformatics Institute (Sha, Martins, Laubenbacher, Mendes, & Shulaev, 2013) contributes an example of a complete data management system for metabolomics together for other genomics data. DOME is anticipated for the management of metabolomic, proteomic, and transcriptomic data. The system is comprised of a subset, B-Net, which functions as a reference database. The B-Net schema is aimed to capture the background information pertaining to the molecular biology of an explicit species that exists in the literature. DOME also inculcates a collection of contrivances for statistical data analysis and data visualization; for example, BROME, standing for BRowser for OMEs. Metabolomics as a research discipline will finally need standards of data management, analysis, and reporting to be implemented by the scientific community.

2.12 CONCLUSIONS

Metabolomics is a comparatively new subject, but it has become an imperative integrative part of genomics and systems biology research. Further, advancements in this area necessitate advancements in both analytical sciences and bioinformatics. Advancement of novel analytical techniques is expansively intended to escalate resolution and comprehensiveness, raising speed and throughput of analytical assays and equipment miniaturization. There are new capable technologies for metabolomics, comprised of microfluidic devices such as UPLC, multidimensional chromatography, with varied detectors and metabolite arrays. To congregate increasing confrontations in metabolomics, a concerted community-wide effort is required. Numerous new research papers, as well as activities of the recently formed Metabolomics Society, are main steps in this course, and pave the way for further steady progress of the discipline.

REFERENCES

Adams, L. D., & Gallagher, S. R. (2004). Two-dimensional gel electrophoresis. *Current Protocols in Molecular Biology*, *67*(1), 10–14.

Afzal, F., Gul, A., & Kazi, A. M. (2016). Next-generation sequencing technologies and plant improvement. In *Plant omics: trends and applications* (pp. 271–294). Cham: Springer.

Agarwal, A., Koppstein, D., Rozowsky, J., Sboner, A., Habegger, L., Hillier, L. W., et al. (2010). Comparison and calibration of transcriptome data from RNA-Seq and tiling arrays. *BMC Genomics*, *11*(1), 383.

Aharoni, A., & O'Connell, A. P. (2002). Gene expression analysis of strawberry achene and receptacle maturation using DNA microarrays. *Journal of Experimental Botany*, *53*(377), 2073–2087.

Ahmadloo, S., Nakaoka, H., Hayano, T., Hosomichi, K., You, H., Utsuno, E., et al. (2017). Rapid and cost-effective high-throughput sequencing for identification of germline mutations of BRCA1 and BRCA2. *Journal of Human Genetics*, *62*(5), 561.

Alkan, C., Coe, B. P., & Eichler, E. E. (2011). Genome structural variation discovery and genotyping. *Nature Reviews Genetics*, *12*(5), 363.

Allwood, J. W., Erban, A., de Koning, S., Dunn, W. B., Luedemann, A., Lommen, A., et al. (2009). Inter-laboratory reproducibility of fast gas chromatography–electron impact–time of flight mass spectrometry (GC–EI–TOF/MS) based plant metabolomics. *Metabolomics*, *5*(4), 479–496.

Allwood, J. W., & Goodacre, R. (2010). An introduction to liquid chromatography–mass spectrometry instrumentation applied in plant metabolomic analyses. *Phytochemical Analysis*, *21*(1), 33–47.

Alon, U. (2007). Network motifs: theory and experimental approaches. *Nature Reviews Genetics*, *8*(6), 450.

Anders, S., & Huber, W. (2010). Differential expression analysis for sequence count data. *Genome Biology*, *11*, R106.

An, F. M., Hsiao, S. R., & Chan, M. T. (2011). Sequencing-based approaches reveal low ambient temperature-responsive and tissue-specific microRNAs in phalaenopsis orchid. *PLoS One*, *6*(5). e18937.

Andrade, F. A. D., Gonçalves, L. S. A., Miglioranza, É., Ruas, C. D. F., Ruas, P. M., & Takahashi, L. S. A. (2016). AFLP analysis of genetic diversity in determinate and indeterminate snap bean accessions. *Acta Scientiarum Agronomy*, *38*(1), 29–34.
Arbona, V., Iglesias, D. J., Talon, M., & Gomez-Cadenas, A. (2009). Plant phenotype demarcation using nontargeted LC-MS and GC-MS metabolite profiling. *Journal of Agricultural and Food Chemistry*, *57*(16), 7338–7347.
Arbona, V., Manzi, M., Ollas, C. D., & Gómez-Cadenas, A. (2013). Metabolomics as a tool to investigate abiotic stress tolerance in plants. *International Journal of Molecular Sciences*, *14*(3), 4885–4911.
Arias, M., Hernandez, M., Remondegui, N., Huvenaars, K., Van Dijk, P., & Ritter, E. (2016). First genetic linkage map of Taraxacum koksaghyz Rodin based on AFLP, SSR, COS and EST-SSR markers. *Scientific Reports*, *6*. 31031.
Arnhard, K., Gottschall, A., Pitterl, F., & Oberacher, H. (2015). Applying 'sequential windowed acquisition of all theoretical fragment ion mass spectra' (SWATH) for systematic toxicological analysis with liquid chromatography-high-resolution tandem mass spectrometry. *Analytical and Bioanalytical Chemistry*, *407*(2), 405–414.
Aslam, B., Basit, M., Nisar, M. A., Khurshid, M., & Rasool, M. H. (2017). Proteomics: technologies and their applications. *Journal of Chromatographic Science*, *55*(2), 182–196.
Atanasova, L., & Druzhinina, I. S. (2010). Global nutrient profiling by Phenotype MicroArrays: a tool complementing genomic and proteomic studies in conidial fungi. *Journal of Zhejiang University Science. B*, *11*(3), 151–168.
Augustsson, P., Karlsen, J. T., Su, H. W., Bruus, H., & Voldman, J. (2016). Iso-acoustic focusing of cells for size-insensitive acousto-mechanical phenotyping. *Nature Communications*, 7. 11556.
Autio, R., Hautaniemi, S., Kauraniemi, P., Yli-Harja, O., Astola, J., Wolf, M., et al. (2003). CGH-Plotter: MATLAB toolbox for CGH-data analysis. *Bioinformatics*, *19*(13), 1714–1715.
Avelar, D. M., & Linardi, P. M. (2010). Use of multiple displacement amplification as pre-polymerase chain reaction (pre-PCR) to amplify genomic DNA of siphonapterids preserved for long periods in scientific collections. *Parasites & Vectors*, *3*(1), 86.
Baird, N. A., Etter, P. D., Atwood, T. S., Currey, M. C., Shiver, A. L., Lewis, Z. A., et al. (2008). Rapid SNP discovery and genetic mapping using sequenced RAD markers. *PLoS One*, *3*(10). e3376.
Baker, M. (2010). MicroRNA profiling: separating signal from noise. *Nature Methods*, 7(9), 687–692.
Ballesteros-Gómez, A., de Boer, J., & Leonards, P. E. (2013). Novel analytical methods for flame retardants and plasticizers based on gas chromatography, comprehensive two-dimensional gas chromatography, and direct probe coupled to atmospheric pressure chemical ionization-high resolution time-of-flight-mass spectrometry. *Analytical Chemistry*, *85*(20), 9572–9580.
Barabási, A. L., Gulbahce, N., & Loscalzo, J. (2011). Network medicine: a network-based approach to human disease. *Nature Reviews Genetics*, *12*(1), 56.
Barbas, C., Moraes, E. P., & Villasenor, A. (2011). Capillary electrophoresis as a metabolomics tool for non-targeted fingerprinting of biological samples. *Journal of Pharmaceutical and Biomedical Analysis*, *55*(4), 823–831.
Barbulovic-Nad, I., Lucente, M., Sun, Y., Zhang, M., Wheeler, A. R., & Bussmann, M. (2006). Bio-microarray fabrication techniques—a review. *Critical Reviews in Biotechnology*, *26*(4), 237–259.
Barderas, M. G., Laborde, C. M., Posada, M., de la Cuesta, F., Zubiri, I., Vivanco, F., et al. (2011). Metabolomic profiling for identification of novel potential biomarkers in cardiovascular diseases. *BioMed Research International*, *2011*.

Barrett, M. T., & Caren, M. P. (2011). *U.S. Patent No. 8,058,055*. Washington, DC: U.S. Patent and Trademark Office.

Barrett, M. T., Scheffer, A., Ben-Dor, A., Sampas, N., Lipson, D., Kincaid, R., et al. (2004). Comparative genomic hybridization using oligonucleotide microarrays and total genomic DNA. *Proceedings of the National Academy of Sciences, 101*(51), 17765–17770.

Battle, K. N. (2013). *Microfluidics for the Analysis of Integral Membrane Proteins: A Top-down Approach*. LSU Doctoral Dissertations Louisiana State University and Agricultural and Mechanical College.

Bentley, D. R., Balasubramanian, S., Swerdlow, H. P., Smith, G. P., Milton, J., Brown, C. G., et al. (2008). Accurate whole human genome sequencing using reversible terminator chemistry. *Nature, 456*(7218), 53.

Bianga, J., Touat-Hamici, Z., Bierla, K., Mounicou, S., Szpunar, J., Chavatte, L., et al. (2014). Speciation analysis for trace levels of selenoproteins in cultured human cells. *Journal of Proteomics, 108*, 316–324.

Bibikova, M., Barnes, B., Tsan, C., Ho, V., Klotzle, B., Le, J. M., et al. (2011). High density DNA methylation array with single CpG site resolution. *Genomics, 98*(4), 288–295.

Bino, R. J., Hall, R. D., Fiehn, O., Kopka, J., Saito, K., Draper, J., et al. (2004). Potential of metabolomics as a functional genomics tool. *Trends in Plant Science, 9*(9), 418–425.

Bloom, G., Yang, I. V., Boulware, D., Kwong, K. Y., Coppola, D., Eschrich, S., et al. (2004). Multi-platform, multi-site, microarray-based human tumor classification. *The American Journal of Pathology, 164*(1), 9–16.

Boeye, G., Gismera, M. J., Sevilla, M. T., & Procopio, J. R. (2017). Selective method based on electrochemical detection coupled HPLC for determination of the allergens atranol and chloroatranol. *Electroanalysis, 29*(1), 116–122.

Bombarely, A., Merchante, C., Csukasi, F., Cruz-Rus, E., Caballero, J. L., Medina-Escobar, N., et al. (2010). Generation and analysis of ESTs from strawberry (*Fragaria xananassa*) fruits and evaluation of their utility in genetic and molecular studies. *BMC Genomics, 11*(1), 503.

Bonasio, R., & Shiekhattar, R. (2014). Regulation of transcription by long noncoding RNAs. *Annual Review of Genetics, 48*, 433–455.

Breadmore, M. C. (2012). Capillary and microchip electrophoresis: challenging the common conceptions. *Journal of Chromatography A, 1221*, 42–55.

Brenner, S., Johnson, M., Bridgham, J., Golda, G., Lloyd, D. H., Johnson, D., et al. (2000). Gene expression analysis by massively parallel signature sequencing (MPSS) on microbead arrays. *Nature Biotechnology, 18*(6), 630.

Brereton, R. G. (2009). *Chemometrics for pattern recognition*. John Wiley & Sons.

Brereton, R. G., & Lloyd, G. R. (2014). Partial least squares discriminant analysis: taking the magic away. *Journal of Chemometrics, 28*(4), 213–225.

Broadhurst, D. I., & Kell, D. B. (2006). Statistical strategies for avoiding false discoveries in metabolomics and related experiments. *Metabolomics, 2*(4), 171–196.

Brookes, A. J. (1999). The essence of SNPs. *Gene, 234*(2), 177–186.

Brotman, Y., Lisec, J., Méret, M., Chet, I., Willmitzer, L., & Viterbo, A. (2012). Transcript and metabolite analysis of the Trichoderma-induced systemic resistance response to *Pseudomonas syringae* in *Arabidopsis thaliana*. *Microbiology, 158*(1), 139–146.

Brown, D. G., Rao, S., Weir, T. L., O'Malia, J., Bazan, M., Brown, R. J., et al. (2016). Metabolomics and metabolic pathway networks from human colorectal cancers, adjacent mucosa, and stool. *Cancer & Metabolism, 4*(1), 11.

Brown, M., Wedge, D. C., Goodacre, R., Kell, D. B., Baker, P. N., Kenny, L. C., et al. (2011). Automated workflows for accurate mass-based putative metabolite identification in LC/MS-derived metabolomic datasets. *Bioinformatics, 27*(8), 1108–1112.

Brückner, A., & Heethoff, M. (2017). A chemo-ecologists' practical guide to compositional data analysis. *Chemoecology, 27*(1), 33–46.

Buermans, H. P. J., & Den Dunnen, J. T. (2014). Next generation sequencing technology: advances and applications. *Biochimica et Biophysica Acta (BBA) - Molecular Basis of Disease, 1842*(10), 1932–1941.

Busconi, M., Colli, L., Sánchez, R. A., Santaella, M., Pascual, M. D. L. M., Santana, O., et al. (2015). AFLP and MS-AFLP analysis of the variation within saffron crocus (*Crocus sativus* L.) germplasm. *PLoS One, 10*(4). e0123434.

Bylesjö, M., Rantalainen, M., Cloarec, O., Nicholson, J. K., Holmes, E., & Trygg, J. (2006). OPLS discriminant analysis: combining the strengths of PLS-DA and SIMCA classification. *Journal of Chemometrics, 20*(8-10), 341–351.

Cajka, T., & Fiehn, O. (2015). Toward merging untargeted and targeted methods in mass spectrometry-based metabolomics and lipidomics. *Analytical Chemistry, 88*(1), 524–545.

Cann, J. (2012). *Interacting macromolecules: the theory and practice of their electrophoresis, ultracentrifugation, and chromatography*. 7. Elsevier.

Capitani, D., Sobolev, A. P., & Mannina, L. (2017). Nuclear Magnetic Resonance–Metabolomics. In *Food Authentication: Management, Analysis and Regulation* (p. 177): Wiley.

Catchpole, G. S., Beckmann, M., Enot, D. P., Mondhe, M., Zywicki, B., Taylor, J., et al. (2005). Hierarchical metabolomics demonstrates substantial compositional similarity between genetically modified and conventional potato crops. *Proceedings of the National Academy of Sciences, 102*(40), 14458–14462.

Catsimpoolas, N. (Ed.), (2012). Isoelectric focusing. In: *Elsevier*.

Cazzonelli, C. I., & Pogson, B. J. (2010). Source to sink: regulation of carotenoid biosynthesis in plants. *Trends in Plant Science, 15*(5), 266–274.

Chaisson, M. J., Wilson, R. K., & Eichler, E. E. (2015). Genetic variation and the de novo assembly of human genomes. *Nature Reviews Genetics, 16*(11), 627.

Chang, C., Xu, G., Bai, Y., Zhang, C., Li, X., Li, M., et al. (2012). Online coupling of capillary electrophoresis with direct analysis in real time mass spectrometry. *Analytical Chemistry, 85*(1), 170–176.

Chang, S. K., & Zhang, Y. (2017). Protein analysis. In *Food analysis* (pp. 315–331). Cham: Springer.

Chauhan, A., Goyal, M. K., & Chauhan, P. (2014). GC-MS technique and its analytical applications in science and technology. *Journal of Analytical & Bioanalytical Techniques, 5*(6), 222.

Chen, J. Q., Lee, J. H., Herrmann, M. A., Park, K. S., Heldman, M. R., Goldsmith, P. K., et al. (2013). Capillary isoelectric-focusing immunoassays to study dynamic oncoprotein phosphorylation and drug response to targeted therapies in non-small cell lung cancer. *Molecular Cancer Therapeutics*,molcanther-0074.

Choy, K. W., Setlur, S. R., Lee, C., & Lau, T. K. (2010). The impact of human copy number variation on a new era of genetic testing. *BJOG: An International Journal of Obstetrics & Gynaecology, 117*(4), 391–398.

Christin, C., Smilde, A. K., Hoefsloot, H. C., Suits, F., Bischoff, R., & Horvatovich, P. L. (2008). Optimized time alignment algorithm for LC−MS data: correlation optimized warping using component detection algorithm-selected mass chromatograms. *Analytical Chemistry, 80*(18), 7012–7021.

Chun, H., & Keleş, S. (2010). Sparse partial least squares regression for simultaneous dimension reduction and variable selection. *Journal of the Royal Statistical Society, Series B: Statistical Methodology, 72*(1), 3–25.

Cloonan, N., Forrest, A. R., Kolle, G., Gardiner, B. B., Faulkner, G. J., Brown, M. K., et al. (2008). Stem cell transcriptome profiling via massive-scale mRNA sequencing. *Nature Methods, 5*(7), 613.

Conesa, A., Madrigal, P., Tarazona, S., Gomez-Cabrero, D., Cervera, A., McPherson, A., et al. (2016). A survey of best practices for RNA-seq data analysis. *Genome Biology, 17*(1), 13.

Consden, R., Gordon, A. H., & Martin, A. J. (1946). Ionophoresis in Silica Jelly; a method for the separation of amino-acids and peptides. *Biochemical Journal, 40*, 33–41.
Cookson, W., Liang, L., Abecasis, G., Moffatt, M., & Lathrop, M. (2009). Mapping complex disease traits with global gene expression. *Nature Reviews Genetics, 10*(3), 184.
Costa, V., Angelini, C., De Feis, I., & Ciccodicola, A. (2010). Uncovering the complexity of transcriptomes with RNA-Seq. *BioMed Research International, 2010*.
Craig, A., Cloarec, O., Holmes, E., Nicholson, J. K., & Lindon, J. C. (2006). Scaling and normalization effects in NMR spectroscopic metabonomic data sets. *Analytical Chemistry, 78*(7), 2262–2267.
Crawford, G. E., Holt, I. E., Whittle, J., Webb, B. D., Tai, D., Davis, S., et al. (2006). Genome-wide mapping of DNase hypersensitive sites using massively parallel signature sequencing (MPSS). *Genome Research, 16*(1), 123–131.
Creek, D. J., Jankevics, A., Breitling, R., Watson, D. G., Barrett, M. P., & Burgess, K. E. (2011). Toward global metabolomics analysis with hydrophilic interaction liquid chromatography–mass spectrometry: improved metabolite identification by retention time prediction. *Analytical Chemistry, 83*(22), 8703–8710.
Curnutte, M. (2012). *Consuming genomes: The coproduction of a new scientific and technological order for genetic testing*. PhD degree in Foundations of the Life Sciences and their Ethical Consequences European School of Molecular Medicine (SEMM) and University of Milan Faculty of Medicine Settore disciplinare: FIL/02
D'Orazio, G., Rocchi, S., & Fanali, S. (2012). Nano-liquid chromatography coupled with mass spectrometry: separation of sulfonamides employing non-porous core–shell particles. *Journal of Chromatography A, 1255*, 277–285.
da Silveira Petruci, J. F., Liebetanz, M. G., Cardoso, A. A., & Hauser, P. C. (2017). Absorbance detector for high performance liquid chromatography based on a deep-UV light-emitting diode at 235 nm. *Journal of Chromatography A, 1512*, 143–146.
Dakna, M., He, Z., Yu, W. C., Mischak, H., & Kolch, W. (2009). Technical, bioinformatical and statistical aspects of liquid chromatography–mass spectrometry (LC–MS) and capillary electrophoresis-mass spectrometry (CE-MS) based clinical proteomics: A critical assessment. *Journal of Chromatography B, 877*(13), 1250–1258.
Daran-Lapujade, P., Rossell, S., van Gulik, W. M., Luttik, M. A., de Groot, M. J., Slijper, M., et al. (2007). The fluxes through glycolytic enzymes in *Saccharomyces cerevisiae* are predominantly regulated at posttranscriptional levels. *Proceedings of the National Academy of Sciences, 104*(40), 15753–15758.
Das, G., Patra, J. K., Lee, S. Y., Kim, C., Park, J. G., & Baek, K. H. (2017). Analysis of metabolomic profile of fermented *Orostachys japonicus* A. Berger by capillary electrophoresis time of flight mass spectrometry. *PLoS One, 12*(7)e0181280.
DeBerardinis, R. J., & Thompson, C. B. (2012). Cellular metabolism and disease: what do metabolic outliers teach us? *Cell, 148*(6), 1132–1144.
Defernez, M., & Colquhoun, I. J. (2003). Factors affecting the robustness of metabolite fingerprinting using 1H NMR spectra. *Phytochemistry, 62*(6), 1009–1017.
Demir, E., Cary, M. P., Paley, S., Fukuda, K., Lemer, C., Vastrik, I., et al. (2010). The BioPAX community standard for pathway data sharing. *Nature Biotechnology, 28*(9), 935.
Desprez, T., Amselem, J., Caboche, M., & Höfte, H. (1998). Differential gene expression in *Arabidopsis* monitored using cDNA arrays. *The Plant Journal, 14*(5), 643–652.
Dettmer, K., Aronov, P. A., & Hammock, B. D. (2007). Mass spectrometry-based metabolomics. *Mass Spectrometry Reviews, 26*(1), 51–78.
Dieterle, F., Riefke, B., Schlotterbeck, G., Ross, A., Senn, H., & Amberg, A. (2011). NMR and MS methods for metabonomics. In *Drug Safety Evaluation* (pp. 385–415): Humana Press.
Dominissini, D., Moshitch-Moshkovitz, S., Salmon-Divon, M., Amariglio, N., & Rechavi, G. (2013). Transcriptome-wide mapping of N 6-methyladenosine by m 6

A-seq based on immunocapturing and massively parallel sequencing. *Nature Protocols*, *8*(1), 176.
Donaghue, C., Davies, N., Ahn, J. W., Thomas, H., Ogilvie, C. M., & Mann, K. (2017). Efficient and cost-effective genetic analysis of products of conception and fetal tissues using a QF-PCR/array CGH strategy; five years of data. *Molecular Cytogenetics*, *10*(1), 12.
Donato, P., Cacciola, F., Tranchida, P. Q., Dugo, P., & Mondello, L. (2012). Mass spectrometry detection in comprehensive liquid chromatography: basic concepts, instrumental aspects, applications and trends. *Mass Spectrometry Reviews*, *31*(5), 523–559.
Duan, S., Zhang, W., Bleibel, W. K., Cox, N. J., & Dolan, M. E. (2008). SNPinProbe_1. 0: A database for filtering out probes in the Affymetrix GeneChip® Human Exon 1.0 ST array potentially affected by SNPs. *Bioinformation*, *2*(10), 469.
Duggan, G. E., Hittel, D. S., Hughey, C. C., Weljie, A., Vogel, H. J., & Shearer, J. (2011). Differentiating short-and long-term effects of diet in the obese mouse using 1H-nuclear magnetic resonance metabolomics. *Diabetes, Obesity and Metabolism*, *13*(9), 859–862.
Dunn, M. J. (Ed.), (2014). *Gel electrophoresis of proteins*: Elsevier.
Dunn, W. B., Bailey, N. J., & Johnson, H. E. (2005). Measuring the metabolome: current analytical technologies. *Analyst*, *130*(5), 606–625.
Dunn, W. B., Broadhurst, D., Begley, P., Zelena, E., Francis-McIntyre, S., Anderson, N., et al. (2011). Procedures for large-scale metabolic profiling of serum and plasma using gas chromatography and liquid chromatography coupled to mass spectrometry. *Nature Protocols*, *6*(7), 1060.
Dunn, W. B., Broadhurst, D. I., Atherton, H. J., Goodacre, R., & Griffin, J. L. (2011). Systems level studies of mammalian metabolomes: the roles of mass spectrometry and nuclear magnetic resonance spectroscopy. *Chemical Society Reviews*, *40*(1), 387–426.
Dunn, W. B., Erban, A., Weber, R. J., Creek, D. J., Brown, M., Breitling, R., et al. (2013). Mass appeal: metabolite identification in mass spectrometry-focused untargeted metabolomics. *Metabolomics*, *9*(1), 44–66.
Easton, D. F., Pooley, K. A., Dunning, A. M., Pharoah, P. D., Thompson, D., Ballinger, D. G., et al. (2007). Genome-wide association study identifies novel breast cancer susceptibility loci. *Nature*, *447*(7148), 1087.
Ekblom, R., & Wolf, J. B. (2014). A field guide to whole-genome sequencing, assembly and annotation. *Evolutionary Applications*, 7(9), 1026–1042.
Elingarami, S., Li, X., & He, N. (2013). Applications of nanotechnology, next generation sequencing and microarrays in biomedical research. *Journal of Nanoscience and Nanotechnology*, *13*(7), 4539–4551.
Ellis, D. I., Harrigan, G. G., & Goodacre, R. (2003). Metabolic fingerprinting with Fourier transform infrared spectroscopy. In *Metabolic Profiling: Its role in biomarker discovery and gene function analysis* (pp. 111–124). Boston, MA: Springer.
El-Mansi, M., Cozzone, A. J., Shiloach, J., & Eikmanns, B. J. (2006). Control of carbon flux through enzymes of central and intermediary metabolism during growth of *Escherichia coli* on acetate. *Current Opinion in Microbiology*, *9*(2), 173–179.
El-Sabrout, K., & El-Raffa, A. (2015). Molecular characterization of Alexandria rabbit line using DNA markers. *Rabbit Genetics*, *5*(1), 1–5.
Fagerquist, C. K., Garbus, B. R., Miller, W. G., Williams, K. E., Yee, E., Bates, A. H., et al. (2010). Rapid identification of protein biomarkers of *Escherichia coli* O157:H7 by matrix-assisted laser desorption ionization-time-of-flight−time-of-flight mass spectrometry and top-down proteomics. *Analytical Chemistry*, *82*(7), 2717–2725.
Farag, M. A., Porzel, A., Schmidt, J., & Wessjohann, L. A. (2012). Metabolite profiling and fingerprinting of commercial cultivars of *Humulus lupulus* L. (hop): a comparison of MS and NMR methods in metabolomics. *Metabolomics*, *8*(3), 492–507.
Farag, M. A., Porzel, A., & Wessjohann, L. A. (2012). Comparative metabolite profiling and fingerprinting of medicinal licorice roots using a multiplex approach of GC–MS, LC–MS and 1D NMR techniques. *Phytochemistry*, *76*, 60–72.

Feist, P., & Hummon, A. B. (2015). Proteomic challenges: sample preparation techniques for microgram-quantity protein analysis from biological samples. *International Journal of Molecular Sciences*, *16*(2), 3537–3563.
Fernie, A. R., Trethewey, R. N., Krotzky, A. J., & Willmitzer, L. (2004). Metabolite profiling: from diagnostics to systems biology. *Nature Reviews Molecular Cell Biology*, *5*(9), 763.
Fiegler, H., Carr, P., Douglas, E. J., Burford, D. C., Hunt, S., Smith, J., et al. (2003). DNA microarrays for comparative genomic hybridization based on DOP-PCR amplification of BAC and PAC clones. *Genes, Chromosomes and Cancer*, *36*(4), 361–374.
Fiehn, O. (2001). Combining genomics, metabolome analysis, and biochemical modelling to understand metabolic networks. *International Journal of Genomics*, *2*(3), 155–168.
Fiehn, O. (2002). Metabolomics—the link between genotypes and phenotypes. In *Functional genomics* (pp. 155–171). Dordrecht: Springer.
Fiehn, O., Kopka, J., Dörmann, P., Altmann, T., Trethewey, R. N., & Willmitzer, L. (2000). Metabolite profiling for plant functional genomics. *Nature Biotechnology*, *18* (11), 1157.
Finotello, F., & Di Camillo, B. (2015). Measuring differential gene expression with RNA-seq: challenges and strategies for data analysis. *Briefings in Functional Genomics*, *14*(2), 130–142.
Forcisi, S., Moritz, F., Kanawati, B., Tziotis, D., Lehmann, R., & Schmitt-Kopplin, P. (2013). Liquid chromatography–mass spectrometry in metabolomics research: mass analyzers in ultra high pressure liquid chromatography coupling. *Journal of Chromatography A*, *1292*, 51–65.
Fortin, T., Salvador, A., Charrier, J. P., Lenz, C., Lacoux, X., Morla, A., et al. (2009). Clinical quantitation of prostate-specific antigen biomarker in the low nanogram/milliliter range by conventional bore liquid chromatography-tandem mass spectrometry (multiple reaction monitoring) coupling and correlation with ELISA tests. *Molecular & Cellular Proteomics*, *8*(5), 1006–1015.
Fotakis, C., Kokkotou, K., Zoumpoulakis, P., & Zervou, M. (2013). NMR metabolite fingerprinting in grape derived products: an overview. *Food Research International*, *54*(1), 1184–1194.
Freeman, W. M., & Hemby, S. E. (2004). Proteomics for protein expression profiling in neuroscience. *Neurochemical Research*, *29*(6), 1065–1081.
Fry, N. K., Savelkoul, P. H., & Visca, P. (2009). Amplified fragment length polymorphism analysis. In *Molecular epidemiology of microorganisms* (pp. 89–104). Totowa, NJ: Humana Press.
Fu, X., Fu, N., Guo, S., Yan, Z., Xu, Y., Hu, H., et al. (2009). Estimating accuracy of RNA-Seq and microarrays with proteomics. *BMC Genomics*, *10*(1), 161.
Fumagalli, E., Baldoni, E., Abbruscato, P., Piffanelli, P., Genga, A., Lamanna, R., et al. (2009). NMR techniques coupled with multivariate statistical analysis: tools to analyse *Oryza sativa* metabolic content under stress conditions. *Journal of Agronomy and Crop Science*, *195*(2), 77–88.
Funk, R. H. (2015). Endogenous electric fields as guiding cue for cell migration. *Frontiers in Physiology*, *6*, 143.
Gaeta, R. T., Yoo, S. Y., Pires, J. C., Doerge, R. W., Chen, Z. J., & Osborn, T. C. (2009). Analysis of gene expression in resynthesized *Brassica napus* allopolyploids using *Arabidopsis* 70mer oligo microarrays. *PLoS One*, *4*(3). e4760.
Galindo-Prieto, B., Eriksson, L., & Trygg, J. (2014). Variable influence on projection (VIP) for orthogonal projections to latent structures (OPLS). *Journal of Chemometrics*, *28*(8), 623–632.

Gallagher, S. R. (2012). One-dimensional SDS gel electrophoresis of proteins. *Current Protocols in Molecular Biology*, *97*(1), 10–12.

Ganal, M. W., Durstewitz, G., Polley, A., Bérard, A., Buckler, E. S., Charcosset, A., et al. (2011). A large maize (*Zea mays* L.) SNP genotyping array: development and germplasm genotyping, and genetic mapping to compare with the B73 reference genome. *PLoS One*, *6*(12). e28334.

Garber, M., Grabherr, M. G., Guttman, M., & Trapnell, C. (2011). Computational methods for transcriptome annotation and quantification using RNA-seq. *Nature Methods*, *8*(6), 469.

García, A., Godzien, J., López-Gonzálvez, Á., & Barbas, C. (2017). Capillary electrophoresis mass spectrometry as a tool for untargeted metabolomics. *Bioanalysis*, *9*(1), 99–130.

Gellatly, K. S., Ash, G. J., & Taylor, J. L. (2001). Development of a method for mRNA differential display in filamentous fungi: comparison of mRNA differential display reverse transcription polymerase chain reaction and cDNA amplified fragment length polymorphism in *Leptosphaeria maculans*. *Canadian Journal of Microbiology*, *47*(10), 955–960.

Gerber, S., Mariette, S., Streiff, R., Bodenes, C., & Kremer, A. (2000). Comparison of microsatellites and amplified fragment length polymorphism markers for parentage analysis. *Molecular Ecology*, *9*(8), 1037–1048.

Gerling, I. C., Solomon, S. S., & Bryer-Ash, M. (2003). Genomes, transcriptomes, and proteomes: molecular medicine and its impact on medical practice. *Archives of Internal Medicine*, *163*(2), 190–198.

Gidhi, A. (2013). *Two-dimensional Protein profiling of Rhizobium isolates of* Pisum sativum *L. (pea) collected from acid soils of Jharkhand* (Doctoral dissertation). Kanke, Ranchi, Jharkhand: Birsa Agricultural University.

Gika, H. G., Theodoridis, G. A., Plumb, R. S., & Wilson, I. D. (2014). Current practice of liquid chromatography–mass spectrometry in metabolomics and metabonomics. *Journal of Pharmaceutical and Biomedical Analysis*, *87*, 12–25.

Gleason, L. U., & Burton, R. S. (2015). RNA-seq reveals regional differences in transcriptome response to heat stress in the marine snail *Chlorostoma funebralis*. *Molecular Ecology*, *24*(3), 610–627.

Godschalk, P. C., Bergman, M. P., Gorkink, R. F., Simons, G., Van den Braak, N., Lastovica, A. J., et al. (2006). Identification of DNA sequence variation in *Campylobacter jejuni* strains associated with the Guillain-Barré syndrome by high-throughput AFLP analysis. *BMC Microbiology*, *6*(1), 32.

González-Ruiz, V., Olives, A. I., & Martín, M. A. (2015). Core-shell particles lead the way to renewing high-performance liquid chromatography. *TrAC Trends in Analytical Chemistry*, *64*, 17–28.

Goodacre, R., Vaidyanathan, S., Dunn, W. B., Harrigan, G. G., & Kell, D. B. (2004). Metabolomics by numbers: acquiring and understanding global metabolite data. *Trends in Biotechnology*, *22*(5), 245–252.

Goodwin, S., McPherson, J. D., & McCombie, W. R. (2016). Coming of age: ten years of next-generation sequencing technologies. *Nature Reviews Genetics*, *17*(6), 333.

Goulitquer, S., Potin, P., & Tonon, T. (2012). Mass spectrometry-based metabolomics to elucidate functions in marine organisms and ecosystems. *Marine Drugs*, *10*(4), 849–880.

Grabherr, M. G., Haas, B. J., Yassour, M., Levin, J. Z., Thompson, D. A., Amit, I., et al. (2011). Full-length transcriptome assembly from RNA-Seq data without a reference genome. *Nature Biotechnology*, *29*(7), 644.

Green, E. D., & Guyer, M. S. (2011). Charting a course for genomic medicine from base pairs to bedside. *Nature*, *470*(7333), 204.

Gresshoff, P. M., & Caetano-Anollés, G. (2018). Commercial applications of DNA profiling by amplification with arbitrary oligonucleotide primers. In *Technology Transfer of Plant Biotechnology*. Boco Raton, FL: CRC Press.

Gromski, P. S., Muhamadali, H., Ellis, D. I., Xu, Y., Correa, E., Turner, M. L., et al. (2015). A tutorial review: metabolomics and partial least squares-discriminant analysis–a marriage of convenience or a shotgun wedding. *Analytica Chimica Acta*, *879*, 10–23.

Grossman, P. D., & Colburn, J. C. (Eds.), (2012). *Capillary electrophoresis: theory and practice*: Academic Press.

Gu, H., Pan, Z., Xi, B., Asiago, V., Musselman, B., & Raftery, D. (2011). Principal component directed partial least squares analysis for combining nuclear magnetic resonance and mass spectrometry data in metabolomics: application to the detection of breast cancer. *Analytica Chimica Acta*, *686*(1-2), 57–63.

Gu, S., & Kay, M. A. (2010). How do miRNAs mediate translational repression? *Silence*, *1*(1), 11.

Gunnarsson, R., Staaf, J., Jansson, M., Ottesen, A. M., Göransson, H., Liljedahl, U., et al. (2008). Screening for copy-number alterations and loss of heterozygosity in chronic lymphocytic leukemia—a comparative study of four differently designed, high resolution microarray platforms. *Genes, Chromosomes and Cancer*, *47*(8), 697–711.

Gygi, S. P., Rist, B., Gerber, S. A., Turecek, F., Gelb, M. H., & Aebersold, R. (1999). Quantitative analysis of complex protein mixtures using isotope-coded affinity tags. *Nature Biotechnology*, *17*(10), 994.

Gzyl, A., Augustynowicz, E., Mosiej, E., Zawadka, M., Gniadek, G., Nowaczek, A., et al. (2005). Amplified fragment length polymorphism (AFLP) versus randomly amplified polymorphic DNA (RAPD) as new tools for inter-and intra-species differentiation within Bordetella. *Journal of Medical Microbiology*, *54*(4), 333–346.

Habib, N., Avraham-Davidi, I., Basu, A., Burks, T., Shekhar, K., Hofree, M., et al. (2017). Massively parallel single-nucleus RNA-seq with DroNc-seq. *Nature Methods*, *14*(10).

Hagel, J. M., & Facchini, P. J. (2008). Plant metabolomics: analytical platforms and integration with functional genomics. *Phytochemistry Reviews*, 7(3), 479–497.

Halligan, D. L., & Keightley, P. D. (2006). Ubiquitous selective constraints in the Drosophila genome revealed by a genome-wide interspecies comparison. *Genome Research*, *16*(7), 875–884.

Han, J., Danell, R. M., Patel, J. R., Gumerov, D. R., Scarlett, C. O., Speir, J. P., et al. (2008). Towards high-throughput metabolomics using ultrahigh-field Fourier transform ion cyclotron resonance mass spectrometry. *Metabolomics*, *4*(2), 128–140.

Harrigan, G. G., & Goodacre, R. (Eds.), (2012). *Metabolic profiling: its role in biomarker discovery and gene function analysis*: Springer Science & Business Media.

He, D., Liu, Y., Cai, M., Pan, H., & Zhang, Q. (2014). The first genetic linkage map of crape myrtle (*L. agerstroemia*) based on amplification fragment length polymorphisms and simple sequence repeats markers. *Plant Breeding*, *133*(1), 138–144.

Head, S. R., Komori, H. K., LaMere, S. A., Whisenant, T., Van Nieuwerburgh, F., Salomon, D. R., et al. (2014). Library construction for next-generation sequencing: overviews and challenges. *Biotechniques*, *56*(2), 61–77.

Hegeman, A. D. (2010). Plant metabolomics—meeting the analytical challenges of comprehensive metabolite analysis. *Briefings in Functional Genomics*, *9*(2), 139–148.

Helm, M., & Motorin, Y. (2017). Detecting RNA modifications in the epitranscriptome: predict and validate. *Nature Reviews Genetics*, *18*(5), 275.

Hennrich, M. L., Groenewold, V., Kops, G. J., Heck, A. J., & Mohammed, S. (2011). Improving depth in phosphoproteomics by using a strong cation exchange-weak anion exchange-reversed phase multidimensional separation approach. *Analytical Chemistry*, *83*(18), 7137–7143.

Hensman, J., Papastamoulis, P., Glaus, P., Honkela, A., & Rattray, M. (2015). Fast and accurate approximate inference of transcript expression from RNA-seq data. *Bioinformatics*, *31*(24), 3881–3889.

Hernández, F., Portolés, T., Pitarch, E., & López, F. J. (2011). Gas chromatography coupled to high-resolution time-of-flight mass spectrometry to analyze trace-level organic compounds in the environment, food safety and toxicology. *TrAC Trends in Analytical Chemistry*, *30*(2), 388–400.
Herzog, C., Poehler, E., Peretzki, A. J., Borisov, S. M., Aigner, D., Mayr, T., et al. (2016). Continuous on-chip fluorescence labelling, free-flow isoelectric focusing and marker-free isoelectric point determination of proteins and peptides. *Lab on a Chip*, *16*(9), 1565–1572.
Hester, S. D., Reid, L., Nowak, N., Jones, W. D., Parker, J. S., Knudtson, K., et al. (2009). Comparison of comparative genomic hybridization technologies across microarray platforms. *Journal of Biomolecular Techniques*, *20*(2), 135.
Hewitt, N. J., Gómez Lechón, M. J., Houston, J. B., Hallifax, D., Brown, H. S., Maurel, P., et al. (2007). Primary hepatocytes: current understanding of the regulation of metabolic enzymes and transporter proteins, and pharmaceutical practice for the use of hepatocytes in metabolism, enzyme induction, transporter, clearance, and hepatotoxicity studies. *Drug Metabolism Reviews*, *39*(1), 159–234.
Hirayama, A., Wakayama, M., & Soga, T. (2014). Metabolome analysis based on capillary electrophoresis-mass spectrometry. *TrAC Trends in Analytical Chemistry*, *61*, 215–222.
Hirst, M., & Marra, M. A. (2010). Next generation sequencing based approaches to epigenomics. *Briefings in Functional Genomics*, *9*(5-6), 455–465.
Hjertén, S. (1985). High-performance electrophoresis: Elimination of electroendosmosis and solute adsorption. *Journal of Chromatography*, *347*, 191–198.
Honaas, L. A., Wafula, E. K., Wickett, N. J., Der, J. P., Zhang, Y., Edger, P. P., et al. (2016). Selecting superior de novo transcriptome assemblies: lessons learned by leveraging the best plant genome. *PLoS One*, *11*(1). e0146062.
Horgan, R. P., & Kenny, L. C. (2011). 'Omic' technologies: genomics, transcriptomics, proteomics and metabolomics. *The Obstetrician & Gynaecologist*, *13*(3), 189–195.
Houzet, L., Deleage, C., Satie, A. P., Merlande, L., Mahe, D., & Dejucq-Rainsford, N. (2015). A new method for rapid screening of end-point PCR products: application to single genome amplified HIV and SIV envelope amplicons. *PLoS One*, *10*(6). e0128188.
Howie, B., Fuchsberger, C., Stephens, M., Marchini, J., & Abecasis, G. R. (2012). Fast and accurate genotype imputation in genome-wide association studies through pre-phasing. *Nature Genetics*, *44*(8), 955.
Hu, S., Xie, Y., Ramachandran, P., Ogorzalek Loo, R. R., Li, Y., Loo, J. A., et al. (2005). Large-scale identification of proteins in human salivary proteome by liquid chromatography/mass spectrometry and two-dimensional gel electrophoresis-mass spectrometry. *Proteomics*, *5*(6), 1714–1728.
Hu, T. T., Pattyn, P., Bakker, E. G., Cao, J., Cheng, J. F., Clark, R. M., et al. (2011). The *Arabidopsis lyrata* genome sequence and the basis of rapid genome size change. *Nature Genetics*, *43*(5), 476.
Huang, X., McLean, R. S., & Zheng, M. (2005). High-resolution length sorting and purification of DNA-wrapped carbon nanotubes by size-exclusion chromatography. *Analytical Chemistry*, 77(19), 6225–6228.
Hühner, J., Lämmerhofer, M., & Neusüß, C. (2015). Capillary isoelectric focusing-mass spectrometry: coupling strategies and applications. *Electrophoresis*, *36*(21-22), 2670–2686.
Hunter, D. J., Kraft, P., Jacobs, K. B., Cox, D. G., Yeager, M., Hankinson, S. E., et al. (2007). A genome-wide association study identifies alleles in FGFR2 associated with risk of sporadic postmenopausal breast cancer. *Nature Genetics*, *39*(7), 870.
Ibáñez, C., Simó, C., García-Cañas, V., Cifuentes, A., & Castro-Puyana, M. (2013). Metabolomics, peptidomics and proteomics applications of capillary electrophoresis-mass spectrometry in foodomics: a review. *Analytica Chimica Acta*, *802*, 1–13.

Ibáñez, C., Simó, C., Martín-Álvarez, P. J., Kivipelto, M., Winblad, B., Cedazo-Mínguez, A., et al. (2012). Toward a predictive model of Alzheimer's disease progression using capillary electrophoresis–mass spectrometry metabolomics. *Analytical Chemistry, 84*(20), 8532–8540.

International HapMap Consortium. (2007). A second generation human haplotype map of over 3.1 million SNPs. *Nature, 449*(7164), 851.

Jain, M., Nilsson, R., Sharma, S., Madhusudhan, N., Kitami, T., Souza, A. L., et al. (2012). Metabolite profiling identifies a key role for glycine in rapid cancer cell proliferation. *Science, 336*(6084), 1040–1044.

Janson, J. C. (Ed.), (2012). Vol. 151. *Protein purification: principles, high resolution methods, and applications*. John Wiley & Sons.

Jhuang, H. Y., Lee, H. Y., & Leu, J. Y. (2017). Mitochondrial–nuclear co-evolution leads to hybrid incompatibility through pentatricopeptide repeat proteins. *EMBO Reports, 18*(1), 87–101.

Jiang, L., Schlesinger, F., Davis, C. A., Zhang, Y., Li, R., Salit, M., et al. (2011). Synthetic spike-in standards for RNA-seq experiments. *Genome Research*.

John, B. B., Joseph, O. O., & Gbolagade, J. S. (2016). Molecular variability of *Celosia argentea* using amplified fragment length polymorphism (AFLP) marker. *Molecular Plant Breeding*, 7.

Jones, F. C., Chan, Y. F., Schmutz, J., Grimwood, J., Brady, S. D., Southwick, A. M., et al. (2012). A genome-wide SNP genotyping array reveals patterns of global and repeated species-pair divergence in sticklebacks. *Current Biology, 22*(1), 83–90.

Jones, O. A., Maguire, M. L., Griffin, J. L., Dias, D. A., Spurgeon, D. J., & Svendsen, C. (2013). Metabolomics and its use in ecology. *Austral Ecology, 38*(6), 713–720.

Joyce, A. R., & Palsson, B. Ø. (2006). The model organism as a system: integrating 'omics' data sets. *Nature Reviews Molecular Cell Biology*, 7(3), 198.

Kanamori-Katayama, M., Itoh, M., Kawaji, H., Lassmann, T., Katayama, S., Kojima, M., et al. (2011). Unamplified cap analysis of gene expression on a single-molecule sequencer. *Genome Research*.

Kanehisa, M., Sato, Y., Kawashima, M., Furumichi, M., & Tanabe, M. (2015). KEGG as a reference resource for gene and protein annotation. *Nucleic Acids Research, 44*(D1), D457–D462.

Kang, J., Choi, M. Y., Kang, S., Kwon, H. N., Wen, H., Lee, C. H., et al. (2008). Application of a 1H nuclear magnetic resonance (NMR) metabolomics approach combined with orthogonal projections to latent structure-discriminant analysis as an efficient tool for discriminating between Korean and Chinese herbal medicines. *Journal of Agricultural and Food Chemistry, 56*(24), 11589–11595.

Kaplan, F., Kopka, J., Sung, D. Y., Zhao, W., Popp, M., Porat, R., et al. (2007). Transcript and metabolite profiling during cold acclimation of *Arabidopsis* reveals an intricate relationship of cold-regulated gene expression with modifications in metabolite content. *The Plant Journal, 50*(6), 967–981.

Karnovsky, A., Weymouth, T., Hull, T., Tarcea, V. G., Scardoni, G., Laudanna, C., et al. (2011). Metscape 2 bioinformatics tool for the analysis and visualization of metabolomics and gene expression data. *Bioinformatics, 28*(3), 373–380.

Katajamaa, M., Miettinen, J., & Orešič, M. (2006). MZmine: toolbox for processing and visualization of mass spectrometry based molecular profile data. *Bioinformatics, 22*(5), 634–636.

Katajamaa, M., & Orešič, M. (2007). Data processing for mass spectrometry-based metabolomics. *Journal of Chromatography A, 1158*(1-2), 318–328.

Keating, B. J., Tischfield, S., Murray, S. S., Bhangale, T., Price, T. S., Glessner, J. T., et al. (2008). Concept, design and implementation of a cardiovascular gene-centric 50 k SNP array for large-scale genomic association studies. *PLoS One, 3*(10). e3583.

Kemsley, E. K., Le Gall, G., Dainty, J. R., Watson, A. D., Harvey, L. J., Tapp, H. S., et al. (2007). Multivariate techniques and their application in nutrition: a metabolomics case study. *British Journal of Nutrition, 98*(1), 1–14.
Keymanesh, K., Darvishi, M. H., & Sardari, S. (2009). Metabolome comparison of transgenic and non-transgenic rice by statistical analysis of FTIR and NMR spectra. *Rice Science, 16* (2), 119–123.
Kim, V. N. (2005). MicroRNA biogenesis: coordinated cropping and dicing. *Nature Reviews Molecular Cell Biology, 6*(5), 376.
Kind, T., Wohlgemuth, G., Lee, D. Y., Lu, Y., Palazoglu, M., Shahbaz, S., et al. (2009). FiehnLib: mass spectral and retention index libraries for metabolomics based on quadrupole and time-of-flight gas chromatography/mass spectrometry. *Analytical Chemistry, 81* (24), 10038–10048.
Kinoshita, E., Kinoshita-Kikuta, E., & Koike, T. (2012). Phos-tag SDS-PAGE systems for phosphorylation profiling of proteins with a wide range of molecular masses under neutral pH conditions. *Proteomics, 12*(2), 192–202.
Kissoudi, M., Sarakatsianos, I., & Samanidou, V. (2018). Isolation and purification of food-grade C-phycocyanin from *Arthrospira platensis* and its determination in confectionery by HPLC with diode array detection. *Journal of Separation Science, 41*(4), 975–981.
Kitano, H. (2002). Systems biology: a brief overview. *Science, 295*(5560), 1662–1664.
Kleffmann, T., Russenberger, D., von Zychlinski, A., Christopher, W., Sjölander, K., Gruissem, W., et al. (2004). The *Arabidopsis thaliana* chloroplast proteome reveals pathway abundance and novel protein functions. *Current Biology, 14*(5), 354–362.
Köcher, T., Swart, R., & Mechtler, K. (2011). Ultra-high-pressure RPLC hyphenated to an LTQ-Orbitrap Velos reveals a linear relation between peak capacity and number of identified peptides. *Analytical Chemistry, 83*(7), 2699–2704.
Kodzius, R., Kojima, M., Nishiyori, H., Nakamura, M., Fukuda, S., Tagami, M., et al. (2006). CAGE: cap analysis of gene expression. *Nature Methods, 3*(3), 211.
Koek, M. M., Jellema, R. H., van der Greef, J., Tas, A. C., & Hankemeier, T. (2011). Quantitative metabolomics based on gas chromatography mass spectrometry: status and perspectives. *Metabolomics*, 7(3), 307–328.
Kogenaru, S., Yan, Q., Guo, Y., & Wang, N. (2012). RNA-seq and microarray complement each other in transcriptome profiling. *BMC Genomics, 13*(1), 629.
Komatsu, S., & Hossain, Z. (Eds.), (2018). *Plant proteomic research*: MDPI.
Kopka, J., Fernie, A., Weckwerth, W., Gibon, Y., & Stitt, M. (2004). Metabolite profiling in plant biology: platforms and destinations. *Genome Biology, 5*(6), 109.
Kozarewa, I., Ning, Z., Quail, M. A., Sanders, M. J., Berriman, M., & Turner, D. J. (2009). Amplification-free Illumina sequencing-library preparation facilitates improved mapping and assembly of (G+C)-biased genomes. *Nature Methods, 6*(4), 291.
Krebs, J. R., & Davies, N. B. (Eds.), (2009). *Behavioural ecology: an evolutionary approach*: John Wiley & Sons.
Kuhlisch, C., & Pohnert, G. (2015). Metabolomics in chemical ecology. *Natural Product Reports, 32*(7), 937–955.
Kuhn, R., & Hoffstetter-Kuhn, S. (2013). *Capillary electrophoresis: principles and practice*. Springer Science & Business Media.
Kumar, R., Ichihashi, Y., Kimura, S., Chitwood, D. H., Headland, L. R., Peng, J., et al. (2012). A high-throughput method for Illumina RNA-Seq library preparation. *Frontiers in Plant Science, 3*, 202.
Kusari, S., & Spiteller, M. (2012). Metabolomics of endophytic fungi producing associated plant secondary metabolites: progress, challenges and opportunities. In *Metabolomics*: InTech.
Łabaj, P. P., Leparc, G. G., Linggi, B. E., Markillie, L. M., Wiley, H. S., & Kreil, D. P. (2011). Characterization and improvement of RNA-Seq precision in quantitative transcript expression profiling. *Bioinformatics, 27*(13), i383–i391.

Lagarde, J., Uszczynska-Ratajczak, B., Carbonell, S., Perez-Lluch, S., Abad, A., Davis, C., et al. (2017). High-throughput annotation of full-length long noncoding RNAs with capture long-read sequencing. *Nature Genetics*, *49*(12), 1731.

Lagarde, J., Uszczynska-Ratajczak, B., Santoyo-Lopez, J., Gonzalez, J. M., Tapanari, E., Mudge, J. M., et al. (2016). Extension of human lncRNA transcripts by RACE coupled with long-read high-throughput sequencing (RACE-Seq). *Nature Communications*, 7. 12339.

Lai, Z. W., Yan, Y., Caruso, F., & Nice, E. C. (2012). Emerging techniques in proteomics for probing nano–bio interactions. *ACS Nano*, *6*(12), 10438–10448.

Lakshmipathy, U., Love, B., Adams, C., Thyagarajan, B., & Chesnut, J. D. (2007). Micro RNA profiling. In *Stem cell assays* (pp. 97–114): Humana Press.

Landt, S. G., Marinov, G. K., Kundaje, A., Kheradpour, P., Pauli, F., Batzoglou, S., et al. (2012). ChIP-seq guidelines and practices of the ENCODE and modENCODE consortia. *Genome Research*, *22*(9), 1813–1831.

Lange, V., Picotti, P., Domon, B., & Aebersold, R. (2008). Selected reaction monitoring for quantitative proteomics: a tutorial. *Molecular Systems Biology*, *4*(1), 222.

Langmead, B., Trapnell, C., Pop, M., & Salzberg, S. L. (2009). Ultrafast and memory-efficient alignment of short DNA sequences to the human genome. *Genome Biology*, *10*(3), R25.

Lapainis, T., Rubakhin, S. S., & Sweedler, J. V. (2009). Capillary electrophoresis with electrospray ionization mass spectrometric detection for single-cell metabolomics. *Analytical Chemistry*, *81*(14), 5858–5864.

Law, K. P., & Lim, Y. P. (2013). Recent advances in mass spectrometry: data independent analysis and hyper reaction monitoring. *Expert Review of Proteomics*, *10*(6), 551–566.

Legendre, M., Pochet, N., Pak, T., & Verstrepen, K. J. (2007). Sequence-based estimation of minisatellite and microsatellite repeat variability. *Genome Research*, *17*(12), 000.

Lehr, M. A., Kilpatrick, C. W., Wilkerson, R. C., & Conn, J. E. (2014). Cryptic species in the Anopheles (Nyssorhynchus) albitarsis (Diptera: Culicidae) complex: incongruence between random amplified polymorphic DNA-polymerase chain reaction identification and analysis of mitochondrial DNA COI gene sequences. *Annals of the Entomological Society of America*, *98*(6), 908–917.

Lei, Z., Huhman, D., & Sumner, L. W. (2011). Mass spectrometry strategies in metabolomics. *Journal of Biological Chemistry*. jbc-R111.

Leiss, K. A., Choi, Y. H., Abdel-Farid, I. B., Verpoorte, R., & Klinkhamer, P. G. (2009). NMR metabolomics of thrips (*Frankliniella occidentalis*) resistance in Senecio hybrids. *Journal of Chemical Ecology*, *35*(2), 219–229.

Leviatan, N., Alkan, N., Leshkowitz, D., & Fluhr, R. (2013). Genome-wide survey of cold stress regulated alternative splicing in *Arabidopsis thaliana* with tiling microarray. *PLoS One*, *8*(6). e66511.

Li, B., & Dewey, C. N. (2011). RSEM: accurate transcript quantification from RNA-Seq data with or without a reference genome. *BMC Bioinformatics*, *12*(1), 323.

Li, J., Ding, Q., Wang, F., Li, H., Zhang, Y., Liu, L., et al. (2018). Genome-wide gene expression profiles in response to downy mildew in Chinese cabbage (*Brassica rapa* L. ssp. *pekinensis*). *European Journal of Plant Pathology*, 1–13.

Li, Y., Wang, D., Li, Z., Wei, J., Jin, C., & Liu, M. (2014). A molecular genetic linkage map of *Eucommia ulmoides* and quantitative trait loci (QTL) analysis for growth traits. *International Journal of Molecular Sciences*, *15*(2), 2053–2074.

Li, Z., Huang, C., Bao, C., Chen, L., Lin, M., Wang, X., et al. (2015). Exon-intron circular RNAs regulate transcription in the nucleus. *Nature Structural & Molecular Biology*, *22*(3), 256.

Liland, K. H. (2011). Multivariate methods in metabolomics–from pre-processing to dimension reduction and statistical analysis. *TrAC Trends in Analytical Chemistry*, *30*(6), 827–841.

Lindon, J. C., & Nicholson, J. K. (2008). Spectroscopic and statistical techniques for information recovery in metabonomics and metabolomics. *Annual Review of Analytical Chemistry*, *1*, 45–69.

Lindon, J. C., Nicholson, J. K., & Holmes, E. (Eds.), (2011). *The handbook of metabonomics and metabolomics*. Elsevier.

Lippman, Z., Gendrel, A. V., Colot, V., & Martienssen, R. (2005). Profiling DNA methylation patterns using genomic tiling microarrays. *Nature Methods*, *2*(3), 219.

Lister, R., Gregory, B. D., & Ecker, J. R. (2009). Next is now: new technologies for sequencing of genomes, transcriptomes, and beyond. *Current Opinion in Plant Biology*, *12*(2), 107–118.

Liu, X., Ahlgren, S., Korthout, H. A., Salomé-Abarca, L. F., Bayona, L. M., Verpoorte, R., et al. (2018). Broad range chemical profiling of natural deep eutectic solvent extracts using a high performance thin layer chromatography–based method. *Journal of Chromatography A*, *1532*, 198–207.

Lu, J., Gosslau, A., Liu, A. Y. C., & Chen, K. Y. (2008). PCR differential display-based identification of regulator of G protein signaling 10 as the target gene in human colon cancer cells induced by black tea polyphenol theaflavin monogallate. *European Journal of Pharmacology*, *601*(1-3), 66–72.

Lu, J. J., Wang, S., Li, G., Wang, W., Pu, Q., & Liu, S. (2012). Chip-capillary hybrid device for automated transfer of sample preseparated by capillary isoelectric focusing to parallel capillary gel electrophoresis for two-dimensional protein separation. *Analytical Chemistry*, *84*(16), 7001–7007.

Lu, W., Clasquin, M. F., Melamud, E., Amador-Noguez, D., Caudy, A. A., & Rabinowitz, J. D. (2010). Metabolomic analysis via reversed-phase ion-pairing liquid chromatography coupled to a stand alone orbitrap mass spectrometer. *Analytical Chemistry*, *82*(8), 3212–3221.

Luo, J., Vijayasankaran, N., Autsen, J., Santuray, R., Hudson, T., Amanullah, A., et al. (2012). Comparative metabolite analysis to understand lactate metabolism shift in Chinese hamster ovary cell culture process. *Biotechnology and Bioengineering*, *109*(1), 146–156.

Luo, W., & Brouwer, C. (2013). Pathview: an R/Bioconductor package for pathway-based data integration and visualization. *Bioinformatics*, *29*(14), 1830–1831.

Macel, M., van Dam, N. M., & Keurentjes, J. J. (2010). Metabolomics: the chemistry between ecology and genetics. *Molecular Ecology Resources*, *10*(4), 583–593.

Macosko, E. Z., Basu, A., Satija, R., Nemesh, J., Shekhar, K., Goldman, M., et al. (2015). Highly parallel genome-wide expression profiling of individual cells using nanoliter droplets. *Cell*, *161*(5), 1202–1214.

Magdeldin, S., Enany, S., Yoshida, Y., Xu, B., Zhang, Y., Zureena, Z., et al. (2014). Basics and recent advances of two dimensional-polyacrylamide gel electrophoresis. *Clinical Proteomics*, *11*(1), 16.

Marguerat, S., & Bähler, J. (2010). RNA-seq: from technology to biology. *Cellular and Molecular Life Sciences*, *67*(4), 569–579.

Marioni, J. C., Mason, C. E., Mane, S. M., Stephens, M., & Gilad, Y. (2008). RNA-seq: an assessment of technical reproducibility and comparison with gene expression arrays. *Genome Research*.

Marshall, A. G., Hendrickson, C. L., & Jackson, G. S. (2006). Fourier transform ion cyclotron resonance mass spectrometry. In *Encyclopedia of analytical chemistry: applications, theory and instrumentation*.

Martinez, N. J., & Gregory, R. I. (2010). MicroRNA gene regulatory pathways in the establishment and maintenance of ESC identity. *Cell Stem Cell*, 7(1), 31–35.

Maštovská, K., & Lehotay, S. J. (2003). Practical approaches to fast gas chromatography–mass spectrometry. *Journal of Chromatography A*, *1000*(1-2), 153–180.

Matukumalli, L. K., Lawley, C. T., Schnabel, R. D., Taylor, J. F., Allan, M. F., Heaton, M. P., et al. (2009). Development and characterization of a high density SNP genotyping assay for cattle. *PLoS One*, *4*(4)e5350.

McCarroll, S. A., Kuruvilla, F. G., Korn, J. M., Cawley, S., Nemesh, J., Wysoker, A., et al. (2008). Integrated detection and population-genetic analysis of SNPs and copy number variation. *Nature Genetics*, *40*(10), 1166.

McCarthy, D. J., Chen, Y., & Smyth, G. K. (2012). Differential expression analysis of multifactor RNA-Seq experiments with respect to biological variation. *Nucleic Acids Research*, *40*, 4288–4297.

Mehl, F., Marti, G., Boccard, J., Debrus, B., Merle, P., Delort, E., et al. (2014). Differentiation of lemon essential oil based on volatile and non-volatile fractions with various analytical techniques: a metabolomic approach. *Food Chemistry*, *143*, 325–335.

Mehmood, T., Liland, K. H., Snipen, L., & Sæbø, S. (2012). A review of variable selection methods in partial least squares regression. *Chemometrics and Intelligent Laboratory Systems*, *118*, 62–69.

Melles, D. C., van Leeuwen, W. B., Snijders, S. V., Horst-Kreft, D., Peeters, J. K., Verbrugh, H. A., et al. (2007). Comparison of multilocus sequence typing (MLST), pulsed-field gel electrophoresis (PFGE), and amplified fragment length polymorphism (AFLP) for genetic typing of Staphylococcus aureus. *Journal of Microbiological Methods*, *69*(2), 371–375.

Mercer, T. R., Gerhardt, D. J., Dinger, M. E., Crawford, J., Trapnell, C., Jeddeloh, J. A., et al. (2012). Targeted RNA sequencing reveals the deep complexity of the human transcriptome. *Nature Biotechnology*, *30*(1), 99.

Mestdagh, P., Van Vlierberghe, P., De Weer, A., Muth, D., Westermann, F., Speleman, F., et al. (2009). A novel and universal method for microRNA RT-qPCR data normalization. *Genome Biology*, *10*(6), R64.

Metallo, C. M., & Vander Heiden, M. G. (2013). Understanding metabolic regulation and its influence on cell physiology. *Molecular Cell*, *49*(3), 388–398.

Metzker, M. L. (2010). Sequencing technologies—the next generation. *Nature Reviews Genetics*, *11*(1), 31.

Meudt, H. M., & Clarke, A. C. (2007). Almost forgotten or latest practice? AFLP applications, analyses and advances. *Trends in Plant Science*, *12*(3), 106–117.

Micheli, M. R., & Bova, R. (Eds.), (2013). *Fingerprinting methods based on arbitrarily primed PCR*: Springer Science & Business Media.

Mitra, K., Carvunis, A. R., Ramesh, S. K., & Ideker, T. (2013). Integrative approaches for finding modular structure in biological networks. *Nature Reviews Genetics*, *14*(10), 719.

Mladenov, P., Finazzi, G., Bligny, R., Moyankova, D., Zasheva, D., Boisson, A. M., et al. (2015). In vivo spectroscopy and NMR metabolite fingerprinting approaches to connect the dynamics of photosynthetic and metabolic phenotypes in resurrection plant Haberlea rhodopensis during desiccation and recovery. *Frontiers in Plant Science*, *6*, 564.

Mohanty, A. K., Yadav, M. L., & Choudhary, S. (2017). Gel electrophoresis of proteins and nucleic acids. In *Protocols in semen biology (comparing assays)* (pp. 233–246). Singapore: Springer.

Molnár-Perl, I. (Ed.), (2005). Vol. 70. *Quantitation of amino acids and amines by chromatography: Methods and protocols*: Elsevier.

Monton, M. R. N., & Soga, T. (2007). Metabolome analysis by capillary electrophoresis–mass spectrometry. *Journal of Chromatography A*, *1168*(1-2), 237–246.

Montpetit, A., Nelis, M., Laflamme, P., Magi, R., Ke, X., Remm, M., et al. (2006). An evaluation of the performance of tag SNPs derived from HapMap in a Caucasian population. *PLoS Genetics*, *2*(3). e27.

Mortazavi, A., Williams, B. A., McCue, K., Schaeffer, L., & Wold, B. (2008). Mapping and quantifying mammalian transcriptomes by RNA-Seq. *Nature Methods*, *5*(7), 621.

Motta, A., Paris, D., Melck, D., De Laurentiis, G., Maniscalco, M., Sofia, M., et al. (2012). Nuclear magnetic resonance-based metabolomics of exhaled breath condensate: methodological aspects. *European Respiratory Journal*, *39*(2), 498–500.

Motto, M. (2017). Genetic tools for crop improvement: past, present, and future. In *More food: road to survival* (p. 39). .

Mullighan, C. G., Phillips, L. A., Su, X., Ma, J., Miller, C. B., Shurtleff, S. A., et al. (2008). Genomic analysis of the clonal origins of relapsed acute lymphoblastic leukemia. *Science*, *322*(5906), 1377–1380.

Muys, C., Thienpont, C. N., Dauchot, N., Maudoux, O., Draye, X., & Cutsem, P. V. (2014). Integration of AFLP s, SSR s and SNP s markers into a new genetic map of industrial chicory (*Cichorium intybus* L. var. *sativum*). *Plant Breeding*, *133*(1), 130–137.

Muzzalupo, I., Pisani, F., Greco, F., & Chiappetta, A. (2015). Direct DNA amplification from virgin olive oil for traceability and authenticity. *European Food Research and Technology*, *241*(1), 151–155.

Nagalakshmi, U., Wang, Z., Waern, K., Shou, C., Raha, D., Gerstein, M., et al. (2008). The transcriptional landscape of the yeast genome defined by RNA sequencing. *Science*, *320* (5881), 1344–1349.

Nathan, R., Getz, W. M., Revilla, E., Holyoak, M., Kadmon, R., Saltz, D., et al. (2008). A movement ecology paradigm for unifying organismal movement research. *Proceedings of the National Academy of Sciences*, *105*(49), 19052–19059.

Neeha, V. S., & Kinth, P. (2013). Nutrigenomics research: a review. *Journal of Food Science and Technology*, *50*(3), 415–428.

Nejat, N., & Mantri, N. (2018). Emerging roles of long non-coding RNAs in plant response to biotic and abiotic stresses. *Critical Reviews in Biotechnology*, *38*(1), 93–105.

Neusüb, C., Römer, J., Höcker, O., & Jooß, K. (2018). Coupling of capillary electromigration techniques to mass spectrometry. In *Capillary electromigration separation methods* (pp. 293–311).

Nielsen, J., & Oliver, S. (2005). The next wave in metabolome analysis. *Trends in Biotechnology*, *23*(11), 544–546.

Nielsen, K. L., Godfrey, P. A., Stegger, M., Andersen, P. S., Feldgarden, M., & Frimodt-Møller, N. (2014). Selection of unique *Escherichia coli* clones by random amplified polymorphic DNA (RAPD): evaluation by whole genome sequencing. *Journal of Microbiological Methods*, *103*, 101–103.

Niu, W., Knight, E., Xia, Q., & McGarvey, B. D. (2014). Comparative evaluation of eight software programs for alignment of gas chromatography–mass spectrometry chromatograms in metabolomics experiments. *Journal of Chromatography A*, *1374*, 199–206.

Nookaew, I., Papini, M., Pornputtapong, N., Scalcinati, G., Fagerberg, L., Uhlén, M., et al. (2012). A comprehensive comparison of RNA-Seq-based transcriptome analysis from reads to differential gene expression and cross-comparison with microarrays: a case study in *Saccharomyces cerevisiae*. *Nucleic Acids Research*, *40*(20), 10084–10097.

Norgaard, R. B. (2006). *Development betrayed: the end of progress and a co-evolutionary revisioning of the future*. Routledge.

Nowrousian, M. (2010). Next-generation sequencing techniques for eukaryotic microorganisms: sequencing-based solutions to biological problems. *Eukaryotic Cell*, *9*(9), 1300–1310.

Nyamundanda, G., Brennan, L., & Gormley, I. C. (2010). Probabilistic principal component analysis for metabolomic data. *BMC Bioinformatics*, *11*(1), 571.

Nybom, H., Weising, K., & Rotter, B. (2014). DNA fingerprinting in botany: past, present, future. *Investigative Genetics*, *5*(1), 1.

Obata, T., & Fernie, A. R. (2012). The use of metabolomics to dissect plant responses to abiotic stresses. *Cellular and Molecular Life Sciences*, *69*(19), 3225–3243.

Okou, D. T., Steinberg, K. M., Middle, C., Cutler, D. J., Albert, T. J., & Zwick, M. E. (2007). Microarray-based genomic selection for high-throughput resequencing. *Nature Methods*, *4*(11), 907.

Oksman-Caldentey, K. M., & Saito, K. (2005). Integrating genomics and metabolomics for engineering plant metabolic pathways. *Current Opinion in Biotechnology*, *16*(2), 174–179.

O'Rourke, J. A., Yang, S. S., Miller, S. S., Bucciarelli, B., Liu, J., Rydeen, A., et al. (2012). An RNA-Seq transcriptome analysis of Pi deficient white lupin reveals novel insights into phosphorus acclimation in plants. *Plant Physiology*, *161*(2), 705–724.

Oshlack, A., Robinson, M. D., & Young, M. D. (2010). From RNA-seq reads to differential expression results. *Genome Biology*, *11*(12), 220.

OuYang, D., Xu, J., Huang, H., & Chen, Z. (2011). Metabolomic profiling of serum from human pancreatic cancer patients using 1 H NMR spectroscopy and principal component analysis. *Applied Biochemistry and Biotechnology*, *165*(1), 148–154.

Ozsolak, F., & Milos, P. M. (2011). RNA sequencing: advances, challenges and opportunities. *Nature Reviews Genetics*, *12*(2), 87.

Paglia, G., Williams, J. P., Menikarachchi, L., Thompson, J. W., Tyldesley-Worster, R., Halldórsson, S., et al. (2014). Ion mobility derived collision cross sections to support metabolomics applications. *Analytical Chemistry*, *86*(8), 3985–3993.

Pareek, C. S., Smoczynski, R., & Tretyn, A. (2011). Sequencing technologies and genome sequencing. *Journal of Applied Genetics*, *52*(4), 413–435.

Parisi, A., Latorre, L., Normanno, G., Miccolupo, A., Fraccalvieri, R., Lorusso, V., et al. (2010). Amplified fragment length polymorphism and multi-locus sequence typing for high-resolution genotyping of Listeria monocytogenes from foods and the environment. *Food Microbiology*, *27*(1), 101–108.

Passari, A. K., Mishra, V. K., Gupta, V. K., Yadav, M. K., Saikia, R., & Singh, B. P. (2015). In vitro and in vivo plant growth promoting activities and DNA fingerprinting of antagonistic endophytic actinomycetes associates with medicinal plants. *PLoS One*, *10*(9). e0139468.

Patterson, A. D., Li, H., Eichler, G. S., Krausz, K. W., Weinstein, J. N., Fornace, A. J., et al. (2008). UPLC-ESI-TOFMS-based metabolomics and gene expression dynamics inspector self-organizing metabolomic maps as tools for understanding the cellular response to ionizing radiation. *Analytical Chemistry*, *80*(3), 665–674.

Patti, G. J., Yanes, O., & Siuzdak, G. (2012). Innovation: metabolomics: the apogee of the omics trilogy. *Nature Reviews Molecular Cell Biology*, *13*(4), 263.

Payyavula, R. S., Navarre, D. A., Kuhl, J. C., Pantoja, A., & Pillai, S. S. (2012). Differential effects of environment on potato phenylpropanoid and carotenoid expression. *BMC Plant Biology*, *12*(1), 39.

Pechlivanis, A., Kostidis, S., Saraslanidis, P., Petridou, A., Tsalis, G., Mougios, V., et al. (2010). 1H NMR-based metabonomic investigation of the effect of two different exercise sessions on the metabolic fingerprint of human urine. *Journal of Proteome Research*, *9*(12), 6405–6416.

Petrakis, E. A., Cagliani, L. R., Polissiou, M. G., & Consonni, R. (2015). Evaluation of saffron (*Crocus sativus* L.) adulteration with plant adulterants by 1H NMR metabolite fingerprinting. *Food Chemistry*, *173*, 890–896.

Pfeifer, D., Pantic, M., Skatulla, I., Rawluk, J., Kreutz, C., Martens, U. M., et al. (2007). Genome-wide analysis of DNA copy number changes and LOH in CLL using high-density SNP arrays. *Blood*, *109*(3), 1202–1210.

Poelchau, M. F., Reynolds, J. A., Denlinger, D. L., Elsik, C. G., & Armbruster, P. A. (2011). A de novo transcriptome of the Asian tiger mosquito, *Aedes albopictus*, to identify candidate transcripts for diapause preparation. *BMC Genomics*, *12*(1), 619.

Poland, J. A., Brown, P. J., Sorrells, M. E., & Jannink, J. L. (2012). Development of high-density genetic maps for barley and wheat using a novel two-enzyme genotyping-by-sequencing approach. *PLoS One*, 7(2). e32253.

Pomastowski, P., & Buszewski, B. (2014). Two-dimensional gel electrophoresis in the light of new developments. *TrAC Trends in Analytical Chemistry*, *53*, 167–177.

Porzel, A., Farag, M. A., Mülbradt, J., & Wessjohann, L. A. (2014). Metabolite profiling and fingerprinting of Hypericum species: a comparison of MS and NMR metabolomics. *Metabolomics*, *10*(4), 574–588.

Pritchard, C. C., Cheng, H. H., & Tewari, M. (2012). MicroRNA profiling: approaches and considerations. *Nature Reviews Genetics*, *13*(5), 358.

Prosser, G. A., Larrouy-Maumus, G., & de Carvalho, L. P. S. (2014). Metabolomic strategies for the identification of new enzyme functions and metabolic pathways. *EMBO Reports*, *15*(6), 657–669.

Putri, S. P., Nakayama, Y., Matsuda, F., Uchikata, T., Kobayashi, S., Matsubara, A., et al. (2013). Current metabolomics: practical applications. *Journal of Bioscience and Bioengineering*, *115*(6), 579–589.

Qiao, L., Tobolkina, E., Liu, B., & Girault, H. H. (2013). Coupling isoelectric focusing gel electrophoresis to mass spectrometry by electrostatic spray ionization. *Analytical Chemistry*, *85*(9), 4745–4752.

Quanbeck, S. M. M., Brachova, L., Campbell, A. A., Guan, X., Perera, A., He, K., et al. (2012). Metabolomics as a hypothesis-generating functional genomics tool for the annotation of *Arabidopsis thaliana* genes of "unknown function". *Frontiers in Plant Science*, *3*, 15.

Rabilloud, T. (Ed.), (2013). *Proteome research: two-dimensional gel electrophoresis and identification methods*: Springer Science & Business Media.

Rabilloud, T., Chevallet, M., Luche, S., & Lelong, C. (2010). Two-dimensional gel electrophoresis in proteomics: past, present and future. *Journal of Proteomics*, *73*(11), 2064–2077.

Rademaker, J. L., Hoste, B., Louws, F. J., Kersters, K., Swings, J., Vauterin, L., et al. (2000). Comparison of AFLP and rep-PCR genomic fingerprinting with DNA-DNA homology studies: Xanthomonas as a model system. *International Journal of Systematic and Evolutionary Microbiology*, *50*(2), 665–677.

Ramautar, R., Busnel, J. M., Deelder, A. M., & Mayboroda, O. A. (2011). Enhancing the coverage of the urinary metabolome by sheathless capillary electrophoresis-mass spectrometry. *Analytical Chemistry*, *84*(2), 885–892.

Ramautar, R., Heemskerk, A. A., Hensbergen, P. J., Deelder, A. M., Busnel, J. M., & Mayboroda, O. A. (2012). CE–MS for proteomics: advances in interface development and application. *Journal of Proteomics*, *75*(13), 3814–3828.

Ramautar, R., Somsen, G. W., & de Jong, G. J. (2013). CE-MS for metabolomics: Developments and applications in the period 2010–2012. *Electrophoresis*, *34*(1), 86–98.

Rasmussen, H. B. (2012). Restriction fragment length polymorphism analysis of PCR-amplified fragments (PCR-RFLP) and gel electrophoresis-valuable tool for genotyping and genetic fingerprinting. In *Gel electrophoresis-principles and basics*: InTech.

Renaut, S., Grassa, C. J., Moyers, B. T., Kane, N. C., & Rieseberg, L. H. (2012). The population genomics of sunflowers and genomic determinants of protein evolution revealed by RNAseq. *Biology*, *1*(3), 575–596.

Reuter, J. A., Spacek, D. V., & Snyder, M. P. (2015). High-throughput sequencing technologies. *Molecular Cell*, *58*(4), 586–597.

Riedelsheimer, C., Lisec, J., Czedik-Eysenberg, A., Sulpice, R., Flis, A., Grieder, C., et al. (2012). Genome-wide association mapping of leaf metabolic profiles for dissecting complex traits in maize. *Proceedings of the National Academy of Sciences*, *109*(23), 8872–8877.

Righetti, P. G., Sebastiano, R., & Citterio, A. (2013). Capillary electrophoresis and isoelectric focusing in peptide and protein analysis. *Proteomics*, *13*(2), 325–340.

Rimmer, A., Phan, H., Mathieson, I., Iqbal, Z., Twigg, S. R., Wilkie, A. O., et al. (2014). Integrating mapping-, assembly-and haplotype-based approaches for calling variants in clinical sequencing applications. *Nature Genetics*, *46*(8), 912.

Risticevic, S., Souza-Silva, E. A., DeEll, J. R., Cochran, J., & Pawliszyn, J. (2015). Capturing plant metabolome with direct-immersion in vivo solid phase microextraction of plant tissues. *Analytical Chemistry, 88*(2), 1266–1274.

Rocha, S. M., Caldeira, M., Carrola, J., Santos, M., Cruz, N., & Duarte, I. F. (2012). Exploring the human urine metabolomic potentialities by comprehensive two-dimensional gas chromatography coupled to time of flight mass spectrometry. *Journal of Chromatography A, 1252*, 155–163.

Rogowska-Wrzesinska, A., Le Bihan, M. C., Thaysen-Andersen, M., & Roepstorff, P. (2013). 2D gels still have a niche in proteomics. *Journal of Proteomics, 88*, 4–13.

Ronald, J., Akey, J. M., Whittle, J., Smith, E. N., Yvert, G., & Kruglyak, L. (2005). Simultaneous genotyping, gene-expression measurement, and detection of allele-specific expression with oligonucleotide arrays. *Genome Research, 15*(2), 284–291.

Rotger, M., Dang, K. K., Fellay, J., Heinzen, E. L., Feng, S., Descombes, P., et al. (2010). Genome-wide mRNA expression correlates of viral control in CD4+T-cells from HIV-1-infected individuals. *PLoS Pathogens, 6*(2). e1000781.

Roux, A., Lison, D., Junot, C., & Heilier, J. F. (2011). Applications of liquid chromatography coupled to mass spectrometry-based metabolomics in clinical chemistry and toxicology: A review. *Clinical Biochemistry, 44*(1), 119–135.

Ruan, Y., Gilmore, J., & Conner, T. (1998). Towards *Arabidopsis* genome analysis: monitoring expression profiles of 1400 genes using cDNA microarrays. *The Plant Journal, 15*(6), 821–833.

Rudd, S. (2003). Expressed sequence tags: alternative or complement to whole genome sequences? *Trends in Plant Science, 8*(7), 321–329.

Rzhetsky, A., Iossifov, I., Koike, T., Krauthammer, M., Kra, P., Morris, M., et al. (2004). GeneWays: a system for extracting, analyzing, visualizing, and integrating molecular pathway data. *Journal of Biomedical Informatics, 37*(1), 43–53.

Safaei-Ghomi, J., Ebrahimabadi, A. H., Djafari-Bidgoli, Z., & Batooli, H. (2009). GC/MS analysis and in vitro antioxidant activity of essential oil and methanol extracts of *Thymus caramanicus* Jalas and its main constituent carvacrol. *Food Chemistry, 115*(4), 1524–1528.

Saini, R. K., & Keum, Y. S. (2018). Significance of genetic, environmental, and pre-and post-harvest factors affecting carotenoid contents in crops: a review. *Journal of Agricultural and Food Chemistry*.

Saito, K., & Matsuda, F. (2010). Metabolomics for functional genomics, systems biology, and biotechnology. *Annual Review of Plant Biology, 61*, 463–489.

Saliba, A. E., Westermann, A. J., Gorski, S. A., & Vogel, J. (2014). Single-cell RNA-seq: advances and future challenges. *Nucleic Acids Research, 42*(14), 8845–8860.

Salzman, R. A., Brady, J. A., Finlayson, S. A., Buchanan, C. D., Summer, E. J., Sun, F., et al. (2005). Transcriptional profiling of sorghum induced by methyl jasmonate, salicylic acid, and aminocyclopropane carboxylic acid reveals cooperative regulation and novel gene responses. *Plant Physiology, 138*(1), 352–368.

Sanford, K., Soucaille, P., Whited, G., & Chotani, G. (2002). Genomics to fluxomics and physiomics—pathway engineering. *Current Opinion in Microbiology, 5*(3), 318–322.

Sasidharan, R., Agarwal, A., Rozowsky, J., & Gerstein, M. (2009). An approach to comparing tiling array and high throughput sequencing technologies for genomic transcript mapping. *BMC Research Notes, 2*(1), 150.

Savelkoul, P. H. M., Aarts, H. J. M., De Haas, J., Dijkshoorn, L., Duim, B., Otsen, M., et al. (1999). Amplified-fragment length polymorphism analysis: the state of an art. *Journal of Clinical Microbiology, 37*(10), 3083–3091.

Scalbert, A., Brennan, L., Fiehn, O., Hankemeier, T., Kristal, B. S., van Ommen, B., et al. (2009). Mass-spectrometry-based metabolomics: limitations and recommendations for future progress with particular focus on nutrition research. *Metabolomics, 5*(4), 435.

Schauer, N., & Fernie, A. R. (2006). Plant metabolomics: towards biological function and mechanism. *Trends in Plant Science, 11*(10), 508–516.

Schauer, N., Semel, Y., Roessner, U., Gur, A., Balbo, I., Carrari, F., et al. (2006). Comprehensive metabolic profiling and phenotyping of interspecific introgression lines for tomato improvement. *Nature Biotechnology*, *24*(4), 447.

Scherling, C., Roscher, C., Giavalisco, P., Schulze, E. D., & Weckwerth, W. (2010). Metabolomics unravel contrasting effects of biodiversity on the performance of individual plant species. *PLoS One*, *5*(9). e12569.

Schripsema, J. (2010). Application of NMR in plant metabolomics: techniques, problems and prospects. *Phytochemical Analysis*, *21*(1), 14–21.

Seki, M., Narusaka, M., Kamiya, A., Ishida, J., Satou, M., Sakurai, T., et al. (2002). Functional annotation of a full-length *Arabidopsis* cDNA collection. *Science*, *296*(5565), 141–145.

Semerikova, S. A., & Semerikov, V. L. (2016). Phylogeny of firs (genus Abies, Pinaceae) based on multilocus nuclear markers (AFLP). *Russian Journal of Genetics*, *52*(11), 1164–1175.

Sha, W., Martins, A. M., Laubenbacher, R., Mendes, P., & Shulaev, V. (2013). The genome-wide early temporal response of *Saccharomyces cerevisiae* to oxidative stress induced by cumene hydroperoxide. *PLoS One*, *8*(9). e74939.

Sharma, D., & Bisht, D. (2016). An efficient and rapid method for enrichment of lipophilic proteins from *Mycobacterium tuberculosis* H37Rv for two-dimensional gel electrophoresis. *Electrophoresis*, *37*(9), 1187–1190.

Shimkets, R. A., Lowe, D. G., Tai, J. T. N., Sehl, P., Jin, H., Yang, R., et al. (1999). Gene expression analysis by transcript profiling coupled to a gene database query. *Nature Biotechnology*, *17*(8), 798.

Shintani, H. (2012). *Handbook of capillary electrophoresis applications*. Springer Science & Business Media.

Shruti, K., Shrey, K., & Vibha, R. (2011). Micro RNAs: tiny sequences with enormous potential. *Biochemical and Biophysical Research Communications*, *407*(3), 445–449.

Shulaev, V. (2006). Metabolomics technology and bioinformatics. *Briefings in Bioinformatics*, *7*(2), 128–139.

Silvente, S., Sobolev, A. P., & Lara, M. (2012). Metabolite adjustments in drought tolerant and sensitive soybean genotypes in response to water stress. *PLoS One*, *7*(6). e38554.

Simmler, C., Napolitano, J. G., McAlpine, J. B., Chen, S. N., & Pauli, G. F. (2014). Universal quantitative NMR analysis of complex natural samples. *Current Opinion in Biotechnology*, *25*, 51–59.

Smart, K. F., Aggio, R. B., Van Houtte, J. R., & Villas-Bôas, S. G. (2010). Analytical platform for metabolome analysis of microbial cells using methyl chloroformate derivatization followed by gas chromatography–mass spectrometry. *Nature Protocols*, *5*(10), 1709.

Smilde, A. K., Jansen, J. J., Hoefsloot, H. C., Lamers, R. J. A., Van Der Greef, J., & Timmerman, M. E. (2005). ANOVA-simultaneous component analysis (ASCA): a new tool for analyzing designed metabolomics data. *Bioinformatics*, *21*(13), 3043–3048.

Smita, S., Singh, K. P., Akhoon, B. A., Gupta, S. K., & Gupta, S. K. (2013). Bioinformatics tools for interpretation of data used in molecular identification. In *Analyzing microbes* (pp. 209–243). Berlin, Heidelberg: Springer.

Smolinska, A., Blanchet, L., Buydens, L. M., & Wijmenga, S. S. (2012). NMR and pattern recognition methods in metabolomics: from data acquisition to biomarker discovery: a review. *Analytica Chimica Acta*, *750*, 82–97.

Snijders, A. M., Nowak, N., Segraves, R., Blackwood, S., Brown, N., Conroy, J., et al. (2001). Assembly of microarrays for genome-wide measurement of DNA copy number. *Nature Genetics*, *29*(3), 263.

Soga, T., & Heiger, D. N. (2000). Amino acid analysis by capillary electrophoresis electrospray ionization mass spectrometry. *Analytical Chemistry*, *72*(6), 1236–1241.

Somervuo, P., Kvist, J., Ikonen, S., Auvinen, P., Paulin, L., Koskinen, P., et al. (2014). Transcriptome analysis reveals signature of adaptation to landscape fragmentation. *PLoS One*, *9*(7). e101467.

Spooner, D. M., McLean, K., Ramsay, G., Waugh, R., & Bryan, G. J. (2005). A single domestication for potato based on multilocus amplified fragment length polymorphism genotyping. *Proceedings of the National Academy of Sciences*, *102*(41), 14694–14699.

Su, Z., Łabaj, P. P., Li, S., Thierry-Mieg, J., Thierry-Mieg, D., Shi, W., et al. (2014). A comprehensive assessment of RNA-seq accuracy, reproducibility and information content by the Sequencing Quality Control Consortium. *Nature Biotechnology*, *32*(9), 903.

Sugimoto, M., Kawakami, M., Robert, M., Soga, T., & Tomita, M. (2012). Bioinformatics tools for mass spectroscopy-based metabolomic data processing and analysis. *Current Bioinformatics*, 7(1), 96–108.

Sugimoto, M., Wong, D. T., Hirayama, A., Soga, T., & Tomita, M. (2010). Capillary electrophoresis mass spectrometry-based saliva metabolomics identified oral, breast and pancreatic cancer-specific profiles. *Metabolomics*, *6*(1), 78–95.

Sultan, M., Schulz, M. H., Richard, H., Magen, A., Klingenhoff, A., Scherf, M., et al. (2008). A global view of gene activity and alternative splicing by deep sequencing of the human transcriptome. *Science*, *321*(5891), 956–960.

Sumner, L. W., Mendes, P., & Dixon, R. A. (2003). Plant metabolomics: large-scale phytochemistry in the functional genomics era. *Phytochemistry*, *62*(6), 817–836.

Sun, X., & Weckwerth, W. (2012). COVAIN: a toolbox for uni-and multivariate statistics, time-series and correlation network analysis and inverse estimation of the differential Jacobian from metabolomics covariance data. *Metabolomics*, *8*(1), 81–93.

Suo, B., He, Y., Paoli, G., Gehring, A., Tu, S. I., & Shi, X. (2010). Development of an oligonucleotide-based microarray to detect multiple foodborne pathogens. *Molecular and Cellular Probes*, *24*(2), 77–86.

Syahir, A., Usui, K., Tomizaki, K. Y., Kajikawa, K., & Mihara, H. (2015). Label and label-free detection techniques for protein microarrays. *Microarrays*, *4*(2), 228–244.

Szymańska, E., Saccenti, E., Smilde, A. K., & Westerhuis, J. A. (2012). Double-check: validation of diagnostic statistics for PLS-DA models in metabolomics studies. *Metabolomics*, *8*(1), 3–16.

Tanaka, H., Miyagawa, Y., Tsujimura, A., Matsumiya, K., Okuyama, A., & Nishimune, Y. (2003). Single nucleotide polymorphisms in the protamine-1 and-2 genes of fertile and infertile human male populations. *Molecular Human Reproduction*, *9*(2), 69–73.

Tang, G., Galili, G., & Zhuang, X. (2007). RNAi and microRNA: breakthrough technologies for the improvement of plant nutritional value and metabolic engineering. *Metabolomics*, *3*(3), 357–369.

Tao, W. A., & Aebersold, R. (2003). Advances in quantitative proteomics via stable isotope tagging and mass spectrometry. *Current Opinion in Biotechnology*, *14*(1), 110–118.

Terova, G., Pisanu, S., Roggio, T., Preziosa, E., Saroglia, M., & Addis, M. F. (2014). Proteomic profiling of sea bass muscle by two-dimensional gel electrophoresis and tandem mass spectrometry. *Fish Physiology and Biochemistry*, *40*(1), 311–322.

Theodoridis, G., Gika, H. G., & Wilson, I. D. (2011). Mass spectrometry-based holistic analytical approaches for metabolite profiling in systems biology studies. *Mass Spectrometry Reviews*, *30*(5), 884–906.

Theodoridis, G. A., Gika, H. G., Want, E. J., & Wilson, I. D. (2012). Liquid chromatography–mass spectrometry based global metabolite profiling: a review. *Analytica Chimica Acta*, *711*, 7–16.

Thomassen, Y. E., van Eikenhorst, G., van der Pol, L. A., & Bakker, W. A. (2013). Isoelectric point determination of live polioviruses by capillary isoelectric focusing with whole column imaging detection. *Analytical Chemistry*, *85*(12), 6089–6094.

Thompson, A., Schäfer, J., Kuhn, K., Kienle, S., Schwarz, J., Schmidt, G., et al. (2003). Tandem mass tags: a novel quantification strategy for comparative analysis of complex protein mixtures by MS/MS. *Analytical Chemistry*, *75*(8), 1895–1904.
Thorisson, G. A., Smith, A. V., Krishnan, L., & Stein, L. D. (2005). The international HapMap project web site. *Genome Research*, *15*(11), 1592–1593.
t'Kindt, R., Scheltema, R. A., Jankevics, A., Brunker, K., Rijal, S., Dujardin, J. C., et al. (2010). Metabolomics to unveil and understand phenotypic diversity between pathogen populations. *PLoS Neglected Tropical Diseases*, *4*(11). e904.
Tohge, T., & Fernie, A. R. (2010). Combining genetic diversity, informatics and metabolomics to facilitate annotation of plant gene function. *Nature Protocols*, *5*(6), 1210.
Tolstikov, V. V., Lommen, A., Nakanishi, K., Tanaka, N., & Fiehn, O. (2003). Monolithic silica-based capillary reversed-phase liquid chromatography/electrospray mass spectrometry for plant metabolomics. *Analytical Chemistry*, *75*(23), 6737–6740.
Tran, T. N., Afanador, N. L., Buydens, L. M., & Blanchet, L. (2014). Interpretation of variable importance in partial least squares with significance multivariate correlation (sMC). *Chemometrics and Intelligent Laboratory Systems*, *138*, 153–160.
Trapnell, C., Hendrickson, D. G., Sauvageau, M., Goff, L., Rinn, J. L., & Pachter, L. (2013). Differential analysis of gene regulation at transcript resolution with RNA-seq. *Nature Biotechnology*, *31*(1), 46.
Trapnell, C., Pachter, L., & Salzberg, S. L. (2009). TopHat: discovering splice junctions with RNA-Seq. *Bioinformatics*, *25*(9), 1105–1111.
Trapnell, C., Williams, B. A., Pertea, G., Mortazavi, A., Kwan, G., Van Baren, M. J., et al. (2010). Transcript assembly and quantification by RNA-Seq reveals unannotated transcripts and isoform switching during cell differentiation. *Nature Biotechnology*, *28*(5), 511.
Trygg, J., & Wold, S. (2002). Orthogonal projections to latent structures (O-PLS). *Journal of Chemometrics*, *16*(3), 119–128.
Trynka, G., Hunt, K. A., Bockett, N. A., Romanos, J., Mistry, V., Szperl, A., et al. (2011). Dense genotyping identifies and localizes multiple common and rare variant association signals in celiac disease. *Nature Genetics*, *43*(12), 1193.
Tsehaye, H., Elameen, A., Tronsmo, A. M., Sundheim, L., Tronsmo, A., Assefa, D., et al. (2016). Genetic variation among Fusarium verticillioides isolates associated with Ethiopian maize kernels as revealed by AFLP analysis. *European Journal of Plant Pathology*, *146* (4), 807–816.
Tuteja, R., & Tuteja, N. (2004). Serial analysis of gene expression (SAGE): unraveling the bioinformatics tools. *Bioessays*, *26*(8), 916–922.
Unoki, H., Takahashi, A., Kawaguchi, T., Hara, K., Horikoshi, M., Andersen, G., et al. (2008). SNPs in KCNQ1 are associated with susceptibility to type 2 diabetes in East Asian and European populations. *Nature Genetics*, *40*(9), 1098.
van Dam, N. M., & van der Meijden, E. (2011). A role for metabolomics in plant ecology. *Biology of Plant Metabolomics: Vol. 43, Annual Plant Reviews* (pp. 87–107).
van den Berg, R. A., Hoefsloot, H. C., Westerhuis, J. A., Smilde, A. K., & van der Werf, M. J. (2006). Centering, scaling, and transformations: improving the biological information content of metabolomics data. *BMC Genomics*, 7(1), 142.
van der Kooy, F., Maltese, F., Choi, Y. H., Kim, H. K., & Verpoorte, R. (2009). Quality control of herbal material and phytopharmaceuticals with MS and NMR based metabolic fingerprinting. *Planta Medica*, *75*(07), 763–775.
Van Emon, J. M. (2015). The omics revolution in agricultural research. *Journal of Agricultural and Food Chemistry*, *64*(1), 36–44.
Velculescu, V. E., Zhang, L., Vogelstein, B., & Kinzler, K. W. (1995). Serial analysis of gene expression. *Science*, *270*(5235), 484–487.
Verlaan, D. J., Berlivet, S., Hunninghake, G. M., Madore, A. M., Larivière, M., Moussette, S., et al. (2009). Allele-specific chromatin remodeling in the ZPBP2/

GSDMB/ORMDL3 locus associated with the risk of asthma and autoimmune disease. *The American Journal of Human Genetics*, *85*(3), 377–393.
Villas-Bôas, S. G., Mas, S., Åkesson, M., Smedsgaard, J., & Nielsen, J. (2005). Mass spectrometry in metabolome analysis. *Mass Spectrometry Reviews*, *24*(5), 613–646.
Villas-Bôas, S. G., Smart, K. F., Sivakumaran, S., & Lane, G. A. (2011). Alkylation or silylation for analysis of amino and non-amino organic acids by GC-MS? *Metabolites*, *1*(1), 3–20.
Vinaixa, M., Schymanski, E. L., Neumann, S., Navarro, M., Salek, R. M., & Yanes, O. (2016). Mass spectral databases for LC/MS-and GC/MS-based metabolomics: state of the field and future prospects. *TrAC Trends in Analytical Chemistry*, *78*, 23–35.
Vuckovic, D. (2012). Current trends and challenges in sample preparation for global metabolomics using liquid chromatography–mass spectrometry. *Analytical and Bioanalytical Chemistry*, *403*(6), 1523–1548.
Vuylsteke, M., Peleman, J. D., & Van Eijk, M. J. (2007). AFLP-based transcript profiling (cDNA-AFLP) for genome-wide expression analysis. *Nature Protocols*, *2*(6), 1399.
Waddell, N. (2008). Microarray-based DNA profiling to study genomic aberrations. *IUBMB Life*, *60*(7), 437–440.
Wang, K., Gan, L., Jeffery, E., Gayle, M., Gown, A. M., Skelly, M., et al. (1999). Monitoring gene expression profile changes in ovarian carcinomas using cDNA microarray. *Gene*, *229*(1), 101–108.
Wang, K., Li, M., Hadley, D., Liu, R., Glessner, J., Grant, S. F., et al. (2007). PennCNV: an integrated hidden Markov model designed for high-resolution copy number variation detection in whole-genome SNP genotyping data. *Genome Research*, *17*(11), 1665–1674.
Wang, X., Li, K., Yao, L., Wang, C., & Van Schepdael, A. (2017). Recent advances in vitamins analysis by capillary electrophoresis. *Journal of Pharmaceutical and Biomedical Analysis*.
Wang, X., Sun, H., Zhang, A., Wang, P., & Han, Y. (2011). Ultra-performance liquid chromatography coupled to mass spectrometry as a sensitive and powerful technology for metabolomic studies. *Journal of Separation Science*, *34*(24), 3451–3459.
Wang, Z., Gerstein, M., & Snyder, M. (2009). RNA-Seq: a revolutionary tool for transcriptomics. *Nature Reviews Genetics*, *10*(1), 57.
Warnke, U., Rappel, C., Meier, H., Kloft, C., Galanski, M., Hartinger, C. G., et al. (2004). Analysis of platinum adducts with DNA nucleotides and nucleosides by capillary electrophoresis coupled to ESI-MS: indications of guanosine 5′-monophosphate O6–N7 chelation. *ChemBioChem*, *5*(11), 1543–1549.
Weckwerth, W., Loureiro, M. E., Wenzel, K., & Fiehn, O. (2004). Differential metabolic networks unravel the effects of silent plant phenotypes. *Proceedings of the National Academy of Sciences*, *101*(20), 7809–7814.
Weckwerth, W., & Morgenthal, K. (2005). Metabolomics: from pattern recognition to biological interpretation. *Drug Discovery Today*, *10*(22), 1551–1558.
Wei, X., Shi, X., Kim, S., Zhang, L., Patrick, J. S., Binkley, J., et al. (2012). Data preprocessing method for liquid chromatography–mass spectrometry based metabolomics. *Analytical Chemistry*, *84*(18), 7963–7971.
Weljie, A. M., Newton, J., Mercier, P., Carlson, E., & Slupsky, C. M. (2006). Targeted profiling: quantitative analysis of 1H NMR metabolomics data. *Analytical Chemistry*, *78*(13), 4430–4442.
Welsh, J., & McClelland, M. (1990). Fingerprinting genomes using PCR with arbitrary primers. *Nucleic Acids Research*, *18*(24), 7213–7218.
Westerhuis, J. A., Hoefsloot, H. C., Smit, S., Vis, D. J., Smilde, A. K., van Velzen, E. J., et al. (2008). Assessment of PLSDA cross validation. *Metabolomics*, *4*(1), 81–89.
Westermeier, R. (2014). Looking at proteins from two dimensions: a review on five decades of 2D electrophoresis. *Archives of Physiology and Biochemistry*, *120*(5), 168–172.

Westermeier, R. (2016). *Electrophoresis in practice: a guide to methods and applications of DNA and protein separations*. John Wiley & Sons.
Williams, J. G., Kubelik, A. R., Livak, K. J., Rafalski, J. A., & Tingey, S. V. (1990). DNA polymorphisms amplified by arbitrary primers are useful as genetic markers. *Nucleic Acids Research*, *18*(22), 6531–6535.
Winkler, M., Escobar García, P., Gattringer, A., Sonnleitner, M., Hülber, K., Schönswetter, P., et al. (2017). A novel method to infer the origin of polyploids from amplified fragment length polymorphism data reveals that the alpine polyploid complex of *Senecio carniolicus* (Asteraceae) evolved mainly via autopolyploidy. *Molecular Ecology Resources*, *17*(5), 877–892.
Wishart, D. S. (2009). Bioinformatics for metabolomics. In *Bioinformatics for systems biology* (pp. 581–599): Humana Press.
Wishart, D. S. (2011). Advances in metabolite identification. *Bioanalysis*, *3*(15), 1769–1782.
Wisman, E., & Ohlrogge, J. (2000). *Arabidopsis* microarray service facilities. *Plant Physiology*, *124*(4), 1468–1471.
Wittenberg, C., & Reed, S. I. (2005). Cell cycle-dependent transcription in yeast: promoters, transcription factors, and transcriptomes. *Oncogene*, *24*(17), 2746.
Wohlfarth, A., Scheidweiler, K. B., Chen, X., Liu, H. F., & Huestis, M. A. (2013). Qualitative confirmation of 9 synthetic cannabinoids and 20 metabolites in human urine using LC–MS/MS and library search. *Analytical Chemistry*, *85*(7), 3730–3738.
Wolfender, J. L., Marti, G., Thomas, A., & Bertrand, S. (2015). Current approaches and challenges for the metabolite profiling of complex natural extracts. *Journal of Chromatography A*, *1382*, 136–164.
Worley, B., & Powers, R. (2013). Multivariate analysis in metabolomics. *Current Metabolomics*, *1*(1), 92–107.
Wright, E. P., Partridge, M. A., Padula, M. P., Gauci, V. J., Malladi, C. S., & Coorssen, J. R. (2014). Top-down proteomics: Enhancing 2D gel electrophoresis from tissue processing to high-sensitivity protein detection. *Proteomics*, *14*(7-8), 872–889.
Wu, W. W., Wang, G., Baek, S. J., & Shen, R. F. (2006). Comparative study of three proteomic quantitative methods, DIGE, cICAT, and iTRAQ, using 2D gel-or LC – MALDI TOF/TOF. *Journal of Proteome Research*, *5*(3), 651–658.
Wu, X., Gong, F., & Wang, W. (2014). Protein extraction from plant tissues for 2DE and its application in proteomic analysis. *Proteomics*, *14*(6), 645–658.
Xia, J., Mandal, R., Sinelnikov, I. V., Broadhurst, D., & Wishart, D. S. (2012). MetaboAnalyst 2.0—a comprehensive server for metabolomic data analysis. *Nucleic Acids Research*, *40*(W1), W127–W133.
Xia, J., Psychogios, N., Young, N., & Wishart, D. S. (2009). MetaboAnalyst: a web server for metabolomic data analysis and interpretation. *Nucleic Acids Research*, *37*(suppl_2), W652–W660.
Xia, J., & Wishart, D. S. (2011). Web-based inference of biological patterns, functions and pathways from metabolomic data using MetaboAnalyst. *Nature Protocols*, *6*(6), 743.
Xiao, D., Liu, S. T., Wei, Y. P., Zhou, D. Y., Hou, X. L., Li, Y., et al. (2016). cDNA-AFLP analysis reveals differential gene expression in incompatible interaction between infected non-heading Chinese cabbage and Hyaloperonospora parasitica. *Horticulture Research*, *3*. 16034.
Xu, Y., & Goodacre, R. (2012). Multiblock principal component analysis: an efficient tool for analyzing metabolomics data which contain two influential factors. *Metabolomics*, *8*(1), 37–51.
Yan, L., Yang, M., Guo, H., Yang, L., Wu, J., Li, R., et al. (2013). Single-cell RNA-Seq profiling of human preimplantation embryos and embryonic stem cells. *Nature Structural & Molecular Biology*, *20*(9), 1131.

Yang, C., Hua, Q., & Shimizu, K. (2002). Integration of the information from gene expression and metabolic fluxes for the analysis of the regulatory mechanisms in Synechocystis. *Applied Microbiology and Biotechnology*, *58*(6), 813–822.

Yu, J., Hu, S., Wang, J., Wong, G. K. S., Li, S., Liu, B., et al. (2002). A draft sequence of the rice genome (*Oryza sativa* L. ssp. *indica*). *Science*, *296*(5565), 79–92.

Yuan, M., Breitkopf, S. B., Yang, X., & Asara, J. M. (2012). A positive/negative ion–switching, targeted mass spectrometry–based metabolomics platform for bodily fluids, cells, and fresh and fixed tissue. *Nature Protocols*, 7(5), 872.

Zargar, S. M., Kurata, R., Rakwal, R., & Fukao, Y. (2015). Peptide separation methodologies for in-depth proteomics. In *Plant cell expansion* (pp. 195–209). New York, NY: Humana Press.

Zeng, H., Deng, Y., & Wu, J. T. (2003). Fast analysis using monolithic columns coupled with high-flow on-line extraction and electrospray mass spectrometric detection for the direct and simultaneous quantitation of multiple components in plasma. *Journal of Chromatography B*, *788*(2), 331–337.

Zeng, W., Hazebroek, J., Beatty, M., Hayes, K., Ponte, C., Maxwell, C., et al. (2014). Analytical method evaluation and discovery of variation within maize varieties in the context of food safety: transcript profiling and metabolomics. *Journal of Agricultural and Food Chemistry*, *62*(13), 2997–3009.

Zhang, A., Sun, H., Wang, P., Han, Y., & Wang, X. (2012). Modern analytical techniques in metabolomics analysis. *Analyst*, *137*(2), 293–300.

Zhang, T., Creek, D. J., Barrett, M. P., Blackburn, G., & Watson, D. G. (2012). Evaluation of coupling reversed phase, aqueous normal phase, and hydrophilic interaction liquid chromatography with Orbitrap mass spectrometry for metabolomic studies of human urine. *Analytical Chemistry*, *84*(4), 1994–2001.

Zhang, X. O., Wang, H. B., Zhang, Y., Lu, X., Chen, L. L., & Yang, L. (2014). Complementary sequence-mediated exon circularization. *Cell*, *159*(1), 134–147.

Zhang, Z., Zhang, W., Tu, J., Li, Y., Xu, G., & Zhang, Y. (2018). Simultaneous detection of protocatechuic acid, chlorogenic acid and caffeic acid in honey by HPLC with ultraviolet and electrochemical detectors. *International Journal of Electrochemical Science*, *13*, 6655–6665.

Zhao, J., Zhang, T., Liu, Y., Wang, X., Zhang, L., Ku, T., et al. (2018). Qualitative and quantitative assessment of DNA quality of frozen beef based on DNA yield, gel electrophoresis and PCR amplification and their correlations to beef quality. *Food Chemistry*, *260*, 160–165.

Zhao, S., Fung-Leung, W. P., Bittner, A., Ngo, K., & Liu, X. (2014). Comparison of RNA-Seq and microarray in transcriptome profiling of activated T cells. *PLoS One*, *9*(1) e78644.

Zhao, X. E., & Suo, Y. R. (2008). Simultaneous determination of monoamine and amino acid neurotransmitters in rat endbrain tissues by pre-column derivatization with high-performance liquid chromatographic fluorescence detection and mass spectrometric identification. *Talanta*, *76*(3), 690–697.

Zheng, C. L., Kawane, S., Bottomly, D., & Wilmot, B. (2014). Analysis considerations for utilizing RNA-Seq to characterize the brain transcriptome. *Vol. 116. International review of neurobiology* (pp. 21–54): Academic Press.

Zhong, S., Joung, J. G., Zheng, Y., Chen, Y. R., Liu, B., Shao, Y., et al. (2011). High-throughput illumina strand-specific RNA sequencing library preparation. *Cold Spring Harbor Protocols*, *2011*(8). pdb-prot5652.

Zhou, J., He, Z., Yang, Y., Deng, Y., Tringe, S. G., & Alvarez-Cohen, L. (2015). High-throughput metagenomic technologies for complex microbial community analysis: open and closed formats. *MBio*, *6*(1). e02288-14.

Zhou, Z., Zhang, C., Zhou, Y., Hao, Z., Wang, Z., Zeng, X., et al. (2016). Genetic dissection of maize plant architecture with an ultra-high density bin map based on recombinant inbred lines. *BMC Genomics*, *17*(1), 178.

Zhu, T., Budworth, P., Han, B., Brown, D., Chang, H. S., Zou, G., et al. (2001). Toward elucidating the global gene expression patternsof developing *Arabidopsis*: parallel analysis of8300 genesby a high-density oligonucleotide probe array. *Plant Physiology and Biochemistry*, *39*(3-4), 221–242.

Zhu, Z. J., Schultz, A. W., Wang, J., Johnson, C. H., Yannone, S. M., Patti, G. J., et al. (2013). Liquid chromatography quadrupole time-of-flight mass spectrometry characterization of metabolites guided by the METLIN database. *Nature Protocols*, *8*(3), 451.

Ziegenhain, C., Vieth, B., Parekh, S., Reinius, B., Guillaumet-Adkins, A., Smets, M., et al. (2017). Comparative analysis of single-cell RNA sequencing methods. *Molecular Cell*, *65*(4), 631–643.

Zimmermann, P., Hirsch-Hoffmann, M., Hennig, L., & Gruissem, W. (2004). GENEVESTIGATOR. *Arabidopsis* microarray database and analysis toolbox. *Plant Physiology*, *136*(1), 2621–2632.

Zou, F. F., Guo, Q., Sun, Y., Zhou, D., Hu, M. X., Hu, H. X., et al. (2016). Identification of protease m1 zinc metalloprotease conferring resistance to deltamethrin by characterization of an AFLP marker in *Culex pipiens* pallens. *Parasites & Vectors*, *9*(1), 172.

FURTHER READING

Abo, R. P., Lin, L., Hunter, S. S., Dolcen, D. N., Paquette, R. R., Laing, A., et al. (2015). Comparative analysis of RNA sequencing methods for characterization of cancer transcriptomics. *Cancer Research*, *75*, 4867.

De Vos, R. C., Moco, S., Lommen, A., Keurentjes, J. J., Bino, R. J., & Hall, R. D. (2007). Untargeted large-scale plant metabolomics using liquid chromatography coupled to mass spectrometry. *Nature Protocols*, *2*(4), 778.

Gachon, C. M., Langlois-Meurinne, M., & Saindrenan, P. (2005). Plant secondary metabolism glycosyltransferases: the emerging functional analysis. *Trends in Plant Science*, *10*(11), 542–549.

Lueders, T., & Friedrich, M. W. (2003). Evaluation of PCR amplification bias by terminal restriction fragment length polymorphism analysis of small-subunit rRNA and mcrA genes by using defined template mixtures of methanogenic pure cultures and soil DNA extracts. *Applied and Environmental Microbiology*, *69*(1), 320–326.

Oshlack, A., & Wakefield, M. J. (2009). Transcript length bias in RNA-seq data confounds systems biology. *Biology Direct*, *4*(1), 14.

Pan, Z., & Raftery, D. (2007). Comparing and combining NMR spectroscopy and mass spectrometry in metabolomics. *Analytical and Bioanalytical Chemistry*, *387*(2), 525–527.

Righetti, P. G. (2013). Detection of neutral hemoglobin mutants by conventional isoelectric focusing and. *TrAC Trends in Analytical Chemistry*, *5*(6), 147.

Swamy, K. B., Lin, C. H., Yen, M. R., Wang, C. Y., & Wang, D. (2014). Examining the condition-specific antisense transcription in *S. cerevisiae* and *S. paradoxus*. *BMC Genomics*, *15*(1), 521.

Tarazona, S., García-Alcalde, F., Dopazo, J., Ferrer, A., & Conesa, A. (2011). Differential expression in RNA-seq: a matter of depth. *Genome Research*. gr-124321.

Xu, D., Xu, D., Yu, X., Liu, Z., He, W., & Ma, Z. (2005). Label-free electrochemical detection for aptamer-based array electrodes. *Analytical Chemistry*, 77(16), 5107–5113.

CHAPTER 3

Metabolomics Studies of Stress in Plants

3.1 INTRODUCTION

Environmental stress is an inevitable process that can result in significant alterations in plant metabolism. To maintain metabolic homeostasis, plants produce numerous metabolites that combat plant stress (Das & Roychoudhury, 2014; Souza, Monteiro, Carvalho, Gratão, & Azevedo, 2017; Ahmad et al., 2018). Environmental stress induces the first activation of networks implicated in signaling pathways, which are followed by signal transduction pathways, resulting in the generation of diverse proteins and metabolites that help achieve the new state, homeostasis (Foyer & Noctor, 2005; Walter & Ron, 2011). From the metabolomic perspective, there are at least three different kinds of metabolites: (1) metabolites implicated in the defense processes, for example, anti-oxidant or osmoprotectants; (2) by-products of stress that appear in the cells because of alteration of normal homeostasis by the alteration in the growth conditions; and (3) signal transduction molecules implicated in arbitrating the acclimation response (Huang et al., 2012). The signal transduction metabolites might be synthesized metabolites that are produced from their conjugated form; for example, the plant hormones' salicylic acid, which could be a diverse by-product of stress metabolism that signals the disruption of signal interruption of cellular homeostasis (Das, Nutan, Singla-Pareek, & Pareek, 2015). The second kind of signaling metabolites could induce by-products of membrane degradation, diverse reactive oxygen species (ROS), or varied oxidized metabolites; for example, phenolic metabolites, or even some antioxidants (Lata, Muthamilarasan, & Prasad, 2015). When plants thrive under adverse growth conditions, abiotic stress reduces plant growth and productivity (Akula & Ravishankar, 2011). Plant metabolism is disturbed due to enzyme inhibition, shortage of substrates, or a higher level of explicit metabolites in plants thriving under abiotic stresses (Bhattacharjee & Saha, 2014). Consequently, in order to maintain the necessary metabolism, the metabolic network must be reconfigured to acclimate it by establishing a new steady state, in light

Ecometabolomics
https://doi.org/10.1016/B978-0-12-814872-3.00003-5

of the prevailing stress conditions (Geigenberger & Fernie, 2014). The metabolic reprogramming is also important for gathering the anti-stress agents, including compatible solutes, antioxidants, and stress-responsive proteins (Gangwar et al., 2014). The accretion of ROS is one more predicament that leads to oxidation and dysfunction of cellular components, which leads to cell death. The optimization of metabolic flux through the organellar electron transport chains is vital to reducing ROS production. In order to offer reducing power for ROS scavenging, the prime task should be maintenance of the redox state (Mailloux & Harper, 2011). Regardless of such significant roles in metabolic regulation under stress conditions, our present perception of this method is fragmented and far from complete.

Metabolomics is an imperative contrivance by which to acquire an inclusive viewpoint of how metabolic networks are regulated, and has been employed by numerous researchers in recent years (Hall, 2006; Prince & Pohnert, 2010). Additionally, it can be employed to elaborate the function of genes as a tool in functional genomics and system biology strategies. The term "metabolomics" refers to an inclusive and quantitative analysis of all small metabolites in a biological system (Kitteringham, Jenkins, Lane, Elliott, & Park, 2009). The plant kingdom is comprised of about 200,000–1,000,000 metabolites, and the KNApSAck database (http://kanaya.naist.jp/KNApSAcK/) includes about 50,000 metabolite entries for plants, demonstrated so far in the literature (Afendi et al., 2011). The most inclusive methods identify only 1000 and 2000 molecular characteristics. Numerous techniques, including gas chromatography-mass spectrometry (GC-MS), liquid chromatography (LC)-MS, capillary electrophoresis (CE)-MS, and nuclear magnetic resonance (NMR) spectroscopy are generally employed in plant metabolomics research (Issaq, Van, Waybright, Muschik, & Veenstra, 2009; Wang, Byun, & Pennathur, 2010). They are employed occasionally in integration because they are expansively complimentary, with each independent method having superior coverage of various kinds of metabolites. Research studies using metabolomic advances to decipher plant responses to abiotic stresses will be illustrated in this chapter. Ultimately, employing data that has already been assembled, we intend to elaborate on common and explicit stress responses. During recent years, the majority of research studies have been focused on analyzing the intricate cascade of events that occur in plants subjected to varied abiotic stresses, and investigating the effects on the gene expression level (Sofo, Scopa, Nuzzaci, & Vitti, 2015). Using the transcriptomics application, along with forward and reverse genetics, hundreds of genes have been linked to environmental stress responses, and regulatory networks of gene expression have been explained. Comparatively less is known about alteration at the metabolomic level, but during the past few years, the widespread metabolite investigation of plant stress response has become a rapidly escalating research field (Joshi et al., 2016). In this chapter, we will summarize all recent research reports illustrating results acquired both in model plants *Arabidopsis thaliana*, and in numerous crop species, investigating the following prime abiotic stresses: low and high temperature, drought, high salinity, and oxidative stress.

3.2 TEMPERATURE STRESS

Temperature is one of the most imperative ecological parameters establishing plant growth and development. Plants are exposed to constant diurnal and seasonal temperature fluctuations, with results depending on whether these divergences from the average values remain within a natural temperature range for each species, or whether extremes of this range are reached (Parmesan & Hanley, 2015; Zhu, 2016). Although the temperature range of survival in some species can be expanded via 'cold acclimation' or 'acquired thermotolerance,' there are adaptive methods whereby plants augment in chilling/freezing or in heat tolerance against preceding low, nonfreezing or high-temperature exposure, respectively. These exceedingly intricate mechanisms are acquired by general reprogramming of the plant transcriptome, proteome, and metabolome. Metabolite profiling was employed to comprehend the dynamics of the *Arabidopsis* response to temperature stress (Grazia Annunziata et al., 2018; Zinta et al., 2018). The researchers investigated GC-MS profiling of *Arabidopsis* plants exposed to heat and cold stress, and recognized an array of identifiable and unidentifiable mass spectral tags used exclusively to denote heat or cold stress, or both.

3.2.1 Low-Temperature Response

Cold stress results in marked alterations in plant metabolism, and the majority of metabolites that are induced by high temperatures are also modulated in response to cold stress. Numerous metabolites induced by both stresses are not correlated to temperature stress (Eremina, Rozhon, & Poppenberger, 2016; Park et al., 2015). Several research reports co-analyzed such metabolomic data, along with transcript profiling data, to unravel the mechanism underlying cold acclimation (Bocian et al., 2015; Pagter et al., 2017; Wu et al., 2016). It has been established that there is no correlation between transcript abundance and metabolite loads, and it was further ascertained that regulatory pathways, independent of transcript abundance, play a significant role in the metabolic adjustment of plants during cold acclimation (Bustamante et al., 2016; Le, Pagter, & Hincha, 2015). Wisniewski, Norelli, Bassett, Artlip, and Macarisin (2011) demonstrated overexpression of C-repeat/dehydration responsive element-binding factor (CBF) 3, and it was revealed that the *Arabidopsis* metabolome is expansively reconfigured in response to low temperatures, which indicated a significant role for the CBF cold response pathway in this process. Targeted profiling revealed that there were marked changes in the carbohydrate metabolism of rice under chilling, salt, and osmotic stress in diverse genotypes exhibiting varying chilling tolerance (Yang, Dai, & Zhang, 2012). It has been estimated that levels of soluble carbohydrates in chilling-tolerant and chilling-sensitive genotypes use quantitative high performance liquid chromatography (HPLC) assays under carbohydrate accretion (Maruyama et al., 2014). The chilling tolerance is correlated with accretion of galactose and raffinose, while sugar decreased in the chilling-sensitive genotypes (Morsy, Jouve, Hausman, Hoffmann, & Stewart, 2007). These genotypes also respond differently to salt and osmotic stress. Morsy et al. (2007) illustrated carbohydrate profiling data integrated with the estimation of oxidative products and antioxidative enzymes, and concluded that chilling-tolerant genotypes possess more effective ROS scavenging systems. Numerous cold-induced metabolites, for example, sugars, profiles and polyamines were recognized via comprehensive investigation, which revealed that they accumulate under cold stress in *Arabidopsis* and other species in studies employing targeted analysis (Kaplan et al., 2007). However, the intensity of metabolite profiling with reference to targeted analysis is comprised of the detection of variations in the metabolite abundance that have not yet been implicated in the process being considered (i.e., cold acclimation) (Maruyama et al., 2009). As illustrated previously, the functional significance of the CBF pathway in the reconstruction of the low-temperature metabolome was established. Cook, Fowler, Fiehn, and Thomashow (2004) also evaluated the metabolic profile of Ws-2 wild-type (wt) plants and transgenic lines ectopically displaying the transcription factor CBF3. In non-acclimated transgenic lines, they observed an 80% increment in metabolites that were cold induced in wt (90% if metabolites that increased more than fivefold were considered); thus ascertaining a significant function of the CBF pathway in the cold response. Maruyama et al. (2009) evaluated the metabolite profile of wt *Arabidopsis* plants thriving under control or stress conditions (cold and drought), and transgenic plants ectopically revealing *DREB1A/CBF3* or an active form of *DREB2A*. The overexpression of *DREB1A/CBF3* has been formerly revealed to confer a higher tolerance to both freezing and dehydration; whereas the overexpression of *DREB2A* significantly enhanced the tolerance to dehydration, but not to freezing, in transgenic plants. Maruyama et al. (2014) illustrated *DREB1A/CBF3* ectopic expression that led to an increased amount of 37 metabolites, 33 of which also accumulated in cold-treated wt plants. Because cold induced the accumulation of 155 metabolites in these experimental conditions, the CBF pathway appeared to be implicated in the increment of only 21% of the cold-induced metabolites, as compared with 80% reported by Cook et al., 2004. In *DREB2A* transgenic plants that are dehydration tolerant, but not freezing tolerant, the level of 28 metabolites increased; 17 of which were also positively affected by *DREB1A/CBF3*. The plants exposed to cold stress and *DREB1A/CBF3* overexpressing plants display seventeen known metabolites, such as myo-inositol, galactinol, raffinose, sucrose, and 13 identified metabolites. Therefore, these metabolites were established as possible candidates, and play a functional role in the cold response. However, the level of the four identifiable metabolites among the seven were also considerably amplified in dehydration-subjected wt plants. Gene expression data acquired by a microarray analysis of the diverse lines and growth conditions were known to be in accordance with metabolite data, specifically the expression level of genes implicated in carbohydrate metabolism that were positively correlated with the accretion of explicit sugars and sugars alcohols under stress (Maruyama et al., 2009). Mattana et al. (2005) reported alterations in the metabolic content of cold-stressed wt and *Osmyb4*-transgenic plants via targeted and profiling analyses. This study revealed higher tolerance levels in transgenic plants that produced higher levels of proline, sucrose, glucose, fructose, glycine betaine, alanine, and sinapoyl malate. The higher levels of metabolic contents in transformed plants were consistent with comprehensive alterations in the mRNA population (Vannini et al., 2006). Kaplan et al. (2004) reported an increase in the levels of metabolites, the majority of which were exhibiting low M_r polarity in response to cold stress. This study further illustrated that there was an even distribution of metabolites across all the temporal stages of cold responses (early, intermediate, and late) in relation to 1–4, 12–24, and 48–96h of cold stress periods, respectively. These results signified that acclimation is a long-term, dynamic method. Gray and Heath (2005) reported that the metabolic profile of *Arabidopsis* leaves transferred to low temperatures constantly fluctuated for up to 49 days; whereas leaves that had grown at low temperatures displayed a stable metabolite composition. But this metabolite profile was quite different than the profiles developed at 4°C. Therefore, the authors anticipated the implication of two divergent metabolic networks against cold stress: one that is ecologically modulated, and another that is developmentally adjusted at low temperatures. The authors conjectured that a similar response may be encountered in response to diverse environmental stimulus. Even though there is quantitative and qualitative disparity among the observations detailed by numerous authors, the assertion that remains constant throughout is the functional role of carbohydrate metabolism. This significant function of carbohydrate metabolism was further ascertained by the comparative metabolite analysis carried out by Hannah et al. (2006). This research study was based on analysis carried out by nine diverse *Arabidopsis* accessions derived from Scandinavia to the Cape Verde Islands, and they differed in their freezing tolerance, integrating transcripts, and metabolite profiling. They revealed the comprehensive variation of transcripts, but not of metabolites associated with the cold acclimation competence. However, the accretion of the metabolites' numerous carbohydrates (i.e., glucose, fructose, sucrose, and raffinose) were closely associated with freezing tolerance. Even though the significant role of soluble sugars, particularly raffinose, has been outlined by numerous studies, it has been illustrated that raffinose accumulation alone is neither indispensable nor adequate for cold acclimation (Zuther et al., 2004). The significant function of carbohydrates has also been illustrated proportionate to heterosis for freezing tolerance (Korn et al., 2010; Korn, Peterek, Mock, Heyer, & Hincha, 2008). Korn et al. (2010) employed comprehensive metabolic profiling to unravel metabolite combinations able to envisage freezing tolerance and its heterosis. They recognized compatible solutes as important predictors for both phenotypes; specifically, metabolites belonging to raffinose biosynthetic pathways, and other, yet-unknown metabolites, in addition to some TCA cycle intermediates that particularly provided only the heterotic phenotype. Korn et al. (2010) attempted to detect a class of metabolites, rather than individual metabolites, that together contain extrapolative potential that seems well adapted for detecting a redundant cellular protection system. For example, those that are symbolized by compatible solutes, where single metabolites could generally alternate with each other as a compensatory mechanisms (Panikulangara, Eggers-Schumacher, Wunderlich, Stransky, & Schöffl, 2004; Zuther et al., 2004).

3.2.2 High-Temperature Stress Response

Although there are enough research reports on cold stress, there are very few reports focused on heat stress in plants. Kaplan et al. (2004) investigated a comprehensive metabolite profiling by GC-MS to detect similarities and disparities in temporal metabolite responses linked with the induction of acquired thermotolerance against heat shock (HS), and acquired freezing tolerance in response to cold shock. The main difference between the plant responses against high and low temperatures was characterized by the temporal dynamics of the induced metabolic changes. Certainly, while during cold exposure, alterations were evenly distributed along the time course of the treatment; however, most of the heat-induced metabolic adjustment occurred within the first 30 minutes (Gray & Heath, 2005). Kaplan et al. (2004) observed 497 low-molecular-weight metabolites, out of which of 143 were affected by HS; among them, numerous pyruvate- and oxaloacetate-derived amino acids, fumarate and malate (oxaloacetate precursors), some amine-containing metabolites (β-alanine, GABA, and putrescine), and several carbohydrates (including maltose, sucrose, raffinose, galactinol, and myo-inositol), which were found to be coordinately amplified. There were 311 metabolites, or mass spectral tags, that were varied against cold stress, which indicated that cold shock affected metabolism more greatly than HS. Furthermore, a higher fraction of HS metabolites was pooled with the cold stress response, with only a very small fraction being HS specific. On the contrary, the majority of metabolites (60%) that were responsive to cold shock were cold-sensitive specific. In both HS and cold stress, 93 metabolites exhibited a generalized response between two thermal stresses. Among these, pyruvate and oxaloacetate-derived amino acids, polyamines, and numerous carbohydrates comprising fructose, sucrose, myo-inositol-phosphate, galactinol, and raffinose increased. Many of these metabolites were either compatible solutes, or precursors for secondary metabolites, with characteristics of defense against pathogens. The functional role of raffinose in the heat-stress response among transgenic plants displayed overexpression of the major HS transcription factor (HSF) 3, and in two *Galactinol synthase1* (*GolS1*) T-DNA knockout mutants. Galactinol synthase is a functional enzyme in the biosynthetic pathway of the raffinose family of oligosaccharides; raffinose, stachyose, and verbascose. The expression of *GolS1* was heat inducible in wt plants, constitutively upregulated in HSF3-overexpressing transgenic plants, and almost completely inhibited in the T-DNA insertion mutants (Krasensky & Jonak, 2012). The wt plants exhibited a basal level of raffinose under control conditions that was amplified in response to heat stress. In transgenic lines, a constitutive remarked accretion was observed that was further induced following heat stress; in contrast, only a basal level devoid of any enhancement after heat stress was measurable in the knockout mutants. Surprisingly, when the heat tolerance phenotype was investigated, no differences between the wt and knockout mutant lines were detected. These observations, together with those acquired by Zuther et al. (2004) pertaining to lack of disparity in the cold-tolerance phenotype among *Arabidopsis* lines accumulating varied levels of raffinose, led to the unanticipated conclusion that varying the raffinose content via t-DNA activated, or transgenic, plants did not affect the stress temperature tolerance phenotype. The most probable explanation is the existence of feedback and or compensative mechanisms whereby a change of raffinose levels is followed by the alteration in sugar/metabolite levels under cold acclimation observed in raffinose synthase knockout mutants (Zuther et al., 2004).

Kaplan et al. (2004) also observed higher levels of signaling molecules, such as salicylic acid, accumulated during both heat and cold stress; although with time, changing courses. This research finding confirmed, in addition to being implicated in the defense mechanism against pathogens, salicylic acid has a significant role in temperature stress responses. Actually, evaluating wt and mutant plants that accumulated diverse salicylic acid (SA) levels, a correlation between prestress levels of endogenous SA, and both basal and acquired thermotolerance was illustrated (Larkindale, Hall, Knight, & Vierling, 2005). Clarke et al. (2009) established the implication of jasmonates in conferring basal thermotolerance to *Arabidopsis* plants. It was further reported that cold-tolerant plants accumulated galactose and raffinose; whereas these sugars decreased in cold-sensitive plants. On the contrary, cold-sensitive plants specifically accumulate higher levels of osmoprotectants, such as mannitol and trehalose, under salt and drought conditions, respectively. Morsy et al. (2007) further demonstrated higher levels of antioxidant metabolites exhibiting higher ROS scavenging activity in cold-tolerant genotype during chilling stress. Cold tolerance in maize, apples, and *Osteospermum* was Myb4 originated, and was positively correlated with higher concentrations of sugars and proline. The transgenic maize plants grown under controlled conditions exhibited no difference in metabolite concentration, with reference to the wt, as the *Osmyb4* gene expression in maize was under the pCOR15a stress-inducible promoter. On the contrary, 6-day-old cold treatments amplified the sugar levels such as fructose, sucrose, glucose, and proline contents in both wt and transgenic maize plants, with the concentration being significantly higher for all of them in the *Osmyb4*-transgenic plants. The utilization of the pCaMV35S constitutive promoter to impel *Osmyb4* expression in *Osteospermum* and apple transgenic plants led to higher levels of sugars and proline concentration before stress exposure (Mattana et al., 2005). Laura et al. (2010) reported the first metabolic profile of *Osteospermum* species under both control and stress growth conditions, thereby confirming and expanding the preceding results on Myb4-driven cold tolerance and metabolic changes. As revealed by the PCA (principal component analysis) score plot, *Osteospermum* samples were dispersed in proportion to genotypes and treatments: control and 2-day cold-treated plants (wt and transgenics) clustered together, while 10-day cold-treated and freezing-treated plants clustered into diverse regions based on genotype. The samples' separation was associated with diverse amounts of sucrose (higher in transgenic plants), inulin, glucose, fructose, and amino acids (higher in wt) that accumulated under stress conditions (Laura et al., 2010). Estimation of proline content and proline/amino acids' ratio reported results on a higher concentration in transgenic apples, consistent with the earlier illustrated overlap between cold and heat stress responses. *Osmyb4* was found to improve the tolerance to both cold and heat stress (Mattana, Carravieri, et al., 2005).

In a more detailed analysis, the authors confirmed that the observed tolerance in transgenic apple plants may be driven by the higher content of sugars (glucose, sucrose, and fructose) and proline present in plants grown under control conditions. In particular, the cold treatment amplified the differences of metabolites' accumulation between wt and transgenics. Moreover, the *Osmyb4*-overexpressing plants showed an improved tolerance phenotype toward drought stress (Pasquali, Biricolti, Locatelli, Baldoni, & Mattana, 2008). The most generalized feature exhibited by all of the Myb4-transgenic plants under cold treatment was higher levels of proline content, whereas wt plants displayed a higher free amino acid pool (Laura et al., 2010; Pasquali et al., 2008). This observation outlined the significance of proline accumulation during stress, consistent with the numerous functions anticipated for this amino acid during abiotic stress (Verbruggen & Hermans, 2008). Du, Wang, Yu, Liu, and Huang (2010) also demonstrated the differential heat tolerance associated with metabolites in two perennial grass species; namely C4 warm-season Bermuda grass (a tolerant species), and C3 cool-season Kentucky bluegrass (a susceptible species) thriving normally under optimum temperature conditions, or exposed to short-term and long-term heat stress. The physiological parameters measured in both species revealed that the Bermuda grass was more heat-tolerant than the Kentucky bluegrass. The metabolic profile also revealed the disparity in the accumulation of numerous metabolites, depending on the species and growth conditions; specifically, Bermuda grass accumulated a higher content of organic acids, amino acids, sugars, and sugar alcohols compared with Kentucky bluegrass, particularly in long-term heat stress.

3.3 WATER AND SALT STRESS

Water stress is the prime constraint for booming agricultural productivity, as it hampers nutrient absorption, alters crop growth, and affects gene expression; thereby influencing yield and quality of crops (Zhao & Running, 2010). In order to combat this stress, plants have evolved innovative adaptive mechanisms comprised of the high accretion of compatible solutes in the cytoplasm (Chemikosova, Pavlencheva, Gur'yanov, & Gorshkova, 2006). Water stress may affect several metabolic pathways, for example, photosynthesis, carbohydrate metabolism, glycolysis, and hormone biosynthesis (Dennison et al., 2003). Stress tolerance is acquired via the accumulation of proline, betaine, and sugars such as fructose and sucrose (Wang et al., 2012). Metabolomic analysis may also contribute to the understanding of drought tolerance, but the data pertaining to drought tolerance associated with metabolomic constituents is still scarce. Metabolomics analysis has been employed to investigate the abiotic stress tolerance of plants; this analysis can establish the explicit responses of biological systems to genetic and ecological alterations (Oliver et al., 2011). Metabolomic investigation is comprised of varied strategies; for example, metabolic fingerprinting, metabolite profiling, targeted analysis, GC-MS, LC-Fourier transform MS, and NMR, to detect small-molecule metabolomic constituents (Kell, 2004; Viant & Sommer, 2013). These approaches can be used to detect metabolomic components precisely (Ruan & Teixeira da Silva, 2011; Barding, Salditos, & Larive, 2012; Barding, 2013). The metabolite profiling in barley exposed to salt stress (Wu et al., 2013), temperature (Kim, Bamba, Harada, Fukusaki, & Kobayashi, 2006), N nutrition (Allwood et al., 2015), and drought stress has been investigated (Bowne et al., 2012). Guo et al. (2018) investigated long-term drought stress in the drought-tolerant (DT) wheat genotype JD17, which stimulated metabolic shifts in the tricarboxylic acid (TCA) cycle, and glycolysis, with the reduction of a γ-aminobutyric acid cycle shut process. The wheat genotype JD17, under drought stress, stimulated the accretion of osmolytes such as proline, and sugars such as sucrose, fructose, mannose, and malic acid. This research extended our perspective on the mechanism implicated in wheat seedlings' drought tolerance. This study also illustrated that GC-MS metabolomics could be a highly efficient strategy for comprehending drought stress effects on plant metabolism. Michaletti, Naghavi, Toorchi, Zolla, and Rinalducci (2018) revealed the integrative biochemical networks of wheat leaves against water stress, and proteomics and metabolomics were used on two spring wheat cultivars (*Bahar*, drought-susceptible; *Kavir*, drought-tolerant). Drought stress stimulates harmful effects on Bahar leaf proteomes, leading to an intensive decline of total protein content, with impairments primarily in photosynthetic proteins and in enzymes implicated in sugar and nitrogen metabolism, along with the detoxification of harmful metabolites. In contrast to this, there are few alterations observed at the protein level in Kavir-stressed leaves. Metabolomes investigated in amino acid pools, organic acids, and sugars are prime metabolites that change tremendously in response to water stress. Bahar cv exhibited higher levels in proline, methionine, arginine, lysine, and aromatic and branched chain amino acids. The accretion of tryptophan via the shikimate pathway appears to preserve auxin production (indoleacrylic acid), while glutamate reduction is associated with polyamine biosynthesis. The kavir metabolome was influenced by drought stress, to a lesser extent, with only two pathways considerably altered; one of them being purine metabolism. These observations methodically contribute to a framework for improved perspectives on the mechanism that controls plant cell response to water stress, with insights into molecules that can be employed for crop enhancement projects. Chmielewska et al. (2016) analyzed proteomics and metabolomic variations in leaves and roots of two barley (*Hordeum vulgare* L.) genotypes with distinct drought tolerance that were exposed to a water deficit. An integrative approach between two-dimensional electrophoresis (2D-PAGE) and matrix-assisted laser desorption time of flight MS (MALDI-TOF and MALDI-TOF/TOF) investigations displayed 121 drought-responsive proteins in leaves, and 182 in roots of both genotypes. Several detected drought-responsive proteins were linked with processes that are generally remarkably influenced during water stress comprised of photosynthesis and carbon metabolism. However, the maximum number of detected leaf and root proteins reflected a typical defense mechanism. Further, the variations in the protein accretion that symbolize a process previously unrelated to drought responses, for example, phenylpropanoids metabolism, were also detected. This study revealed about 100 drought-affected, low-molecular-weight metabolites symbolizing varied metabolite types, with amino acid being the most affected metabolite class, as confirmed via tandem GC-TOF MS. This study compared the data procured from proteomic and metabolomic investigations to explore the current association between these levels of molecular organization. This study also revealed organ specificity of recorded variations, and displayed divergence in response to water stress of drought susceptible and tolerant barley lines. This study signified that numerous detected proteins and metabolites whose accretion levels were amplified with drought in the investigated susceptible barley variety displayed higher constitutive accretion of levels in the drought-resistant line. This might indicate that constitutive biochemical predisposition symbolizes a more-improved drought tolerance mechanism than inducible responses. Sánchez, Erban, Udvardi, and Kopka (2012) evaluated comparative variations via GC/electron impact-TOF MS (GC/EI-TOF-MS) in metabolite levels of *Lotus japonicus* ecotype Gifu plants subjected to water stress. There was an increase in fructose, glucose, galactose, and maltose levels. Obata et al. (2015) reported GC-MS-based metabolite profiling to investigate the metabolite levels in the leaf blades of 10 hybrids in maize subjected to water stress. This study illustrated an expansive assortment that was comprised of numerous sugar alcohols that decreased, for example, maltose and erythritol, or did not change (e.g., raffinose and xylitol). Further, there was pronounced accretion of an extensive assortment of amino acids such as tryptophan, alanine, serine, glycine, leucine, and γ-aminobutyric acid (GABA). Sun et al. (2016) investigated the metabolic variations via NMR in maize plants exposed to drought conditions. This study revealed that water stress levels enhance the concentration of organic acids implicated in the TCA cycle, such as malate and citrate, and amino acids such as proline and alanine, and sugars, particularly fructose. Sucrose and glucose act as osmolytes to regulate cell homeostasis. Nyarukowaa, Koecha, Lootsc, and Apostolidesa (2016) detected and estimated metabolites via targeted analysis by GC-MS/MS and UPLC-MS in tolerant and drought-sensitive (DS) *Camellia sinensis* cultivars exposed to varied water stress levels. This study reported the significant alteration in metabolite levels between DT and DS tea cultivars. The amino acids such as asparagine, glycine, valine, isoleucine, proline, and leucine were significantly higher in the DT cultivars than in the DS cultivars. The high amino acid levels in DT cultivars justify their function as osmolytes (Shi & Chan, 2014). Further, this inclination toward higher amino acid levels results in lowered protein levels due to degradation of proteins to amino acids. Moreover, a significant increase in the polyols level, such as mannitol, was also reported. This study further reported that mannitol accumulated at a lesser rate in comparison with

DT, than in DS tea cultivars. This metabolite renders stress tolerance via actively scavenging hydroxyl radicals, and is also observed in lesser concentrations in DT cultivars than in DS wheat crops. Accumulation of polyphenols is the first line of defense against varied water stress levels (Di Ferdinando, Brunetti, Agati, & Tattini, 2014), due to their ability to donate hydrogen and scavenge singlet oxygen (Labanowska et al., 2013). Scalabrin et al. (2015) demonstrated higher polyamine levels in *Nicotiana langsdorffii* thriving under a water deficit, as investigated via metabolomic approaches such as LC-MS. Sánchez-Rodríguez et al. (2011) reported an inclination toward a higher pool of hydroxycinnamic acid and flavonoids in diverse species, as documented in leaves of *Solanum lycopersicum* cultivars exposed to a water deficit.

Metabolomic approaches are extensively employed for investigating the salt-induced accumulation of major metabolites in specific tissues and organs implicated in N metabolism, such as allantoin, urea, and glutamine, among others (Slama, Abdelly, Bouchereau, Flower, & Savouré, 2015). GC-MS analysis of the perennial ryegrass (*Lolium perenne*) demonstrated a remarkable increase in carbohydrates, and a 71.9% rise in total metabolite content in leaves (Hu, Chen, Hu, Amombo, & Fu, 2016). This study revealed 42.2% sucrose among the total carbohydrate content, which was related to its imperative function as major signal molecule in source-sink regulation, and in plant structure and metabolism. Hu et al. (2013) illustrated metabolic variations in ryegrass plants exposed to 255 mM NaCl for 12 h or 4 days, and revealed higher levels of carbohydrates such as tagatose, psicose, glucoheptose, idose, allose, talose, lactulose, and cellobiose; and one sugar alcohol (inositol). Hence, carbohydrate accumulation is a predominant feature of plants exposed to varied salt stress levels, due to their potential role in ROS scavenging and the aversion of cell lipid peroxidation (Keunen, Peshev, Vangronsveld, Van Den Ende, & Cuypers, 2013). Sanchez, Siahpoosh, Roessner, Udvardi, and Kopka (2008) compared tolerance levels in DT plants in *Thellungiella* halophytes and salt-sensitive *Arabidopsis thaliana*, which was investigated through metabolic fingerprinting. This revealed similar NaCl dose-dependent patterns comprised of the steady shifting of metabolomes with increasing salt contents in the medium until a division at higher stress intensities activated a remarkable metabolic reconfiguration. There was a high accumulation of soluble sugars such as fructose, glucose, sucrose, raffinose, and other unidentified putative complex glycans under salt stress; although metabolite levels were more prominent in *Thellungiella* (Lugan et al., 2010). *Thellungiella* plants accumulated higher levels of proline and exhibited lower mobilization as compared with *Arabidopsis* plants. Similar research studies were carried out in the halophyte *Prosopis strombulifera* grown under varied salt concentrations, and seedlings were analyzed via GC and UPLC-q-TOF-MS/MS (Llanes, Arbona, Gómez-Cadenas, & Luna, 2016). This study revealed an accumulation of amino acids, organic acids under salt stress, and the cell damage as result of the high respiration rate induced by sulfate anion in the medium. In contrast to this, there was an upsurge pool of explicit amino acids such as leucine in roots, and peptides such as cysteinyl-glycine, in leaves, as recorded in NaCl-treated plants, which could be an adaptive strategy to maintain cell homeostasis during salt stress. Wang, Liu, Li, and Han (2015) investigated metabolomic approaches such as GC-MS in the shoots and roots of two similar rice genotypes with distinct salt tolerance levels thriving under varied salt and abscisic acid (ABA) treatments and their combinations. Their study illustrated organ-specific transcriptome reprogramming and primary metabolite alterations in the two genotypes against varied treatments. Metabolite profiling integrated with transcript data provided the evidence that ABA could activate a readjustment of carbohydrate metabolism and carbon mobilization in salt stress seedlings. But there is scant data available pertaining to metabolome profiling that demonstrates the association between ABA and salt stress responses in plants. Metabolomic profiling of *Prosopis strombulifera* investigated the concentration of the oxylipin 13 hydroxyoctadecadienoic acid identified via UPLC-qTOF-MS/MS, which declined in the roots of salt-treated plants. Llanes et al. (2016) identified the jasmonic acid methylester in the same plants by GC-MS, and confirmed the concentration was organ-specific under varied salinity stress. Bendaly et al. (2016) illustrated variations in leaf metabolomes via GC-flame ionization detector (GC-FID) of xerohalophyte species *Atriplex halimus* subjected to salinity (400 mM NaCl). The observations demonstrated high antioxidant activity, which was related to numerous metabolites, such as sucrose, threonine, phenylalanine, α-alanine, and proline. Lee et al. (2016) carried out the metabolomic studies using UPLC-MS in the roots of *Salicornia herbacea* subjected to varied NaCl levels, and it was revealed that structural and functional analyses exhibited 53 metabolites as NaCl concentration increased in the medium. Muscolo et al. (2015) compared the metabolic profiles of two species with distinct tolerance levels to water and salt stress, and the study revealed overlapping biochemical pathways, along with explicit metabolic adjustments to diverse environmental conditions. Keeping the previous research reports in mind, we may conclude that metabolomics is required to promote the information of plant physiology, due to its elaborate data pertaining to dynamic metabolic networks. The divergence and expansive number of metabolites identified in each study may make it difficult to understand the function of each one of them during abiotic stress response conditions in the plants. Despite the diversity of the metabolites, they have to be depicted as a snapshot of the metabolic pathways present in the tissue under particular conditions. The comparative metabolite profiling among species and cultivars varying in their stress tolerance, together with the metabolite content of wt plants and mutant and transgenic species, were documented to emphasize qualitative and quantitative discrepancies associated with the phenotypes. The identification and characterization of the explicit metabolites can also symbolize good indicators for stress tolerance phenotypes in the screening of varieties, such as in plant breeding programs. Currently, omics approaches are emerging as an imperative strategy for illustrating plant responses to diverse stress conditions. The integrated approaches of metabolomics, proteomics, transcriptomics, and statistical tools will provide us with an entire snapshot of how plants respond to stresses, and help with developing strategies to amplify drought and salt stress tolerance in varied plant species. Furthermore, the inculcation of these tolerance traits in response to varied stresses in breeding programs of agriculturally important crops is a promising development for the future.

3.4 SULFUR AND PHOSPHORUS STRESS

Nutrient limitation also significantly influences plant growth and metabolism, particularly deficit levels of macronutrients such as carbon (C), nitrogen (N), phosphorus (P), and sulfur (S) (Bielecka et al., 2015; Obata & Fernie, 2012). Nutrient limitation has a direct influence on metabolism, because a majority of organic metabolites includes the amalgamation of these elements (Arbona, Manzi, Ollas, & Gómez-Cadenas, 2013; Sanchez et al., 2011). The varying environment has constantly altered the balance between carbon absorption and carbon assimilation (Fatichi, Leuzinger, & Körner, 2014). The carbon limitation for even shorter periods of time resulted in growth inhibition, which is not instantly reversed when C becomes available again (Bauerle et al., 2012). The metabolite profile of *Arabidopsis* seedlings grown under limited carbon levels was investigated, and it was observed that carbon-starved seedlings exhibited lower levels of carbohydrates, and carbon-containing metabolites, such as myo-inositol, raffinose, glycerate, and fatty acids (Nikiforova et al., 2005). Moreover, inhibition in N and

S assimilation also led to a decrease in S-containing amino acids, such as methionine and glutamine, glutamate, aspartate, and alanine (Nikiforova et al., 2006). The higher level of amino acids may also be due to proteolysis. A majority of the changes are regressed instantly following re-addition of sucrose into the media. Usadel et al. (2008) employed an extended dark period to induce C starvation under more natural conditions in the *Arabidopsis* rosettes. Interestingly, the variation in metabolite levels was mainly comparable to those observed in liquid culture seedlings. The significant decline in carbohydrate levels within first 4 hours of the prolonged dark period signifies that the treatment induced C starvation significantly, and that carbohydrates commenced to severely limit metabolism (Gibon et al., 2009). However, organic acids and other carbon-containing metabolites exhibited a somewhat slow decline. The dark treatment for a prolonged period induces acute C starvation and leaf senescence at the end of the experiment. The metabolite profile of *Arabidopsis* leaves exposed to a prolonged dark period has been investigated in various research studies to elaborate the metabolic bases of dark induced senescence, and the function of the mitochondrial substitutive electron transport pathway (Skirycz et al., 2010). Even though these two research studies illustrate similar metabolic responses for the first few days of dark treatment, a subset of metabolites displays a biphasic profile for longer dark periods. This is specifically recognized for some TCA cycle intermediates comprised of fumarate, isocitrate, malate, and succinate, which pooled following 7 days of dark treatment, regardless of reductions during the first 3 days of treatment. Moreover, accumulations of amino acids such as GABA, particularly, are pronounced (Fait, Fromm, Walter, Galili, & Fernie, 2008). Metabolite profiles were also evaluated, and an assortment of mutant deficits were observed in the genes implicated in mitochondrial alternative electron transport flavoprotein/electron transfer, including a flavoprotein:ubiquinone oxidoreductase (ETF/ETFQO) complex, such as *ETFQO* (Yang & Swenson, 2007) and *ETFβ* (Frerman, 2013), plus enzymes implicated in the stipulation of its substrates, such as *IVDH*, *D2HGDH*, and *PSHX* (Kirma, Araújo, Fernie, & Galili, 2012). Moreover, the elaborate investigation displayed the implication of the ETF/ETFQO complex in the catabolic pathways of branched chain amino acids and lysine, and functioned as an electron donor to the mitochondrial ubiquinol pool (Andreyev, Kushnareva, Murphy, & Starkov, 2015). These investigations indicated that more integrative investigation of the function of all characteristics of protein degradation and resultant remobilization should be evaluated within the framework of comprehending metabolic response to stress (Zhou et al., 2016). Urbanczyk-Wochniak and Fernie (2004) investigated the influence of N deficiency on the metabolite levels in tomato leaves, and observed lower levels of amino acids. The prime regulators of carbon and nitrogen metabolism, namely 2-oxoglutarate, along with TCA intermediates such as citrate, isocitrate, succinate, fumarate, and malate declined (Sweetlove, Beard, Nunes-Nesi, Fernie, & Ratcliffe, 2010). Tschoep et al. (2009) investigated the limited N supply in *Arabidopsis*, and observed that malate and fumarate levels were remarkably reduced in a low N environment. However, their rosette protein content was consistent, and total amino acid levels increased in comparison with the N-replete plants. The results displayed that *Arabidopsis* responds adaptively to low N conditions. Morcuende et al. (2007) evaluated the metabolite profile of *Arabidopsis* seedlings thriving in a liquid culture under P starvation. The levels of sugar phosphate were very low, but metabolites further down in glycolysis, glycerate 3-phosphate, glycerate-2-phosphate, and phosphenol pyruvate increased in P-deficient seedlings. P-deficient seedlings exhibited a significant accretion of starch, sucrose, and reducing sugars, together with organic acids such as citrate, fumarate, malate, and oxoglutarate. The content majority of amino acids remains the same or increases slightly, while those of numerous amino acids comprised of aromatic acids such as histidine, arginine, and threonine declined. Transcriptomic data, along with metabolite analysis, reveals that P starvation results in a shift toward the accretion of carbohydrates, organic acids, and amino acids (Farag et al., 2018). The effects of P starvation on crops such as common beans and barley were also investigated. Hernández et al. (2007) employed metabolite profiling to monitor the effect of P deficiency in the roots and nodules of the common bean. The majority of the amino acids were increased in P-stressed roots. The accretion of numerous sugars anticipated that sugars may be partitioned preferentially to P-stressed roots to support the expression of P-stress-induced genes (Hajlaoui, El Ayeb, Garrec, & Denden, 2010). The declined levels of organic acids possibly reflect exudation of metabolites from the roots into the rhizosphere. Amino acids and other N-containing metabolites were reduced, together with sugars, while organic acids were accumulated in P-deficient nodules. Such a divergent response may be because of N deficiency in P-limited nodules, in which the sole N supply from fixed N_2 might be restrained under environmental limitations, such as

P starvation (Jeudy et al., 2010). Huang et al. (2008) investigated metabolites from both shoots and roots of P-deficient barley. Limited P amplified the levels of phosphorylated intermediates such as glucose-6-P, fructose-6-P, inositol-1-P, and glycerol-3-P; and organic acids such as 2-oxoglutarate, succinate, fumarate, and malate. The observations displayed that P-deficient plants altered carbohydrate metabolism originally to decrease P consumption and salvage P from small P-containing metabolites, which subsequently decreases the levels of organic acid in TCA cycle (Wasaki et al., 2003).

Sulfur is another macronutrient indispensable to the biosynthesis of the S-containing amino acids cysteine and methionine, together with an extensive range of S-containing metabolites, including glutathione (Anjum et al., 2012; Kopriva, Mugford, Matthewman, & Koprivova, 2009). There are few metabolomic studies on the responses to S starvation in *Arabidopsis* (Hoefgen & Nikiforova, 2008; Nikiforova, Kopka, et al., 2005; Ostaszewska, Juszczuk, Kołodziejek, & Rychter, 2014). During the course of S-stress response, two metabolic states can be differentiated. The instant metabolic response is comprised of the decline of organic S-containing metabolites on the S assimilation; for example, cysteine and glutathione, resulting in the accumulation of their precursor *O*-acetyl-serine (OAS) together with serine, and to the successive re-channeling of the metabolic flow to glycine and tryptophan. Glucosinolate catabolism is induced to retrieve S from it. After a long-term response, the lipid contents and S-containing molecules (*S*-adenosyl methionine) declined. Low S supply results in N imbalance, and further to the variations in carbon (C1) metabolism that link photorespiration, S assimilation and N dumping (Patel, Huang, Low-Décarie, & Lefsrud, 2015). The results of the *Arabidopsis* plants with altered OAS levels indicated the significance of this metabolite, because OAS functions signaled the role for an explicit sulfate response, as well as for the regulation of the transcript levels of an explicit gene set, irrespective of the sulfur status of the plants.

Potassium is an indispensable organic metabolite that is required as a cofactor of enzymes implicated in numerous biosynthetic reactions (Hasanuzzaman et al., 2018). Armengaud, Breitling, and Amtmann (2010) employed metabolite profiling to detect metabolic targets of K stress. Metabolite profiles of limiting K *Arabidopsis* plants were identified by high metabolite profiling to characterize metabolic targets of K stress. Metabolite profiles of low K *Arabidopsis* were identified by high enhancement of soluble sugars such as sucrose, fructose, and glucose, and less enhancement in total proteins and the general amino acid level. Numerous basic or neutral amino acids accumulated during K deficiency, while acidic amino acids declined. Further, an immense decline of pyruvate and organic acids was observed, but only in the roots, and not in the shoots. They also estimated enzyme activities and determined that the main consequence of K deficiency induces an inhibition of glycolysis by the direct enzyme inhibition.

Sulfur stress was investigated via a metabolomic approach. Nikiforova, Daub, Hesse, Willmitzer, and Hoefgen (2005) employed untargeted GC-MS and LC-MS profiling to assess the response of 134 identified metabolites, and 6023 unidentified nonredundant ions traced to sulfur limitation. The reconstruction of the coordinated network of metabolic regulation was induced by sulfur stress via metabolic profiling (Nikiforova, Daub, et al., 2005). This metabolomic data was therefore investigated along with transcriptomics data to reorganize gene-metabolite link networks implicated in *Arabidopsis* response to sulfur starvation (Nikiforova, Kopka, et al., 2005). Integration of transcriptomics and metabolomics approaches was employed to analyze transcriptional and metabolic responses of bean plants growing under P-deficient and P-sufficient conditions (Hernández et al., 2007). GS-TOF-MS profiling of common beans thriving under phosphorus stress conditions detected an array of metabolites considerably varied in P-deficient roots. A majority of metabolites such as amino acids, polyols, and sugars were increased in P-stressed plants (Hernández et al., 2007).

3.5 OXIDATIVE STRESS

Numerous research studies have ascertained the functional role of the metabolic system, such as the defensive role of the ascorbate glutathione system in combating oxidative stress in plants (Akram, Shafiq, & Ashraf, 2017; Nakabayashi & Saito, 2015; Noctor, Lelarge-Trouverie, & Mhamdi, 2015). Jiménez-Arias et al. (2015) illustrated the dynamics of the metabolic flux against oxidative stress induced by menadione (MD) in heterotrophic *Arabidopsis* cells. This study employed GC-MS profiling to estimate the levels of 50 polar metabolites after stress treatment, and linked metabolic changes to corresponding changes in mRNA levels estimated in the sample. This study illustrated that oxidative stress primarily led to marked inhibition of the TCA cycle, and expansive sectors of amino acids metabolism, followed by the backing up of glycolysis, and routing of the carbon flow toward the oxidative pentose phosphate pathway (OPPP). Transcriptomic investigation of similar samples also exhibited a coordinated transcriptional response to tackle with stress with main triggers from anabolic, to catabolic, metabolism (Kumar, Sreeharsha, Mudalkar, Sarashetti, & Reddy, 2017). The generalized result of the majority of abiotic stresses is enhanced generation of ROS, which are extremely toxic, and lead to degradation of proteins, lipids, carbohydrates, chlorophyll, and DNA; thereby leading to oxidative stress (Gill & Tuteja, 2010; Kundu et al., 2018). ROS are prime by-products of processes such as photosynthetic or respiratory electron transport. There is generally equilibrium between the generation and scavenging of ROS, but abiotic stress factors may hamper this equilibrium, resulting in an abrupt rise in intracellular levels of ROS (Apel & Hirt, 2004). The majority of research studies have investigated ROS-scavenging enzymatic antioxidants that symbolize the first defense mechanism, while very few research studies have been reported pertaining to the direct results of oxidative stress on the plant metabolome (Ahmad, Sarwat, & Sharma, 2008; Li, Liu, & Zhang, 2010). Baxter et al. (2007), employed redox-active quinone MD for inducing oxidative stress in *Arabidopsis* cell suspension cultures, and identified the dynamics of metabolic alteration as reflected by changing metabolite fluxes, and in ^{13}C-labeling kinetics. The combined investigation of such metabolic fluxes, such as an increase in hexose and triose phosphate, gluconate ribose; a decline in malate, and few amino acids, signified a marked inhibition of the TCA and routing of carbon flux into the OPPP. The decline of prime antioxidants such as ascorbic acid associated with the accretion of threonate formed following its degradation, signifying continuing intense oxidative stress with a failure to recycle the oxidized ascorbate completely (Baxter et al., 2007). In contrast, these studies performed on *Arabidopsis* roots from hydroponically grown plants outlined comparable metabolic variations in immediate MD responses (30 min), following prolonged oxidative stress after 2 and 6 h. Changes recorded in *Arabidopsis* roots and cultured cells obviously differentiated (Lehmann et al., 2009). The roots treated with MD displayed 56 polar metabolites, 33 among which were immensely affected within 30 min, and 39 were altered in as little as two time points. The primary variations parallel to interpretations in cell cultures were comprised of a decline in the TCA cycle metabolites and linked amino acids, and a rise in OPPP intermediates such as ribose-5-P and ribulose-5-P, and some glycolytic intermediates (Genga et al., 2011). With the passage of time, the metabolite content (e.g., TCA cycle) intermediates and some amino acids reverted to actual values, and increased in the roots, contrary to the response of cultured cells, in which the majority of metabolites remained lessened during this time (Li, Wang, Pidatala, Chang, & Cao, 2014). However, there was significant disparity among cells growing in culture to those of roots in regard to glycolysis, while cultured cells displayed a continuous rise in hexose 6-phosphate, and transient enhancement in 3-PGA, while cells in the roots exhibited a marked decline in hexose 6-phosphate, and linear enhancement in pyruvate was observed (Shingaki-Wells et al., 2011). Furthermore, the menadione-treated roots exhibited higher proline levels, with an associated decline in its precursor, glutamate. This was followed by an alteration in polyamine metabolism, with a decline in putrescine, and accretion of some methionine-derived aliphatic glucosinolates (Lehmann et al., 2009). Higher tolerance to MD-induced oxidative stress was exhibited by a rise in cultured cells overexpressing the *Arabidopsis Bax Inhibitor-1* (*AtBI-1*) gene (Ishikawa et al., 2010). Bax Inhibitor-1 is an endoplasmic reticulum membrane protein, which functions as a cell death suppression factor that is comprehensively conserved in animals and higher plants. In this study, the authors employed CE-MS to investigate metabolic responses in rice against cell death-inducible oxidative stress. Metabolomic data revealed the shift in carbon flow from the central pathway to OPPP in control rice cells due to higher demand for NADPH as the reducing power, along with data acquired in roots and cultured cells of *Arabidopsis* (Lehmann et al., 2009). Although the decline of carbon metabolism took place in the central pathway, a significant accretion of the majority of amino acids originated from the PEP. Pyruvate and oxaloacetate was found in MD-treated rice cells. This result was inconsistent with data acquired in MD-treated *Arabidopsis*, in which reduced levels of numerous amino acids were associated with a decline in their precursors (Lehmann et al., 2009). But the overexpression of *AtBI-1* did not generate any remarkable effect on the primary metabolism in control cultured cells. However, obvious differences between *AtBI*-1 overexpressing and control cells was observed following a 24 h exposure to stress, primarily in some metabolic pathways (i.e., glycolysis, amino acids of the glutamate and aspartate families), and the machinery of redox and energy metabolism. These observations indicated that tolerance to oxidative stress was conferred by the AtBI-1 factor because of the higher capacity of metabolic acclimation, with the recovery of metabolite composition that declined during the early response (Petrov, Hille, Mueller-Roeber, & Gechev, 2015). Metabolomic approaches were also employed in the plants exposed to O_3. Due to global warming in recent years, ozone depletion has significantly impacted plant growth, development, and productivity

(Cho et al., 2008; Lankadurai, Nagato, & Simpson, 2013). Ozone is the photochemically produced air pollutant that is released into the intercellular space of leaves via stomata, which reacts with water and instantly produces ROS (Mills et al., 2018). The ozone has varying effects on plants, depending on its concentration and length of exposure, and on the susceptibility of the plants, and signs of damage may range from visible chlorosis and necrosis in the leaves, to inhibition of photosynthesis and lower yields (Emberson et al., 2018; Ghosh, Singh, Agrawal, & Agrawal, 2018). The effects of the ozone have been illustrated with two main methods: by subjecting plants to acute or chronic doses of ozone, which creates further stress conditions. The plant responses to the ozone have been investigated via omics tools in diverse species, for example, *Arabidopsis*, rice, and birch (Bison et al., 2018; Cho et al., 2008; Li, Muthuramalingam, & Mahalingam, 2015; Vainonen & Kangasjärvi, 2015).

Researchers have combined transcriptomics, proteomics, and metabolomics approaches in rice seedlings subjected to O_3 exposure. CE-MS-based metabolomic profiling revealed a rise in the content of numerous amino acids, such as GABA, glutathione, and sakuranetin, a main rice secondary metabolite (Cho et al., 2008). Integrating diverse omics approaches enabled the authors to indicate glutamate, GABA, and glutamate dehydrogenase as potent biomarkers for O_3 damages in rice. Long-term ozone exposure in white birches (*Betula pendula* Roth) with varied O_3 sensitivity was investigated using leaf metabolomic data, which revealed 339 low-molecular-weight metabolites. Ozone enhancement led to a high flux of phenolic metabolites, for example, chlorogenic acid, quercetin glycosides, and lipophilic compounds linked to leaf cuticular wax formation (Yamaji, Julkunen-Tiitto, Rousi, Freiwald, & Oksanen, 2003). In contrast, a decline in the concentration of several carbohydrates and chlorophyll-associated metabolites was induced by higher ozone levels (Oksanen, Freiwald, Prozherina, & Rousi, 2005).

3.6 HEAVY METAL STRESS

The heavy metal stress induced in plants can be investigated via metabolomic approaches such as NMR-based fingerprinting and metabolite profiling (Ariza, García-Barrera, García-Sevillano, González-Fernández, & Gómez-Jacinto, 2013; Bundy et al., 2007). Metabolite fingerprinting via NMR spectroscopy integrated with multivariate statistics investigation was employed to differentiate between control and cadmium treated *Silene cucubalus* cell cultures (Bailey, Oven, Holmes, Nicholson, & Zenk, 2003). Metabolites that exhibited enhancement in cadmium treated cells were detected as malic acid and acetate, while glutamate and branched chain amino acids declined (Hughes, Bundy, Want, Kille, & Stürzenbaum, 2009). Metabolite profiling of *Arabidopsis* cells subjected to cesium stress via NMR exhibited that metabolite alteration due to Cs stress is comprised primarily of sugar metabolism and glycolytic fluxes, and is dependent on potassium levels in the cells (Le Lay et al., 2006). With the advent of the industrial revolution, the biosphere was polluted with toxic metals at an alarming rate (Förstner & Wittmann, 2012), and pollution of agricultural soils via heavy metals has become an imperative ecological issue because of its significant ecological effects (Wu et al., 2010). Plants growing in soil contaminated with heavy metals exhibit visible symptoms such as chlorosis, growth retardation, browning of root tips, impaired ionic transfer of micronutrients from roots to shoots, generation of oxidative stress and ROS, altered germination, disruption of lipids in biomembranes together with harmful effects on plant physiological processes such as photosynthesis, water relations, and mineral nutrition (Jan & Parray, 2016). Bailey et al. (2003) employed high field ^{1}H NMR spectroscopy combined with chemometric investigation to acquire data pertaining to variations in endogenous profiles for *Crotalaria cobalticola* plant cells subjected to cobalt chloride. Control samples displayed a higher flux of branched chain amino acids and secondary metabolite precursors phenylalanine and tyrosine, while choline, glutamate, alanine, and lactate were higher in the dosed samples. Li, Gao, Gjetvaj, Westcott, and Gruber (2009) subjected *Brassica carinata* seedlings to higher concentrations of lithium ions. The seedlings displayed pronounced accumulation of lithium, and the lipid and phenolic composition dramatically changed in brown-seeded seedlings. The sinapic acid esters and chloroplast lipids were substituted with benzoate derivatives such as resveratol and oxylipins following lithium exposure. On the contrary, the yellow-seeded plants maintained the same phenolic and lipid composition prior and subsequent to lithium, and did not tolerate the high metal concentration that was tolerated by the brown-seeded line. Metabolomic approaches confirmed the accumulation of some amino acids implicated in the cell compartmentation and cadmium detoxification in tomato seedlings subjected to cadmium stress (Sharma & Dietz, 2006). Roessner et al. (2006) have also employed the metabolomic approach to compare metabolite profiles in root and leaf tissues of an intolerant, commercial cultivar of barley (cv *Clipper*), and a boron-tolerant Algerian landrace (cv Sahara) subjected to high boron concentrations (200 and 1000 μmol/dm^3). The authors investigated the number and amplitude of metabolite changes in roots that were greater in Clipper than in Sahara. Metabolomic approaches were also employed for the investigation of nickel-rich latex of the hyperaccumulating *Sebertia acuminata* via GC-MS analysis. There were more than 120 metabolites identified, and 57 among them were subsequently detected (Callahan et al., 2008). The first major metabolite reported in the latex was methylated aldaric acid (2,4,5-trihydroxy-3-methoxy-1,6-hexan-dioic acid), and its structure was validated by 1D and 2D NMR spectroscopy. Following this, citric acid is the next major small organic metabolite present in the latex. Nickel(II) complexes of stoichiometry Ni(II):acid = 1:2 were identified for these two acids, along with malic, itaconic, erythronic, galacturonic, tartaric, aconitic, and saccharic acids. These observations contribute further validation that organic acids may play an important role in the transport, and probably in the storage of metal ions in hyperaccumulating plants. Metabolomic response in tomato plants subjected to long-term cadmium toxicity was demonstrated in hydroponic conditions wherein cadmium accumulation was found to be pronounced in the roots and lower parts of the plants (Mazzei et al., 2016). Metabolomic approaches such as ^{1}H NMR, HPLC-PDA and colorimetric approaches yielded the absolute quantification of 28 metabolites. This study further demonstrated lower levels of ascorbic acid and proline, and increased levels of α-tocopherol, and asparagine and tyrosine accumulation increased, as observed in 100 μmol/dm^3 cadmium-treated leaves. Pigment concentration, like that of caroteniods and chlorophyll content, declined with 100 μmol/dm^3 cadmium-mature leaves, which reinforces that reduced gene expression is indispensable for isoprenoid and carotenoid accretion. Hédiji et al. (2015) illustrated results that display marked metabolic and physiological changes in seedlings exposed to long-term exposure of 20 μmol/dm^3 cadmium using GC-MS. This study revealed identification of 80 metabolites by their retention time indices, and specific mass fragments were detected. The levels of carbohydrates, organic acids, amino acids and other stress-responsive metabolites altered under cadmium stress. Cd-treated plants exhibited higher levels of alanine, β-alanine, proline, serine, putrescine, sucrose, and other metabolites with compatible solute-like properties, particularly 4-aminobutyric acid, glycerol, raffinose, and trehalose in comparison with control plants (Parida, Panda, & Rangani, 2018). Moreover, concentration of antioxidants such as α-tocopherol, campesterol, β-sitosterol, and isoflavone also increased significantly. These observations further validate the significant function of antioxidant metabolites in the defense mechanism against cadmium stress, and indicate that the metabolite profile is a potent contrivance that can assist in instantly categorizing ecologically altered plants, and unravelling the process of cadmium stress responses (Xie et al., 2014). The metabolomic investigation in rice roots was carried out to link differential transcriptome data to a biological method. This study described the intricate network of regulatory pathways that adapts chromium responses of rice. The combined matrix of both transcriptome and metabolome data, following appropriate normalization and preliminary computation, created a visual snapshot of the association between metabolites (Xia & Wishart, 2016). The prevalence of diverse motifs in the gene subsets indicated the implication of motif-explicit, transcription-altering proteins in chromium stress. So as to evaluate the metabolomic changes induced by the metal ions in Brassica, plants were exposed to varied concentrations of copper, iron, and manganese in separate treatments. ^{1}H NMR and 2D ^{1}H NMR spectra, coupled with PCA and partial least square-discriminant analysis (PLS-DA), were employed to evaluate the metabolic change in *Brassica rapa* (var. Raapstelen) (Jahangir, Abdel-Farid, Choi, & Verpoorte, 2008). Diverse chemometric procedures, followed by ^{1}H NMR evaluation, exhibited numerous metabolic consequences (Fonville et al., 2010). Among the metabolites that revealed disparity, glucoinolates and hydroxycinnamic acids conjugated with malates were found to be the differentiating metabolites, as were main metabolites such as carbohydrates and amino acids. This study revealed that the effects of copper and iron on plant metabolism were more amplified than that of manganese, and that the metabolomics varied per kind of metal and per concentration (Bundy et al., 2008). Le Lay et al. (2006) analyzed the integration and localization of ^{133}Cs in a plant cellular model, and the metabolic response induced was investigated as a function of external potassium concentration. Metabolomic analysis revealed cesium stress as a function of potassium nutrition effects via ^{13}C and ^{31}P NMR analysis of organic acid extracts. These investigations indicated that sugar metabolism and glycolytic fluxes were affected depending upon the levels of K^{+} ions in the medium (Armengaud et al., 2009).

3.7 METABOLOMIC ANALYSIS OF PLANT-STRESS INTERACTIONS

Plants thriving under diverse stress conditions reveal different self-defense mechanisms, which are induced by altered gene expression patterns. This leads to qualitative and quantitative divergences in protein synthesis, resulting in the modulation of certain metabolic and defensive pathways (Cramer, Urano, Delrot, Pezzotti, & Shinozaki, 2011; Wingler & Roitsch, 2008).

The different metabolic profiling technologies contribute imperative insights to biologists to help them comprehend the defense mechanisms of plants thriving under stress conditions (Ahsan, Renaut, & Komatsu, 2009). Currently, metabolomic methods contribute diverse multivariate analyses derived from varied hyphenated and chromatographic detection systems, such as gas or LC coupled with MS, or NMR-based approaches (Gika, Theodoridis, Plumb, & Wilson, 2014; Lindon & Nicholson, 2008). The metabolomic data mining and analysis can be carried out via an integrated mode of statistical methods, for example, independent component analysis (ICA) and analysis of variances (Ritchie, Holzinger, Li, Pendergrass, & Kim, 2015). Metabolomics integrated with gene expression, protein interaction, and other diverse regulatory pathways can be useful to different organisms with inconsequential variations (Alonso, Marsal, & Julià, 2015). Currently, this technology can be employed to complete metabolite accumulation in the cellular localization at a particular stage of growth in the cell. Metabolomics can be employed as a functional instrument to comprehend ecological responses in plants (Bundy et al., 2007). In recent years, metabolomics has been used as a tool for illustrating the biogenic volatile metabolites released by diverse plant species (Iijima, 2014). Metabolomics are more flexible in response to variations, compared with transcriptomes and proteomes. There are numerous regulatory genes induced in response to environmental stresses controlling diverse secondary metabolites in plants that are final products of diverse biochemical pathways. These metabolites can be identified as primary or secondary, and associations among the two classes are widespread. Primary metabolism is comprised of significant metabolites, for example, those that are implicated in prime carbohydrate metabolism (Rontein, Dieuaide-Noubhani, Dufourc, Raymond, & Rolin, 2002). Secondary metabolism might be associated with environmental signals consisting of cell signaling and interspecies communication. The biosynthetic pathway of prime metabolites is well conserved among species that illustrate the growth characteristics of diverse species. Research has provided demonstrated the significance and mechanism of metabolites for plant development. This is referred to as primary metabolism, instead of metabolism at the secondary level, because of the disparity in metabolic content. Most of the data pertaining to secondary metabolism is primarily acquired from proceeding forward in the genetic methods where explicit phenotypic traits may be associated with variations in specified metabolites. Naturally, such analysis is based on targeted methods where only a single metabolite might be analyzed (Yonekura-Sakakibara et al., 2008). Metabolomics studies in rice have illustrated that varied metabolites can promote seed germination (Moongngarm & Saetung, 2010). The distinctiveness of metabolites among wts and mutants plants (Wakasa et al., 2006), the metabolome profiling at diverse stages of development (Tarpley, Duran, Kebrom, & Sumner, 2005), and the investigation of regular metabolite differences among modified rice varieties has been documented (Kusano et al., 2011). Crop production has been significantly reduced due to synergistic effects of metabolites on stress tolerant drought, heat, and salinity, either individually or combined. Breeding for developing stress tolerance in crops is the most significant method to manage growing ecological challenges in agriculture. Significant analysis has been done in exploring plant responses to environmental stresses (Obata & Fernie, 2012; Urano et al., 2009), but the associations among varied stresses, and the mechanisms that regulate inert stress tolerance mechanisms have been less frequently studied (Cairns et al., 2013; Suzuki et al., 2013). Because of the intricate stereochemical orientation of atoms, the structure of metabolites is less understood than that of genes and proteins (Lewis & Gorrod, 2002). Genotypic and ecological interactions also affect the levels and composition of plant metabolites. Metabolomics is an important factor for consideration in the study of stress biology in plants and animals via the detection of varied metabolites. These metabolites are comprised of by-products of stress metabolism and stress signal transduction molecules that are implicated in the adaptation of plants to stresses. These could be confirmed via direct estimations, variations in proteomes and transcriptome expressions, and authenticated by mutant analysis (Ideker, Dutkowski, & Hood, 2011). When plants perceive changes in the environment, the signal transduction pathways instigate and activate diverse protein metabolites that restore homeostasis (Mittler, Vanderauwera, Gollery, & Van Breusegem, 2004). Complete metabolomic analysis for every plant would be quite expensive, but via metabolic profiling, a subclass of metabolites can be investigated at any time, or growth stage of the plant (Arbona et al., 2013). Plants exhibit diverse responses when they perceive biotic/abiotic stimuli that are associated to a specific stress tolerance mechanism. Alteration in primary metabolism is recognized as generally evident in all metabolic responses, and is comprised of modifications in the levels of sugar/sugar alcohol, amino acids, and TCA intermediates exhibiting general propensities in ecological stress responses. However, alterations in the secondary metabolism are specific to a particular stress, and are species-specific as well. For example, polyamines and sugars might be responsible for adaptation to extreme cold temperatures in *Thellungiella* accessions (Colinet, Larvor, Laparie, & Renault, 2012). Some of the metabolites associated with abiotic stress tolerance are comprised of polyols such as mannitols, sorbitol, dimethyldimethylsulfoniopropionate, sucrose, trehalose, fructan, proline, and ectoine that function as osmolytes and osmoprotectants to protect plants under varied stresses. An assortment of epicuticular waxes defend plants from further water loss via drought stress, and offer protection against the divergent functions of varied pathogens. Numerous low-molecular-weight metabolites include ascorbic acids, carotenoids, tocopherols, anthocyanin, and so forth. These protect plants from oxidative damage associated with a diversity of stresses, and defend them by tracking down active oxygen intermediates released during diverse stress conditions. The generation of phytolexins, and the initiation of the phenylpropaniods pathway and lignin biosynthesis are linked with plant stress defense mechanisms. Plant biosynthetic metabolites such as salicylic acid, methyl salicylate, and jasmonic acid and methyl jesmonate are generated under stress. They function as signaling metabolites, stimulating systemic defense in crops (Shulaev, 2006). In recent years, metabolomics have been employed for several applications, such as evaluation of stress effects in plants, following the contribution of particular metabolites within particular biosynthetic or secondary deprivation pathways, and organization of varied plant samples (Saito & Matsuda, 2010). The targeted or non-targeted method can be employed, depending on the particular objectives. The technique employed for investigation might affect the assembly of metabolites in the metabolomic platform. The majority of the currently employed metabolomics platforms are based on targeted methods. Overall, the actual metabolic organization of a given plant species is the result of a specific gene expression profile. Precursors and intermediate metabolites are activated to release bioactive metabolites when a certain metabolic pathway is activated. Plant hormones are fundamentally implicated in regulating the release of these precursors and intermediates, which can activate or deactivate metabolic steps (Zhao & Verpoorte, 2007). The stability, defense, and signaling of metabolites can be used to estimate plant tolerance to diverse abiotic stress conditions (Kerchev, Fenton, Foyer, & Hancock, 2012). It has been illustrated that sugars or sugar alcohols and amino acids are the primary metabolites that accumulate in plant parts when they are exposed to stresses (Krasensky & Jonak, 2012; Valerio et al., 2011). However, because of the variations in concentration, fluctuation in secondary metabolite levels is typically because of complex regulatory process (Arbona et al., 2013).

3.8 METABOLIC NETWORK IMPLICATED IN STRESS SIGNALING

Plants are exposed to an array of abiotic stresses, such as drought, cold, high temperature, salinity, and high radiation flux, which leads to reduced plant growth and development, finally resulting in plant death (Petrov et al., 2015; Singh et al., 2016). Plant stress responses are dynamic, and implicate intricate cross-talk between diverse regulatory levels, including modification of metabolism and gene expression for physiological and morphological adaptation (Krasensky & Jonak, 2012: Kuromori, Mizoi, Umezawa, Yamaguchi-Shinozaki, & Shinozaki, 2015). There are numerous research articles that describe the metabolic regulation underlying the drought, extreme temperature, and salinity stress, along with the signaling events implicated in arbitrating stress-induced metabolic variations (Sewelam, Kazan, & Schenk, 2016). To simplify the expansive metabolomic data, we can cite an example where we assume two groups, such as the control group (wt), and treated group (mutant type), generally result in '*m*' independent biological replicates, each with '*n*' technical replicates for each group. Let p represent the number of metabolites to be estimated in the metabolomics experiment. Then the outline of all the experimental data within this experimental design ends up in two data arrays, each with *pxnxm* dimensions. Suppose that the technical variance of the procedure has been displayed to be much less than the biological variance (vartech$\ll$varbiol); a technical mean value can then be drafted for each biological replicate, and the data arrays decrease to data matrices with *pxm* dimensions.

Even though fundamental statistical approaches and evaluations instantly facilitate a direct comparison of diverse data sets, and contribute data about the underlying biological mechanism, the produced hypotheses can become indistinct because of varied biochemical elucidations for one metabolomic characteristic. The covariance matrix of a metabolomics data set assists in estimating such an ambiguity. Following is a statistical estimate for how components (i.e., metabolites) are associated with each other

$$\text{Cov}(x, y) = 1m - 1\sum i = 1m(xi - x^{-})(yi - y^{-})$$

x and y are variables representing metabolite levels, with a sample size m and mean value $-x$ and $-y$. Therefore, a *pxp* covariance matrix is derived from a *pxm* data matrix. This quadratic association between the number of metabolites and probable metabolic associations exemplarily illustrates why even small metabolomic data sets (i.e., $p < 10$) are difficult to understand without suitable mathematical methods. This underscores the fact that the metabolomic coverage of experimental methods is automatically related to an exponential increase in possible elucidation that has to be considered. Hence, to explore the prime practical and all-inclusive biochemical elucidation for all determined experimental data, the instant relevance of varied statistical methods that substantiate or even harmonize each other has become a usual method. The fundamental statistical methods are well established, and several platforms and graphical user interfaces have emerged to enable and aid such procedures for summarizing the accessible tools and software (Sheth & Thaker, 2014; Sugimoto, Kawakami, Robert, Soga, & Tomita, 2012). The majority of these statistical policies even go beyond this scope, and are capable of evaluating, determining, and representing multivariate problems, and using approaches such as PCA, the ICA (Steuer, Morgenthal, Weckwerth, & Selbig, 2007), or the independent PCA (Yao, Coquery, & Le Cao, 2012). Depending on the covariance data, these multivariate approaches reduce the dimensionality of the data set while preserving as much data variance as possible. This produces abstract variables (i.e., components) that apply the most prominent and typical features of multivariate data sets, therefore enabling and assisting the generation of a biological hypothesis. These are complemented with other unsupervised and supervised statistical methods, for example, hierarchical clustering, *k*-means clustering, and PLS-DA. Molecular compounds provide the separation of the sample by varying degrees, which can be identified (Bellaire et al., 2014; Korman, Oh, Raskind, & Banks, 2012; Le Cao, Boitard, & Besse, 2011; Uarrota et al., 2014). These approaches of multivariate data investigation are an imperative part of most system biology methods, not only for metabolomics, but mainly also other data sets derived from varied omics techniques, such as transcriptomics and proteomics (Weckwerth, 2008). The integrative evaluation of varied experimental high throughput data can be carried out through special software packages that could be intended for an inclusive regression analysis, correlation, and visualization. An outstanding outline of software tool, platforms, and workflows of computational tasks in system biology was contributed earlier (Ghosh, Matsuoka, Asai, Hsin, & Kitano, 2011). Even though these platforms contribute an extensive assortment of statistical and mathematical methods, the ultimate result that a systems biologist is interested in might always be similar: the detection of particular metabolic clusters and patterns (i.e., samples) with similar statistical features separating the control from the treatment group. This ultimately helps us understand which metabolic steps or pathways are probably affected by the established perturbation, and reveals conclusions pertaining to biochemical or molecular biological mechanisms that can lead to a testable hypothesis. Several research studies from varied fields of biological research have determined the relevance and utility of such computational workflows (Altaf-Ul-Amin, Afendi, Kiboi, & Kanaya, 2014). Parallel to this study, procuring data on the regulatory associations between levels of molecular organizations has been known to be available only to a very limited degree. This is chiefly because of intricacies and nonlinear associations that are present between transcript levels, proteins, and metabolites. Therefore, if metabolite content varies as a result of genomics or ecological perturbations, reasons for this could include:

(i) Feedback/forward regulation of enzymes with nonlinear kinetics.
(ii) Post-translational modification of enzymes/proteins implicated in the metabolite biosynthesis, its interconversion, and transport.
(iii) Variations in the rate of protein biosynthesis, or degradation, and so on.

Therefore, determining a certain result from an array of metabolite levels by regulatory approaches in a metabolic network is cumbersome, excluding a few cases in which (a) all components of the metabolic networks are recognized and biochemically identified, or (b) all association between levels of molecular organization can, because of simplification, be known as linear. In recent years, an enormous amount of data pertaining to genome-wide metabolic network structures and regulatory principles has been explored related to, for example, molecular systems of prokaryotes (Carrera et al., 2014), yeast (Sanchez, Perez-Correa, & Agosin, 2014), algae (Chang et al., 2011), and higher plants (Hill et al., 2013; Mintz-Oron et al., 2012). However, we are still far from being able to complete all of the requirements for these methods. On the contrary, mathematical modeling enables linearization methods. Plant acclimation in response to changes in the environment needs a new state of cellular homeostasis acquired by a subtle balance between multiple pathways. The significant metabolites are hormones, secondary messengers, phosphatases, and protein kinases within the stress-induced signaling network that execute an array of biochemical and physiological processes (Hirayama & Shinozaki, 2010). Metabolic alteration in response to stress is imperative for procuring stress tolerance. Transcriptional investigation of stress signaling mutants exhibited that in several mutants with modified stress tolerance; genes implicated in the synthesis of stress-linked metabolites are modified, and thus can be directly or indirectly implicated in metabolic adjustment upon stress (Joshi et al., 2016). However, because of post-transcriptional and post-translational alterations, compartmentalization, metabolite stability, substrate availability, and so forth, changes in transcript abundance are not essentially translated to any change in metabolite level. A process has been developed to associate signal transduction with metabolite responses against drought, salt, and temperature stress conditions comprised of the targeted analysis of metabolites and the metabolite profiling of mutants in signaling components (Prasch & Sonnewald, 2013).

3.8.1 Functional Significance of ABA in Metabolic Adjustments

ABA is an imperative regulator of abiotic stress signaling, and is implicated in varied environmental stresses (Guajardo, Correa, & Contreras-Porcia, 2016; Hossain et al., 2010). ABA instantly accumulates against diverse environment stress conditions, and ABA-deficient plants have an altered stress response (Jones, 2016). ABA enhances stomatal closure, hampers stomatal opening to decrease water loss by transpiration, and induces the expression of several stress-associated genes. Recent studies have signified a functional role in the regulation of stress-induced metabolic adjustments (Trivedi, Gill, & Tuteja, 2016; Yoshida, Mogami, & Yamaguchi-Shinozaki, 2015). Comparative analysis of metabolic profiles of *Arabidopsis* plants subjected to ABA and high soil salinity displayed ABA-induced and ABA-independent steps of salt stress–induced metabolic reorganizations (Kempa, Krasensky, Dal Santo, Kopka, & Jonak, 2008). ABA and salt stress results led to a reduction of starch and accretion of maltose. However, the consequent carbon flux appears to be differentially regulated. While glucose-6-phosphate and fructose-6-phosphate levels, and the glucose/fructose ratio, are reduced while thriving under salt stress conditions, glucose-6-phosphate and fructose-6-phosphate levels and the glucose/fructose ratio are enhanced against ABA treatment. Fascinatingly, ABA did not induce galactinol and raffinose accretion (Kempa et al., 2008). Urano et al. (2009) illustrated an accumulation of sugars in *Arabidopsis* mutants deficient in ABA, and induced galactinol and raffinose accretion under drought conditions. Numerous research studies have described stress-induced accretion of the citrate, malate, succinate, and GABA independent of ABA intervention. On the contrary, accretion of numerous osmotic stress–induced proteinogenic amino acids comprised of proline were induced and depended on ABA (Urano et al., 2009).

3.8.2 Role of Proline Accumulation in Signaling

Proline has multiple functions in stress defense. Numerous protein kinases that are significant for salt, drought, and cold tolerance have revealed the regulatory role of proline accumulation. Such as in osmotic stress, SNF-related protein kinases (SnRK2s) are activated (McLoughlin et al., 2012). Evaluation of *Arabidopsis* mutants displayed that ABA-responsive SnRKs 2.2, 2.3, and 2.6 are significant for osmotic stress and ABA-induced proline accretion, while SnRK2.9 appears to function as a negative regulator in ABA-induced proline accumulation (Fujii, Verslues, & Zhu, 2011). CBL interacting kinases (CIPKs), which are actually SnRK3s that are associated with calcineurin B-like (CBL) calcium binding proteins, are implicated in adjusting proline levels. Overexpression of genes OsCIPK03 and OsCIPK12 in rice exposed to cold and drought increased proline levels under stress conditions. Though overexpression of the closely associated OsCIPK15 increased salt tolerance irrespective of any effects on proline accumulation, which explains the implication of only a subset of SnRK3s in proline accumulation against stress (Xiang, Huang, & Xiong, 2007). Furthermore, calcium-responsive protein kinases, *Arabidopsis* calcium-dependent protein kinase 6 (CDPK6) (Wang et al., 2016) and soybean calmodulin GmCAM4 (Cao, Li, Liu, Kong, & Tran, 2018) upregulate stress tolerance and proline levels in *Arabidopsis*, anticipating an essential role for intracellular calcium signals in proline metabolism. MAPK (mitogen-activated protein kinase) cascades execute numerous methods comprised of abiotic stress responses. Numerous MAPKs are activated in response to cold, salt, and drought in monocot and dicot plants (Ichimura et al., 2002; Xiong and Yang, 2003), and genetic manipulation of MAPK signaling modifies plant tolerance to these stresses (Muchate, Nikalje, Rajurkar, Suprasanna, & Nikam, 2016). Jan and Parray (2016) also demonstrated the positive function of MAPK-based signaling in stress-induced proline accretion. While ABA and numerous stress-associated protein kinases can induce proline accumulation, a maize protein phosphatase type 2C down-regulated stress-induced proline accumulation, and conferred tolerance to hyperosmotic stress (You et al., 2014). PP2C type phosphatases PP2Cs are implicated in controlling varied processes, including development and responses to environmental stress, and have been revealed to control stress-induced MAPK and SnRK2 protein kinases negatively (Umezawa et al., 2009). Phospholipid metabolism intercedes in hyperosmotic-stress responses (Testerink & Munnik, 2011). Using a pharmacological advance,

PLC-based signaling has been implicated in salt-induced proline accumulation (Parre et al., 2007). Protein kinases are involved in stress-induced changes in carbon metabolism. Carbohydrate metabolism is instantly modulated against environmental stress. Plant glycogen synthase kinase 3 (GSK-3)/shaggy-like kinases is known to be associated with stress signaling and carbon metabolism (Kempa et al., 2007). The positive regulator MsK4 for salt stress tolerance was identified in *Medicago sativa*, and plants that exhibit overexpression of MsK4 display higher sugar levels, such as raffinose- and galactinol-increased putrescine levels and amino acids under normal growth conditions. The gene expression of MsK4 is associated with starch granules and higher starch and sugar levels in *Arabidopsis* thriving in salt stress conditions (Kempa et al., 2007). These observations are exciting from an evolutionary point of view, as GSK was basically characterized as a regulator of storage carbohydrate glycogen (Woodgett & Cohen, 1984). Proline accumulation is basically a result of the upregulation of OsCIPK03, OsCIPK12, and the MAPK kinase ZmMKK4 under stress conditions, which ultimately leads to high-soluble sugar content in rice and maize seedlings (Ahmad & Wani, 2013). Moreover, protein kinase *ZmMKK4* is a negative regulator of drought tolerance, while as *SAL1* has been correlated with sugar and starch metabolism (Kong et al., 2011).

3.8.3 Transcription Factors Regulating Metabolite Levels

The varied array of plant stresses such as salinity, drought, and temperature have a remarkable effect on gene expression. Numerous genes for enzymes implicated in cellular metabolism are differentially expressed upon stress and alteration of stress-associated transcription factors, and have been shown to stimulate alterations in stress-related metabolite levels (Donati, Sander, & Link, 2018; Kochanowski et al., 2017). The family of transcriptional factors implicated in osmotic stress include AP2/EREBP (Apetala2/ethylene-responsive element binding protein), which regulates AP2/EREBP (Apetala2/ethylene-responsive element binding protein), that plays a significant role in osmotic stress (Licausi, Ohme-Takagi, & Perata, 2013). Higher tolerance levels in response to freezing, drought, and salt stress are concomitant with overexpression levels of transcription factors in *Arabidopsis*. The plants that display higher levels of expression of CBF/DREB1 accumulate higher levels of proline and soluble sugars (glucose, fructose, sucrose, and raffinose), particularly during cold acclimation (Achard et al., 2008). The entire metabolic profile of CBF3/DREB1A overexpressers grown at normal growth temperatures is similar to that of plants thriving under cold stress (Maruyama et al., 2004). Osmotic stress stimulates DREB2A (dehydration responsive element binding protein 2A) and overexpression of a constitutive active version of DREB2A without the inhibitory domain-enhanced tolerance to dehydration, but to a somewhat lesser extent, freezing temperatures (Song, Je, Hong, & Lim, 2014). Corresponding to this tolerance outline, the metabolite profile of plants displays overexpression of the constitutive active form DREB2A, more similar to that of dehydration-exposed, rather than that of cold-treated plants (Maruyama et al., 2004). Though the comprehensive metabolite profile of plants shows overexpression of constitutive active forms of DREB2A, which is similar to plants under stress, but without enhancing levels of galactinol, raffinose, and proline. Proline accumulation is prominent in response to drought, and not heat stress, which is further validated by the higher expression of DREB2A in response to heat stress tolerance (Bita & Gerats, 2013).

HSF are prime regulators of heat-stress-responsive genes, and these are encoded by an expansive gene family with diverse expression patterns and functions (von Koskull-Döring, Scharf, & Nover, 2007). Similarly, overexpression of HsFA2 results in the constitutive accretion of galactinol and raffinose, and enhances the tolerance of *Arabidopsis* plants exposed to diverse environmental stresses (Nishizawa, Yabuta, & Shigeoka, 2008). There have been reports of overexpression of HSF3/HsfA1b improved thermotolerance (Prandl, Hinderhofer, Eggers-Schumacher, & Schofû, 1998) and raffinose levels under normal and heat stress conditions (Panikulangara et al., 2004).

The NAC domain family of plant-specific transcriptional regulators is implicated in developmental processes together with hormonal control and stress defense (Olsen, Ernst, Leggio, & Skriver, 2005). The NAC family includes OsNAC5 activated in response to osmotic stress and ABA. Enhanced expression of OsNAC5 leads to higher levels of proline, sugar, and high tolerance levels to cold, salt, and drought stress (Song, Chen, Chen, Dai, & Zhang, 2011). Knock-down lines exhibited lower levels of proline and soluble sugar as soon as they were subjected to stress, and were more sensitive to stress (Song et al., 2011). *Arabidopsis* that exhibit overexpression of Myb transcription factors such as Osmyb4 displayed high levels of soluble sugars such as glucose, fructose, and sucrose, along with glycine betaine (Wani et al., 2012). Accumulation of these metabolites was associated with high drought and freezing tolerance (Qin et al., 2012; Zandalinas, Mittler, Balfagón, Arbona, & Gómez-Cadenas, 2018).

3.9 CONCLUSION

Diverse metabolites and intricate metabolic pathways are required for natural stress tolerance. So in order to understand this natural tolerance, we need to decipher the analysis of the metabolic adjustments of plants with varied levels of stress tolerance. The elucidation of metabolic adjustments of plants with varied metabolite levels exhibit diverse stress tolerance and transgenic strategies that can contribute significantly to complementing indications for in-depth perspectives into the function of diverse metabolites adapting to severe environmental conditions. Regardless of these significant inputs, numerous queries need to be answered, such as the developmental stage of plant is significant for its potential to tolerate an adverse conditions and cellular metabolism changes during development. Hence, it will be exciting to investigate how the developmental stage influences the metabolic adjustment to stress conditions. Metabolic networks are extremely dynamic, and metabolites can migrate between diverse cellular compartments. Metabolic profiling only displays the steady-state level of metabolites; comprehensive kinetics and flux analyses will be instrumental for an improved perception of metabolic fluctuations against stresses. Metabolic analysis at the subcellular level in explicit tissues is an extended challenge. Stress-induced signaling networks are illustrated comprehensively. However, the signaling processes needed for homeostasis of fundamental cellular and metabolic processes in undesirable environments are beginning to appear. Improvements in how environmental changes are communicated via cellular signal transduction to provoke a coordinated metabolic response, and how the function of metabolic enzymes are altered by transcriptional and post-translational modifications, are of fundamental scientific interest, and will contribute to meeting the objective of augmented plant stress tolerance and efficiency in an ever-changing environment.

REFERENCES

Achard, P., Gong, F., Cheminant, S., Alioua, M., Hedden, P., & Genschik, P. (2008). The cold-inducible CBF1 factor-dependent signaling pathway modulates the accumulation of the growth repressing DELLA proteins via its effect on gibberellin metabolism. *The Plant Cell*, *20*, 2117–2129.

Afendi, F. M., Okada, T., Yamazaki, M., Hirai-Morita, A., Nakamura, Y., Nakamura, K., et al. (2011). KNApSAcK family databases: integrated metabolite–plant species databases for multifaceted plant research. *Plant and Cell Physiology*, *53*(2), e1.

Ahmad, P., & Wani, M. R. (Eds.), (2013). *Physiological mechanisms and adaptation strategies in plants under changing environment.* In Vol. 2. Springer Science & Business Media.

Ahmad, P., Ahanger, M. A., Singh, V. P., Tripathi, D. K., Alam, P., & Alyemeni, M. N. (Eds.), (2018). *Plant metabolites and regulation under environmental stress*: Academic Press.

Ahmad, P., Sarwat, M., & Sharma, S. (2008). Reactive oxygen species, antioxidants and signaling in plants. *Journal of Plant Biology, 51*(3), 167–173.

Ahsan, N., Renaut, J., & Komatsu, S. (2009). Recent developments in the application of proteomics to the analysis of plant responses to heavy metals. *Proteomics, 9*(10), 2602–2621.

Akram, N. A., Shafiq, F., & Ashraf, M. (2017). Ascorbic acid—a potential oxidant scavenger and its role in plant development and abiotic stress tolerance. *Frontiers in Plant Science, 8*, 613.

Akula, R., & Ravishankar, G. A. (2011). Influence of abiotic stress signals on secondary metabolites in plants. *Plant Signaling & Behavior, 6*(11), 1720–1731.

Allwood, J. W., Chandra, S., Xu, Y., Dunn, W. B., Correa, E., Hopkins, L., et al. (2015). Profiling of spatial metabolite distributions in wheat leaves under normal and nitrate limiting conditions. *Phytochemistry, 115*, 99–111.

Alonso, A., Marsal, S., & Julià, A. (2015). Analytical methods in untargeted metabolomics: state of the art in 2015. *Frontiers in Bioengineering and Biotechnology, 3*, 23.

Altaf-Ul-Amin, M., Afendi, F. M., Kiboi, S. K., & Kanaya, S. (2014). Systems biology in the context of big data and networks. *BioMed Research International, 2014*, 428570. https://doi.org/10.1155/2014/428570.

Andreyev, A. Y., Kushnareva, Y. E., Murphy, A. N., & Starkov, A. A. (2015). Mitochondrial ROS metabolism: 10 years later. *Biochemistry (Moscow), 80*(5), 517–531.

Anjum, N. A., Gill, S. S., Umar, S., Ahmad, I., Duarte, A. C., & Pereira, E. (2012). Improving growth and productivity of oleiferous Brassicas under changing environment: significance of nitrogen and sulphur nutrition, and underlying mechanisms. *The Scientific World Journal, 2012*.

Apel, K., & Hirt, H. (2004). Reactive oxygen species: metabolism, oxidative stress, and signal transduction. *Annual Review of Plant Biology, 55*, 373–399.

Arbona, V., Manzi, M., Ollas, C. D., & Gómez-Cadenas, A. (2013). Metabolomics as a tool to investigate abiotic stress tolerance in plants. *International Journal of Molecular Sciences, 14*(3), 4885–4911.

Ariza, J. G., García-Barrera, T., García-Sevillano, M. A., González-Fernández, M., & Gómez-Jacinto, V. (2013). Metallomics and metabolomics of plants under environmental stress caused by metals. In *Heavy metal stress in plants* (pp. 173–201). Berlin, Heidelberg: Springer.

Armengaud, P., Breitling, R., & Amtmann, A. (2010). Coronatine-insensitive 1 (COI1) mediates transcriptional responses of *Arabidopsis thaliana* to external potassium supply. *Molecular Plant, 3*(2), 390–405.

Armengaud, P., Sulpice, R., Miller, A. J., Stitt, M., Amtmann, A., & Gibon, Y. (2009). Multilevel analysis of primary metabolism provides new insights into the role of potassium nutrition for glycolysis and nitrogen assimilation in *Arabidopsis* roots. *Plant Physiology, 150*(2), 772–785.

Bailey, N. J., Oven, M., Holmes, E., Nicholson, J. K., & Zenk, M. H. (2003). Metabolomic analysis of the consequences of cadmium exposure in *Silene cucubalus* cell cultures via 1H NMR spectroscopy and chemometrics. *Phytochemistry, 62*(6), 851–858.

Barding, G. A. (2013). *Metabolomics of complex biological systems to uncover molecular mechanisms in rice and other organisms*. Doctoral dissertationUC Riverside.

Barding, G. A., Salditos, R., & Larive, C. K. (2012). Quantitative NMR for bioanalysis and metabolomics. *Analytical and Bioanalytical Chemistry, 404*(4), 1165–1179.

Bauerle, W. L., Oren, R., Way, D. A., Qian, S. S., Stoy, P. C., Thornton, P. E., et al. (2012). Photoperiodic regulation of the seasonal pattern of photosynthetic capacity and the

implications for carbon cycling. *Proceedings of the National Academy of Sciences*, *109*(22), 8612–8617.
Baxter, C. J., Redestig, H., Schauer, N., Repsilber, D., Patil, K. R., Nielsen, J., et al. (2007). The metabolic response of heterotrophic *Arabidopsis* cells to oxidative stress. *Plant Physiology*, *143*(1), 312–325.
Bellaire, A., Ischebeck, T., Staedler, Y., Weinhaeuser, I., Mair, A., Parameswaran, S., et al. (2014). Metabolism and development—integration of micro computed tomography data and metabolite profiling reveals metabolic reprogramming from floral initiation to silique development. *The New Phytologist*, *202*, 322–335. https://doi.org/10.1111/nph.12631.
Bendaly, A., Messedi, D., Smaoui, A., Ksouri, R., Bouchereau, A., & Abdelly, C. (2016). Physiological and leaf metabolome changes in the xerohalophyte species *Atriplex halimus* induced by salinity. *Plant Physiology and Biochemistry*, *103*, 208–218.
Bhattacharjee, S., & Saha, A. K. (2014). Plant water-stress response mechanisms. In *Approaches to plant stress and their management* (pp. 149–172). New Delhi: Springer.
Bielecka, M., Watanabe, M., Morcuende, R., Scheible, W. R., Hawkesford, M. J., Hesse, H., et al. (2015). Transcriptome and metabolome analysis of plant sulfate starvation and resupply provides novel information on transcriptional regulation of metabolism associated with sulfur, nitrogen and phosphorus nutritional responses in *Arabidopsis*. *Frontiers in Plant Science*, *5*, 805.
Bison, J. V., Cardoso-Gustavson, P., de Moraes, R. M., da Silva Pedrosa, G., Cruz, L. S., Freschi, L., et al. (2018). Volatile organic compounds and nitric oxide as responses of a Brazilian tropical species to ozone: the emission profile of young and mature leaves. *Environmental Science and Pollution Research*, *25*(4), 3840–3848.
Bita, C., & Gerats, T. (2013). Plant tolerance to high temperature in a changing environment: scientific fundamentals and production of heat stress-tolerant crops. *Frontiers in Plant Science*, *4*, 273.
Bocian, A., Zwierzykowski, Z., Rapacz, M., Koczyk, G., Ciesiołka, D., & Kosmala, A. (2015). Metabolite profiling during cold acclimation of *Lolium perenne* genotypes distinct in the level of frost tolerance. *Journal of Applied Genetics*, *56*(4), 439–449.
Bowne, J. B., Erwin, T. A., Juttner, J., Schnurbusch, T., Langridge, P., Bacic, A., et al. (2012). Drought responses of leaf tissues from wheat cultivars of differing drought tolerance at the metabolite level. *Molecular Plant*, *5*(2), 418–429.
Bundy, J. G., Keun, H. C., Sidhu, J. K., Spurgeon, D. J., Svendsen, C., Kille, P., et al. (2007). Metabolic profile biomarkers of metal contamination in a sentinel terrestrial species are applicable across multiple sites. *Environmental Science & Technology*, *41*(12), 4458–4464.
Bundy, J. G., Sidhu, J. K., Rana, F., Spurgeon, D. J., Svendsen, C., Wren, J. F., et al. (2008). 'Systems toxicology' approach identifies coordinated metabolic responses to copper in a terrestrial non-model invertebrate, the earthworm *Lumbricus rubellus*. *BMC Biology*, *6*(1), 25.
Bustamante, C. A., Monti, L. L., Gabilondo, J., Scossa, F., Valentini, G., Budde, C. O., et al. (2016). Differential metabolic rearrangements after cold storage are correlated with chilling injury resistance of peach fruits. *Frontiers in Plant Science*, 7, 1478.
Cairns, J. E., Crossa, J., Zaidi, P. H., Grudloyma, P., Sanchez, C., Araus, J. L., et al. (2013). Identification of drought, heat, and combined drought and heat tolerant donors in maize. *Crop Science*, *53*(4), 1335–1346.
Callahan, D. L., Roessner, U., Dumontet, V., Perrier, N., Wedd, A. G., Richard, A. J., et al. (2008). LC–MS and GC–MS metabolite profiling of nickel (II) complexes in the latex of the nickel-hyperaccumulating tree *Sebertia acuminata* and identification of methylated aldaric acid as a new nickel (II) ligand. *Phytochemistry*, *69*(1), 240–251.
Cao, D., Li, Y., Liu, B., Kong, F., & Tran, L. S. P. (2018). Adaptive mechanisms of soybean grown on salt-affected soils. *Land Degradation & Development*, *29*(4), 1054–1064.

Carrera, J., Estrela, R., Luo, J., Rai, N., Tsoukalas, A., & Tagkopoulos, I. (2014). An integrative, multi-scale, genome-wide model reveals the phenotypic landscape of *Escherichia coli*. *Molecular Systems Biology*, *10*, 735. https://doi.org/10.15252/msb.20145108.
Chang, R. L., Ghamsari, L., Manichaikul, A., Hom, E. F., Balaji, S., Fu, W., et al. (2011). Metabolic network reconstruction of Chlamydomonas offers insight into light-driven algal metabolism. *Molecular Systems Biology*, *7*, 518. https://doi.org/10.1038/msb.2011.52.
Chemikosova, S. B., Pavlencheva, N. V., Gur'yanov, O. P., & Gorshkova, T. A. (2006). The effect of soil drought on the phloem fiber development in long-fiber flax. *Russian Journal of Plant Physiology*, *53*(5), 656–662.
Chmielewska, K., Rodziewicz, P., Swarcewicz, B., Sawikowska, A., Krajewski, P., Marczak, Ł., et al. (2016). Analysis of drought-induced proteomic and metabolomic changes in barley (*Hordeum vulgare* L.) leaves and roots unravels some aspects of biochemical mechanisms involved in drought tolerance. *Frontiers in Plant Science*, *7*, 1108.
Cho, K., Shibato, J., Agrawal, G. K., Jung, Y. H., Kubo, A., Jwa, N. S., et al. (2008). Integrated transcriptomics, proteomics, and metabolomics analyses to survey ozone responses in the leaves of rice seedling. *Journal of Proteome Research*, *7*(7), 2980–2998.
Clarke, S. M., Cristescu, S. M., Miersch, O., Harren, F. J., Wasternack, C., & Mur, L. A. (2009). Jasmonates act with salicylic acid to confer basal thermotolerance in Arabidopsis thaliana. *New Phytologist*, *182*(1), 175–187.
Colinet, H., Larvor, V., Laparie, M., & Renault, D. (2012). Exploring the plastic response to cold acclimation through metabolomics. *Functional Ecology*, *26*, 711–722.
Cook, D., Fowler, S., Fiehn, O., & Thomashow, M. F. (2004). A prominent role for the CBF cold response pathway in configuring the low-temperature metabolome of *Arabidopsis*. *Proceedings of the National Academy of Sciences*, *101*(42), 15243–15248.
Cramer, G. R., Urano, K., Delrot, S., Pezzotti, M., & Shinozaki, K. (2011). Effects of abiotic stress on plants: a systems biology perspective. *BMC Plant Biology*, *11*(1), 163.
Das, K., & Roychoudhury, A. (2014). Reactive oxygen species (ROS) and response of antioxidants as ROS-scavengers during environmental stress in plants. *Frontiers in Environmental Science*, *2*, 53.
Das, P., Nutan, K. K., Singla-Pareek, S. L., & Pareek, A. (2015). Oxidative environment and redox homeostasis in plants: dissecting out significant contribution of major cellular organelles. *Frontiers in Environmental Science*, *2*, 70.
Dennison, P. E., Roberts, D. A., Thorgusen, S. R., Regelbrugge, J. C., Weise, D., & Lee, C. (2003). Modeling seasonal changes in live fuel moisture and equivalent water thickness using a cumulative water balance index. *Remote Sensing of Environment*, *88*(4), 442–452.
Di Ferdinando, M., Brunetti, C., Agati, G., & Tattini, M. (2014). Multiple functions of polyphenols in plants inhabiting unfavorable Mediterranean areas. *Environmental and Experimental Botany*, *103*, 107–116.
Donati, S., Sander, T., & Link, H. (2018). Crosstalk between transcription and metabolism: how much enzyme is enough for a cell? *Wiley Interdisciplinary Reviews: Systems Biology and Medicine*, *10*(1)e1396.
Du, H., Wang, Z., Yu, W., Liu, Y., & Huang, B. (2010). Differential metabolic responses of perennial grass *Cynodon transvaalensis* × *Cynodon dactylon* (C4) and *Poa pratensis* (C3) to heat stress. *Physiologia Plantarum*, *141*(3), 251–264.
Emberson, L. D., Pleijel, H., Ainsworth, E. A., van den Berg, M., Ren, W., Osborne, S., et al. (2018). Ozone effects on crops and consideration in crop models. *European Journal of Agronomy*.
Eremina, M., Rozhon, W., & Poppenberger, B. (2016). Hormonal control of cold stress responses in plants. *Cellular and Molecular Life Sciences*, *73*(4), 797–810.
Fait, A., Fromm, H., Walter, D., Galili, G., & Fernie, A. R. (2008). Highway or byway: the metabolic role of the GABA shunt in plants. *Trends in Plant Science*, *13*(1), 14–19.

Farag, M. A., Meyer, A., Ali, S. E., Salem, M. A., Giavalisco, P., Westphal, H., et al. (2018). Comparative metabolomics approach detects stress-specific responses during coral bleaching in soft corals. *Journal of Proteome Research, 17*(6), 2060–2071.

Fatichi, S., Leuzinger, S., & Körner, C. (2014). Moving beyond photosynthesis: from carbon source to sink-driven vegetation modeling. *New Phytologist, 201*(4), 1086–1095.

Fonville, J. M., Richards, S. E., Barton, R. H., Boulange, C. L., Ebbels, T. M., Nicholson, J. K., et al. (2010). The evolution of partial least squares models and related chemometric approaches in metabonomics and metabolic phenotyping. *Journal of Chemometrics, 24*(11-12), 636–649.

Förstner, U., & Wittmann, G. T. (2012). *Metal pollution in the aquatic environment*. Springer Science & Business Media.

Foyer, C. H., & Noctor, G. (2005). Redox homeostasis and antioxidant signaling: a metabolic interface between stress perception and physiological responses. *The Plant Cell, 17* (7), 1866–1875.

Frerman, F. E. (2013). Electron transfer flavoproteins. In *Encyclopedia of biophysics* (pp. 606–611). Berlin, Heidelberg: Springer.

Fujii, H., Verslues, P. E., & Zhu, J. K. (2011). *Arabidopsis* decuple mutant reveals the importance of SnRK2 kinases in osmotic stress responses in vivo. *Proceedings of the National Academy of Sciences, 108*(4), 1717–1722.

Gangwar, S., Singh, V. P., Tripathi, D. K., Chauhan, D. K., Prasad, S. M., & Maurya, J. N. (2014). Plant responses to metal stress: the emerging role of plant growth hormones in toxicity alleviation. *Emerging technologies and management of crop stress tolerance* (pp. 215–248). Vol. 2 (pp. 215–248). Elsevier.

Geigenberger, P., & Fernie, A. R. (2014). Metabolic control of redox and redox control of metabolism in plants. *Antioxidants & Redox Signaling, 21*(9), 1389–1421.

Genga, A., Mattana, M., Coraggio, I., Locatelli, F., Piffanelli, P., & Consonni, R. (2011). Plant metabolomics: A characterisation of plant responses to abiotic stresses. In *Abiotic stress in plants—mechanisms and adaptations*. InTech.

Ghosh, A., Singh, A. A., Agrawal, M., & Agrawal, S. B. (2018). Ozone toxicity and remediation in crop plants. *Sustainable agriculture reviews* (pp. 129–169). Vol. 27 (pp. 129–169). Cham: Springer.

Ghosh, S., Matsuoka, Y., Asai, Y., Hsin, K. -Y., & Kitano, H. (2011). Software for systems biology: from tools to integrated platforms. *Nature Reviews Genetics, 12*, 821–832. https://doi.org/10.1038/nrg3096.

Gibon, Y., PYL, E. T., Sulpice, R., Lunn, J. E., Hoehne, M., Guenther, M., et al. (2009). Adjustment of growth, starch turnover, protein content and central metabolism to a decrease of the carbon supply when *Arabidopsis* is grown in very short photoperiods. *Plant, Cell & Environment, 32*(7), 859–874.

Gika, H. G., Theodoridis, G. A., Plumb, R. S., & Wilson, I. D. (2014). Current practice of liquid chromatography–mass spectrometry in metabolomics and metabonomics. *Journal of Pharmaceutical and Biomedical Analysis, 87*, 12–25.

Gill, S. S., & Tuteja, N. (2010). Reactive oxygen species and antioxidant machinery in abiotic stress tolerance in crop plants. *Plant Physiology and Biochemistry, 48*(12), 909–930.

Gray, G. R., & Heath, D. (2005). A global reorganization of the metabolome in *Arabidopsis* during cold acclimation is revealed by metabolic fingerprinting. *Physiologia Plantarum, 124*(2), 236–248.

Grazia Annunziata, M., Apelt, F., Carillo, P., Krause, U., Feil, R., Koehl, K., et al. (2018). Response of *Arabidopsis* primary metabolism and circadian clock to low night temperature in a natural light environment. *Journal of Experimental Botany*,.

Guajardo, E., Correa, J. A., & Contreras-Porcia, L. (2016). Role of abscisic acid (ABA) in activating antioxidant tolerance responses to desiccation stress in intertidal seaweed species. *Planta, 243*(3), 767–781.

Guo, R., Shi, L., Jiao, Y., Li, M., Zhong, X., Gu, F., et al. (2018). Metabolic responses to drought stress in the tissues of drought-tolerant and drought-sensitive wheat genotype seedlings. *AoB Plants*, *10*(2)ply016.

Hajlaoui, H., El Ayeb, N., Garrec, J. P., & Denden, M. (2010). Differential effects of salt stress on osmotic adjustment and solutes allocation on the basis of root and leaf tissue senescence of two silage maize (*Zea mays* L.) varieties. *Industrial Crops and Products*, *31*(1), 122–130.

Hall, R. D. (2006). Plant metabolomics: from holistic hope, to hype, to hot topic. *New Phytologist*, *169*(3), 453–468.

Hannah, M. A., Wiese, D., Freund, S., Fiehn, O., Heyer, A. G., & Hincha, D. K. (2006). Natural genetic variation of freezing tolerance in *Arabidopsis*. *Plant Physiology*, *142*(1), 98–112.

Hasanuzzaman, M., Bhuyan, M. H. M., Nahar, K., Hossain, M. S., Mahmud, J. A., Hossen, M. S., et al. (2018). Potassium: a vital regulator of plant responses and tolerance to abiotic stresses. *Agronomy*, *8*(3), 31.

Hédiji, H., Djebali, W., Belkadhi, A., Cabasson, C., Moing, A., Rolin, D., et al. (2015). Impact of long-term cadmium exposure on mineral content of *Solanum lycopersicum* plants: consequences on fruit production. *South African Journal of Botany*, *97*, 176–181.

Hernández, G., Ramírez, M., Valdés-López, O., Tesfaye, M., Graham, M. A., Czechowski, T., et al. (2007). Phosphorus stress in common bean: root transcript and metabolic responses. *Plant Physiology*, *144*(2), 752–767.

Hill, K., Porco, S., Lobet, G., Zappala, S., Mooney, S., Draye, X., et al. (2013). Root systems biology: integrative modeling across scales, from gene regulatory networks to the rhizosphere. *Plant Physiology*, *163*, 1487–1503. https://doi.org/10.1104/pp.113.227215.

Hirayama, T., & Shinozaki, K. (2010). Research on plant abiotic stress responses in the post-genome era: past, present and future. *The Plant Journal*, *61*, 1041–1052.

Hoefgen, R., & Nikiforova, V. J. (2008). Metabolomics integrated with transcriptomics: assessing systems response to sulfur-deficiency stress. *Physiologia Plantarum*, *132*(2), 190–198.

Hossain, M. A., Cho, J. I., Han, M., Ahn, C. H., Jeon, J. S., An, G., et al. (2010). The ABRE-binding bZIP transcription factor OsABF2 is a positive regulator of abiotic stress and ABA signaling in rice. *Journal of Plant Physiology*, *167*(17), 1512–1520.

Hu, T., Chen, K., Hu, L., Amombo, E., & Fu, J. (2016). H_2O_2 and Ca^{2+} signals integrated to enhance perennial ryegrass acclimation to salt stresses. *Scientific Reports*. *6*, https://doi.org/10.1038/srep36396.

Hu, Y., Chen, L., Wang, H., Zhang, L., Wang, F., & Yu, D. (2013). *Arabidopsis* transcription factor WRKY8 functions antagonistically with its interacting partner VQ9 to modulate salinity stress tolerance. *The Plant Journal*, *74*(5), 730–745.

Huang, C. Y., Roessner, U., Eickmeier, I., Genc, Y., Callahan, D. L., Shirley, N., et al. (2008). Metabolite profiling reveals distinct changes in carbon and nitrogen metabolism in phosphate-deficient barley plants (*Hordeum vulgare* L.). *Plant and Cell Physiology*, *49*(5), 691–703.

Huang, G. T., Ma, S. L., Bai, L. P., Zhang, L., Ma, H., Jia, P., et al. (2012). Signal transduction during cold, salt, and drought stresses in plants. *Molecular Biology Reports*, *39*(2), 969–987.

Hughes, S. L., Bundy, J. G., Want, E. J., Kille, P., & Stürzenbaum, S. R. (2009). The metabolomic responses of *Caenorhabditis elegans* to cadmium are largely independent of metallothionein status, but dominated by changes in cystathionine and phytochelatins. *Journal of Proteome Research*, *8*(7), 3512–3519.

Ichimura, K., Shinozaki, K., Tena, G., Sheen, J., Henry, Y., Champion, A., et al. (2002). Mitogen-activated protein kinase cascades in plants: a new nomenclature. *Trends in Plant Science*, 7(7), 301–308.

Ideker, T., Dutkowski, J., & Hood, L. (2011). Boosting signal-to-noise in complex biology: prior knowledge is power. *Cell, 144*(6), 860–863.

Iijima, Y. (2014). Recent advances in the application of metabolomics to studies of biogenic volatile organic compounds (BVOC) produced by plant. *Metabolites, 4*(3), 699–721.

Ishikawa, T., Takahara, K., Hirabayashi, T., Matsumura, H., Fujisawa, S., Terauchi, R., et al. (2010). Metabolome analysis of response to oxidative stress in rice suspension cells over-expressing cell death suppressor Bax inhibitor-1. *Plant & Cell Physiology, 51*(1), 9–20.

Issaq, H. J., Van, Q. N., Waybright, T. J., Muschik, G. M., & Veenstra, T. D. (2009). Analytical and statistical approaches to metabolomics research. *Journal of Separation Science, 32* (13), 2183–2199.

Jahangir, M., Abdel-Farid, I. B., Choi, Y. H., & Verpoorte, R. (2008). Metal ion-inducing metabolite accumulation in *Brassica rapa*. *Journal of Plant Physiology, 165*(14), 1429–1437.

Jan, S., & Parray, J. A. (2016). Heavy metal stress signalling in plants. In *Approaches to heavy metal tolerance in plants* (pp. 33–55). Singapore: Springer.

Jeudy, C., Ruffel, S., Freixes, S., Tillard, P., Santoni, A. L., Morel, S., et al. (2010). Adaptation of *Medicago truncatula* to nitrogen limitation is modulated via local and systemic nodule developmental responses. *New Phytologist, 185*(3), 817–828.

Jiménez-Arias, D., Pérez, J. A., Luis, J. C., Martín-Rodríguez, V., Valdés-González, F., & Borges, A. A. (2015). Treating seeds in menadione sodium bisulphite primes salt tolerance in *Arabidopsis* by inducing an earlier plant adaptation. *Environmental and Experimental Botany, 109*, 23–30.

Jones, A. M. (2016). A new look at stress: abscisic acid patterns and dynamics at high-resolution. *New Phytologist, 210*(1), 38–44.

Joshi, R., Wani, S. H., Singh, B., Bohra, A., Dar, Z. A., Lone, A. A., et al. (2016). Transcription factors and plants response to drought stress: current understanding and future directions. *Frontiers in Plant Science, 7*, 1029.

Kaplan, F., Kopka, J., Haskell, D. W., Zhao, W., Schiller, K. C., Gatzke, N., et al. (2004). Exploring the temperature-stress metabolome of *Arabidopsis*. *Plant Physiology, 136*(4), 4159–4168.

Kaplan, F., Kopka, J., Sung, D. Y., Zhao, W., Popp, M., Porat, R., et al. (2007). Transcript and metabolite profiling during cold acclimation of *Arabidopsis* reveals an intricate relationship of cold-regulated gene expression with modifications in metabolite content. *The Plant Journal, 50*(6), 967–981.

Kell, D. B. (2004). Metabolomics and systems biology: making sense of the soup. *Current Opinion in Microbiology, 7*(3), 296–307.

Kempa, S., Krasensky, J., Dal Santo, S., Kopka, J., & Jonak, C. (2008). A central role of abscisic acid in stress-regulated carbohydrate metabolism. *PLoS One, 3*(12)e3935.

Kempa, S., Rozhon, W., Šamaj, J., Erban, A., Baluška, F., Becker, T., et al. (2007). A plastid-localized glycogen synthase kinase 3 modulates stress tolerance and carbohydrate metabolism. *The Plant Journal, 49*(6), 1076–1090.

Kerchev, P. I., Fenton, B., Foyer, C. H., & Hancock, R. D. (2012). Infestation of potato (Solanum tuberosum L.) by the peach-potato aphid (Myzus persicae Sulzer) alters cellular redox status and is influenced by ascorbate. *Plant, Cell & Environment, 35*(2), 430–440.

Keunen, E. L. S., Peshev, D., Vangronsveld, J., Van Den Ende, W. I. M., & Cuypers, A. N. N. (2013). Plant sugars are crucial players in the oxidative challenge during abiotic stress: extending the traditional concept. *Plant, Cell & Environment, 36*(7), 1242–1255.

Kim, J. K., Bamba, T., Harada, K., Fukusaki, E., & Kobayashi, A. (2006). Time-course metabolic profiling in *Arabidopsis thaliana* cell cultures after salt stress treatment. *Journal of Experimental Botany, 58*(3), 415–424.

Kirma, M., Araújo, W. L., Fernie, A. R., & Galili, G. (2012). The multifaceted role of aspartate-family amino acids in plant metabolism. *Journal of Experimental Botany, 63* (14), 4995–5001.

Kitteringham, N. R., Jenkins, R. E., Lane, C. S., Elliott, V. L., & Park, B. K. (2009). Multiple reaction monitoring for quantitative biomarker analysis in proteomics and metabolomics. *Journal of Chromatography B, 877*(13), 1229–1239.

Kochanowski, K., Gerosa, L., Brunner, S. F., Christodoulou, D., Nikolaev, Y. V., & Sauer, U. (2017). Few regulatory metabolites coordinate expression of central metabolic genes in Escherichia coli. *Molecular Systems Biology, 13*(1), 903.

Kong, X., Pan, J., Zhang, M., Xing, X. I. N., Zhou, Y. A. N., Liu, Y., et al. (2011). ZmMKK4, a novel group C mitogen-activated protein kinase kinase in maize (*Zea mays*), confers salt and cold tolerance in transgenic *Arabidopsis*. *Plant, Cell & Environment, 34*(8), 1291–1303.

Kopriva, S., Mugford, S. G., Matthewman, C., & Koprivova, A. (2009). Plant sulfate assimilation genes: redundancy versus specialization. *Plant Cell Reports, 28*(12), 1769–1780.

Korman, A., Oh, A., Raskind, A., & Banks, D. (2012). Statistical methods in metabolomics. *Methods in Molecular Biology, 856*, 381–413. https://doi.org/10.1007/978-1-61779-585-5_16.

Korn, M., Gärtner, T., Erban, A., Kopka, J., Selbig, J., & Hincha, D. K. (2010). Predicting *Arabidopsis* freezing tolerance and heterosis in freezing tolerance from metabolite composition. *Molecular Plant, 3*(1), 224–235.

Korn, M., Peterek, S., Mock, H. P., Heyer, A. G., & Hincha, D. K. (2008). Heterosis in the freezing tolerance, and sugar and flavonoid contents of crosses between *Arabidopsis thaliana* accessions of widely varying freezing tolerance. *Plant, Cell and Environment, 31*(6), 813–827.

Krasensky, J., & Jonak, C. (2012). Drought, salt, and temperature stress-induced metabolic rearrangements and regulatory networks. *Journal of Experimental Botany, 63*(4), 1593–1608.

Kumar, S., Sreeharsha, R. V., Mudalkar, S., Sarashetti, P. M., & Reddy, A. R. (2017). Molecular insights into photosynthesis and carbohydrate metabolism in *Jatropha curcas* grown under elevated CO_2 using transcriptome sequencing and assembly. *Scientific Reports*, 7(1), 11066.

Kundu, P., Gill, R., Ahlawat, S., Anjum, N. A., Sharma, K. K., Ansari, A. A., et al. (2018). Targeting the redox regulatory mechanisms for abiotic stress tolerance in crops. In *Biochemical, physiological and molecular avenues for combating abiotic stress tolerance in plants* (pp. 151–220). .

Kuromori, T., Mizoi, J., Umezawa, T., Yamaguchi-Shinozaki, K., & Shinozaki, K. (2015). Stress signaling networks: drought stress. *Molecular Biology*, 1–23.

Kusano, M., Tohge, T., Fukushima, A., Kobayashi, M., Hayashi, N., Otsuki, H., et al. (2011). Metabolomics reveals comprehensive reprogramming involving two independent metabolic responses of Arabidopsis to UV-B light. *The Plant Journal, 67*(2), 354–369.

Labanowska, M., Filek, M., Kurdziel, M., Bidzińska, E., Miszalski, Z., & Hartikainen, H. (2013). EPR spectroscopy as a tool for investigation of differences in radical status in wheat plants of various tolerances to osmotic stress induced by NaCl and PEG-treatment. *Journal of Plant Physiology, 170*, 136–145.

Lankadurai, B. P., Nagato, E. G., & Simpson, M. J. (2013). Environmental metabolomics: an emerging approach to study organism responses to environmental stressors. *Environmental Reviews, 21*(3), 180–205.

Larkindale, J., Hall, J. D., Knight, M. R., & Vierling, E. (2005). Heat stress phenotypes of *Arabidopsis* mutants implicate multiple signaling pathways in the acquisition of thermotolerance. *Plant Physiology, 138*(2), 882–897.

Lata, C., Muthamilarasan, M., & Prasad, M. (2015). Drought stress responses and signal transduction in plants. In *Elucidation of abiotic stress signaling in plants* (pp. 195–225). New York, NY: Springer.

Laura, M., Consonni, R., Locatelli, F., Fumagalli, E., Allavena, A., Coraggio, I., et al. (2010). Metabolic response to cold and freezing of *Osteospermum ecklonis* overexpressing Osmyb4. *Plant Physiology and Biochemistry*, *48*(9), 764–771.

Le Cao, K. A., Boitard, S., & Besse, P. (2011). Sparse PLS discriminant analysis: biologically relevant feature selection and graphical displays for multiclass problems. *BMC Bioinformatics*, *12*, 253. https://doi.org/10.1186/1471-2105-12-253.

Le Lay, P., Isaure, M. P., Sarry, J. E., Kuhn, L., Fayard, B., Le Bail, J. L., et al. (2006). Metabolomic, proteomic and biophysical analyses of *Arabidopsis thaliana* cells exposed to a caesium stress. Influence of potassium supply. *Biochimie*, *88*(11), 1533–1547.

Le, M. Q., Pagter, M., & Hincha, D. K. (2015). Global changes in gene expression, assayed by microarray hybridization and quantitative RT-PCR, during acclimation of three *Arabidopsis thaliana* accessions to sub-zero temperatures after cold acclimation. *Plant Molecular Biology*, *87*(1-2), 1–15.

Lee, S. J., Jeong, E. M., Ki, A. Y., Oh, K. S., Kwon, J., Jeong, J. H., et al. (2016). Oxidative defense metabolites induced by salinity stress in roots of *Salicornia herbacea*. *Journal of Plant Physiology*, *206*, 133–142.

Lehmann, M., Schwarzländer, M., Obata, T., Sirikantaramas, S., Burow, M., Olsen, C. E., et al. (2009). The metabolic response of *Arabidopsis* roots to oxidative stress is distinct from that of heterotrophic cells in culture and highlights a complex relationship between the levels of transcripts, metabolites, and flux. *Molecular Plant*, *2*(3), 390–406.

Lewis, D. F. V., & Gorrod, J. W. (2002). Molecular orbital calculations and nicotine metabolism: a rationale for experimentally observed metabolite ratios. *Drug Metabolism and Drug Interactions*, *19*(1), 29–40.

Li, K., Wang, X., Pidatala, V. R., Chang, C. P., & Cao, X. (2014). Novel quantitative metabolomic approach for the study of stress responses of plant root metabolism. *Journal of Proteome Research*, *13*(12), 5879–5887.

Li, X., Gao, P., Gjetvaj, B., Westcott, N., & Gruber, M. Y. (2009). Analysis of the metabolome and transcriptome of *Brassica carinata* seedlings after lithium chloride exposure. *Plant Science*, *177*(1), 68–80.

Li, Y., Liu, Y., & Zhang, J. (2010). Advances in the research on the AsA-GSH cycle in horticultural crops. *Frontiers of Agriculture in China*, *4*(1), 84–90.

Li, Y., Muthuramalingam, M., & Mahalingam, R. (2015). Plant responses to tropospheric ozone. In *Genetic manipulation in plants for mitigation of climate change* (pp. 1–14). New Delhi: Springer.

Licausi, F., Ohme-Takagi, M., & Perata, P. (2013). APETALA 2/ethylene responsive factor (AP 2/ERF) transcription factors: mediators of stress responses and developmental programs. *New Phytologist*, *199*(3), 639–649.

Lindon, J. C., & Nicholson, J. K. (2008). Spectroscopic and statistical techniques for information recovery in metabonomics and metabolomics. *Annual Review of Analytical Chemistry*, *1*, 45–69.

Llanes, A., Arbona, V., Gómez-Cadenas, A., & Luna, V. (2016). Metabolomic profiling of the halophyte *Prosopis strombulifera* shows sodium salt-specific response. *Plant Physiology and Biochemistry*, *108*, 145–157.

Lugan, R., Niogret, M. F., Leport, L., Guegan, J. P., Larher, F., Savoure, A., et al. (2010). Metabolome and water homeostasis analysis of *Thellungiella salsuginea* suggests that dehydration tolerance is a key response to osmotic stress in this halophyte. *The Plant Journal*, *64*, 215), 229.

Mailloux, R. J., & Harper, M. E. (2011). Uncoupling proteins and the control of mitochondrial reactive oxygen species production. *Free Radical Biology and Medicine*, *51*(6), 1106–1115.

Maruyama, K., Sakuma, Y., Kasuga, M., Ito, Y., Seki, M., Goda, H., et al. (2004). Identification of cold-inducible downstream genes of the *Arabidopsis* DREB1A/CBF3 transcriptional factor using two microarray systems. *The Plant Journal*, *38*(6), 982–993.

Maruyama, K., Takeda, M., Kidokoro, S., Yamada, K., Sakuma, Y., Urano, K., et al. (2009). Metabolic pathways involved in cold acclimation identified by integrated analysis of metabolites and transcripts regulated by DREB1A and DREB2A. *Plant Physiology*, *150*(4), 1972–1980.

Maruyama, K., Urano, K., Yoshiwara, K., Morishita, Y., Sakurai, N., Suzuki, H., et al. (2014). Integrated analysis of the effects of cold and dehydration on rice metabolites, phytohormones, and gene transcripts. *Plant Physiology*, *164*(4), 1759–1771.

Mattana, M., Biazzi, E., Consonni, R., Locatelli, F., Vannini, C., Provera, S., et al. (2005). Overexpression of Osmyb4 enhances compatible solute accumulation and increases stress tolerance of *Arabidopsis thaliana*. *Physiologia Plantarum*, *125*(2), 212–223.

Mattana, M., Carravieri, S., Vannini, C., Bracale, M., Locatelli, F., Baldoni, E., et al. (2005). *Osmyb4: a tool to improve multiple stress tolerance in crops. In R. E. Evenson & V. Santaniello (Eds.), Agricultural biotechnology: ten years after, 9th ICABR international conference.* http://www.economia.uniroma2.it/conferenze/icabr2005.

Mazzei, P., Vinale, F., Woo, S. L., Pascale, A., Lorito, M., & Piccolo, A. (2016). Metabolomics by proton high-resolution magic-angle-spinning nuclear magnetic resonance of tomato plants treated with two secondary metabolites isolated from trichoderma. *Journal of Agricultural and Food Chemistry*, *64*(18), 3538–3545.

McLoughlin, F., Galvan-Ampudia, C. S., Julkowska, M. M., Caarls, L., van der Does, D., Lauriere, C., et al. (2012). The Snf1-related protein kinases SnRK2.4 and SnRK2.10 are involved in maintenance of root system architecture during salt stress. *The Plant Journal*, *72*(3), 436–449.

Michaletti, A., Naghavi, M. R., Toorchi, M., Zolla, L., & Rinalducci, S. (2018). Metabolomics and proteomics reveal drought-stress responses of leaf tissues from spring-wheat. *Scientific Reports*, *8*(1), 5710.

Mills, G., Sharps, K., Simpson, D., Pleijel, H., Broberg, M., Uddling, J., et al. (2018). Ozone pollution will compromise efforts to increase global wheat production. *Global Change Biology*. https://doi.org/10.1111/gcb.14157.

Mintz-Oron, S., Meir, S., Malitsky, S., Ruppin, E., Aharoni, A., & Shlomi, T. (2012). Reconstruction of *Arabidopsis* metabolic network models accounting for subcellular compartmentalization and tissue-specificity. *Proceedings of the National Academy of Sciences of the United States of America*, *109*, 339–344. https://doi.org/10.1073/pnas.1100358109.

Mittler, R., Vanderauwera, S., Gollery, M., & Van Breusegem, F. (2004). Reactive oxygen gene network of plants. *Trends in Plant Science*, *9*(10), 490–498.

Moongngarm, A., & Saetung, N. (2010). Comparison of chemical compositions and bioactive compounds of germinated rough rice and brown rice. *Food Chemistry*, *122*(3), 782–788.

Morcuende, R., Bari, R., Gibon, Y., Zheng, W., Pant, B. D., Bläsing, O., et al. (2007). Genome-wide reprogramming of metabolism and regulatory networks of *Arabidopsis* in response to phosphorus. *Plant, Cell & Environment*, *30*(1), 85–112.

Morsy, M. R., Jouve, L., Hausman, J. F., Hoffmann, L., & Stewart, J. M. (2007). Alteration of oxidative and carbohydrate metabolism under abiotic stress in two rice (*Oryza sativa* L.) genotypes contrasting in chilling tolerance. *Journal of Plant Physiology*, *164*(2), 157–167.

Muchate, N. S., Nikalje, G. C., Rajurkar, N. S., Suprasanna, P., & Nikam, T. D. (2016). Plant salt stress: adaptive responses, tolerance mechanism and bioengineering for salt tolerance. *The Botanical Review*, *82*(4), 371–406.

Muscolo, A., Junker, A., Klukas, C., Weigelt-Fischer, K., Riewe, D., & Altmann, T. (2015). Phenotypic and metabolic responses to drought and salinity of four contrasting lentil accessions. *Journal of Experimental Botany*, *66*(18), 5467–5480.

Nakabayashi, R., & Saito, K. (2015). Integrated metabolomics for abiotic stress responses in plants. *Current Opinion in Plant Biology*, *24*, 10–16.
Nikiforova, V. J., Bielecka, M., Gakiere, B., Krueger, S., Rinder, J., Kempa, S., et al. (2006). Effect of sulfur availability on the integrity of amino acid biosynthesis in plants. *Amino Acids*, *30*(2), 173–183.
Nikiforova, V. J., Daub, C. O., Hesse, H., Willmitzer, L., & Hoefgen, R. (2005). Integrative gene-metabolite network with implemented causality deciphers informational fluxes of sulphur stress response. *Journal of Experimental Botany*, *56*(417), 1887–1896.
Nikiforova, V. J., Kopka, J., Tolstikov, V., Fiehn, O., Hopkins, L., Hawkesford, M. J., et al. (2005). Systems rebalancing of metabolism in response to sulfur deprivation, as revealed by metabolome analysis of *Arabidopsis* plants. *Plant Physiology*, *138*(1), 304–318.
Nishizawa, A., Yabuta, Y., & Shigeoka, S. (2008). Galactinol and raffinose constitute a novel function to protect plants from oxidative damage. *Plant Physiology*, *147*(3), 1251–1263.
Noctor, G., Lelarge-Trouverie, C., & Mhamdi, A. (2015). The metabolomics of oxidative stress. *Phytochemistry*, *112*, 33–53.
Nyarukowaa, C., Koecha, R., Lootsc, T., & Apostolidesa, Z. (2016). SWAPDT: a method for short-time withering assessment of probability for drought tolerance in *Camellia sinensis* validated by targeted metabolomics. *Journal of Plant Physiology*, *198*, 39–48.
Obata, T., & Fernie, A. R. (2012). The use of metabolomics to dissect plant responses to abiotic stresses. *Cellular and Molecular Life Sciences*, *69*(19), 3225–3243.
Obata, T., Witt, S., Lisec, J., Palacios-Rojas, N., Florez-Sarasa, I., Yousfi, S., et al. (2015). Metabolite profiles of maize leaves in drought, heat, and combined stress field trials reveal the relationship between metabolism and grain yield. *Plant Physiology*, *169*, 2665–2683.
Oksanen, E., Freiwald, V., Prozherina, N., & Rousi, M. (2005). Photosynthesis of birch (*Betula pendula*) is sensitive to springtime frost and ozone. *Canadian Journal of Forest Research*, *35*(3), 703–712.
Oliver, M. J., Guo, L., Alexander, D. C., Ryals, J. A., Wone, B. W., & Cushman, J. C. (2011). A sister group contrast using untargeted global metabolomic analysis delineates the biochemical regulation underlying desiccation tolerance in *Sporobolus stapfianus*. *The Plant Cell*. tpc-110.
Olsen, A. N., Ernst, H. A., Leggio, L. L., & Skriver, K. (2005). NAC transcription factors: structurally distinct, functionally diverse. *Trends in Plant Science*, *10*(2), 79–87.
Ostaszewska, M., Juszczuk, I. M., Kołodziejek, I., & Rychter, A. M. (2014). Long-term sulphur starvation of *Arabidopsis thaliana* modifies mitochondrial ultrastructure and activity and changes tissue energy and redox status. *Journal of Plant Physiology*, *171*(7), 549–558.
Pagter, M., Alpers, J., Erban, A., Kopka, J., Zuther, E., & Hincha, D. K. (2017). Rapid transcriptional and metabolic regulation of the deacclimation process in cold acclimated *Arabidopsis thaliana*. *BMC Genomics*, *18*(1), 731.
Panikulangara, T. J., Eggers-Schumacher, G., Wunderlich, M., Stransky, H., & Schöffl, F. (2004). Galactinol synthase1. A novel heat shock factor target gene responsible for heat-induced synthesis of raffinose family oligosaccharides in *Arabidopsis*. *Plant Physiology*, *136*(2), 3148–3158.
Parida, A. K., Panda, A., & Rangani, J. (2018). Metabolomics-guided elucidation of abiotic stress tolerance mechanisms in plants. In *Plant metabolites and regulation under environmental stress* (pp. 89–131). Elsevier.
Park, S., Lee, C. M., Doherty, C. J., Gilmour, S. J., Kim, Y., & Thomashow, M. F. (2015). Regulation of the *Arabidopsis* CBF regulon by a complex low-temperature regulatory network. *The Plant Journal*, *82*(2), 193–207.
Parmesan, C., & Hanley, M. E. (2015). Plants and climate change: complexities and surprises. *Annals of Botany*, *116*(6), 849–864.
Parre, E., Ghars, M. A., Leprince, A. S., Thiery, L., Lefebvre, D., Bordenave, M., et al. (2007). Calcium signaling via phospholipase C is essential for proline accumulation upon

ionic but not nonionic hyperosmotic stresses in *Arabidopsis*. *Plant Physiology*, *144*(1), 503–512.

Pasquali, G., Biricolti, S., Locatelli, F., Baldoni, E., & Mattana, M. (2008). Osmyb4 expression improves adaptive responses to drought and cold stress in transgenic apples. *Plant Cell Reports*, *27*(10), 1677–1686.

Patel, A. K., Huang, E. L., Low-Décarie, E., & Lefsrud, M. G. (2015). Comparative shotgun proteomic analysis of wastewater-cultured microalgae: nitrogen sensing and carbon fixation for growth and nutrient removal in *Chlamydomonas reinhardtii*. *Journal of Proteome Research*, *14*(8), 3051–3067.

Petrov, V., Hille, J., Mueller-Roeber, B., & Gechev, T. S. (2015). ROS-mediated abiotic stress-induced programmed cell death in plants. *Frontiers in Plant Science*, *6*, 69.

Prandl, R., Hinderhofer, K., Eggers-Schumacher, G., & Schofû, F. (1998). HSF3, a new heat shock factor from *Arabidopsis thaliana*, derepresses the heat shock response and confers thermo tolerance when over expressed in transgenic plants. *Molecular & General Genetics*, *258*, 269–278.

Prasch, C. M., & Sonnewald, U. (2013). Simultaneous application of heat, drought and virus to *Arabidopsis thaliana* plants reveals significant shifts in signaling networks. *Plant Physiology*, *162*(4), 1849–1866.

Prince, E. K., & Pohnert, G. (2010). Searching for signals in the noise: metabolomics in chemical ecology. *Analytical and Bioanalytical Chemistry*, *396*(1), 193–197.

Qin, Y., Wang, M., Tian, Y., He, W., Han, L., & Xia, G. (2012). Over-expression of TaMYB33 encoding a novel wheat MYB transcription factor increases salt and drought tolerance in *Arabidopsis*. *Molecular Biology Reports*, *39*(6), 7183–7192.

Ritchie, M. D., Holzinger, E. R., Li, R., Pendergrass, S. A., & Kim, D. (2015). Methods of integrating data to uncover genotype–phenotype interactions. *Nature Reviews Genetics*, *16*(2), 85.

Roessner, U., Patterson, J. H., Forbes, M. G., Fincher, G. B., Langridge, P., & Bacic, A. (2006). An investigation of boron toxicity in barley using metabolomics. *Plant Physiology*, *142*(3), 1087–1101.

Rontein, D., Dieuaide-Noubhani, M., Dufourc, E. J., Raymond, P., & Rolin, D. (2002). The metabolic architecture of plant cells – stability of central metabolism and flexibility of anabolic pathways during the growth cycle of tomato cells. *Journal of Biological Chemistry*, *277*, 43948–43960.

Ruan, C. J., & Teixeira da Silva, J. A. (2011). Metabolomics: creating new potentials for unraveling the mechanisms in response to salt and drought stress and for the biotechnological improvement of xero-halophytes. *Critical Reviews in Biotechnology*, *31*(2), 153–169.

Saito, K., & Matsuda, F. (2010). Metabolomics for functional genomics, systems biology, and biotechnology. *Annual Review of Plant Biology*, *61*, 463–489.

Sanchez, B. J., Perez-Correa, J. R., & Agosin, E. (2014). Construction of robust dynamic genome-scale metabolic model structures of *Saccharomyces cerevisiae* through iterative re-parameterization. *Metabolic Engineering*, *25C*, 159–173. https://doi.org/10.1016/j.ymben.2014.07.004.

Sanchez, D. H., Pieckenstain, F. L., Szymanski, J., Erban, A., Bromke, M., Hannah, M. A., et al. (2011). Comparative functional genomics of salt stress in related model and cultivated plants identifies and overcomes limitations to translational genomics. *PLoS One*, *6* (2)e17094.

Sanchez, D. H., Siahpoosh, M. R., Roessner, U., Udvardi, M., & Kopka, J. (2008). Plant metabolomics reveals conserved and divergent metabolic responses to salinity. *Physiologia Plantarum*, *132*(2), 209–219.

Sánchez, F. S., Erban, A., Udvardi, M. K., & Kopka, J. (2012). Comparative metabolomics of drought acclimation in model and forage legumes. *Plant, Cell & Environment, 35*, 136–149.
Sánchez-Rodríguez, E., Rubio-Wilhelmi, M., del, M., Ríos, J. J., Blasco, B., Rosales, M. Á., et al. (2011). Ammonia production and assimilation: its importance as a tolerance mechanism during moderate water deficit in tomato plants. *Journal of Plant Physiology, 168*, 816–823.
Scalabrin, E., Radaelli, M., Rizzato, G., Bogani, P., Buiatti, M., Gambaro, A., et al. (2015). Metabolomic analysis of wild and transgenic *Nicotiana langsdorffii* plants exposed to abiotic stresses: unraveling metabolic responses. *Analytical and Bioanalytical Chemistry, 407*(21), 6357–6368.
Sewelam, N., Kazan, K., & Schenk, P. M. (2016). Global plant stress signaling: reactive oxygen species at the cross-road. *Frontiers in Plant Science, 7*, 187.
Sharma, S. S., & Dietz, K. J. (2006). The significance of amino acids and amino acid-derived molecules in plant responses and adaptation to heavy metal stress. *Journal of Experimental Botany, 57*(4), 711–726.
Sheth, B., & Thaker, V. (2014). Plant systems biology: insights, advances and challenges. *Planta, 240*, 33–54. https://doi.org/10.1007/s00425-014-2059-5.
Shi, H., & Chan, Z. (2014). Improvement of plant abiotic stress tolerance through modulation of the polyamine pathway. *Journal of Integrative Plant Biology, 56*, 114–121.
Shingaki-Wells, R. N., Huang, S., Taylor, N. L., Carroll, A. J., Zhou, W., & Millar, A. H. (2011). Differential molecular responses of rice and wheat coleoptiles to anoxia reveal novel metabolic adaptations in amino acid metabolism for tissue tolerance. *Plant Physiology, 156*(4), 1706–1724.
Shulaev, V. (2006). Metabolomics technology and bioinformatics. *Briefings in Bioinformatics, 7* (2), 128–139.
Singh, R., Singh, S., Parihar, P., Mishra, R. K., Tripathi, D. K., Singh, V. P., et al. (2016). Reactive oxygen species (ROS): beneficial companions of plants' developmental processes. *Frontiers in Plant Science, 7*, 1299.
Skirycz, A., De Bodt, S., Obata, T., De Clercq, I., Claeys, H., De Rycke, R., et al. (2010). Developmental stage specificity and the role of mitochondrial metabolism in the response of *Arabidopsis* leaves to prolonged mild osmotic stress. *Plant Physiology, 152*(1), 226–244.
Slama, I., Abdelly, C., Bouchereau, A., Flower, T., & Savouré, A. (2015). Diversity, distribution and roles of osmoprotective compounds accumulated in halophytes under abiotic stress. *Annals of Botany, 115*, 433–447.
Sofo, A., Scopa, A., Nuzzaci, M., & Vitti, A. (2015). Ascorbate peroxidase and catalase activities and their genetic regulation in plants subjected to drought and salinity stresses. *International Journal of Molecular Sciences, 16*(6), 13561–13578.
Song, C., Je, J., Hong, J. K., & Lim, C. O. (2014). Ectopic expression of an *Arabidopsis* dehydration-responsive element-binding factor DREB2C improves salt stress tolerance in crucifers. *Plant Cell Reports, 33*(8), 1239–1254.
Song, S. Y., Chen, Y., Chen, J., Dai, X. Y., & Zhang, W. H. (2011). Physiological mechanisms underlying OsNAC5-dependent tolerance of rice plants to abiotic stress. *Planta, 234*(2), 331–345.
Souza, L. A., Monteiro, C. C., Carvalho, R. F., Gratão, P. L., & Azevedo, R. A. (2017). Dealing with abiotic stresses: an integrative view of how phytohormones control abiotic stress-induced oxidative stress. *Theoretical and Experimental Plant Physiology, 29*(3), 109–127.
Steuer, R., Morgenthal, K., Weckwerth, W., & Selbig, J. (2007). A gentle guide to the analysis of metabolomic data. *Methods in Molecular Biology, 358*, 105–126. https://doi.org/10.1007/978-1-59745-244-1_7.

Sugimoto, M., Kawakami, M., Robert, M., Soga, T., & Tomita, M. (2012). Bioinformatics tools for mass spectroscopy-based metabolomic data processing and analysis. *Current Bioinformatics*, 7, 96–108. https://doi.org/10.2174/157489312799304431.

Sun, C. X., Li, M. Q., Gao, X. X., Liu, L. N., Wu, X. F., & Zhou, J. H. (2016). Metabolic response of maize plants to multi-factorial abiotic stresses. *Plant Biology*, *18*, 120–129.

Suzuki, N., Miller, G., Salazar, C., Mondal, H. A., Shulaev, E., Cortes, D. F., et al. (2013). Temporal-spatial interaction between reactive oxygen species and abscisic acid regulates rapid systemic acclimation in plants. *The Plant Cell*, *25*(9), 3553–3556.

Sweetlove, L. J., Beard, K. F., Nunes-Nesi, A., Fernie, A. R., & Ratcliffe, R. G. (2010). Not just a circle: flux modes in the plant TCA cycle. *Trends in Plant Science*, *15*(8), 462–470.

Tarpley, L., Duran, A. L., Kebrom, T. H., & Sumner, L. W. (2005). Biomarker metabolites capturing the metabolite variance present in a rice plant developmental period. *BMC Plant Biology*, *5*(1), 8.

Testerink, C., & Munnik, T. (2011). Molecular, cellular, and physiological responses to phosphatidic acid formation in plants. *Journal of Experimental Botany*, *62*(7), 2349–2361.

Trivedi, D. K., Gill, S. S., & Tuteja, N. (2016). *Abscisic acid (ABA): biosynthesis, regulation, and role in abiotic stress tolerance* (pp. 315–326). Weinheim, Germany: Wiley-VCH.

Tschoep, H., Gibon, Y., Carillo, P., Armengaud, P., Szecowka, M., Nunes-Nesi, A., et al. (2009). Adjustment of growth and central metabolism to a mild but sustained nitrogen-limitation in *Arabidopsis*. *Plant, Cell & Environment*, *32*, 300–318.

Uarrota, V. G., Moresco, R., Coelho, B., Nunes Eda, C., Peruch, L. A., Neubert Ede, O., et al. (2014). Metabolomics combined with chemometric tools (PCA, HCA, PLS-DA and SVM) for screening cassava (*Manihot esculenta* Crantz) roots during postharvest physiological deterioration. *Food Chemistry*, *161*, 67–78. https://doi.org/10.1016/j.foodchem.2014.03.110.

Umezawa, T., Sugiyama, N., Mizoguchi, M., Hayashi, S., Myouga, F., Yamaguchi-Shinozaki, K., et al. (2009). Type 2C protein phosphatases directly regulate abscisic acid-activated protein kinases in *Arabidopsis*. *Proceedings of the National Academy of Sciences*, *106*(41), 17588–17593.

Urano, K., Maruyama, K., Ogata, Y., Morishita, Y., Takeda, M., Sakurai, N., et al. (2009). Characterization of the ABA-regulated global responses to dehydration in *Arabidopsis* by metabolomics. *The Plant Journal*, *57*(6), 1065–1078.

Urbanczyk-Wochniak, E., & Fernie, A. R. (2004). Metabolic profiling reveals altered nitrogen nutrient regimes have diverse effects on the metabolism of hydroponically-grown tomato (*Solanum lycopersicum*) plants. *Journal of Experimental Botany*, *56*(410), 309–321.

Usadel, B., Bläsing, O. E., Gibon, Y., Retzlaff, K., Höhne, M., Günther, M., et al. (2008). Global transcript levels respond to small changes of the carbon status during progressive exhaustion of carbohydrates in *Arabidopsis* rosettes. *Plant Physiology*, *146*(4), 1834–1861.

Vainonen, J. P., & Kangasjärvi, J. (2015). Plant signalling in acute ozone exposure. *Plant, Cell & Environment*, *38*(2), 240–252.

Valerio, C., Costa, A., Marri, L., Issakidis-Bourguet, E., Pupillo, P., Trost, P., et al. (2011). Thioredoxin-regulated beta-amylase (BAM1) triggers diurnal starch degradation in guard cells, and in mesophyll cells under osmotic stress. *Journal of Experimental Botany*, *62*, 545–555.

Vannini, C., Iriti, M., Bracale, M., Locatelli, F., Faoro, F., Croce, P., et al. (2006). The ectopic expression of the rice Osmyb4 gene in *Arabidopsis* increases tolerance to abiotic, environmental and biotic stresses. *Physiological and Molecular Plant Pathology*, *69*(1-3), 26–42.

Verbruggen, N., & Hermans, C. (2008). Proline accumulation in plants: a review. *Amino Acids*, *35*(4), 753–759.

Viant, M. R., & Sommer, U. (2013). Mass spectrometry based environmental metabolomics: a primer and review. *Metabolomics*, *9*(1), 144–158.

von Koskull-Döring, P., Scharf, K. D., & Nover, L. (2007). The diversity of plant heat stress transcription factors. *Trends in Plant Science, 12*(10), 452–457.
Wakasa, K., Hasegawa, H., Nemoto, H., Matsuda, F., Miyazawa, H., Tozawa, Y., et al. (2006). High-level tryptophan accumulation in seeds of transgenic rice and its limited effects on agronomic traits and seed metabolite profile. *Journal of Experimental Botany, 57*, 3069–3078.
Walter, P., & Ron, D. (2011). The unfolded protein response: from stress pathway to homeostatic regulation. *Science, 334*(6059), 1081–1086.
Wang, C. J., Yang, W., Wang, C., Gu, C., Niu, D. D., Liu, H. X., et al. (2012). Induction of drought tolerance in cucumber plants by a consortium of three plant growth-promoting rhizobacterium strains. *PLoS One*, 7(12)e52565.
Wang, C., Liu, Y., Li, S. S., & Han, G. Z. (2015). Insights into the origin and evolution of the plant hormone signaling machinery. *Plant Physiology, 167*(3), 872–886.
Wang, J. H., Byun, J., & Pennathur, S. (2010). Analytical approaches to metabolomics and applications to systems biology. *Seminars in Nephrology, 30*(5), 500–511.
Wang, J. L., Huang, S. Y., Li, T. T., Chen, K., Ning, H. R., & Zhu, X. Q. (2016). Evaluation of the basic functions of six calcium-dependent protein kinases in *Toxoplasma gondii* using CRISPR-Cas9 system. *Parasitology Research, 115*(2), 697–702.
Wani, S. H., Singh, N. B., Jeberson, S. M., Sanghera, G. S., Haribhushan, A., Chaudhury, B. U., et al. (2012). Molecular strategies for identification and deployment of gene (s) for abiotic stress tolerance in crop plants. *International Journal of Life Sciences, 1*(2), 128–142.
Wasaki, J., Yonetani, R., Kuroda, S., Shinano, T., Yazaki, J., Fujii, F., et al. (2003). Transcriptomic analysis of metabolic changes by phosphorus stress in rice plant roots. *Plant, Cell & Environment, 26*(9), 1515–1523.
Weckwerth, W. (2008). Integration of metabolomics and proteomics in molecular plant physiology-coping with the complexity by data-dimensionality reduction. *Physiologia Plantarum, 132*, 176–189. https://doi.org/10.1111/j.1399-3054.2007.01011.x.
Wingler, A., & Roitsch, T. (2008). Metabolic regulation of leaf senescence: interactions of sugar signalling with biotic and abiotic stress responses. *Plant Biology, 10*, 50–62.
Wisniewski, M., Norelli, J., Bassett, C., Artlip, T., & Macarisin, D. (2011). Ectopic expression of a novel peach (*Prunus persica*) CBF transcription factor in apple (Malus × domestica) results in short-day induced dormancy and increased cold hardiness. *Planta, 233*(5), 971–983.
Woodgett, J. R., & Cohen, P. (1984). Multisite phosphorylation of glycogen synthase. Molecular basis for the substrate specificity of glycogen synthase kinase-3 and casein kinase-II (glycogen synthase kinase-5). *Biochimica et Biophysica Acta, 788*, 339–347.
Wu, D., Shen, Q., Cai, S., Chen, Z. H., Dai, F., & Zhang, G. (2013). Ionomic responses and correlations between elements and metabolites under salt stress in wild and cultivated barley. *Plant and Cell Physiology, 54*(12), 1976–1988.
Wu, G., Kang, H., Zhang, X., Shao, H., Chu, L., & Ruan, C. (2010). A critical review on the bio-removal of hazardous heavy metals from contaminated soils: issues, progress, eco-environmental concerns and opportunities. *Journal of Hazardous Materials, 174*(1-3), 1–8.
Wu, Z. G., Jiang, W., Chen, S. L., Mantri, N., Tao, Z. M., & Jiang, C. X. (2016). Insights from the cold transcriptome and metabolome of *Dendrobium officinale*: global reprogramming of metabolic and gene regulation networks during cold acclimation. *Frontiers in Plant Science*, 7, 1653.
Xia, J., & Wishart, D. S. (2016). Using MetaboAnalyst 3.0 for comprehensive metabolomics data analysis. *Current Protocols in Bioinformatics, 55*(1), 14.10.1–14.10.19.
Xiang, Y., Huang, Y., & Xiong, L. (2007). Characterization of stress-responsive CIPK genes in rice for stress tolerance improvement. *Plant Physiology, 144*(3), 1416–1428.

Xie, Y., Hu, L., Du, Z., Sun, X., Amombo, E., Fan, J., et al. (2014). Effects of cadmium exposure on growth and metabolic profile of bermudagrass [*Cynodon dactylon* (L.) Pers.]. *PLoS One*, *9*(12)e115279.

Xiong, L., & Yang, Y. (2003). Disease resistance and abiotic stress tolerance in rice are inversely modulated by an abscisic acid–inducible mitogen-activated protein kinase. *The Plant Cell*, *15*(3), 745–759.

Yamaji, K., Julkunen-Tiitto, R., Rousi, M., Freiwald, V., & Oksanen, E. (2003). Ozone exposure over two growing seasons alters root-to-shoot ratio and chemical composition of birch (*Betula pendula* Roth). *Global Change Biology*, *9*(10), 1363–1377.

Yang, A., Dai, X., & Zhang, W. H. (2012). A R2R3-type MYB gene, OsMYB2, is involved in salt, cold, and dehydration tolerance in rice. *Journal of Experimental Botany*, *63*(7), 2541–2556.

Yang, K. Y., & Swenson, R. P. (2007). Modulation of the redox properties of the flavin cofactor through hydrogen-bonding interactions with the N (5) atom: Role of αSer254 in the electron-transfer flavoprotein from the methylotrophic bacterium W3A1. *Biochemistry*, *46*(9), 2289–2297.

Yao, F., Coquery, J., & Le Cao, K. A. (2012). Independent principal component analysis for biologically meaningful dimension reduction of large biological data sets. *BMC Bioinformatics*, *13*, 24. https://doi.org/10.1186/1471-2105-13-24.

Yonekura-Sakakibara, K., Tohge, T., Matsuda, F., Nakabayashi, R., Takayama, H., Niida, R., et al. (2008). Comprehensive flavonol profiling and transcriptome coexpression analysis leading to decoding gene–metabolite correlations in Arabidopsis. *The Plant Cell*, *20*(8), 2160–2176.

Yoshida, T., Mogami, J., & Yamaguchi-Shinozaki, K. (2015). Omics approaches toward defining the comprehensive abscisic acid signaling network in plants. *Plant and Cell Physiology*, *56*(6), 1043–1052.

You, J., Zong, W., Hu, H., Li, X., Xiao, J., & Xiong, L. (2014). A STRESS-RESPONSIVE NAC1-regulated protein phosphatase gene rice protein phosphatase18 modulates drought and oxidative stress tolerance through abscisic acid-independent reactive oxygen species scavenging in rice. *Plant Physiology*, *166*(4), 2100–2114.

Zandalinas, S. I., Mittler, R., Balfagón, D., Arbona, V., & Gómez-Cadenas, A. (2018). Plant adaptations to the combination of drought and high temperatures. *Physiologia Plantarum*, *162*(1), 2–12.

Zhao, M., & Running, S. W. (2010). Drought-induced reduction in global terrestrial net primary production from 2000 through 2009. *Science*, *329*(5994), 940–943.

Zhao, J., & Verpoorte, R. (2007). Manipulating indole alkaloid production by Catharanthus roseus cell cultures in bioreactors: from biochemical processing to metabolic engineering. *Phytochemistry Reviews*, *6*(2–3), 435–457.

Zhou, Q., Guo, J. J., He, C. T., Shen, C., Huang, Y. Y., Chen, J. X., et al. (2016). Comparative transcriptome analysis between low-and high-cadmium-accumulating genotypes of pakchoi (*Brassica chinensis* L.) in response to cadmium stress. *Environmental Science & Technology*, *50*(12), 6485–6494.

Zhu, J. K. (2016). Abiotic stress signaling and responses in plants. *Cell*, *167*(2), 313–324.

Zinta, G., AbdElgawad, H., Peshev, D., Weedon, J. T., Van den Ende, W., Nijs, I., et al. (2018). Dynamics of metabolic responses to periods of combined heat and drought in *Arabidopsis thaliana* under ambient and elevated atmospheric CO_2. *Journal of Experimental Botany*, *69*(8), 2159–2170.

Zuther, E., Büchel, K., Hundertmark, M., Stitt, M., Hincha, D. K., & Heyer, A. G. (2004). The role of raffinose in the cold acclimation response of *Arabidopsis thaliana*. *FEBS Letters*, *576*(1-2), 169–173.

FURTHER READING

Ara, H., & Sinha, A. K. (2015). Role of mitogen-activated protein kinase cascade in combating abiotic stress in plants. In *Elucidation of abiotic stress signaling in plants* (pp. 207–229). New York, NY: Springer.
Kumar, S., Stecher, G., & Tamura, K. (2016). MEGA7: molecular evolutionary genetics analysis version 7.0 for bigger datasets. *Molecular Biology and Evolution*, *33*(7), 1870–1874.
Majumder, A. L. (2017). Phosphoinositides and phospholipase C signalling in plant stress response: a revisit. *Proceedings of the Indian National Science Academy*, *83*(4), 845–863.
Wang, X. (2005). Regulatory functions of phospholipase D and phosphatidic acid in plant growth, development, and stress responses. *Plant Physiology*, *139*(2), 566–573.

CHAPTER 4

Climatic Change and Metabolome Fluxes

4.1 INTRODUCTION

Plants have to endure climate change, ecological threats, and even sometimes natural catastrophes in order to survive. The multiple environmental stresses include limited nutrient stress, high temperature, and elevated CO_2 levels, either individually or in combination (Meena et al., 2017; Xu, Jiang, Jia, & Zhou, 2016). These environmental stresses affect overall plant structure and signify a high risk for promoting sustainable agriculture at a time of important growth in the global population (Amedie, 2013). In order to combat environmental stresses, plants have an extensive range of molecular pathways to identify changes instantly, and consequently make adjustments (Kensler, Wakabayashi, & Biswal, 2007; Rejeb, Pastor, & Mauch-Mani, 2014). The objective of reconstituting steps prevailing under persistently varying ecological conditions has been an intriguing topic for several decades. However, there is still a considerable knowledge gap, and we are usually incapable of predicting exactly how plants will combat environmental challenges. Such perspectives are needed to breed crops or produce transgenic varieties with high tolerance levels against numerous ecological stress factors, because plants are concurrently subjected to varied ecological disturbances. For future efforts, we can advance the potential of systems biology strategies to evaluate biological networks to comprehend and promote plant responses to ecological stress. Higher levels of atmospheric greenhouse gasses have altered the global climate, increasing the earth's surface temperature by 0.74°C in the past century (IPCC, 2007). This rise in temperature is parallel with the rise in the concentration of atmospheric CO_2 from 280 to 394 ppm (Monastersky, 2013; Wang, Jiang, Zhang, & Zhou, 2011). The present increase continues to rise exponentially in spite of global policies intended to impede it (Fang et al., 2009). Along with an increase in atmospheric CO_2 concentrations, there is a concomitant rise in other greenhouse gasses; for example, methane and nitrogen oxides that are extensively released by anthropogenic activities (Portmann, Daniel, & Ravishankara, 2012). These instant changes might surpass the competence

https://doi.org/10.1016/B978-0-12-814872-3.00004-7

of organisms, populations, and communities to address them. Hence, an instant query in the biology of global change is how current vegetation or plant life is combatting this new challenge. In this chapter, we will summarize the present consequences of climate change; the competence of terrestrial organisms, populations, communities and ecosystems to tackle current climatic change; and the upscaling of their rejoinders from the community, ecosystem, and biosphere at molecular and genetic levels (Peñuelas et al., 2013; Sterner & Elser, 2002). We will also address some unraveled issues necessitating more research for an enhanced perspective on the competence at the organism, population, community, and ecosystem levels to combat climatic change, encompassing the association with other triggers of global change, and for enhanced understanding of the likely reaction on climate of these changes in organisms, populations, communities, and ecosystems.

4.2 SEASONAL STOICHIOMETRIC AND METABOLOMIC CHANGES

The prime metabolic shift in plant life can be determined by monitoring the changing metabolic flux in response to seasonal variation. Metabolomic changes are often associated with shifts in elemental stoichiometry of organisms thriving under changing environmental conditions and diverse developmental stages, because elements function mainly as components of molecular compounds (Elser, Fagan, Kerkhoff, Swenson, & Enquist, 2010; Rivas-Ubach, Sardans, Pérez-Trujillo, Estiarte, & Peñuelas, 2012). Several research studies have established the strong association between changes in elemental stoichiometry C/N/P/K and variation in leaf metabolomes throughout the growth cycle and ecological stress (Brunetti, George, Tattini, Field, & Davey, 2013; Gargallo-Garriga et al., 2015). In *Erica multiflora* exposed to drought and global warming, there was a decline in the N/P ratio, with a concomitant rise in primary metabolites of leaves, particularly during active growth seasons (Bernal, Estiarte, & Peñuelas, 2011; Prieto, Peñuelas, Ogaya, & Estiarte, 2008). This research finding established the growth rate hypothesis (GRH), which affirms the decline of N/P rations in growing organisms as result of their allocation into macromolecules such as RNA (Sardans, Peñuelas, Prieto, & Estiarte, 2008; Rajkumar, Prasad, Swaminathan, & Freitas, 2013). In the summer, and under drought stress, a lower level of N/K and P/K in leaves is in line with the function of K in osmotic protection, which corresponds with a rise in osmoprotectant levels to avert water stress. These research reports represent a step in the perspective pertaining to the association between stoichiometry and organism metabolism. The elemental stoichiometry ratio of C/N/P levels in environment and biomass exhibit a statistically significant association with organism traits that in turn modulate the structure and function of ecosystems (Sardans, Peñuelas, Coll, Vayreda, & Rivas-Ubach, 2012; Yuan & Chen, 2015). The GRH is one of the main themes of the environmental stoichiometry, and it suggests that growing organisms must amplify their distribution of P to RNA to convene rising demands for the protein synthesis necessary for growth (Sistla & Schimel, 2012). Higher growth rates are exhibited by species displaying low ratios of ecological N/P and C/P, and they may trigger shifts in species communities (Fanin, Fromin, Buatois, & Hättenschwiler, 2013; Taylor & Townsend, 2012). The GRH has been reestablished in freshwater ecosystems, with exclusions pertaining to the distribution of nutrients to function, excluding growth (Elser et al., 2003; Matzek & Vitousek, 2009). Conversely, the relevance of GRH has no, or little, significance in terrestrial ecosystems (Hall et al., 2011). In terrestrial ecosystems, other than the allocation of P and N in growth, these nutrients can function in storage, defending against abiotic and biotic stress. Hence, the phenotypic responses in these other fundamental organismal functions must be taken into account when investigating the association of C/N/P ratio with the metabolomes of organism and ecosystem function and structure (Peñuelas & Sardans, 2009). So in order to comprehend the relationship between the environment and elemental stoichiometry, the primary step will be the establishment of a link between elemental stoichiometry to the metabolomes of organisms and ecosystem structure and function (Yu et al., 2010; Zhang et al., 2013). Numerous research studies have been illustrated to document significant phenotypic responses of organisms to drought and global warming at the molecular level. Ecometabolomic research studies illustrate effective strategies for obtaining perspectives pertaining to molecular plasticity of the organism to drought and global warming. The ecometabolomic research reports illustrate the identification of major metabolic pathways accountable for organism responses, and advance assistance to identify the genes implicated in the response (Guy, Kaplan, Kopka, Selbig, & Hincha, 2008; Sardans, Peñuelas, & Rivas-Ubach, 2011). The relevance of growing omics techniques to ecophysiological research has established the efficiency of plants and animals in displaying plastic molecular reactions to drought and warming (Blackman, Aspinwall, Tissue, & Rymer, 2017). Plants respond to drought stress by increased sugar solublization and enhanced gene expression implicated in abscisic acid

(ABA) and proline biosynthetic pathways at the molecular level. The associated changes in gene expression are subsequently associated with metabolomic shifts (Krugman, Peleg, Quansah, et al., 2011; Spieb, Oufir, Matusikova, et al., 2012; Warren, Aranda, & Cano, 2012). The genes implicated in drought tolerance are generally associated with the continuation of turgor and cell integrity (Aranjuelo, Molero, Erice, Avice, & Nogues, 2011; Erxleben, Gewssler, Vervliet-Scheebaum, & Reski, 2012). The prime metabolites accumulated under drought stress include proline, ABA, and gamma aminobutyric acid (GABA). Drought stress stimulates a diverse class of metabolites that are determined by genotypes and ecological factors (Sanchez, Schwabe, Erban, Udvardi, & Kopka, 2012; Saxena et al., 2011). In concurrence with drought stress, global warming also suggests modulation in organisms at the molecular level. The relevance of omics technologies demonstrates enhanced gene expression of heat stress proteins and other metabolites (Sardans et al., 2011). Numerous research studies anticipate an increase in protein catabolism allowing the allocation of proteins into anti-stress factors (William et al., 2013). The mechanistic response of plants to global warming and climatic changes varies, even in species of the same genus (Bagley et al., 2015; Urban, 2015). Numerous research studies illustrate the varied responses of plants to global climate change, which do not entail significant variation in secondary metabolism (Nelson et al., 2014). Nybakken, Sandvik, and Klanderud (2011) demonstrated meager variation in the carbon-based secondary metabolic flux in response to global warming in sub-alpine ecosystems (Craine et al., 2013; Zamboni, Fendt, Rühl, & Sauer, 2009). Drought and global warming reveal physiological, phenological, and anatomical changes in organisms (Aspinwall et al., 2015; Buckley & Kingsolver, 2012). These physiological changes are manifested as a result of varied molecular composition associated with altered elemental stoichiometry with diverse responses within the same community (Theocharis, Clément, & Barka, 2012; Clément and Barka, 2012). Variation in elemental stoichiometry C: N: P affects the nutrient cycling in ecosystems, and energy transfer via food webs as observed in herbivorous communities. The shifts in elemental stoichiometry of plants have advanced inferences on the functionality of trophic webs (Sardans, Peñuelas, et al., 2012; Verschoor, Van Dijk, Huisman, & Van Donk, 2013). Similar research studies demonstrating drought and global warming effects on terrestrial animals are still scarce. Aphids subjected to high temperatures exhibit low growth rate, and low levels of enzymes implicated in metabolic pathways that lead to biosynthesis of exoskeletal proteins (Peñuelas et al., 2013). Metabolomic research establishes that high temperature increases amino acids and protein pools, and reduces soluble sugar levels (Bagri et al., 2017; Khan, Bano, Rahman, Rathinasabapathi, & Babar, 2018). Hence, it can be conjectured that omics technologies exhibit high sensitivity for identifying metabolic shifts of organisms subjected to drought or global warming. These techniques reveal instant enhancements in the enzymatic synthesis associated with metabolism and metabolites that are associated with osmotic adjustment and anti-stress processes. Although there is still a dearth of research reports documenting the coupling of climatic change and genomic metabolomics with nutrient cycles, there are reports available on elemental stoichiometry, along with physiological and seasonal shifts, with alterations in the ecosystem's structure (Sardans et al., 2011; Trivedi, Anderson, & Singh, 2013). The interdisciplinary research studies should offer improved perspectives on the methods and developments based on the variation in resource utilization, in intraspecies and interspecies competition, and intraspecies substitution and selection under global change.

Seasonal variations induce the corresponding change in elemental stoichiometry, such as favorable climatic conditions of spring, along with high soil water availability, higher photosynthetic rates, and increased N and P uptake. Enhanced elemental concentration of N and P levels allocate increased amino acids and protein synthesis, which subsequently leads to increased RNA synthesis (Frigstad et al., 2011; Rivas-Ubach et al., 2012). A declining N/P ratio throughout growing season has been observed in terrestrial plants, as well as in animals (Gusewell, 2004; Matzek & Vitousek, 2009). Furthermore, the spring season exhibits higher carbon allocation to growth and energy supplies contributing to primary metabolism, while less carbon was incorporated into the defensive line for secondary metabolism (Hartman et al., 2018; Liu et al., 2015). Actually, C/N, C/P, and C/K ratios were also positively associated with non-polar metabolites; for example, fatty acids, terpenoids, polyphenols, and other C-rich components. On the contrary, summer leaves exhibit high sugar levels due to high sugar accretion during spring, or as result of an increase in cellular osmotic potential (Sardans, Rivas-Ubach, & Peñuelas, 2012a, 2012b). Conversely, the highest levels of K/P and the lowest N/K and C/K ratios were observed in summer leaves. Potassium has an imperative function in maintaining the plant-water continuum through regulation of osmotic potential and stomatal movement (Sardans & Peñuelas, 2015; Sardans, Peñuelas, et al., 2012). Global warming enhances fatty acid levels associated with drought stress, as demonstrated in several research studies (Bennett, Bell, Davy, Webster, & Francis, 2018;

Kachroo & Kachroo, 2009; Upchurch, 2008). In a water deficit habitat such as the Mediterranean ecosystem, elemental stoichiometry and metabolomes of leaves are significantly affected due to oxidative stress induced by drought conditions (Peñuelas et al., 2018; Suseela & Tharayil, 2018). Similarly, in the autumn season, under drought stress, leaf metabolomes are predominated by polyphenolic compounds, quinic acid, and tartaric acid exhibiting antioxidant activity (Sardans et al., 2011; Sánchez-Rodríguez, Ruiz, Ferreres, & Moreno, 2012). These metabolites have high antioxidant potential, resulting from their high competence as electron donors, and their ability to change the kinetics of peroxidation. Similarly, a higher accumulation of choline was observed in drought stressed leaves in summer. These results are supported by research studies that observed an increased release of aromatic amino acids in drought-stressed plants (Ahmad et al., 2008; Gou et al., 2015). This metabolomic disparity was followed by an increment in K levels leading to low C/K and N/K ratios and high K/P ratios, as observed in the summer during drought season; which is likely associated with water use efficiency (WUE; Slama, Abdelly, Bouchereau, Flowers, & Savouré, 2015; Kabbadj et al., 2017). On the contrary, there are research studies that have not been able to demonstrate a link between N/P and protein/RNA ratios (Matzek & Vitousek, 2009). Due to immense environmental threats, terrestrial plants and animals assimilate huge levels of N and P in defense against biotic and abiotic stress. Further, C: N: P biomass can be associated with significant ecological processes such as litter, decomposition, nitrogen fixation, plant-herbivore-predator association (Sardans, Rivas-Ubach, Estiarte, Ogaya, & Peñuelas, 2013; Rivas-Ubach, Peñuelas i Reixach, & Sardans, 2014), ecosystem diversity and structure, and ecological adaptability to varied stresses (Damiani, Colombo, Di Filippo, Pescini, & Mauri, 2016; Schwahn, Beleggia, Omranian, & Nikoloski, 2017). Moreover, it was confirmed that the N/P and C/P content ratios declined in the growing season, following the GRH, and this elemental stoichiometry was associated with metabolomic shifts in the plants displaying high sugars and amino acid content (Brunetti et al., 2013; Srivastava, Kowalski, Callahan, Meikle, & Creek, 2016). We need to integrate stoichiometry and ecometabolomics in order to advance our knowledge of ontogenetically and ecologically associated shifts in C/N/P/K levels, and of how these levels can change to acquire an optimal allocation for growth and other physiological activities. The coupling advances our perspective on the impact of stoichiometry on the living organism, and on the structure, function, and evolution of ecosystems (Hessen, Elser, Sterner, & Urabe, 2013; Meunier et al., 2017).

4.3 CLIMATIC PARAMETERS AND METABOLOMIC SHIFT

There are numerous functional genomics research studies that demonstrate the alteration in regulatory pathways in response to climatic change and metabolic pathways linked with factors for plant stress responses (Fester, 2015; Lee et al., 2010; Oksanen, Keski-Saari, Kontunen-Soppela, & Keinänen, 2013). However, relevant data on agricultural crops remains inadequate. In order to acquire food security, we need more generic solutions for crop protection. In this context, the relevance of omics technologies employing system biology could prove advantageous and facilitate deciphering the implication of diverse signaling pathways in summarizing the responses in response to climate change effects. To facilitate coverage of plant responses to climate change, and to expand molecular engineering approaches for enhancing the plant tolerance to diverse stresses, it will be an imperative step to combine omics data with bio-informatics based system biology/system level modeling and to acquire computational models. Using system biology evaluation, it was possible to follow the changing regulation of photosynthetic pathways in C3 plants in response to global climate change (Reddy, Rasineni, & Raghavendra, 2010; Yu, Chen, Xu, & Huang, 2012). Using a bio-informatics analysis of *Arabidopsis* microarray data, a new regulatory method was anticipated that integrates transcriptional and post-translation regulation, and manipulates the amino-acid flux against abiotic stress (Kilian, Peschke, Berendzen, Harter, & Wanke, 2012; Steinfath, Repsilber, Scholz, Walther, & Selbig, 2007). Detection of independent alterations can be drafted effectively by simulating stomatal movement via a guard cell ABA-signaling model (Albert et al., 2017; Huntingford et al., 2015). This may provide the most promising prospect for conferring high drought tolerance, and proceeding prospective wet-bench evaluation (Abdurakhmonov et al., 2016). Some research studies demonstrate that a compendium of genomic signatures can be employed to classify the plant abiotic stress phenotype in *Arabidopsis* as per transcriptome procedures, and then be associated with gene co-expression system evaluation to resolve genes controlling the phenotype response (Barah, Winge, Kusnierczyk, Tran, & Bones, 2013). This strategy can resolve the incidence of identified stress-induced pathways, and marker genes will be ascertained, and a generalized abiotic stress-induced transcriptome and associated phenotypic classification to stress duration will be provided (Shinozaki & Yamaguchi-Shinozaki, 2007; Atkinson & Urwin, 2012). Integration of genomic data analysis with reverse engineering network modeling facilitates the evaluation of the transcriptional control of *Arabidopsis* (Cho et al., 2007; di Bernardo et al., 2005). These studies anticipated that *Arabidopsis* have advanced higher integration in terms of transcriptional control among cellular metabolic activities implicated in responses and adjustment to varying surroundings; while gene networks constitutively expressed, or less linked to stress responses, are differentiated by lower networking. Recently the novel computational model AraNet was launched, which anticipates the function of unidentified plant genes with exclusive speed and precision (Margolin et al., 2006; Martinez, 2016). The AraNet software has facilitated the discovery of a regulator of drought sensitivity, *At1g80710*, now named DROUGHT SENSITIVE1 *DRS1* (Loha, Lowa, & Mohamada, 2015). Current research focuses on a system biology-type strategy to dig out issues pertaining to plant ecophysiology in response to the changing climate via the modeling of diverse plants' stress responses (Chapin III et al., 2012; Nicotra et al., 2010). However, comprehending the prospective way toward unraveling challenges posed by instantaneous global climate change can be repressive. There is still a need to ascertain an inclusive predictive model to comprehend varied biological networks in plants' responses to integrated climate change devastations (Franks, Weber, & Aitken, 2014; Anderegg, 2015). Such a model could be valuable for manipulating plant metabolism to varying climates. Finally, this could lead to sustainable agriculture that is more resilient to varying climates. Climatic changes have also induced shifts in plant distribution that are directly associated with changes in plant ecophysiology, growth, and productivity (Becklin et al., 2016; Parmesan & Hanley, 2015). High mountain biomes are most sensitive to increased temperatures resulting from global warming (Kohler, Wehrli, & Jurek, 2014; Petitpierre et al., 2016). These types of distribution shifts are reported in European and American mountains, while the largest and highest mountainous areas are located in Asia (Körner et al., 2017). Similar studies have been carried out on Himalaya herbs, wherein it is indispensible to identify climate-induced shifts in elevation, and ecophysiological characteristics of endemic herbs (Jan, Kamili, Parray, & Bedi, 2016; Jan, Kamili, Parray, Bedi, & Ahmad, 2016). This study illustrated a considerably increasing elevational shift of 55.2 m/decade, in contrast to average global shifting of 6.1 m/decade, and impromptu changes in secondary metabolite concentration (Agnihotri et al., 2017). Further, mountainous ecosystems are found to be more susceptible to biodiversity loss because of ecosystem fragmentation (Newbold et al., 2015). Moreover, an ecologically vulnerable ecosystem is characterized by upward shifting of vegetation zones, range expansion, and contraction together with shifts in phenology against climatic change (Bellard, Bertelsmeier, Leadley, Thuiller, & Courchamp, 2012; Kelly & Goulden, 2008). A climate-induced shift in species is driven by species' distributional pattern changes, varied regional climates, and diverse physiological mechanisms (Somero, 2012; Garcia, Cabeza, Rahbek, & Araújo, 2014). It has been observed that for every 0.5°C decline in temperature per 100 m rise in altitude, there is an upward shift of vegetation belts at a rate of 8–10 m per decade, and an added risk of plant extinction in response to climatic change in mountainous summits (Tranquillini, 2012). This upward shift in vegetation will advance upper vegetation zones toward much higher altitudes in the mountains (Cannone, Sgorbati, & Guglielmin, 2007; Pauli et al., 2012). This vegetational shift can induce structural changes in plants, along with the modulation of metabolite levels (Graff, Aguiar, & Chaneton, 2007; Turkington, 2009). The prime climactic factor for stimulating the metabolic and physiological changes includes elevated levels of carbon dioxide, which alters stomatal density, WUE, and also secondary metabolites. Although stomatal density is under genetic control, the comparative stomatal density remains impervious under prevailing ecological conditions; but elevated levels of carbon dioxide and high temperatures affect stomatal density (Lake & Woodward, 2008; Tiwari, Kumar, Yadav, & Chauhan, 2013). Secondary metabolites are significant as phytochemicals and nutraceuticals, and their concentrations are affected by diverse ecological factors that are predominant at varied altitudes (Robinson, Ryan, & Newman, 2012; Sampaio, Edrada-Ebel, & Da Costa, 2016). Increased atmospheric CO2 levels increase the carbon supply necessary for development and growth, and even its allocation to C-based secondary metabolites; for example, fatty acids, sterols, flavonoids, and other phenolics (Jia, Wang, Chen, He, & Liu, 2014; Ncube, Finnie, & Van Staden, 2012). Himalaya is the richest global biodiversity hotspot, and an important resource of the eighth-largest Asian River, supplied with 1500 glaciers crucial to the sustenance of more than two billion people (Sharma, Chettri, & Oli, 2010; Mittermeier, Turner, Larsen, Brooks, & Gascon, 2011). Climatic change has an immense effect on Himalaya, and its impact on biodiversity is more devastating here than in any other spot across the world (Xu et al., 2009). Further, climatic changes at regional levels are dissimilar from climate trends at the global level. Moreover, with the impetus for instant economic progress in India during the past two decades, the existing environment of Himalaya has been significantly degraded (Telwala, Brook, Manish, & Pandit, 2013). Increased anthropogenic activities as result of expansion in modern infrastructure have amplified massive deforestation, over-urbanization, increased soil erosion, and pollution of natural resources such as soil and water (Wei & Ye, 2014). These mountains are rapidly losing their tolerance level to withstand regional climatic variations (Pepin et al., 2015). The ecophysiological alterations induced in response to climatic change in local flora and fauna are highly inconsistent. With some meager examples on few observations, there are no systematic studies on climate change effects on endemic flora and fauna of Himalaya. However, some research studies have been carried out on herbarium specimens of three representative plants, namely, *Geranium nepalense* Sweet (Geraniaceae), *Inula racemosa* Hook. f. (Asteraceae), and *Lavatera kashmiriana* Camb. (Malvaceae) collected during the past century to inspect the response of Himalayan plants to climate change. The old herbarium data (1900–1960) was evaluated against newer herbarium records (1961–2010), and investigated for measurement of stomatal density and metabolic composition of herbarium specimens (Agnihotri et al., 2017; Gairola, Sharma, & Bedi, 2014). This study was carried out in temperate and alpine regions of Kashmir Himalaya to record the effects of climatic change during the past 110 years on distribution patterns and ecophysiological attributes. The observations demonstrate alterations in distribution patterns and stomatal density in response to elevated temperatures as a result of higher levels of carbon dioxide, and a higher concentration of carbon-based secondary metabolites within diverse herbaceous species (Caldera, De Costa, Woodward, Lake, & Ranwala, 2017; Casson & Hetherington, 2010; Nippert, Fay, Carlisle, Knapp, & Smith, 2009). There was significant altitudinal shift in all three herbaceous species, as represented by distinct scattered plots. There was an average rise of 55.2 m/decade between 1900 and 2010 that was greater than the global average rate of 6.1 m/decade. Larger altitudinal shifts were reported at the maximum height for *G. nepalense* and *I. racemosa*, more so than for *L. kashmiriana*, which is a high-altitude species (Gairola et al., 2014). During the years 1900–2010 in Kashmir Himalaya, the average annual maximum temperature increased by 0.31°C/100 years, and the mean annual average minimum temperature increased by 0.79°C/100 years (Singh, Bassignana-Khadka, Singh Karky, & Sharma, 2011). Along with elevation of temperature, there has been a consequent rise in precipitation levels by 8.15 mm in Kashmir Himalaya, which led to variations in ecophysiological traits of the aforementioned herbs.

There was a significant reduction in stomatal density of 21.64%, 17.58%, and 16.84% in *G. nepalense, I. racemosa*, and *L. kashmiriana*, respectively ($P < 0.01$) in all three plant species from the beginning of the 20th century.

The gas chromatography of these herbs revealed higher concentrations of palmitic acid, stearic acid, and linolenic acid; while other fatty acids were present in small amounts (Gairola et al., 2014). The pronounced accumulation of palmitic acid was found during 1900–1960, more so than for 1961–2010 for these plant species, while stearic acid and linoleic acid showed similar trends (Agnihotri et al., 2017). Conversely, polyunsaturated fatty acid (18: 3) exhibited an inverse drift to that of monounsaturated (18: 1) and saturated fatty acids (16: 0 and 18: 0). A marked shift of 0.51°C/100 years has been reported for the Indian subcontinent during the period 1901–2007. Several ecophysiological characteristics demonstrated a species maximum temperature limit to be 32°C, and a rise of 72.94 ppm in the atmospheric global CO_2 concentration (316.90–389.84 ppm) has been recorded in the past 50 years, which significantly alters the ecophysiology of herbaceous species (Barua et al., 2014). Higher levels of CO_2 and elevated temperatures can affect stomatal density, cell differentiation, and physiological metabolism (Reddy et al., 2010; Rogiers, Hardie, & Smith, 2011). Forests thrive at lower CO_2 partial pressures, and are more responsive to the variations in CO_2 partial pressures than the flora from the higher CO_2 environment (Martz, Vuosku, Ovaskainen, Stark, & Rautio, 2016). Hence, elevated levels of CO_2 will lead to lower stomatal density and carbon isotope inequity values at high altitudes (Du et al., 2016; Hu et al., 2015). This inclination toward reduced stomatal density might be due to the adaptation of plants to preserve water levels at high CO_2 (Franks, Doheny-Adams T, Britton-Harper, & Gray, 2015). Following diverse climactic observations, it was established that there is a strong positive association between stomatal density and carbon-based secondary metabolites, and both are inversely proportional to increasing CO_2 and altitude (Chater et al., 2015; Huang, Reichelt, Chowdhury, Hammerbacher, & Hartmann, 2017). Diverse secondary metabolites exhibiting imperative medicinal properties are synthesized, such as gallic acid and quercetin isolated from *G. nepalense*, and inulin from *I. racemosa* (Chauhan, 1999; Gamage et al., 2018). The metabolic flux of these metabolites was affected by diverse ecological factors comprised of temperature, water retention, CO_2 concentration, and elevational clines (Choudhary, 2015; Huang et al., 2011). Significant variations in fatty acids were observed during the past 100 years in the Himalayas. An increase in phytosterols are affected by a rise in temperature, CO_2 concentration, and precipitation, thereby modulating phytochemicals (Sah, 2013). There are several other research reports that demonstrate that Himalayan flora is shifting at rapid speeds (55.2 m/decade) toward higher altitudes than the general postulation of altitudinal shifting of 8–10 m/decade in global mountain ecosystems (Pandey, Irulappan, Bagavathiannan, & Senthil-Kumar, 2017; Pandey et al., 2017; Mishra, Stobdan, Thangaraj, Seth, & Singh, 2018). Increased CO_2 levels and high temperatures led to an increased decline in stomatal density at high altitudes, and a marked disparity in stable carbon isotopes and secondary metabolite accumulation, signifying the tolerance tactics of plants to combat environmental stress (Bose, Choudhury, Tandon, & Kumaria, 2017; Jan, Majeed, Andrabi, & John, 2018). Thus, it can be conjectured that there is a concurrence in the rate of upward shifting of mountainous flora and the level of climatic changes that can be further extended by other human activities that function differentially in each high-altitude ecosystem (Pauli, Gottfried, & Grabherr, 2014; Sproull, Quigley, Sher, & González, 2015)

4.3.1 Increases in Atmospheric CO_2

Elevated levels of CO_2 modulate elemental stoichiometry, such as high C:P ratios in C3 plants; for example, C3 grass plants exhibit, on average, a 38% rise in C:P ratios. While a terrestrial ecosystem exhibited almost a 22% rise in C:N ratios in photosynthetic tissues. On the contrary, aquatic ecosystems, such as oceans, demonstrate an inclination toward higher phosphorus demands in response to elevated levels of CO_2. The high CO_2 levels raise the phosphorus uptake demands, global warming augments water column stratification, and nitrogen fixation ability and atmospheric inputs amplify N:P ratios (Li, Jiang, & Liu, 2016; Verspagen, Van de Waal, Finke, Visser, & Huisman, 2014). Elevated CO_2 increases crop productivity by enhancing the net photosynthetic rate, and modulates the impacts of climate change in C3 and C4 plants (Wang et al., 2015; Wang et al., 2015; Killi, Bussotti, Raschi, & Haworth, 2017). Previous research reports demonstrated differential responses of C3 and C4 plants to elevated levels of CO_2, and C3 accumulated more biomass due to higher photosynthetic activity than that of C4 plants (Kumar, Sharma, Soni, & Pawar, 2017; Srinivasarao et al., 2016). Elevated levels of CO_2 reduce the WUE per unit area of vegetation for both C3 and C4 plants (Kurepin et al., 2018). It has been confirmed that C4 plants are originated from C3 plants, as C4 photosynthesis functions as a CO_2 pump and prevents the oxygenation reaction by effectively raising the intracellular CO_2 to the O_2 ratio at the site of Rubisco (da Silva, Mazumdar, & Naik, 2016; Gamage et al., 2018). Elevated atmospheric CO_2 generally leads to a decline in nutrients such as calcium and magnesium due to a lowering of the transpiration-driven mass flow of nutrients, as a result of decreased stomatal conductance (Houshmandfar et al., 2017). Houshmandfar, Fitzgerald, Armstrong, Macabuhay, & Tausz, 2015 illustrated that in addition to the decline in transpiration rate, high CO_2 reduced calcium and magnesium in xylem sap. Taub and Wang (2008) first proposed the restrained uptake of nutrients under high CO_2 levels. This decreased nutrient uptake by roots resulted from a lower nutrient demand compared with growth, and the restrained transpiration-driven mass flow of nutrients decreased because of lower stomatal conductance under CO_2 (McGrath & Lobell, 2013). Elevated levels of CO_2 induce the accumulation of oxalates in leaves of *Rumex obtusifolius*, which results in cell proliferation (Miyagi et al., 2011). A high CO_2 level not only reduces photorespiration rates, but also stimulates defense-related metabolites and pathogen-related genes that are followed by up-regulation of reactive oxygen species (ROS) signaling (Noctor & Mhamdi, 2017; Smith, 2018). The genes are implicated in CO_2 signaling and stomatal regulation, but the link between CO_2 availability and biotic stress remains unexplained. Current research illustrates that metabolite flux and ROS signaling are known to be imperative in connecting CO_2 availability to activate defense systems (Engineer et al., 2016; Sierla, Waszczak, Vahisalu, & Kangasjärvi, 2016). Recent research illustrates the stimulation of a phenylpropaniod pathway in response to high levels of CO_2 in tobacco (*Nicotiana tabacum*), poplars (*Populus euramericana*), and *A. thaliana* (Becker & Kläring, 2016; Rajashekar, 2018). The main evidence of a relationship between high CO_2 and secondary metabolite flux can be demonstrated by increased substrate availability for photosynthetic carbon fixation (Abadie, Bathellier, & Tcherkez, 2018; Liu, Shen, Xin, & Wang, 2016). The major effect of elevated CO_2 levels is arbitrated by Ribulose-1,5-bisphosphate carboxylase/oxygenase (Rubisco). CO_2 fixation can also be catalyzed by cytosolic phosphoenolpyruvate carboxylase (PEPc), which instigates C4 acid biosynthesis (Temme, Liu, Cornwell, Cornelissen, & Aerts, 2015). PEP is the connecting link, as cytosolic phosphoenolpyruvate carboxylase could collide with a degree of biosynthesis of defense-linked compounds (Caretto, Linsalata, Colella, Mita, & Lattanzio, 2015). *Arabidopsis* mutants demonstrating high levels of 2-methylerythritol 2,4-cyclopyrophosphate (MEcPP) can be linked with a pronounced accumulation of salicylic acid and higher tolerance levels to biotrophic pathogens (Noctor & Mhamdi, 2017). The majority of carbon resources are consumed in isoprene emission, which is implicated in tolerance strategies against oxidative stress. However, elevated levels of CO_2 are known to hamper isoprene synthesis, likely via activation of PEPc activity (Trowbridge et al., 2012;

Werner & Gessler, 2011). The interrelationship between elevated levels of CO_2, MEP pathway, and PR responses remains yet unraveled. Further high levels of CO_2 can result in increased sugar accumulation, but the amino acid biosynthesis in response to high CO_2 can be ambiguous. Like the high influx of salicylic acid in response to high CO_2, amino acids' glutamine synthesis was stimulated, as evidenced by glutamine receptors and proline levels that are involved in PR-related responses (Trowbridge et al., 2012). Glycine is the only amino acid that declines in response to high CO_2 levels, likely due to the inactivation of photorespiration (Smith, 2018). Numerous simple organic acids are involved in tolerance strategies against biotic stress (Choi et al., 2018). High levels of CO_2 in *A. thaliana* demonstrate the accumulation of citrate and salicylic acid (Valegård, Hasse, Andersson, & Gunn, 2018), and the accumulation of organic acid in an apoplastic space of barley subject to bacterial challenges (Rico, Jones, & Preston, 2009). Citric acid and other enzymes are implicated in signaling pathways against biotic stress in *A. thaliana* (Coolen et al., 2016; Rejeb et al., 2014). Malic acid is another metabolite that demonstrates the association between carbon levels to PR responses. Stimulation of salicylic acid and PR-linked tolerance in response to high CO_2 levels was attenuated by upstream PEPc and downstream malate in mutants of *Arabidopsis* (Caldwell, Bornman, Ballaré, Flint, & Kulandaivelu, 2007; Noctor, 2006). High levels of CO_2 induce a defense tolerance that is associated with metabolites in azelaic acid and pipecolic acid that are activated against oxidative stress (Kissoudis, van de Wiel, Visser, & van der Linden, 2014; Uzuner, Sağlam, & Kadıoğlu, 2017). These metabolites increase in response to oxidative stress, but their biosynthesis under high CO_2 is still unrevealed. A steady rise of CO_2 levels over many generations may not lead to significant metabolic alterations, as reported in C4 plants that maintain CO_2 levels in bundle sheath cells that are comparatively greater than that of mesophyll cells of both C3 and C4 plants. In spite of this, cell-specific transcriptomes of C4 plants have not revealed over-expression of PR linked components in the bundle sheath (Ding et al., 2015). However, a defense-associated pathway might be comparatively tolerant to CO_2 in maize and other monocots, which needs to be addressed. Moreover, it could be conjectured that some CO_2 signaling metabolites could be inactivated through the evolution of C4 photosynthesis (Bauwe, Hagemann, Kern, & Timm, 2012; Driever & Kromdijk, 2013). However, it could be more exciting to investigate whether biotechnological tailoring of C4 pathways into C3 plants has any consequence on biotic stress response. Higher production of PR-related metabolites under high CO_2 can be valuable only against pathogens; but at lower biotic stress risks, the biosynthesis of defense-related metabolites would curtail the yield at the maximum limit (Mhamdi & Noctor, 2016; Pedreschi & Lurie, 2015). Explication of the proccess that relates CO_2 levels to PR responses could supply data that could be modulated to optimize plant development and tolerance strategies under high CO_2 levels.

4.3.2 Cascade Effect

There is an instantaneous proclivity toward research studies that pertain to the effect of climate change on biogeochemical cycling, trophic association, and species diversity (Greaver et al., 2016; Shibata et al., 2015; Thakur et al., 2015). The climatic factors affect diversity through direct, as well as indirect, pathways; signifying the implication of varied cascades against explicit and expansive climatic variables (Gohli & Jordal, 2017; Rosenzweig et al., 2014). Research studies demonstrating the integrative impacts of variation in climatic variables; for example precipitation, temperature, CO_2, species interactions, or diversity cascades are still scarce (Andresen et al., 2016; Fine, 2015). The hypothetical outline for demonstrating the climatic change and varied cascades, and even direct biotic interactions, are still underway, and need to advance further. We need to investigate the significant association between varied climatic factors in order to determine the combined effect of climactic factors on diversity cascades. Significant variations in temperature and atmospheric CO_2 hampers the association between plants, plant biochemistry, and the insect-herbivore association via intricate links (Lämke & Unsicker, 2018). Climate change can provoke top-down effects on global diversity via the alteration of herbivores and their natural enemy progression and survival that could hasten a massive decline in plant richness and evenness (Greaver et al., 2016; Thomson, Macfadyen, & Hoffmann, 2010). Climatic variables such as high temperature, elevated levels of CO_2, and severe weather conditions could have direct, depressing impacts on the survival of a significantly imperative group of insect herbivores; parasitoid wasps (Moreira, Abdala-Roberts, Parra-Tabla, & Mooney, 2015; Block, Vaughan, Christensen, Alborn, & Tumlinson, 2017). Due to these implications, the effects on parasitoids, insect herbivore outbursts, can amplify, leading to a decline in herbivore evenness, and a consequent decline in plant diversity (Pureswaran et al., 2015; Rasmann, Pellissier, Defossez, Jactel, & Kunstler, 2014). Such types of environmental episodes occurring as result of varied climatic parameters can possibly be implicated in expansive effects, such as Cohen's statistical effect size (McPherson, Jetz, & Rogers,

2004). The negative effects of CO_2 and temperature were reported in alfa, which revealed a synergistic increment in developmental asynchrony between parasitoids and herbivores (Elad & Pertot, 2014). Numerous reviews and empirical research studies demonstrating the direct effects of climate change variables on plant quality, biomass, and diversity have been conducted. These studies have shown, for example, that increased CO_2 levels amplify plant growth directly via activating the photosynthetic rate, and indirectly by declining plant quality for herbivores (Meza-Canales, Meldau, Zavala, & Baldwin, 2017; Züst & Agrawal, 2017). The altered plant quality has varied effects on herbivore competence, consequently leading to modulation in consumer-resource dynamics, which leads to a shift in herbivore richness and evenness (Hillebrand, Bennett, & Cadotte, 2008; Sasaki & Lauenroth, 2011). Research studies have investigated the impacts of climate change on tritrophic associations, signifying that effects on these links will facilitate incidences, and in some cases, a related decline in diversity within wild and cultivated ecosystems comprised mainly of agriculture (Dyer, Richards, Short, & Dodson, 2013). The most predominant factor for generating diversity cascades will be amplified climatic variability and severe weather events linked with climatic change, because they hamper regulation of herbivores via modulation of predator and parasitoid interaction, consequently resulting in massive shifts in herbivore and plant diversity (Scherber et al., 2010; Tylianakis, Didham, Bascompte, & Wardle, 2008). Climatic variability-induced cascades can induce varied changes, such as increased pest herbivore incidence, interference of normal pest control via innate enemies, high accumulation of C-based defense-related metabolites, and random diversity patterns in response to global warming and elevated CO_2 (Rzanny, Kuu, & Voigt, 2013; Tylianakis et al., 2008). The Indian Himalaya region demonstrates a distinct cascading effect on ecosystems, resulting in fundamental changes that are intricate and significant (Chakraborty, Saha, Sachdeva, & Joshi, 2018; Sandhu & Sandhu, 2014). The cascade effect can be assessed at different levels, such as ecological, regional livelihood, downstream watersheds, and global feedback with some level of ambiguity (Kulkarni, Patwardhan, Kumar, Ashok, & Krishnan, 2013; Xu et al., 2009). However, the ecological effects of high-elevation species to climatic variables is particularly missing in Greater Himalaya, but it is usually anticipated that instant responses by species to climate change might hamper associations (Shrestha, Gautam, & Bawa, 2012; Whipple, 2009). Significant ecological cascading effects comprise secondary extinctions activated by a decline of key species in the alpine ecosystem.

The endemic-rich Himalaya region is comprised of numerous plant species that could not respond effectively to anticipated rates and the extent of climate change (Kumar, 2012). One of the apparent threats is species extinction from inaccessible mountain ranges in case of upward shifts of taxa (Forero-Medina, Terborgh, Socolar, & Pimm, 2011; Telwala et al., 2013). Generally, the climatic change results in intricate changes in natural vegetation, while species reveal hermetic responses (Seinfeld & Pandis, 2016). Species invasion by weeds and exotic species from lower elevations are expected. The varied response of cascading effects of climate change on the regional level need to be recognized, anticipated, and filtered through several cultural contexts (Weber, 2017).

High levels of CO_2 can modulate plant stoichiometry, which may have amplified impacts at the ecosystem level. High CO_2 levels can lead to carbon enrichment, which affects trophic interactions, resulting in increased growth, amplified nitrogen fixation capacity in legumes, and high symbiotic fungal associations (Bi, Ismar, Sommer, & Zhao, 2018; Tobita, Yazaki, Harayama, & Kitao, 2016). Further, it has an indirect effect on herbivore feeding rates, which anticipates compensatory behavior by herbivores against inadequate food quality (La Pierre & Smith, 2016; Urabe, Shimizu, & Yamaguchi, 2018). Several research studies, however, illustrate an increase in feeding rates to compensate for the inadequate food quality under high CO_2 levels (Guo, Kainz, Sheldon, & Bunn, 2016; Li et al., 2018). High CO_2 levels in the shrub oak community for prolonged periods of time led to a decline in food quality, which consequently altered insect-herbivore interactions (Jamieson, Schwartzberg, Raffa, Reich, & Lindroth, 2015). However, low-quality food supports some herbivore insects, which significantly modulates community composition. High CO_2 levels further aggravate fungal virulence, and alter the biosynthesis of pathogen-related metabolites (Martin, Pohlers, Mühlschlegel, & Kurzai, 2017; Meunier et al., 2017; Mikkelsen, Jørgensen, & Lyngkjær, 2015). Elevated levels of CO_2 reduces the C:N ratio by decreasing litter, as observed in *Betula pendula* (Godbold et al., 2014). A decline in litter quality, and a higher C:N ratio, can decelerate nutrient cycling, which favors the formation of recalcitrant soil formation predominated by a high C:N ratio (Knops, Wedin, & Naeem, 2010; Delgado-Baquerizo, García-Palacios, Milla, Gallardo, & Maestre, 2015). However, a high litter C:N ratio is not in concurrence with N mineralization rates (Rahman, Tsukamoto, Tokumoto, Shuvo, & Ashikur, 2013; Xiong, Zeng, Xia, & Guo, 2014). Further, high CO_2 levels induce altered C:N and C:P ratios, and usually herbivores have lower

C:N and C:P ratios than the plants they consume. This disparity in the ratios of C:N in plants might support the expansion of consumer communities with elevated C:N ratios (Yue et al., 2016). The focus of diversity cascades research should be on utilizing a combination of the best available methods to create a broader synthesis and an improved understanding of mechanisms behind cascades.

4.3.3 Eutrophication Effects

Eutrophication has been a predominant issue that was primarily controlled by treatment with algicides, and then by managing nutrient acquisition (Foley, Jones, Maberly, & Rippey, 2012; Moss et al., 2011). Numerous studies demonstrate the intensifying effect of climate change on eutrophication, and concurrence was observed between climate change and eutrophication (Jeppesen et al., 2010, 2011). Climate change has intensified the eutrophication by affecting the rate of precipitation, and glacier melting, which subsequently results in increased diffusion of nutrient inflow (Moss, 2012; Sommaruga, 2015). Increasing nutrient inflow and rising temperatures together amplify eutrophication levels that favor the predominance of cyanobacterial populations, floating plants, and the ultimate loss of underground vegetation in response to high temperatures (Jeppesen et al., 2009; Paerl & Paul, 2012). High temperatures favor eutrophication by augmenting the rate of mineralization in catchment soils, leading to low oxygen availability (Allen, Hook, Biederman, & Stein, 2002; Jones & Brett, 2014). Eutrophication enables the organism to be less dependent on imported organic matter, and enables a greater ability for autotrophic carbon dioxide fixation, resulting in high productivity and respiration rates, and bulk production of deoxygenated waters, and methane gas and sediment (Jeppesen et al., 2016). A climate change-induced rise in temperature reduces the competence of sediment in carbon storage, producing more methane, and elevating respiration to the net photosynthetic ratio at the community level (Trolle, Hamilton, Pilditch, Duggan, & Jeppesen, 2011; Xu et al., 2015). Climate change-induced eutrophication has led to a significant shift in the availability of soil nutrients, such as altered N:P ratio (Sardans, Peñuelas, et al., 2012). Plants respond to climate through their chemical shifts, and altered utilization of nutrient resources (Tipping, Somerville, & Luster, 2016), which modulates nutrient cycling such as C, N, and P that subsequently yield effects on the community structure (Finzi et al., 2011). Eutrophication decreases the activity of some enzymes implicated in the turnover, and decreases nutrient availability (Bloor & Bardgett, 2012), which significantly alters the elemental composition, the extent of which varies in different communities (Liao et al., 2008). For example, cold and wet temperate areas exhibit high soil organic matter in response to climate change (Butler et al., 2012), leading to a high influx of soil nutrients (Aerts, Van Bodegom, & Cornelissen, 2012), which amplifies community soil biomass, leading to varied species richness (Yergeau et al., 2012). Such variations offer new competence platforms for both interspecific and intraspecific competition among plants and microbes (Elgersma, Yu, Vor, & Ehrenfeld, 2012).

Climate change can alter the association of interspecific competition by altering the abiotic component structure. Numerous reports illustrate that present climate changes have an immense effect on organisms, populations, communities, and terrestrial ecosystems by varying phenotypes, genotypes, growth, phenology, geographical distribution, species competence, ecological associations, and risk of extinction at the community level (Gonella et al., 2015; Lee, Flory, & Phillips, 2012). Ecosystems are dynamic, with constant changing structures and functions, and have remarkable effects on climate change itself. There are very few research reports that illustrate the competence of organisms, populations, communities, and ecosystems to respond to the associations with other instantaneous stresses triggered by various factors of global change, and to the new links that are produced (Flint, 2012; Torossian, Kordas, & Helmuth, 2016). Current changes in nutrient dynamics, such as in the N:P ratios of organisms and environments that some ecosystem are bearing as consequence of the unstable input to the biosphere by humans (Peñuelas, Sardans, Rivas-Ubach, & Janssens, 2012), can be strongly linked with climate change. It has been speculated from various diverse studies in the climatic change era, that the rise in N:P ratio can be associated with the decline in water resource availability supporting species with low productivity rates, and more judicial utilization of resources (Finzi et al., 2011; Stott et al., 2010). Nutrient availability and elemental stoichiometry can alter the rate and direction of the rejoinder at the organism, population, and community level to climate change (Moore et al., 2013; Rivas-Ubach et al., 2012). Current research reports documenting the interactions between climatic manipulations with eutrophication, or elevated CO_2 levels, can decipher the impacts of climate change on metabolic flux. There is an immediate need to integrate the diverse disciplines covering wider time ranges and space in which the experiments are conducted, so as to cover more buffered alterations (Leuzinger et al., 2011). These research studies need to be complemented with inventories (Carnicer et al., 2011), remote sensing data (Zhao & Running, 2010), paleoecological data (Estiarte, Peñuelas, López-Martínez, & Pérez-Obiol, 2008), and expansive data sets (Kattge et al., 2011) to reveal the basic effects of climate change on the earth. The integration of omics technologies with nutrient cycling, elemental stoichiometry, physiological and phenological changes, and ecosystem structure shifts will pave the way toward an integrative perception of mechanism and method underlying biological effects of climate change (Agathos & Boon, 2015; Song et al., 2016; Song et al., 2016).

4.4 VARIABLE ELEMENTAL STOICHIOMETRY AND GLOBAL CHANGE

The prime objective of ecological stoichiometry is to illustrate organism survival strategies, and ecosystem structure and function in response to C:N:P stoichiometric ratios of abiotic components such as water, soil, and organisms. Research studies illustrating the association between significant environmental variables and C:N:P ratios of the environment and organisms have increased over the past 20 years (Sardans, Rivas-Ubach, & Peñuelas, 2012a, 2012b). The nutrients C, N, and P are the most significant with respect to their imperative role in ecosystem structure and organism stoichiometry, with the aim of identifying the usual links between stoichiometry and structure, and the function of both aquatic and terrestrial ecosystems. Aquatic ecosystems possess varied C:N:P ratios in oceans, in a confined range similar to the Redfield ratio in deep water and inner-oceanic areas (Galbraith & Martiny, 2015; Talarmin et al., 2016). Terrestrial ecosystems exhibit the usual proclivity toward a rise in soil and plant N:P ratios, particularly in cold and temperate to tropical ecosystems, but with immense alterations within each climatic belt (Smith et al., 2014; Song, Li, et al., 2016; Song, Yao, et al., 2016). The nutrient or elemental ratio of C:N:P is more highly limited in organisms than in their abiotic components, such as water and soil, in which they thrive well (Gulis et al., 2017; Sistla, Appling, Lewandowska, Taylor, & Wolf, 2015). The competence to adapt this ratio implicates numerous strategies, ranging from leaf re-absorption in plants, to the excretion control in animals (Rivas-Ubach et al., 2014). Numerous disparities in C:N:P ratios illustrated in diverse taxa were compared, such as fresh water ecosystems, revealing steady support that indicates a high growth rate in response to a low N:P ratio in a trophic web; but terrestrial ecosystem have not yet yielded any conclusive results in GRH

(Daines, Clark, & Lenton, 2014; Delgado-Baquerizo, Reich, García-Palacios, & Milla, 2016). Current research studies demonstrate the imperative role of the N:P ratio in the evolution of organism genomes (Guignard et al., 2017; Meunier et al., 2017). Moreover, research is required to interpret the stoichiometry of diverse trophic levels under diverse C:N:P, along with the K ecological ratio in long-term ecosystem scale studies. Integration of elemental stoichiometry with metabolomics and genomics should enhance our perspective on integrating diverse levels of biological associations, from elemental stoichiometry, to the structure and evolution of the ecosystem, through cellular metabolism and nutrient cycling (Chen et al., 2016; Chen et al., 2016; Zhang et al., 2018).

The GRH is the prime theme of ecological stoichiometry, which suggests that growing organisms must amplify their nutrient distribution, such as P to RNA, to acquire enough proteins for growth (Cross, Hood, Benstead, Huryn, & Nelson, 2015; Fan et al., 2015). Thus, GRH has been evaluated, principally by investigating the associations of N:P ratios of organisms and the environment, and community dynamics (Sitters, Atkinson, Guelzow, Kelly, & Sullivan, 2015). The GRH is supported by freshwater ecosystems (Sardans, Peñuelas, et al., 2012), but it has not been validated in terrestrial plants. The intricate relevance of GRH to terrestrial ecosystems is because of their ability to invest a considerable portion of N and P to functions other than growth; for example, storage, a defensive mechanism against stress (Giordano, Palmucci, & Raven, 2015; Zhao et al., 2014). The association between N:P and the protein:RNA ratio is insignificant, and the addition of the N and P ratio with K, associated with water relations, can advance our perspective on the links between stoichiometric ratios and ecological variables (Leal, Seehausen, & Matthews, 2017; Zechmeister-Boltenstern et al., 2015). Hence, research studies on K could assist us in comprehending the collapse of GRH in "terrestrial ecosystems" (Niu et al., 2016). Thus, it can be conjectured that the distribution of C, N, P, and K, within and between plant tissues, must be targeted on improving the WUE in dry environments, while as in the case of wet habitats, the allocation of C, N, P, and K could be associated with growth and survival, instead of WUE (Huang et al., 2016; Sardans & Peñuelas, 2014). On the contrary, in forest biomes, K is restrained, and the N:K ratio of plants can explain the variation in species diversity and ecosystem structure (Busch et al., 2018). Moreover, alterations in K concentration may vary against biotic associations (Tatarko & Knops, 2018). Moreover, like N and P, K stoichiometry has significant effects on trophic webs by affecting the growth of herbivores, and the community structure in herbivores (Lurgi, Montoya, & Montoya, 2016). Elevated K levels in aerial parts of plants can reduce carbohydrates and amino acids, and lead to an increment in the sclerenchymatous parenchyma layer (Wakeel, Gul, & Zörb, 2016). Elemental stoichiometry has a remarkable effect on the main ecological processes; for example, litter decomposition (Delgado-Baquerizo et al., 2016; Zhang & Elser, 2017), N_2 fixation competence, plant herbivore links, tolerance strategies against varied stresses, and ecosystem dynamics (Savage et al., 2016; Singh & Chauhan, 2017).

In marine ecosystems, disparity in the C:N:P ratio can be expansive, such as in the Redfield ocean N:P ratio of 16:1, which is highly invariable and species specific at the community level (Gan et al., 2014; Clare, 2015). This disparity in nutrients is because of the exchange between the competitive equilibrium, which allows massive allocation to P-deficit reserve acquisition machinery, and consequently to a greater N:P ratio, and exponential growth that assists higher P-enriched assemblage machinery, and therefore lesser N: P ratios (Longphuirt, O'Boyle, Wilkes, Dabrowski, & Stengel, 2016). Thus, the N:P ratio is variable, and can change with respect to a change in organisms growing exponentially, and those in equilibrium among the community. The marine ecosystem maintains N:P ratios, even when the N:P ratio drops by amplifying N_2 biomass by nitrogen fixation, via nitrogen fixers, which, under decomposition, release high N organic matter and nitrates that consequently increase N:P ratios (Hao, Gao, & Yang, 2017; Marleau, Guichard, & Loreau, 2015). On the contrary, when the N:P ratio drops, nitrogen fixation is reduced by a limited supply of P, consequently refurbishing the N:P ratio in the water (Hao et al., 2017), signifying that phosphorus has a main regulatory role in marine productivity (Shen et al., 2015). Conversely, denitrification negates N_2 fixation, and signifies the process of excess N discharge, because higher N_2 needs to be necessitated with high O_2 supply, which consequently leads to O_2 depletion and denitrification (Su, Pahlow, Wagner, & Oschlies, 2015). But these processes occur at variable rates in diverse marine ecosystems, such as the Mediterranean and northern tropical Atlantic, which function as sources, and the Arabian Sea and eastern tropical Pacific, which act as sinks (Lejeusne, Chevaldonné, Pergent-Martini, Boudouresque, & Perez, 2010). However, nitrogen fixation can be significantly altered by the availability of microelements such as Fe and Mo, which are significant for enzymatic fixation of N_2 (Weisany, Raei, & Allahverdipoor, 2013). The impact of micronutrients on nitrogen fixation is diverse in freshwater and marine ecosystems (Harrison, 2016; Hobbie & Villéger, 2015). The nitrogen fixation in eutrophic lakes occurs only when the N:P ration of nutrient loading is equal or below the Redfield ratio, while as in estuaries, there is low nitrogen fixation due to insignificant levels of Fe and Mo, which consequently reduces nitrogenase activity, and hence blocks the passage to approach the Redfield ratio (Camenzind, Hättenschwiler, Treseder, Lehmann, & Rillig, 2017; Marino & Howarth, 2016).

Conversely, terrestrial ecosystems are exposed to extensive climatic changes that significantly affect the elemental stoichiometry of both ecosystems and organisms. Numerous research reports illustrate C:N, C:P, and N:P rations against varied climatic and ecological gradients. Research studies illustrate higher P reserves in young soils of higher altitudes than that of older soil, which displays depleted P levels due to prolonged meteorization of surface parental materials (Bremer, Farley, Chadwick, & Harden, 2016; Huang et al., 2015). Elser et al. (2010) demonstrated high and low latitudinal gradients in P and N, respectively, while investigating foliar N and P composition of woody plants thriving near the equator. Numerous other research reports demonstrated the decline in P levels of plants inhabiting tropical climates (Batterman, Wurzburger, & Hedin, 2013). Further, litter decomposition was regulated by nitrogen content, which modulates microbial activity, and the high N:P ratio of litter in tropical ecosystems allows more rapid nutrient cycling, and instant uptake and averting of nutrient depletion via leaching in the ecosystem (Grandawa, 2014). Tropical forests display high C:P ratios, which are main determinant factors in litter decomposition. Tropical forests are known to accumulate more nitrogen in contrast to temperate forests, because limited P levels, high N_2 fixers, and a continuous exchange of bioavailable N at the ecosystem level (Cleveland et al., 2011; Fanin, Hättenschwiler, Schimann, & Fromin, 2015). Some of the tropical plants display a carnivorous mode of nutrition due to P-deficit soils (Bloomfield, Farquhar, & Lloyd, 2014). While in some cases, such as *Escherichia coli*, plants exhibit high organic cellular C, C:P, and N:P biomass ratios at elevated temperatures via distribution of a high portion of the P-cell quota toward membrane formation (Cross et al., 2015). However, there is no consistent pattern followed by the N:P ratio between and within varied climatic belts (Zhang, Xue, Lyu, Tong, & Jiang, 2017). Tripathi and Raghubanshi (2013) demonstrated a higher limitation of N than P in a dry, tropical ecosystem of India, against tropical region; while Australian soils display low P levels, as compared with other continents. N availability can be augmented by high P levels via activation of phosphatase activity in all plants, or by augmenting plant growth (Marklein & Houlton, 2012; Richardson,

Barea, McNeill, & Prigent-Combaret, 2009). Research studies carried out in agronomic fields demonstrate a decline in growth rate in response to limited N:P ratio in leaves. A comparative analysis of elemental stoichiometry in terrestrial and aquatic ecosystems demonstrate higher C:P and C:N ratios in terrestrial plants than in aquatic ecosystems, demonstrating a higher allocation toward low nutrient structural organization (Scharler et al., 2015; Van Der Werff, 2011). This disparity in biogeochemical cycling could be a strategy to avert interspecific competition, signifying a less variable N:P ratio in comparison with C:N and C:P ratios (Dalling, Heineman, González, & Ostertag, 2016; Julian, Gerber, King, & Wright, 2017). Herbaceous plants display lower N:P ratios than woody plants (An & Shangguan, 2010), and graminoids display higher ratios than forbs (Gusewell, 2004).

Although there are numerous research reports that illustrate varied elemental stoichiometry in different ecosystems, there is still a need to acquire a more critical perception pertaining to the fate of C in organisms, food webs, and ecosystems; and it is principally controlled by the mass-balance theory and key nutrient availability (McLauchlan, Williams, & Engstrom, 2013). The elucidations of nutrient stoichiometry widen our perception underlying the evolution of food via trophic levels. As we can observe, there is a higher level of homeostasis in heterotrophs than autotrophs, instead of disparity among diverse heterotrophic groups (Persson et al., 2010). The high variability in autotrophs is because of nutrient richness in its surroundings, turnover rates, and balance between light and nutrients (Hofmann, Lamprecht, Pauli, & Illmer, 2016; Wirtz & Kerimoglu, 2016). Thus, it can be conjectured that elemental stoichiometry and species evolution run concurrently, signifying that ecological and organismic stoichiometry can affect the molecular structure and evolution of organisms at varied levels (Bailey et al., 2009; Sterner & Elser, 2002). Environmental stoichiometry can impact transcriptional expression and metabolomic structure, thereby affecting phenotypic screening, and the final genomic snapshot. Some research studies illustrate the possibility of integrating elemental stoichiometry with an omics approach from an ecological point of view. There are numerous research reports documenting the significant effects of P levels on the gene expression of highly conserved genes, such as phosphate transporter systems and P biogeochemical cycling, comprised of acquisition, assimilation, and P allocation; which is usually discreet (Hinsinger et al., 2011; Jeyasingh & Weider, 2007). Advanced research studies, such as transcriptomics in algae (Cock et al., 2010), microbes (Richardson et al., 2011), and terrestrial plants (Veneklaas et al., 2012), have displayed significant changes in gene expression against available P levels. These kinds of effects can intervene with species fitness and its evolutionary drive. Comparative genomics of *Oryza sativa* and *A. thaliana* using a TIGR database and protein sequences revealed higher N content in genomes and proteomes of cultivated plants than in wild plants (Singh, Jethva, Singla-Pareek, Pareek, & Kushwaha, 2016). These studies demonstrated the prospective function of N and P levels in the genomic evolution of plants, and related genomic evolution with environmental stoichiometry, within which the organism is evolved. Hence, it can be conjectured that the N:P ratio is a determinant factor that impacts ecosystem structure at moderate and prolonged time scales. Thus, it can be summarized that the natural gradients of N:P availability ratios can impart chemical control over species evolution.

4.5 BACKGROUND RADIATION AND METABOLOMIC RESPONSE

Solar radiation is the prime energy resource for driving the photosynthetic carbon fixation in plants. Ultraviolet radiation accounts for 6.3% (UV)-A (λ = 315–400 nm), and UV-B (λ = 280–315 nm) represents 1.5% of incoming solar radiation. Climatic change has led to stratospheric ozone loss, which amplifies the UV-B radiation flux reaching the earth's surface; but with almost negligible effects on levels of UV-A (Bornman et al., 2015; Menon et al., 2016). Both UV-A and UV-B are biologically detrimental to terrestrial plants, but a higher rate of emissions of UV-B can penetrate the stratosphere, and induce DNA damage via dimerization of thymidine residues, and membrane damage via lipid and protein oxidation (Cadet, Douki, & Ravanat, 2015; Cadet, Grand, & Douki, 2014). Therefore, plants have to cope with exchanges between irradiance requirements, and regulating the harmful effects of UV radiation.

The problem may have been especially severe in the geological past, when environmental chemistry models signified that the stability of the stratospheric ozone layer protecting earth's surface biota from high UV-B flux could be degraded by geological processes (Zepp, Erickson Iii, Paul, & Sulzberger, 2011). The depletion of the stratospheric ozone is caused by indiscriminate anthropogenic activities, such as high chlorofluorocarbon (CFC) emissions, increased UV-B radiation flux from high southern (Bais et al., 2015; Duffy, 2017) and northern latitude ecosystems, with depressing effects on photosynthetic primary production. Implementation of the Montreal Protocol has reduced CFC emissions; however, research investigations illustrated a high influx of UV-B flux by 2020, relative to 1979–1992. Additionally, an increased efflux of halogenated compounds and melting of glacier Arctic and Antarctic caps has been observed as a result of climatic change (Madronich et al., 2015). Plants respond to radiation through two routes: one, through biosynthesis of UV-absorbing compounds; and by DNA repair (Häder et al., 2015; Huché-Thélier et al., 2016). Elevated UV-B radiation results in declined growth due to a higher allocation of nutrient resources in defense-related measures (Caretto et al., 2015). These defense-related strategies mainly include the biosynthesis of UV-absorbing phenolics compounds and their accumulation on photosynthetic tissues, along with deposition on pollen spores (Wallace, Fleming, Wellman, & Beerling, 2011). However, short-term UV-B radiation modulates plant morphology and physiology by inducing a decline in leaf area (Chen, Chen, et al., 2016; Chen, Huang, et al., 2016), root and hypocotyls length, epidermal cell expansion (Verdaguer, Díaz-Guerra, Font, González, & Llorens, 2018), and high trichome density (Bickford, 2016). Increased UV-B flux can reduce or increase the photosynthetic rate by inducing intricate changes in stomatal behavior and lower plant fitness (Devi et al., 2017). Most of the UV-associated research studies have investigated the plant response to UV-B doses that are proportionate with anthropogenic ozone depletion. However, the ecophysiological responses of plants to prolonged doses of UV-B ($>20\,kJ\,m^{-2}\,d^{-1}$ BE) beyond those induced by anthropogenic ozone depletion might also be significant on evolutionary scales (Millett, Foot, Thompson, & Svensson, 2017). Research studies have been conducted to describe effects of UV-B flux on two accessions, Columbia-0 (Col-0) and the UV-B-sensitive mutant (*fah-1*) *A. thaliana* L., metabolism. UV-B radiation for prolonged periods resulted in significant changes in morphology and physiology that are proportionate with the production of defense-related secondary metabolites (Booij-James, Dube, Jansen, Edelman, & Mattoo, 2000). These defense-related metabolites include phenylpropaniods, which incur high costs related to the growth and development of plants, followed by adaptation and recovery, which is manifested through life expansion in *Arabidopsis* (Huang et al., 2010). The instant and steady production of phenylpropaniods, recognized as exclusive defense responses, incurred constant expenditures in terms of plant fitness, with meager effects on seed germination capacity (Wargent, Nelson, McGhie, & Barnes, 2015; Yu et al., 2018). The most significant morphological response to high radiation flux includes an increase in leaf trichome density, which provides protection through interception, absorption, and scattering of UV-B radiation (Kumari, Prasad, & Agrawal, 2015; Jan, Kamili,

Parray, & Bedi, 2016; Jan, Kamili, Parray, Bedi, & Ahmad, 2016; Ma & Upadhyaya, 2018). Further, trichome offers several other protection strategies against herbivory, including alterations to boundary layer dynamics, dispersal of water vapor, and the absorption of radiant energy (Crocoll, Asbach, Novak, Gershenzon, & Degenhardt, 2010). Plants thriving under high radiation flux reveal elevated levels of defense-related secondary metabolites, and a high tolerance level against herbivory (Jan, Kamili, Parray, & Bedi, 2016; Jan, Kamili, Parray, Bedi, & Ahmad, 2016). Gene expression of UV-B-related metabolites and herbivory response overlap both instigate the activation of signaling mitogen-activated protein kinases (MAPKs) (Sinha, Jaggi, Raghuram, & Tuteja, 2011). However, there is no research report that demonstrates the UV-induced signal transduction against wounding, but the function of leaf trichomes during early leaf development of oaks and olives are implicated in the protection strategy (Barnes et al., 2015). The change in stomatal density is also the main tolerance strategy of plants against high radiation flux, such as a decline in stomatal density against UV-B flux, which has been observed in *Glycine max* (Kataria, Baghel, & Guruprasad, 2017). The general responses of stomatal density to high levels of CO_2 levels and humidity are strongly linked to, and mainly regulated by, ABA concentration. UV radiation absorbs ABA, and is consequently inactivated by photolysis, and a decline in ABA would correspond to this mechanistic link (Chater et al., 2015). Other phytohormones such as jasmonic acid (JA), ethylene, and ROS are all enhanced in response to UV-B. These phytohormones and ROS are implicated in the decline of stomatal density in the systemic response to high CO_2 levels (Xia et al., 2015). High UV radiation induces MAPK activity in *Lycopersicon peruvianum*, and some of *MAPK*, such as *MPK3* and *MPK6*, are implicated in stomatal development (Raja, Majeed, Kang, Andrabi, & John, 2017); while MKK4 and MPK6 are imperative regulators of stomatal development (Jan & Parray, 2016). Loss of MAPKs function leads to the surplus of stomata and the disruption of the one-cell spacing pattern. On the contrary, the gain in the function of MAPK leads to stomatal development. Hence, it can anticipated that a UV-induced decline in stomatal density is stimulated via MAPKs activation executing upstream of JA, leading to the inhibition of the stomatal opening (Xu & Zhang, 2015). UV radiation induces an increase in sinapic acid, its ester, and salt sinapoyl malate in wild-type *A. thaliana*, which absorbs UV-B radiation (Neugart et al., 2014). Pronounced accumulation of sinapic acid, its ester, and salt sinapoyl malate function as sunscreen, absorbing the surplus of UV-B flux; hence, averting the cellular disorder (Kaling et al., 2015). In addition to an increment in sinapic acid, its ester, and salt sinapoyl malate; there is a concurrent increase of ferulic acid and 5-hydroxy-coniferaldehyde. In addition to these metabolites, there is an increment in flavonoid compounds, quercetin–glycoside, and kaempferol–glycoside. These flavonoid metabolites function as antioxidants, conferring protection against high UV-B radiation flux (Wargent et al., 2015). Accumulation of flavonoids and its glycosides in initial developmental stages of the *Betula pubescens* seedling demonstrate its protective function. However, there is a disparity in the extent of protection conferred by flavonoids against UV radiation; for example, quercetin displays high absorbance compared with kaempferol in response to high UV-B flux (Dillon, Chludil, & Zavala, 2017). Both quercetin and kaempferol are expensive in terms of carbon allocation, each containing 15 C in contrast to 11 C for sinapic acid, and 10 C for ferulic acid (Deng & Lu, 2017). This increase in secondary metabolite influx incurs a high cost on plant fitness without affecting the germination rate significantly. The localization of pronounced accumulations in tissues is highly diverse, such as in *Brassica rapa*, in which the accretion of metabolites occurs in seed testa without affecting the seed testa biomass. There are various other physiogonomic and metabolic responses of plants that are subjected to varied UV-B flux, but these responses are more consistent with short-term dose research studies resulting in acclimation, which is considered to be multiphasic. However, the majority of research studies conjectured that the overall increase in phenylpropaniods function in adaptation strategies. There is significant experimental evidence that demonstrates that the lack of a phenylpropanoid pathway can result in low plasticity in plants under a high UV-B radiation flux. Thus, it can determine that a high influx of secondary metabolites can facilitate the repair and protection of seedlings under a high radiation flux (Verdaguer, Jansen, Llorens, Morales, & Neugart, 2017). Further, these studies illustrate explicit tolerance strategies provided by these metabolites, and such multiphasic adaption approaches, occur at diverse developmental stages, and may facilitate carbon acquisition. The expenses incurred include a decline in flower number, and total biomass; although the allocation of resources in the UV protection mechanism regulates germination, leading to alleviation of UV-induced effects.

4.6 NUTRIENT LIMITATIONS AND HOMEOSTATIC COMPETENCE

The flow of nutrients within different trophic levels is regulated through the food chain or food web, which demonstrates energy transfer in ecosystems

(Julian et al., 2017; Peng, Chen, Yan, & Han, 2016). However, the elemental stoichiometry of food such as C:N and C:P ratio limits the food chain efficiency transfer through the food web, as illustrated in aquatic ecosystems (Manzoni, Taylor, Richter, Porporato, & Ågren, 2012; Pujo-Pay et al., 2011). A food web exhibiting a high C:P ratio in particulate matter at the base reduces the matter transfer within the food web (Fink & Von Elert, 2006). On the contrary, a decline in C:P and N:P ration water enhances the input of consumers to C, N, and P cycling, modulating the proportions of C inputs metabolized via consumers, and likely resulting in decreased ecosystem level C storage (Rivas-Ubach et al., 2012). The P levels of herbivores in freshwater ecosystems are functional aspects for the growth of carnivores (Duffy et al., 2007). In case the C:P ratio of herbivores is high, carnivores can compensate for the C surplus by decreasing growth, leading to meager variations in C:N:P ratios, steady with GRH. Significant data maintains such top-down effects by grazers and predators on N:P rations in water, and producers. These effects are because of the ability of grazers and predators to accumulate N and P, and also due to excretion. High pressure by grazers has been illustrated to increase N and P availability by excretion, signifying that increased regeneration of N by grazers results in lower C:N and C:P rations in grazed, than in ungrazed algae; hence enhancing algal nutritional quality (Frigstad et al., 2011) and producing more rapid N and P cycling. In spite of recurrently signifying only a fraction of ecosystem biomass, predators significantly influence the stoichiometry of lower trophic levels and the environment by their expansive trophic activities, and ability to structure lower trophic levels (Prater et al., 2017). Terrestrial ecosystems demonstrate the intensive effects of a decline in C:N and C:P ratios greater than that of aquatic ecosystems (Wang, Sardans, et al., 2015; Wang, Xie, et al., 2015). Research reports have also illustrated nutrient shifts such as grazer-stimulated N deficit plants to P-deficit plants by amplifying N levels in leaves (Singh, Baer, Riebesell, Martiny, & Lomas, 2015). A foliar C:P ratio is associated with a C:P ratio of soil via the impact of the soil nutrient availability on foliar nutrient content, which can limit herbivore investment in P-enriched metabolites dispensable for growth (Talarmin et al., 2016). These studies demonstrate the broader drift toward a homeostatic control of the C:N:P stoichiometry of soils via soil-plant-herbivore associations that substantiate more investigation (Singh et al., 2015). Sequestration of grazers leads to a reduction of P levels, as the biomass of grazers increases at a low N:P ratio. These changes result in reduced plant growth, and ultimately in a bottom-up effect in herbivore growth

(Erga et al., 2017). Specific grazers maintain their C:N:P stoichiometry by ingesting plants with varied compositions of these elements (Kou et al., 2018). These types of mechanisms operate in diverse ecosystems to maintain C:N:P stoichiometry, even within different trophic levels (Delgado-Baquerizo et al., 2018; Hillebrand, Cowles, Lewandowska, Van de Waal, & Plum, 2014). There are numerous research studies on human consumption and litter that provide facts substantiating this premise; although uptake and nutrient cycling with covalent chemistry such as C, N, and P are predominantly under biological control, and that subsequently regulates and diversifies stoichiometry at a minimal range (Bi et al., 2018; Sperfeld, Wagner, Halvorson, Malishev, & Raubenheimer, 2017). However, we need to supplement such facts with more research emphasis on climatic factors to ascertain any trends. The ability to alter C:P ratio can have an imperative function in an organism's response subjected to herbivory and predation with more significance in both the organism's structure, and its biotic relationship with the environment (Ngatia, Turner, Njoka, Young, & Reddy, 2015). In order to regulate homeostasis in C:P and C:N ratios within each trophic level requires the implication of resorption, digestive-absorptive methods, higher grazing activity, and alterations in respiration (Sardans, Rivas-Ubach, & Peñuelas, 2012a, 2012b). Hence, this homeostatic competence has an expansive energy cost that reduces plant growth and reproductive ability (Elser et al., 2010; Rivas-Ubach et al., 2012).

4.6.1 Functional Significance of Plant Metabolite Levels in Studying Ecosystem Buffering Capacities

Ecometabolomics has been identified as the most elite approach being used for particular organisms sampled from given environments (Gargallo-Garriga et al., 2016; Iriti & Vitalini, 2018; Peters et al., 2018; Snart, Hardy, & Barrett, 2015). Despite numerous intrinsic constraints, metabolite profiling is becoming remarkably eminent under field conditions (Fernie, Trethewey, Krotzky, & Willmitzer, 2004; Fiehn et al., 2000; Johnson, Ivanisevic, & Siuzdak, 2016). It has been successfully employed to tackle topics such as species interactions, associations between growth and chemical stoichiometry, or the plant stress response (Gargallo-Garriga et al., 2017; Hasler-Sheetal, Castorani, Glud, Canfield, & Holmer, 2016; Meena et al., 2017; Sun, Gao, Fu, Zhou, & Wu, 2015). As illustrated in the preceding sections, plant stress has a remarkable effect on plant metabolomes, and metabolite profiling has become exceedingly significant for deciphering the interaction between organisms and the environment (van Dam & Bouwmeester, 2016;

Verma et al., 2018; Wolfender, Marti, Thomas, & Bertrand, 2015). In comparison with indoor studies, external stress is masked, at least by the environment, to some degree. Such stress buffering capacities of agroecosystems are significant for keeping in view the ever-expanding anthropogenic effect on ecosystems (Ekbladh et al., 2002; Eksvärd & Rydberg, 2010; Lawrence, 2015). In this section, we review the concept of using metabolite profiles for relevant measurements. Several research reports anticipate the utilization of parameters of the response of the key plant species to the given stress treatment as proxies for determining and comparing stress buffering capacities of ecosystems (Fester, 2015; Kéfi, Holmgren, & Scheffer, 2016; Medlyn, De Kauwe, & Duursma, 2016; Suseela & Tharayil, 2018). Investigation of ecosystem stress buffering capacities may advance our perspective on how ecosystems cope with stress, and might improve our ability to predict ecosystem variations (Baert, Janssen, Sabbe, & De Laender, 2016; Forsman & Wennersten, 2016; Hilderbrand & Utz, 2015; Musavi et al., 2017).

The survival of plants in a given environment is not only determined by the degree of stress only, but also by the protection conferred by the environment (Peterman, Waring, Seager, & Pollock, 2013). These protective effects can be determined by diverse abiotic parameters such as edaphic factors, or biotic factors, such as composition and diversity of the plant community or rhizosphere (Philippot, Raaijmakers, Lemanceau, & Van Der Putten, 2013). Steffen et al. (2011) ascertained the significance of protective environmental effects under the prevailing global climate change era. However, the protective potential of the environment has been reduced significantly by anthropogenic activities, due to a decline in biodiversity (Folke et al., 2004). In order to avert this ecological deterioration, an advanced perspective on the stress buffering capacities of ecosystems is indispensable (Ossipov et al., 2008; Wallenstein, Hess, Lewis, Steltzer, & Ayers, 2010).

Fester (2015) assumed the plant's reaction to explicit stress treatment as an alternative for the stress buffering capacities of ecosystems because a plant's primary production is a dominant process in all ecosystems. But like any other proxy, the strategy selected will include only specific features of the ecosystem studied. This strategy is entirely dependent on the general hypothesis that plants from ecosystems with high buffering capacities will experience a milder or more transient effect of a given stress treatment in comparison with plants from ecosystems with weaker buffering capacities (Dauwe, Holliday, Aitken, & Mansfield, 2012; Zhao et al., 2013). Fester (2015) also used above-ground metabolite profiling for estimating the plant stress responses. As illustrated in previous sections of this chapter, plant metabolite profiles are explicitly responsive to stress, and include data pertaining to the degree of the stress, duration of the stress, the plant's competence to recover from stress, or to acclimatize to the new situation. Hence, the intensity of the effect would be a fundamental parameter, and can be inferred from the numerous important alterations in the metabolite levels (Rivas-Ubach et al., 2012, 2014). The duration of a given impact is available by kinetic experiments. In order to demonstrate the competence of plants to cope with stress, we will require marker metabolites, which signify the health status of the plant under stress (Dannenmann et al., 2009; Genga et al., 2011). Several research reports validated that the metabolite profiling is more appropriate for determining the plant fitness under stress in contrast to the conventional parameter for plant performance, such as biomass production, photosynthetic activity, and so forth (Pessarakli, 2016; Zhao et al., 2013). Metabolite profiling would provide a more inclusive snapshot of the plant metabolome than the conventional parameters (Dunn, Broadhurst, Atherton, Goodacre, & Griffin, 2011). They will facilitate high resolution referring to kinetic estimations, and provide high sensitivity for estimation (Larive, Barding Jr, & Dinges, 2014; Robinson, Ukrainetz, Kang, & Mansfield, 2007). Vogel et al., 2013 illustrated the effects of roof treatments in an open field experiment using metabolite profiling, which detects metabolite level, but not more conventional parameters. This approach proposed objectives for estimating ecosystem characteristics by recording factors from individual organisms. Because of this focus, respective experimental designs would have to vary in a few ways from existing ecometabolomics experiments:

(a) In several ecosystems, various treatments and measurements have to be employed on the pilot level, with various experimental repetitions. This is due to the fact that major interest in organism characteristics and specified technical issues with plot-level repetitions, such an experimental design, is not frequently found in metabolite profiling. Although the plant's environment was included as an experimental factor by numerous authors (Patra, Schluttenhofer, Wu, Pattanaik, & Yuan, 2013; Poff & Zimmerman, 2010; Poorter et al., 2012), the majority of them employed individual plants (and not diverse experimental plots) as replicates, with the exception of Sehring (2010) and Rivas-Ubach et al. (2014).

(b) Stress treatments have to be applied to complete ecosystems instead of individual plants. Most of the studies from metabolite profiling (Pandey, Irulappan, et al., 2017; Pandey, Kumar, et al., 2017;

Okazaki & Saito, 2012) indeed applied treatments to the area surrounding experimental plants, with the exception of tree girdling by Dannenmann et al. (2009) and Kim, Baek, Park, Im, and Kim (2017). The size of the treated area should be considered carefully when aiming for assertions on ecosystem properties. This theme has become less significant when relying on spatial or temporal differences as starting points of the experiment (Valares Masa, Alias Gallego, Chaves Lobon, & Sosa Diaz, 2016; Yang et al., 2012).

(c) Though the plant's stress response could be employed as a proxy for ecosystem buffering capacities, insights into the nature of buffering mechanisms are not possible this way. For such objectives, factors from the ecosystem investigated have to be correlated with metabolite data, and parameters correlated with buffering capacities need to be characterized. Respective analyses, such as "phenotypic anchoring" (Bundy, Davey, & Viant, 2009; Gargallo-Garriga et al., 2015; Peñuelas et al., 2013), were carried out in the field of toxicology, and for associating particular phenotypic traits such as growth traits, morphological and biochemical characteristics with specific metabolic characteristics such as carbohydrate and lignin precursors (Arbona, Manzi, Ollas, Gómez-Cadenas, 2013; Elmendorf et al., 2012; Fernández-Martínez et al., 2014; Robinson et al., 2007). Environmental anchoring necessary for detecting ecosystem parameters significant for buffering stress in a similar way will be possible in the context of expansive research consortia, where several ecosystem factors can be assessed and employed for statistical analyses (Hennion et al., 2012; Webb, Freeman, & Broeckling, 2016). The detection of parameters significant for stress buffering capacities of ecosystems would permit devising rational management measures aimed at regulating and amplifying ecosystem stress elasticity.

In summary, we anticipate employing metabolite profiles representing the plant's response to a given effect of stress as a proxy for estimating and comparing an ecosystem's alleviating effects on this effect. Such an approach is based on the statement that plants from ecosystems with high stress buffering capacity will experience lessened effects of given treatments when compared with the plants from ecosystems with low stress buffering capacity (Döring et al., 2015; McCormack & Guo, 2014). On a more theoretical level, a decline in the capacity of an ecosystem to recover from small perturbations has been signified as one of the major causes for "critical slowing down" of the system when approaching a critical point of transition (Scheffer et al., 2009). As summarized herein, metabolite profiles of the key plant species challenged by specific stress parameters applied to an ecosystem might be employed as an estimate for this capacity, and might therefore offer the possibility of producing respective early warning signals. Environmental anchoring can also be employed for the detection of characteristics significant for ecosystem buffering capacities that would be supportive in advancing the management of agroecosystems. Metabolite profiles, as summarized, have the potential to contribute significant data on ecosystems subjected to stress; and significant interdisciplinary links between ecologists and analytical chemists should form the basis for comprehending this perspective.

4.7 CONCLUSION

There are numerous research reports that demonstrate current climate changes that are significantly impacting metabolite flux at both the organism and ecosystem level. Climate-induced metabolic responses can alter the structure and function of ecosystems manifested through the modifications of elemental stoichiometry. Changes in elemental stoichiometry can have discreet effects on the biosphere, allowing humans to positively impact climate change. Climate change has consequents effects on nutrient resources ultimately alter the growth rate and directive response of organisms. Although present research studies illustrating integrative responses of climatic variables and associated eutrophication or high CO_2 levels can facilitate our perceptive pertaining to the effects of climate change on metabolic flux. We need to assemble expansive research covering more extensive experimental data and encompassing more buffered changes for prolonged periods of time to procure reproducible results. Further, we need to complement our facts on remote sensing data, paleoecological data, and extensive data sets to unravel the effects of climate change on life on earth.

REFERENCES

Abadie, C., Bathellier, C., & Tcherkez, G. (2018). Carbon allocation to major metabolites in illuminated leaves is not just proportional to photosynthesis when gaseous conditions (CO_2 and O_2) vary. *New Phytologist.*

Abdurakhmonov, I. Y., Ayubov, M. S., Ubaydullaeva, K. A., Buriev, Z. T., Shermatov, S. E., Ruziboev, H. S., et al. (2016). RNA interference for functional genomics and improvement of cotton (*Gossypium* sp.). *Frontiers in Plant Science*, 7, 202.

Aerts, R., Van Bodegom, P. M., & Cornelissen, J. H. C. (2012). Litter stoichiometric traits of plant species of high-latitude ecosystems show high responsiveness to global change without causing strong variation in litter decomposition. *New Phytologist*, *196*(1), 181–188.

Agathos, S. N., & Boon, N. (2015). Editorial overview: environmental biotechnology. *Current Opinion in Biotechnology*, *33*, 5–7.

Agnihotri, P., Husain, T., Shirke, P. A., Sidhu, O. P., Singh, H., Dixit, V., et al. (2017). Climate change-driven shifts in elevation and ecophysiological traits of Himalayan plants during the past century. *Current Science*, *112*(3), 595–601.

Ahmad, R., Kim, M. D., Back, K. H., Kim, H. S., Lee, H. S., Kwon, S. Y., et al. (2008). Stress-induced expression of choline oxidase in potato plant chloroplasts confers enhanced tolerance to oxidative, salt, and drought stresses. *Plant Cell Reports*, *27*(4), 687–698.

Albert, R., Acharya, B. R., Jeon, B. W., Zañudo, J. G., Zhu, M., Osman, K., et al. (2017). A new discrete dynamic model of ABA-induced stomatal closure predicts key feedback loops. *PLoS Biology*, *15*(9)e2003451.

Allen, W. C., Hook, P. B., Biederman, J. A., & Stein, O. R. (2002). Temperature and wetland plant species effects on wastewater treatment and root zone oxidation. *Journal of Environmental Quality*, *31*(3), 1010–1016.

Amedie, F. A. (2013). *Impacts of Climate Change on Plant Growth, Ecosystem Services, Biodiversity, and Potential Adaptation Measure.* Doctoral dissertation, MS thesis in Atmospheric ScienceSweden: University of Gothenburg.

An, H., & Shangguan, Z. (2010). Leaf Stoichiometric trait and specific leaf area of dominant species in the secondary succession of the Loess Plateau. *Polish Journal of Ecology*, *58*, 103–113.

Anderegg, W. R. (2015). Spatial and temporal variation in plant hydraulic traits and their relevance for climate change impacts on vegetation. *New Phytologist*, *205*(3), 1008–1014.

Andresen, L. C., Müller, C., de Dato, G., Dukes, J. S., Emmett, B. A., Estiarte, M., et al. (2016). Shifting impacts of climate change: long-term patterns of plant response to elevated CO2, drought, and warming across ecosystems. *Advances in ecological research* (pp. 437–473). Vol. 55 (pp. 437–473). Academic Press.

Aranjuelo, I., Molero, G., Erice, G., Avice, J. C., & Nogues, S. (2011). Plant physiology and proteomics reveals the leaf response to drought in alfalfa (*Medicago sativa* L.). *Journal of Experimental Botany*, *62*, 111–123.

Arbona, V., Manzi, M., Ollas, C. De., & Gómez-Cadenas, A. (2013). Metabolomics as a tool to investigate abiotic stress tolerance in plants. *International Journal of Molecular Sciences*, *14* (3), 4885–4911. https://doi.org/10.3390/ijms14034885.

Aspinwall, M. J., Loik, M. E., Resco de Dios, V., Tjoelker, M. G., Payton, P. R., & Tissue, D. T. (2015). Utilizing intraspecific variation in phenotypic plasticity to bolster agricultural and forest productivity under climate change. *Plant, Cell & Environment*, *38* (9), 1752–1764.

Atkinson, N. J., & Urwin, P. E. (2012). The interaction of plant biotic and abiotic stresses: from genes to the field. *Journal of Experimental Botany*, *63*(10), 3523–3543.

Baert, J. M., Janssen, C. R., Sabbe, K., & De Laender, F. (2016). Per capita interactions and stress tolerance drive stress-induced changes in biodiversity effects on ecosystem functions. *Nature Communications*, 7, 12486.

Bagley, J., Rosenthal, D. M., Ruiz-Vera, U. M., Siebers, M. H., Kumar, P., Ort, D. R., et al. (2015). The influence of photosynthetic acclimation to rising CO_2 and warmer temperatures on leaf and canopy photosynthesis models. *Global Biogeochemical Cycles*, *29*(2), 194–206.

Bagri, J., Yadav, A., Anwar, K., Dkhar, J., Singla-Pareek, S. L., & Pareek, A. (2017). Metabolic shift in sugars and amino acids regulates sprouting in Saffron corm. *Scientific Reports*, 7(1), 11904.

Bailey, J. K., Schweitzer, J. A., Ubeda, F., Koricheva, J., LeRoy, C. J., Madritch, M. D., et al. (2009). From genes to ecosystems: a synthesis of the effects of plant genetic factors across levels of organization. *Philosophical Transactions of the Royal Society of London B: Biological Sciences*, *364*(1523), 1607–1616.

Bais, A. F., McKenzie, R. L., Bernhard, G., Aucamp, P. J., Ilyas, M., Madronich, S., et al. (2015). Ozone depletion and climate change: impacts on UV radiation. *Photochemical & Photobiological Sciences*, *14*(1), 19–52.
Barah, P., Winge, P., Kusnierczyk, A., Tran, D. H., & Bones, A. M. (2013). Molecular signatures in *Arabidopsis thaliana* in response to insect attack and bacterial infection. *PLoS One*, *8*(3)e58987.
Barnes, P. W., Flint, S. D., Ryel, R. J., Tobler, M. A., Barkley, A. E., & Wargent, J. J. (2015). Rediscovering leaf optical properties: new insights into plant acclimation to solar UV radiation. *Plant Physiology and Biochemistry*, *93*, 94–100.
Barua, A., Katyaini, S., Mili, B., Gooch, P., & for Integrated Mountain Development (ICIMOD) (2014). Climate change and poverty: building resilience of rural mountain communities in South Sikkim, Eastern Himalaya, India. *Regional Environmental Change*, *14*(1), 267–280.
Batterman, S. A., Wurzburger, N., & Hedin, L. O. (2013). Nitrogen and phosphorus interact to control tropical symbiotic N2 fixation: a test in Inga punctata. *Journal of Ecology*, *101* (6), 1400–1408.
Bauwe, H., Hagemann, M., Kern, R., & Timm, S. (2012). Photorespiration has a dual origin and manifold links to central metabolism. *Current Opinion in Plant Biology*, *15*(3), 269–275.
Becker, C., & Kläring, H. P. (2016). CO_2 enrichment can produce high red leaf lettuce yield while increasing most flavonoid glycoside and some caffeic acid derivative concentrations. *Food Chemistry*, *199*, 736–745.
Becklin, K. M., Anderson, J. T., Gerhart, L. M., Wadgymar, S. M., Wessinger, C. A., & Ward, J. K. (2016). Examining plant physiological responses to climate change through an evolutionary lens. *Plant Physiology*, *172*(2), 635–649.
Bellard, C., Bertelsmeier, C., Leadley, P., Thuiller, W., & Courchamp, F. (2012). Impacts of climate change on the future of biodiversity. *Ecology Letters*, *15*(4), 365–377.
Bennett, H., Bell, J. J., Davy, S. K., Webster, N. S., & Francis, D. S. (2018). Elucidating the sponge stress response; lipids and fatty acids can facilitate survival under future climate scenarios. *Global Change Biology*.
Bernal, M., Estiarte, M., & Peñuelas, J. (2011). Drought advances spring growth phenology of the Mediterranean shrub *Erica multiflora*. *Plant Biology*, *13*(2), 252–257.
Bi, R., Ismar, S. M., Sommer, U., & Zhao, M. (2018). Simultaneous shifts in elemental stoichiometry and fatty acids of *Emiliania huxleyi* in response to environmental changes. *Biogeosciences*, *15*(4), 1029.
Bickford, C. P. (2016). Ecophysiology of leaf trichomes. *Functional Plant Biology*, *43*(9), 807–814.
Blackman, C. J., Aspinwall, M. J., Tissue, D. T., & Rymer, P. D. (2017). Genetic adaptation and phenotypic plasticity contribute to greater leaf hydraulic tolerance in response to drought in warmer climates. *Tree Physiology*, *37*(5), 583–592.
Block, A., Vaughan, M. M., Christensen, S. A., Alborn, H. T., & Tumlinson, J. H. (2017). Elevated carbon dioxide reduces emission of herbivore-induced volatiles in *Zea mays*. *Plant, Cell & Environment*, *40*(9), 1725–1734.
Bloomfield, K. J., Farquhar, G. D., & Lloyd, J. (2014). Photosynthesis–nitrogen relationships in tropical forest tree species as affected by soil phosphorus availability: a controlled environment study. *Functional Plant Biology*, *41*(8), 820–832.
Bloor, J. M., & Bardgett, R. D. (2012). Stability of above-ground and below-ground processes to extreme drought in model grassland ecosystems: interactions with plant species diversity and soil nitrogen availability. *Perspectives in Plant Ecology, Evolution and Systematics*, *14*(3), 193–204.
Booij-James, I. S., Dube, S. K., Jansen, M. A., Edelman, M., & Mattoo, A. K. (2000). Ultraviolet-B radiation impacts light-mediated turnover of the photosystem II reaction center

heterodimer in *Arabidopsis* mutants altered in phenolic metabolism. *Plant Physiology, 124*(3), 1275–1284.

Bornman, J. F., Barnes, P. W., Robinson, S. A., Ballare, C. L., Flint, S. D., & Caldwell, M. M. (2015). Solar ultraviolet radiation and ozone depletion-driven climate change: effects on terrestrial ecosystems. *Photochemical & Photobiological Sciences, 14*(1), 88–107.

Bose, B., Choudhury, H., Tandon, P., & Kumaria, S. (2017). Studies on secondary metabolite profiling, anti-inflammatory potential, in vitro photoprotective and skin-aging related enzyme inhibitory activities of *Malaxis acuminata*, a threatened orchid of nutraceutical importance. *Journal of Photochemistry and Photobiology B: Biology, 173*, 686–695.

Bremer, L. L., Farley, K. A., Chadwick, O. A., & Harden, C. P. (2016). Changes in carbon storage with land management promoted by payment for ecosystem services. *Environmental Conservation, 43*(4), 397–406.

Brunetti, C., George, R. M., Tattini, M., Field, K., & Davey, M. P. (2013). Metabolomics in plant environmental physiology. *Journal of Experimental Botany, 64*(13), 4011–4020.

Buckley, L. B., & Kingsolver, J. G. (2012). Functional and phylogenetic approaches to forecasting species' responses to climate change. *Annual Review of Ecology, Evolution, and Systematics, 43*.

Bundy, J. G., Davey, M. P., & Viant, M. R. (2009). Environmental metabolomics: a critical review and future perspectives. *Metabolomics, 5*, 3–21.

Busch, V., Klaus, V. H., Penone, C., Schäfer, D., Boch, S., Prati, D., et al. (2018). Nutrient stoichiometry and land use rather than species richness determine plant functional diversity. *Ecology and Evolution, 8*(1), 601–616.

Butler, S. M., Melillo, J. M., Johnson, J. E., Mohan, J., Steudler, P. A., Lux, H., et al. (2012). Soil warming alters nitrogen cycling in a New England forest: implications for ecosystem function and structure. *Oecologia, 168*(3), 819–828.

Cadet, J., Douki, T., & Ravanat, J. L. (2015). Oxidatively generated damage to cellular DNA by UVB and UVA radiation. *Photochemistry and Photobiology, 91*(1), 140–155.

Cadet, J., Grand, A., & Douki, T. (2014). Solar UV radiation-induced DNA bipyrimidine photoproducts: formation and mechanistic insights. *Photoinduced phenomena in nucleic acids* (pp. 249–275). Vol. II (pp. 249–275). Cham: Springer.

Caldera, H. I. U., De Costa, W. A., Woodward, F. I., Lake, J. A., & Ranwala, S. M. (2017). Effects of elevated carbon dioxide on stomatal characteristics and carbon isotope ratio of *Arabidopsis thaliana* ecotypes originating from an altitudinal gradient. *Physiologia Plantarum, 159*(1), 74–92.

Caldwell, M. M., Bornman, J. F., Ballaré, C. L., Flint, S. D., & Kulandaivelu, G. (2007). Terrestrial ecosystems, increased solar ultraviolet radiation, and interactions with other climate change factors. *Photochemical & Photobiological Sciences, 6*(3), 252–266.

Camenzind, T., Hättenschwiler, S., Treseder, K., Lehmann, A., & Rillig, M. C. (2017). Nutrient limitation of soil microbial processes in tropical forests. *Ecological Monographs, 88*(1), 4–21.

Cannone, N., Sgorbati, S., & Guglielmin, M. (2007). Unexpected impacts of climate change on alpine vegetation. *Frontiers in Ecology and the Environment, 5*(7), 360–364.

Caretto, S., Linsalata, V., Colella, G., Mita, G., & Lattanzio, V. (2015). Carbon fluxes between primary metabolism and phenolic pathway in plant tissues under stress. *International Journal of Molecular Sciences, 16*(11), 26378–26394.

Carnicer, J., Coll, M., Ninyerola, M., Pons, X., Sanchez, G., & Peñuelas, J. (2011). Widespread crown condition decline, food web disruption, and amplified tree mortality with increased climate change-type drought. *Proceedings of the National Academy of Sciences, 108*(4), 1474–1478.

Casson, S. A., & Hetherington, A. M. (2010). Environmental regulation of stomatal development. *Current Opinion in Plant Biology, 13*(1), 90–95.

Chakraborty, A., Saha, S., Sachdeva, K., & Joshi, P. K. (2018). Vulnerability of forests in the Himalayan region to climate change impacts and anthropogenic disturbances: a systematic review. *Regional Environmental Change*, 1–17.

Chapin, III,, F. S., Jefferies, R. L., Reynolds, J. F., Shaver, G. R., Svoboda, J., & Chu, E. W. (Eds.), (2012). *Arctic ecosystems in a changing climate: an ecophysiological perspective*. Academic Press.

Chater, C., Peng, K., Movahedi, M., Dunn, J. A., Walker, H. J., Liang, Y. K., et al. (2015). Elevated CO_2-induced responses in stomata require ABA and ABA signaling. *Current Biology*, *25*(20), 2709–2716.

Chauhan, N. S. (1999). *Medicinal and aromatic plants of Himachal Pradesh*. Indus publishing.

Chen, M., Huang, Y., Liu, G., Qin, F., Yang, S., & Xu, X. (2016). Effects of enhanced UV-B radiation on morphology, physiology, biomass, leaf anatomy and ultrastructure in male and female mulberry (*Morus alba*) saplings. *Environmental and Experimental Botany*, *129*, 85–93.

Chen, Y. L., Chen, L. Y., Peng, Y. F., Ding, J. Z., Li, F., Yang, G. B., et al. (2016). Linking microbial C:N:P stoichiometry to microbial community and abiotic factors along a 3500-km grassland transect on the Tibetan Plateau. *Global Ecology and Biogeography*, *25*(12), 1416–1427.

Cho, K. H., Choo, S. M., Jung, S. H., Kim, J. R., Choi, H. S., & Kim, J. (2007). Reverse engineering of gene regulatory networks. *IET Systems Biology*, *1*(3), 149–163.

Choi, J., Ko, J., Ng, C. T., Jeong, S., Tenhunen, J., Xue, W., et al. (2018). Quantification of CO_2 fluxes in paddy rice based on the characterization and simulation of CO_2 assimilation approaches. *Agricultural and Forest Meteorology*, *249*, 348–366.

Choudhary, K. K. (2015). *Effects of enhanced ultraviolet-B irradiation on selected leguminous plants*. Doctoral dissertationBanaras Hindu University.

Clare, M. (2015). *A comparative study examining the limiting nutrients within four lake basins of the English Lake District*. Lancashire: Lancaster Environment Centre, University of Lancaster.

Cleveland, C. C., Townsend, A. R., Taylor, P., Alvarez-Clare, S., Bustamante, M., Chuyong, G., et al. (2011). Relationships among net primary productivity, nutrients and climate in tropical rain forest: a pan-tropical analysis. *Ecology Letters*, *14*(9), 939–947.

Cock, J. M., Sterck, L., Rouzé, P., Scornet, D., Allen, A. E., Amoutzias, G., et al. (2010). The Ectocarpus genome and the independent evolution of multicellularity in brown algae. *Nature*, *465*(7298), 617.

Coolen, S., Proietti, S., Hickman, R., Davila Olivas, N. H., Huang, P. P., Van Verk, M. C., et al. (2016). Transcriptome dynamics of *Arabidopsis* during sequential biotic and abiotic stresses. *The Plant Journal*, *86*(3), 249–267.

Craine, J. M., Ocheltree, T. W., Nippert, J. B., Towne, E. G., Skibbe, A. M., Kembel, S. W., et al. (2013). Global diversity of drought tolerance and grassland climate-change resilience. *Nature Climate Change*, *3*(1), 63.

Crocoll, C., Asbach, J., Novak, J., Gershenzon, J., & Degenhardt, J. (2010). Terpene synthases of oregano (*Origanum vulgare* L.) and their roles in the pathway and regulation of terpene biosynthesis. *Plant Molecular Biology*, *73*(6), 587–603.

Cross, W. F., Hood, J. M., Benstead, J. P., Huryn, A. D., & Nelson, D. (2015). Interactions between temperature and nutrients across levels of ecological organization. *Global Change Biology*, *21*(3), 1025–1040.

da Silva, R., Mazumdar, A., & Naik, B. (2016). Relative contribution of C3 and C4 type terrestrial organic matter in the Mahanadi offshore (Bay of Bengal) sediments and climatic implication. *In 19th EGU General Assembly, EGU2017, proceedings from the conference, 23–28 April 2017*, Austria: Vienna (p. 1232).

Daines, S. J., Clark, J. R., & Lenton, T. M. (2014). Multiple environmental controls on phytoplankton growth strategies determine adaptive responses of the N:P ratio. *Ecology Letters*, *17*(4), 414–425.

Dalling, J. W., Heineman, K., González, G., & Ostertag, R. (2016). Geographic, environmental and biotic sources of variation in the nutrient relations of tropical montane forests. *Journal of Tropical Ecology*, *32*(5), 368–383.

Damiani, C., Colombo, R., Di Filippo, M., Pescini, D., & Mauri, G. (2016, October). Linking alterations in metabolic fluxes with shifts in metabolite levels by means of kinetic modeling. In *Italian Workshop on Artificial Life and Evolutionary Computation* (pp. 138–148). Cham: Springer.

Dannenmann, M., Simon, J., Gasche, R., Holst, J., Naumann, P. S., Kogel, K. I., et al. (2009). Tree girdling provides insights on the role of labile carbon in nitrogen partitioning between soil microorganisms and adult European beech. *Soil Biology and Biochemistry*, *41*, 1622–1631.

Dauwe, R., Holliday, J. A., Aitken, S. N., & Mansfield, S. D. (2012). Metabolic dynamic during cold acclimation within and among population of Sitka spruce (Picea sitchensis). *New Phytologist*, *194*, 192–205.

Delgado-Baquerizo, M., Eldridge, D. J., Maestre, F. T., Ochoa, V., Gozalo, B., Reich, P. B., et al. (2018). Aridity decouples C:N:P stoichiometry across multiple trophic levels in terrestrial ecosystems. *Ecosystems*, *21*(3), 459–468.

Delgado-Baquerizo, M., García-Palacios, P., Milla, R., Gallardo, A., & Maestre, F. T. (2015). Soil characteristics determine soil carbon and nitrogen availability during leaf litter decomposition regardless of litter quality. *Soil Biology and Biochemistry*, *81*, 134–142.

Delgado-Baquerizo, M., Reich, P. B., García-Palacios, P., & Milla, R. (2016). Biogeographic bases for a shift in crop C:N:P stoichiometries during domestication. *Ecology Letters*, *19*(5), 564–575.

Deng, Y., & Lu, S. (2017). Biosynthesis and regulation of phenylpropanoids in plants. *Critical Reviews in Plant Sciences*, *36*(4), 257–290.

Devi, E. L., Kumar, S., Singh, T. B., Sharma, S. K., Beemrote, A., Devi, C. P., et al. (2017). Adaptation strategies and defence mechanisms of plants during environmental stress. In *Medicinal plants and environmental challenges* (pp. 359–413). Cham: Springer.

di Bernardo, D., Thompson, M. J., Gardner, T. S., Chobot, S. E., Eastwood, E. L., Wojtovich, A. P., et al. (2005). Chemogenomic profiling on a genome-wide scale using reverse-engineered gene networks. *Nature Biotechnology*, *23*(3), 377.

Dillon, F. M., Chludil, H. D., & Zavala, J. A. (2017). Solar UV-B radiation modulates chemical defenses against *Anticarsia gemmatalis* larvae in leaves of field-grown soybean. *Phytochemistry*, *141*, 27–36.

Ding, Z., Weissmann, S., Wang, M., Du, B., Huang, L., Wang, L., et al. (2015). Identification of photosynthesis-associated C4 candidate genes through comparative leaf gradient transcriptome in multiple lineages of C3 and C4 species. *PLoS One*, *10* (10)e0140629.

Döring, T. F., Vieweger, A., Pautasso, M., Vaarst, M., Finckh, M. R., & Wolfe, M. S. (2015). Resilience as a universal criterion of health. *Journal of the Science of Food and Agriculture*, *95*(3), 455–465.

Driever, S. M., & Kromdijk, J. (2013). Will C3 crops enhanced with the C4 CO_2-concentrating mechanism live up to their full potential (yield)? *Journal of Experimental Botany*, *64*(13), 3925–3935.

Du, B., Sun, B., Zhang, M., Yang, G., Xing, L., Tang, F., et al. (2016). Atmospheric palaeo-CO_2 estimates based on the carbon isotope and stomatal data of Cheirolepidiaceae from the Lower Cretaceous of the Jiuquan Basin, Gansu Province. *Cretaceous Research*, *62*, 142–153.

Duffy, J. E., Cardinale, B. J., France, K. E., McIntyre, P. B., Thébault, E., & Loreau, M. (2007). The functional role of biodiversity in ecosystems: incorporating trophic complexity. *Ecology Letters*, *10*(6), 522–538.

Duffy, S. J. (2017). *Environmental chemistry: a global perspective*. Oxford University Press.

Dunn, W. B., Broadhurst, D. I., Atherton, H. J., Goodacre, R., & Griffin, J. L. (2011). Systems level studies of mammalian metabolomes: the roles of mass spectrometry and nuclear magnetic resonance spectroscopy. *Chemical Society Reviews*, *40*(1), 387–426.

Dyer, L. A., Richards, L. A., Short, S. A., & Dodson, C. D. (2013). Effects of CO_2 and temperature on tritrophic interactions. *PLoS One*, *8*(4)e62528.

Ekbladh, G., Grönlund, E., Ingemarson, F., Karlsson, L., Nilsson, S., & Strid Eriksson, I. (2002). *Ecosystem properties and principles of living systems as foundation for sustainable agriculture—critical reviews of environmental assessment tools, key findings and questions from a course process*. Uppsala: Centre for Sustainable Agriculture, Swedish University of Agricultural Sciences.

Eksvärd, K., & Rydberg, T. (2010). Integrating participatory learning and action research and systems ecology: a potential for sustainable agriculture transitions. *Systemic Practice and Action Research*, *23*(6), 467–486.

Elad, Y., & Pertot, I. (2014). Climate change impacts on plant pathogens and plant diseases. *Journal of Crop Improvement*, *28*(1), 99–139.

Elgersma, K. J., Yu, S., Vor, T., & Ehrenfeld, J. G. (2012). Microbial-mediated feedbacks of leaf litter on invasive plant growth and interspecific competition. *Plant and Soil*, *356*(1-2), 341–355.

Elmendorf, S. C., Henry, G. H., Hollister, R. D., Björk, R. G., Boulanger-Lapointe, N., Cooper, E. J., et al. (2012). Plot-scale evidence of tundra vegetation change and links to recent summer warming. *Nature Climate Change*, *2*(6), 453.

Elser, J. J., Acharya, K., Kyle, M., Cotner, J., Makino, W., Markow, T., et al. (2003). Growth rate–stoichiometry couplings in diverse biota. *Ecology Letters*, *6*(10), 936–943.

Elser, J. J., Fagan, W. F., Kerkhoff, A. J., Swenson, N. G., & Enquist, B. J. (2010). Biological stoichiometry of plant production: metabolism, scaling and ecological response to global change. *New Phytologist*, *186*(3), 593–608.

Engineer, C. B., Hashimoto-Sugimoto, M., Negi, J., Israelsson-Nordström, M., Azoulay-Shemer, T., Rappel, W. J., et al. (2016). CO_2 sensing and CO_2 regulation of stomatal conductance: advances and open questions. *Trends in Plant Science*, *21*(1), 16–30.

Erga, S. R., Haugen, S. B., Bratbak, G., Egge, J. K., Heldal, M., Mork, K. A., et al. (2017). Seasonal variations in C:N:Si:Ca:P:Mg:S:K:Fe relationships of seston from Norwegian coastal water: impact of extreme offshore forcing during winter-spring 2010. *Marine Chemistry*, *196*, 1–12.

Erxleben, A., Gewssler, A., Vervliet-Scheebaum, M., & Reski, R. (2012). Metabolite profiling of the moss *Physcomitrella patens* reveals evolutionary conservation of osmoprotective substances. *Plant Cell Reports*, *31*, 427–436.

Estiarte, M., Peñuelas, J., López-Martínez, C., & Pérez-Obiol, R. (2008). Holocene palaeoenvironment in a former coastal lagoon of the arid south eastern Iberian Peninsula: salinization effects on $\delta^{15}N$. *Vegetation History and Archaeobotany*, *17*(6), 667.

Fan, H., Wu, J., Liu, W., Yuan, Y., Hu, L., & Cai, Q. (2015). Linkages of plant and soil C:N:P stoichiometry and their relationships to forest growth in subtropical plantations. *Plant and Soil*, *392*(1-2), 127–138.

Fang, H. J., Yu, G. R., Cheng, S. L., Mo, J. M., Yan, J. H., & Li, S. (2009). ^{13}C abundance, water soluble and microbial biomass carbon as potential indicators of soil organic carbon dynamics in subtropical forest at different successional stages and subject to different nitrogen loads. *Plant and Soil*, *320*, 243–254.

Fanin, N., Fromin, N., Buatois, B., & Hättenschwiler, S. (2013). An experimental test of the hypothesis of non-homeostatic consumer stoichiometry in a plant litter–microbe system. *Ecology Letters*, *16*(6), 764–772.

Fanin, N., Hättenschwiler, S., Schimann, H., & Fromin, N. (2015). Interactive effects of C, N and P fertilization on soil microbial community structure and function in an Amazonian rain forest. *Functional Ecology*, *29*(1), 140–150.

Fernández-Martínez, M., Vicca, S., Janssens, I. A., Sardans, J., Luyssaert, S., Campioli, M., et al. (2014). Nutrient availability as the key regulator of global forest carbon balance. *Nature Climate Change*, *4*(6), 471.

Fernie, A. R., Trethewey, R. N., Krotzky, A. J., & Willmitzer, L. (2004). Metabolite profiling: from diagnostics to systems biology. *Nature Reviews Molecular Cell Biology*, *5*(9), 763.

Fester, T. (2015). Plant metabolite profiles and the buffering capacities of ecosystems. *Phytochemistry*, *110*, 6–12.

Fiehn, O., Kopka, J., Dormann, P., Altman, T., Trethewey, R., & Willmitzer, L. (2000). Metabolite profiling for functional genomics. *Nature Biotechnology*, *18*, 1157–1161.

Fine, P. V. (2015). Ecological and evolutionary drivers of geographic variation in species diversity. *Annual Review of Ecology, Evolution, and Systematics*, *46*, 369–392.

Fink, P., & Von Elert, E. (2006). Physiological responses to stoichiometric constraints: nutrient limitation and compensatory feeding in a freshwater snail. *Oikos*, *115*(3), 484–494.

Finzi, A. C., Austin, A. T., Cleland, E. E., Frey, S. D., Houlton, B. Z., & Wallenstein, M. D. (2011). Responses and feedbacks of coupled biogeochemical cycles to climate change: examples from terrestrial ecosystems. *Frontiers in Ecology and the Environment*, *9*(1), 61–67.

Flint, R. W. (2012). *Practice of sustainable community development: a participatory framework for change*. Springer Science & Business Media.

Foley, B., Jones, I. D., Maberly, S. C., & Rippey, B. (2012). Long-term changes in oxygen depletion in a small temperate lake: effects of climate change and eutrophication. *Freshwater Biology*, *57*(2), 278–289.

Folke, C., Carpenter, S., Walker, B., Scheffer, M., Elmqvist, T., Gunderson, L., et al. (2004). Regime shift, resilience and biodiversity in ecosystem management. *Annual Review of Ecology, Evolution, and Systematics*, *35*, 557–581.

Forero-Medina, G., Terborgh, J., Socolar, S. J., & Pimm, S. L. (2011). Elevational ranges of birds on a tropical montane gradient lag behind warming temperatures. *PLoS One*, *6*(12) e28535.

Forsman, A., & Wennersten, L. (2016). Inter-individual variation promotes ecological success of populations and species: evidence from experimental and comparative studies. *Ecography*, *39*(7), 630–648.

Franks, P. J., Doheny-Adams T, W., Britton-Harper, Z. J., & Gray, J. E. (2015). Increasing water-use efficiency directly through genetic manipulation of stomatal density. *New Phytologist*, *207*(1), 188–195.

Franks, S. J., Weber, J. J., & Aitken, S. N. (2014). Evolutionary and plastic responses to climate change in terrestrial plant populations. *Evolutionary Applications*, 7(1), 123–139.

Frigstad, H., Andersen, T., Hessen, D. O., Naustvoll, L. J., Johnsen, T. M., & Bellerby, R. G. (2011). Seasonal variation in marine C:N:P stoichiometry: can the composition of seston explain stable Redfield ratios? *Biogeosciences*, *8*(10), 2917–2933.

Gairola, S., Sharma, J., & Bedi, Y. S. (2014). A cross-cultural analysis of Jammu, Kashmir and Ladakh (India) medicinal plant use. *Journal of Ethnopharmacology*, *155*(2), 925–986.

Galbraith, E. D., & Martiny, A. C. (2015). A simple nutrient-dependence mechanism for predicting the stoichiometry of marine ecosystems. *Proceedings of the National Academy of Sciences*, *112*(27), 8199–8204.

Gamage, D., Thompson, M., Sutherland, M., Hirotsu, N., Makino, A., & Seneweera, S. (2018). New insights into the cellular mechanisms of plant growth at elevated atmospheric carbon dioxide. *Plant, Cell & Environment*, *41*(6), 1233–1246.

Gan, J., Lu, Z., Cheung, A., Dai, M., Liang, L., Harrison, P. J., et al. (2014). Assessing ecosystem response to phosphorus and nitrogen limitation in the Pearl River plume using the regional ocean modeling system (ROMS). *Journal of Geophysical Research, Oceans*, *119* (12), 8858–8877.

Garcia, R. A., Cabeza, M., Rahbek, C., & Araújo, M. B. (2014). Multiple dimensions of climate change and their implications for biodiversity. *Science*, *344*(6183), 1247579.
Gargallo-Garriga, A., Sardans, J., Pérez-Trujillo, M., Oravec, M., Urban, O., Jentsch, K. J., et al. (2015). Warming differentially influences the effects of drought on stoichiometry and metabolomics in shoots and roots. *New Phytologist*, *207*, 591–603.
Gargallo-Garriga, A., Sardans, J., Pérez-Trujillo, M., Guenther, A., Llusià, J., Rico, L., et al. (2016). Shifts in plant foliar and floral metabolomes in response to the suppression of the associated microbiota. *BMC Plant Biology*, *16*(1), 78.
Gargallo-Garriga, A., Wright, S. J., Sardans, J., Pérez-Trujillo, M., Oravec, M., Večeřová, K., et al. (2017). Long-term fertilization determines different metabolomic profiles and responses in saplings of three rainforest tree species with different adult canopy position. *PLoS One*, *12*(5)e0177030.
Genga, A., Mattana, M., Coraggio, I., Locatelli, F., Piffanelli, P., & Consonni, R. (2011). Plant metabolomics: a characterisation of plant responses to abiotic stresses. In *Abiotic stress in plants-mechanisms and adaptations*. InTech.
Giordano, M., Palmucci, M., & Raven, J. A. (2015). Growth rate hypothesis and efficiency of protein synthesis under different sulphate concentrations in two green algae. *Plant, Cell & Environment*, *38*(11), 2313–2317.
Godbold, D., Tullus, A., Kupper, P., Sõber, J., Ostonen, I., Godbold, J. A., et al. (2014). Elevated atmospheric CO_2 and humidity delay leaf fall in *Betula pendula*, but not in *Alnus glutinosa* or *Populus tremula* × *tremuloides*. *Annals of Forest Science*, *71*(8), 831–842.
Gohli, J., & Jordal, B. H. (2017). Explaining biogeographical range size and measuring its effect on species diversification in bark beetles. *Journal of Biogeography*, *44*(9), 2132–2144.
Gonella, E., Pajoro, M., Marzorati, M., Crotti, E., Mandrioli, M., Pontini, M., et al. (2015). Plant-mediated interspecific horizontal transmission of an intracellular symbiont in insects. *Scientific Reports*, *5*, 15811.
Gou, W., Tian, L., Ruan Zh, Z. P., Chen, F., Zhang, L., Cui Zh, Z. P., et al. (2015). Accumulation of choline and glycinebetaine and drought stress tolerance induced in maize (*Zea mays*) by tree plant growth promoting Rhizobacteria (PGPR) strains. *Pakistan Journal of Botany*, *47*(2), 581–586.
Graff, P., Aguiar, M. R., & Chaneton, E. J. (2007). Shifts in positive and negative plant interactions along a grazing intensity gradient. *Ecology*, *88*(1), 188–199.
Grandawa, M. M. (2014, May). Characterisation of physicochemical properties of *Arachis hypogaea* L shells (groundnut) as environmental remidation. In *International Conference on Chemical, Biological and Environmental Sciences (ICCBES'14)* (pp. 12–13).
Greaver, T. L., Clark, C. M., Compton, J. E., Vallano, D., Talhelm, A. F., Weaver, C. P., et al. (2016). Key ecological responses to nitrogen are altered by climate change. *Nature Climate Change*, *6*(9), 836.
Guignard, M. S., Leitch, A. R., Acquisti, C., Eizaguirre, C., Elser, J. J., Hessen, D. O., et al. (2017). Impacts of nitrogen and phosphorus: from genomes to natural ecosystems and agriculture. *Frontiers in Ecology and Evolution*, *5*, 70.
Gulis, V., Kuehn, K. A., Schoettle, L. N., Leach, D., Benstead, J. P., & Rosemond, A. D. (2017). Changes in nutrient stoichiometry, elemental homeostasis and growth rate of aquatic litter-associated fungi in response to inorganic nutrient supply. *The ISME Journal*, *11*(12), 2729.
Guo, F., Kainz, M. J., Sheldon, F., & Bunn, S. E. (2016). The importance of high-quality algal food sources in stream food webs–current status and future perspectives. *Freshwater Biology*, *61*(6), 815–831.
Gusewell, S. (2004). N:P ratios in terrestrial plants: variation and functional significance. *The New Phytologist*, *164*, 243–266.
Guy, C., Kaplan, F., Kopka, J., Selbig, J., & Hincha, D. K. (2008). Metabolomics of temperature. *Physiologia Plantarum*, *132*(2), 220–235.

Häder, D. P., Williamson, C. E., Wängberg, S. Å., Rautio, M., Rose, K. C., Gao, K., et al. (2015). Effects of UV radiation on aquatic ecosystems and interactions with other environmental factors. *Photochemical & Photobiological Sciences*, *14*(1), 108–126.

Hall, E. K., Maixner, F., Franklin, O., Daims, H., Richter, A., & Battin, T. (2011). Linking microbial and ecosystem ecology using ecological stoichiometry: a synthesis of conceptual and empirical approaches. *Ecosystems*, *14*(2), 261–273.

Hao, Z., Gao, Y., & Yang, T. (2017). Seasonal variation of DOM and associated stoichiometry for freshwater ecosystem in the subtropical watershed: indicating the optimal C:N:P ratio. *Ecological Indicators*, *78*, 37–47.

Harrison, M. D. (2016). Nutrient dynamics. In *Encyclopedia of estuaries* (pp. 462–463). Netherlands: Springer.

Hartman, S. E., Humphreys, M. P., Kivimäe, C., Woodward, E. M. S., Kitidis, V., McGrath, T., et al. (2018). Seasonality and spatial heterogeneity of the surface ocean carbonate system in the northwest European continental shelf. *Progress in Oceanography*.

Hasler-Sheetal, H., Castorani, M. C., Glud, R. N., Canfield, D. E., & Holmer, M. (2016). Metabolomics reveals cryptic interactive effects of species interactions and environmental stress on nitrogen and sulfur metabolism in seagrass. *Environmental Science & Technology*, *50*(21), 11602–11609.

Hennion, F., Bouchereau, A., Gauthier, C., Hermant, M., Vernon, P., & Prinzing, A. (2012). Variation in amine composition in plant species: how it integrates macroevolutionary and environmental signals. *American Journal of Botany*, *99*(1), 36–45.

Hessen, D. O., Elser, J. J., Sterner, R. W., & Urabe, J. (2013). Ecological stoichiometry: an elementary approach using basic principles. *Limnology and Oceanography*, *58*(6), 2219–2236.

Hilderbrand, R. H., & Utz, R. M. (2015). Ecological thresholds and resilience in streams. In *Rivers–physical, fluvial and environmental processes* (pp. 461–478). Cham: Springer.

Hillebrand, H., Bennett, D. M., & Cadotte, M. W. (2008). Consequences of dominance: a review of evenness effects on local and regional ecosystem processes. *Ecology*, *89*(6), 1510–1520.

Hillebrand, H., Cowles, J. M., Lewandowska, A., Van de Waal, D. B., & Plum, C. (2014). Think ratio! A stoichiometric view on biodiversity–ecosystem functioning research. *Basic and Applied Ecology*, *15*(6), 465–474.

Hinsinger, P., Betencourt, E., Bernard, L., Brauman, A., Plassard, C., Shen, J., et al. (2011). P for two, sharing a scarce resource: soil phosphorus acquisition in the rhizosphere of intercropped species. *Plant Physiology*, *156*(3), 1078–1086.

Hobbie, S. E., & Villéger, S. (2015). Interactive effects of plants, decomposers, herbivores, and predators on nutrient cycling. In *Trophic ecology: bottom-up and top-down interactions across aquatic and terrestrial systems* (p. 233). Cambridge University Press.

Hofmann, K., Lamprecht, A., Pauli, H., & Illmer, P. (2016). Distribution of prokaryotic abundance and microbial nutrient cycling across a high-alpine altitudinal gradient in the Austrian Central Alps is affected by vegetation, temperature, and soil nutrients. *Microbial Ecology*, *72*(3), 704–716.

Houshmandfar, A., Fitzgerald, G. J., Armstrong, R., Macabuhay, A. A., & Tausz, M. (2015). Modelling stomatal conductance of wheat: an assessment of response relationships under elevated CO_2. *Agricultural and Forest Meteorology*, *214*, 117–123.

Houshmandfar, A., Fitzgerald, G. J., O'Leary, G., Tausz-Posch, S., Fletcher, A., & Tausz, M. (2017). The relationship between transpiration and nutrient uptake in wheat changes under elevated atmospheric CO_2. *Physiologia Plantarum*.

Hu, J. J., Xing, Y. W., Turkington, R., Jacques, F. M., Su, T., Huang, Y. J., et al. (2015). A new positive relationship between p CO2 and stomatal frequency in *Quercus guyavifolia* (Fagaceae): a potential proxy for palaeo-CO_2 levels. *Annals of Botany*, *115*(5), 777–788.

Huang, H., Zhang, J., Meng, P., Fu, Y., Zheng, N., & Gao, J. (2011). Seasonal variation and meteorological control of CO_2 flux in a hilly plantation in the mountain areas of North China. *Acta Meteorologica Sinica*, *25*(2), 238–248.

Huang, J., Gu, M., Lai, Z., Fan, B., Shi, K., Zhou, Y. H., et al. (2010). Functional analysis of the *Arabidopsis* PAL gene family in plant growth, development, and response to environmental stress. *Plant Physiology*, *153*(4), 1526–1538.

Huang, J., Reichelt, M., Chowdhury, S., Hammerbacher, A., & Hartmann, H. (2017). Increasing carbon availability stimulates growth and secondary metabolites via modulation of phytohormones in winter wheat. *Journal of Experimental Botany*, *68*(5), 1251–1263.

Huang, L., Zhang, Y., Shi, Y., Liu, Y., Wang, L., & Yan, N. (2015). Comparison of phosphorus fractions and phosphatase activities in coastal wetland soils along vegetation zones of Yancheng National Nature Reserve, China. *Estuarine, Coastal and Shelf Science*, *157*, 93–98.

Huang, Z., Liu, B., Davis, M., Sardans, J., Peñuelas, J., & Billings, S. (2016). Long-term nitrogen deposition linked to reduced water use efficiency in forests with low phosphorus availability. *New Phytologist*, *210*(2), 431–442.

Huché-Thélier, L., Crespel, L., Le Gourrierec, J., Morel, P., Sakr, S., & Leduc, N. (2016). Light signaling and plant responses to blue and UV radiations—perspectives for applications in horticulture. *Environmental and Experimental Botany*, *121*, 22–38.

Huntingford, C., Smith, D. M., Davies, W. J., Falk, R., Sitch, S., & Mercado, L. M. (2015). Combining the [ABA] and net photosynthesis-based model equations of stomatal conductance. *Ecological Modelling*, *300*, 81–88.

IPCC AR4 WG3 (2007). Climate change 2007: mitigation of climate change. In B. Metz, O. R. Davidson, P. R. Bosch, R. Dave, & L. A. Meyer (Eds.), *Contribution of Working Group III to the fourth assessment report of the intergovernmental panel on climate change*. Cambridge University Press. ISBN 978-0-521-88011-4 pb: 978-0-521-70598-1.

Iriti, M., & Vitalini, S. (2018). Plant metabolomics in the global scenario of food security: a systems-biology approach for sustainable crop production. *International Journal of Molecular Sciences*, *19*(7), 2094.

Jamieson, M. A., Schwartzberg, E. G., Raffa, K. F., Reich, P. B., & Lindroth, R. L. (2015). Experimental climate warming alters aspen and birch phytochemistry and performance traits for an outbreak insect herbivore. *Global Change Biology*, *21*(7), 2698–2710.

Jan, N., Majeed, U., Andrabi, K. I., & John, R. (2018). Cold stress modulates osmolytes and antioxidant system in *Calendula officinalis*. *Acta Physiologiae Plantarum*, *40*(4), 73.

Jan, S., Kamili, A. N., Parray, J. A., & Bedi, Y. S. (2016). Differential response of terpenes and anthraquinones derivatives in *Rumex dentatus* and *Lavandula officinalis* to harsh winters across north-western Himalaya. *Natural Product Research*, *30*(5), 608–612.

Jan, S., Kamili, A. N., Parray, J. A., Bedi, Y. S., & Ahmad, P. (2016). Microclimatic variation in UV perception and related disparity in tropane and quinolizidine alkaloid composition of *Atropa acuminata*, *Lupinus polyphyllus* and *Hyoscyamus niger*. *Journal of Photochemistry and Photobiology B: Biology*, *161*, 230–235.

Jan, S., & Parray, J. A. (2016). Heavy metal stress signalling in plants. In *Approaches to heavy metal tolerance in plants*. Singapore: Springer.

Jeppesen, E., Kronvang, B., Meerhoff, M., Søndergaard, M., Hansen, K. M., Andersen, H. E., et al. (2009). Climate change effects on runoff, catchment phosphorus loading and lake ecological state, and potential adaptations. *Journal of Environmental Quality*, *38*(5), 1930–1941.

Jeppesen, E., Kronvang, B., Olesen, J. E., Audet, J., Søndergaard, M., Hoffmann, C. C., et al. (2011). Climate change effects on nitrogen loading from cultivated catchments in Europe: implications for nitrogen retention, ecological state of lakes and adaptation. *Hydrobiologia*, *663*(1), 1–21.

Jeppesen, E., Meerhoff, M., Holmgren, K., González-Bergonzoni, I., Teixeira-de Mello, F., Declerck, S. A., et al. (2010). Impacts of climate warming on lake fish community structure and potential effects on ecosystem function. *Hydrobiologia*, *646*(1), 73–90.

Jeppesen, E., Trolle, D., Davidson, T. A., Bjerring, R., Søndergaard, M., Johansson, L. S., et al. (2016). Major changes in CO_2 efflux when shallow lakes shift from a turbid to a clear water state. *Hydrobiologia*, *778*(1), 33–44.

Jeyasingh, P. D., & Weider, L. J. (2007). Fundamental links between genes and elements: evolutionary implications of ecological stoichiometry. *Molecular Ecology*, *16*(22), 4649–4661.

Jia, X., Wang, W., Chen, Z., He, Y., & Liu, J. (2014). Concentrations of secondary metabolites in tissues and root exudates of wheat seedlings changed under elevated atmospheric CO_2 and cadmium-contaminated soils. *Environmental and Experimental Botany*, *107*, 134–143.

Johnson, C. H., Ivanisevic, J., & Siuzdak, G. (2016). Metabolomics: beyond biomarkers and towards mechanisms. *Nature Reviews Molecular Cell Biology*, *17*(7), 451.

Jones, J., & Brett, M. T. (2014). Lake nutrients, eutrophication, and climate change. In *Global environmental change* (pp. 273–279). Netherlands: Springer.

Julian, P., Gerber, S., King, J., & Wright, A. L. (2017). Nutrient homeostasis and mechanisms related to nutrient retention by wetland macrophytes in a subtropical wetland. *bioRxiv*. https://doi.org/10.1101/221465.

Kabbadj, A., Makoudi, B., Mouradi, M., Pauly, N., Frendo, P., & Ghoulam, C. (2017). Physiological and biochemical responses involved in water deficit tolerance of nitrogen-fixing *Vicia faba*. *PLoS One*, *12*(12)e0190284.

Kachroo, A., & Kachroo, P. (2009). Fatty acid–derived signals in plant defense. *Annual Review of Phytopathology*, *47*, 153–176.

Kaling, M., Kanawati, B., Ghirardo, A., Albert, A., Winkler, J. B., Heller, W., et al. (2015). UV-B mediated metabolic rearrangements in poplar revealed by non-targeted metabolomics. *Plant, Cell & Environment*, *38*(5), 892–904.

Kataria, S., Baghel, L., & Guruprasad, K. N. (2017). Alleviation of adverse effects of ambient UV stress on growth and some potential physiological attributes in soybean (*Glycine max*) by seed pre-treatment with static magnetic field. *Journal of Plant Growth Regulation*, *36*(3), 550–565.

Kattge, J., Ogle, K., Bönisch, G., Díaz, S., Lavorel, S., Madin, J., et al. (2011). A generic structure for plant trait databases. *Methods in Ecology and Evolution*, *2*(2), 202–213.

Kéfi, S., Holmgren, M., & Scheffer, M. (2016). When can positive interactions cause alternative stable states in ecosystems? *Functional Ecology*, *30*(1), 88–97.

Kelly, A. E., & Goulden, M. L. (2008). Rapid shifts in plant distribution with recent climate change. *Proceedings of the National Academy of Sciences*, *105*(33), 11823–11826.

Kensler, T. W., Wakabayashi, N., & Biswal, S. (2007). Cell survival responses to environmental stresses via the Keap1-Nrf2-ARE pathway. *Annual Review of Pharmacology and Toxicology*, *47*, 89–116.

Khan, N., Bano, A., Rahman, M. A., Rathinasabapathi, B., & Babar, M. A. (2018). UPLC-HRMS based untargeted metabolic profiling reveals changes in chickpea (*Cicer arietinum*) metabolome following long-term drought stress. *Plant, Cell & Environment*.

Kilian, J., Peschke, F., Berendzen, K. W., Harter, K., & Wanke, D. (2012). Prerequisites, performance and profits of transcriptional profiling the abiotic stress response. *Biochimica et Biophysica Acta (BBA)-Gene Regulatory Mechanisms*, *1819*(2), 166–175.

Killi, D., Bussotti, F., Raschi, A., & Haworth, M. (2017). Adaptation to high temperature mitigates the impact of water deficit during combined heat and drought stress in C3 sunflower and C4 maize varieties with contrasting drought tolerance. *Physiologia Plantarum*, *159*(2), 130–147.

Kim, M. S., Baek, S. H., Park, S. U., Im, K. H., & Kim, J. K. (2017). Targeted metabolite profiling to evaluate unintended metabolic changes of genetic modification in resveratrol-enriched rice (Oryza sativa L.). *Applied Biological Chemistry, 60*(2), 205–214.
Kissoudis, C., van de Wiel, C., Visser, R. G., & van der Linden, G. (2014). Enhancing crop resilience to combined abiotic and biotic stress through the dissection of physiological and molecular crosstalk. *Frontiers in Plant Science, 5*, 207.
Knops, J. M., Wedin, D. A., & Naeem, S. (2010). The role of litter quality feedbacks in terrestrial nitrogen and phosphorus cycling. *Open Ecology Journal, 3*, 14–25.
Kohler, T., Wehrli, A., & Jurek, M. (Eds.), (2014). *Mountains and climate change: A global concern. Sustainable Mountain Development Series* (p. 136). Bern, Switzerland: Centre for Development and Environment (CDE), Swiss Agency for Development and Cooperation (SDC) and Geographica Bernensia.
Körner, C., Jetz, W., Paulsen, J., Payne, D., Rudmann-Maurer, K., & Spehn, E. M. (2017). A global inventory of mountains for bio-geographical applications. *Alpine Botany, 127* (1), 1–15.
Kou, L., Chen, W., Jiang, L., Dai, X., Fu, X., Wang, H., et al. (2018). Simulated nitrogen deposition affects stoichiometry of multiple elements in resource-acquiring plant organs in a seasonally dry subtropical forest. *Science of the Total Environment, 624*, 611–620.
Krugman, T., Peleg, Z., Quansah, L., et al. (2011). Alteration in expression of hormonerelated genes in wild emmer wheat roots associated with drought adaptation mechanisms. *Functional & Integrative Genomics, 11*, 565–583.
Kulkarni, A., Patwardhan, S., Kumar, K. K., Ashok, K., & Krishnan, R. (2013). Projected climate change in the Hindu Kush–Himalayan region by using the high-resolution regional climate model PRECIS. *Mountain Research and Development, 33*(2), 142–151.
Kumar, P. (2012). Assessment of impact of climate change on Rhododendrons in Sikkim Himalayas using Maxent modelling: limitations and challenges. *Biodiversity and Conservation, 21*(5), 1251–1266.
Kumar, V., Sharma, A., Soni, J. K., & Pawar, N. (2017). Physiological response of C3, C4 and CAM plants in changeable climate. *The Pharma Innovation, 6*(9, Part B), 70.
Kumari, R., Prasad, M. N. V., & Agrawal, S. B. (2015). Growth and defense metabolism of plants exposed to ultraviolet-B radiation. In *Sustainable agriculture reviews* (pp. 263–305). Cham: Springer.
Kurepin, L. V., Stangl, Z. R., Ivanov, A. G., Bui, V., Mema, M., Hüner, N., et al. (2018). Contrasting acclimation abilities of two dominant boreal conifers to elevated CO_2 and temperature. *Plant, Cell & Environment.*
La Pierre, K. J., & Smith, M. D. (2016). Soil nutrient additions increase invertebrate herbivore abundances, but not herbivory, across three grassland systems. *Oecologia, 180*(2), 485–497.
Lake, J. A., & Woodward, F. I. (2008). Response of stomatal numbers to CO_2 and humidity: control by transpiration rate and abscisic acid. *New Phytologist, 179*(2), 397–404.
Lämke, J. S., & Unsicker, S. B. (2018). Phytochemical variation in treetops: causes and consequences for tree-insect herbivore interactions. *Oecologia*, 1–12.
Larive, C. K., Barding, G. A., Jr., & Dinges, M. M. (2014). NMR spectroscopy for metabolomics and metabolic profiling. *Analytical Chemistry, 87*(1), 133–146.
Lawrence, P. G. (2015). *Resilience of Montana's agroecosystems to economic and climatic change.* Doctoral dissertationMontana State University-Bozeman, College of Agriculture.
Leal, M. C., Seehausen, O., & Matthews, B. (2017). The ecology and evolution of stoichiometric phenotypes. *Trends in Ecology & Evolution, 32*(2), 108–117.
Lee, J. E., Lee, B. J., Chung, J. O., Hwang, J. A., Lee, S. J., Lee, C. H., et al. (2010). Geographical and climatic dependencies of green tea (*Camellia sinensis*) metabolites: a ^1H NMR-based metabolomics study. *Journal of Agricultural and Food Chemistry, 58*(19), 10582–10589.

Lee, M. R., Flory, S. L., & Phillips, R. P. (2012). Positive feedbacks to growth of an invasive grass through alteration of nitrogen cycling. *Oecologia, 170*(2), 457–465.

Lejeusne, C., Chevaldonné, P., Pergent-Martini, C., Boudouresque, C. F., & Perez, T. (2010). Climate change effects on a miniature ocean: the highly diverse, highly impacted Mediterranean Sea. *Trends in Ecology & Evolution, 25*(4), 250–260.

Leuzinger, S., Luo, Y., Beier, C., Dieleman, W., Vicca, S., & Körner, C. (2011). Do global change experiments overestimate impacts on terrestrial ecosystems? *Trends in Ecology & Evolution, 26*(5), 236–241.

Li, X., Jiang, D., & Liu, F. (2016). Soil warming enhances the hidden shift of elemental stoichiometry by elevated CO_2 in wheat. *Scientific Reports, 6*, 23313.

Li, Z., Song, Y., Li, L., Wang, L., Liu, B., Parajulee, M. N., et al. (2018). Influence of elevated CO_2 on development and food utilization of target armyworm *Mythimna separata* fed on transgenic Bt maize infected by nitrogen-fixing bacteria. *bioRxiv*. https://doi.org/10.1101/292565.

Liao, C., Peng, R., Luo, Y., Zhou, X., Wu, X., Fang, C., et al. (2008). Altered ecosystem carbon and nitrogen cycles by plant invasion: a meta-analysis. *New Phytologist, 177*(3), 706–714.

Liu, F., Liu, Y., Wang, G., Song, Y., Liu, Q., Li, D., et al. (2015). Seasonal variations of C:N: P stoichiometry and their trade-offs in different organs of *Suaeda salsa* in Coastal Wetland of Yellow River Delta, China. *PLoS One, 10*(9)e0138169.

Liu, L., Shen, F., Xin, C., & Wang, Z. (2016). Multi-scale modeling of *Arabidopsis thaliana* response to different CO_2 conditions: from gene expression to metabolic flux. *Journal of Integrative Plant Biology, 58*(1), 2–11.

Loha, S. K., Lowa, S. T., & Mohamada, M. S. (2015). A review of software for predicting gene function. *International Journal of Bio-Science and Bio-Technology*, 7(2), 57–70.

Longphuirt, S. N., O'Boyle, S., Wilkes, R., Dabrowski, T., & Stengel, D. B. (2016). Influence of hydrological regime in determining the response of macroalgal blooms to nutrient loading in two Irish estuaries. *Estuaries and Coasts, 39*(2), 478–494.

Lurgi, M., Montoya, D., & Montoya, J. M. (2016). The effects of space and diversity of interaction types on the stability of complex ecological networks. *Theoretical Ecology, 9*(1), 3–13.

Ma, L., & Upadhyaya, M. K. (2018). Effects of leaf position on reflectance, transmittance and absorption of red and far-red light in tomato, *Chenopodium album* and *Amaranthus retroflexus* leaves. *Weed Research, 58*(1), 17–24.

Madronich, S., Shao, M., Wilson, S. R., Solomon, K. R., Longstreth, J. D., & Tang, X. Y. (2015). Changes in air quality and tropospheric composition due to depletion of stratospheric ozone and interactions with changing climate: implications for human and environmental health. *Photochemical & Photobiological Sciences, 14*(1), 149–169.

Manzoni, S., Taylor, P., Richter, A., Porporato, A., & Ågren, G. I. (2012). Environmental and stoichiometric controls on microbial carbon-use efficiency in soils. *New Phytologist, 196*(1), 79–91.

Margolin, A. A., Wang, K., Lim, W. K., Kustagi, M., Nemenman, I., & Califano, A. (2006). Reverse engineering cellular networks. *Nature Protocols, 1*(2), 662.

Marino, R., & Howarth, R. W. (2016). Why is planktonic nitrogen fixation so rare in coastal marine ecosystems? Insights from a cross-systems approach. In *Aquatic microbial ecology and biogeochemistry: a dual perspective* (pp. 127–139). Cham: Springer.

Marklein, A. R., & Houlton, B. Z. (2012). Nitrogen inputs accelerate phosphorus cycling rates across a wide variety of terrestrial ecosystems. *New Phytologist, 193*(3), 696–704.

Marleau, J. N., Guichard, F., & Loreau, M. (2015). Emergence of nutrient co-limitation through movement in stoichiometric meta-ecosystems. *Ecology Letters, 18*(11), 1163–1173.

Martin, R., Pohlers, S., Mühlschlegel, F. A., & Kurzai, O. (2017). CO_2 sensing in fungi: at the heart of metabolic signaling. *Current Genetics*, *63*(6), 965–972.
Martinez, M. (2016). Computational tools for genomic studies in plants. *Current Genomics*, *17* (6), 509–514.
Martz, F., Vuosku, J., Ovaskainen, A., Stark, S., & Rautio, P. (2016). The snow must go on: ground ice encasement, snow compaction and absence of snow differently cause soil hypoxia, CO_2 accumulation and tree seedling damage in boreal forest. *PLoS One*, *11* (6), e0156620.
Matzek, V., & Vitousek, P. M. (2009). N:P stoichiometry and protein: RNA ratios in vascular plants: an evaluation of the growth-rate hypothesis. *Ecology Letters*, *12*(8), 765–771.
McCormack, M. L., & Guo, D. (2014). Impacts of environmental factors on fine root lifespan. *Frontiers in Plant Science*, *5*, 205.
McGrath, J. M., & Lobell, D. B. (2013). Reduction of transpiration and altered nutrient allocation contribute to nutrient decline of crops grown in elevated CO_2 concentrations. *Plant, Cell & Environment*, *36*, 697–705.
McLauchlan, K. K., Williams, J. J., & Engstrom, D. R. (2013). Nutrient cycling in the palaeorecord: fluxes from terrestrial to aquatic ecosystems. *The Holocene*, *23*(11), 1635–1643.
McPherson, J. A. N. A., Jetz, W., & Rogers, D. J. (2004). The effects of species' range sizes on the accuracy of distribution models: ecological phenomenon or statistical artefact? *Journal of Applied Ecology*, *41*(5), 811–823.
Medlyn, B. E., De Kauwe, M. G., & Duursma, R. A. (2016). New developments in the effort to model ecosystems under water stress. *New Phytologist*, *212*(1), 5–7.
Meena, K. K., Sorty, A. M., Bitla, U. M., Choudhary, K., Gupta, P., Pareek, A., et al. (2017). Abiotic stress responses and microbe-mediated mitigation in plants: the omics strategies. *Frontiers in Plant Science*, *8*, 172.
Menon, S. S., Uppal, M., Randhawa, S., Cheema, M. S., Aghdam, N., Usala, R. L., et al. (2016). Radiation metabolomics: current status and future directions. *Frontiers in Oncology*, *6*, 20.
Meunier, C. L., Boersma, M., El-Sabaawi, R., Halvorson, H. M., Herstoff, E. M., Van de Waal, D. B., et al. (2017). From elements to function: toward unifying ecological stoichiometry and trait-based ecology. *Frontiers in Environmental Science*, *5*, 18.
Meza-Canales, I. D., Meldau, S., Zavala, J. A., & Baldwin, I. T. (2017). Herbivore perception decreases photosynthetic carbon assimilation and reduces stomatal conductance by engaging 12-oxo-phytodienoic acid, mitogen-activated protein kinase 4 and cytokinin perception. *Plant, Cell & Environment*, *40*(7), 1039–1056.
Mhamdi, A., & Noctor, G. (2016). High CO_2 primes plant biotic stress defences through redox-linked pathways. *Plant Physiology*, *172*(2), 929–942.
Mikkelsen, B. L., Jørgensen, R. B., & Lyngkjær, M. F. (2015). Complex interplay of future climate levels of CO_2, ozone and temperature on susceptibility to fungal diseases in barley. *Plant Pathology*, *64*(2), 319–327.
Millett, J., Foot, G. W., Thompson, J. C., & Svensson, B. M. (2017). Geographic variation in Sundew (Drosera) leaf colour: plant–plant interactions counteract expected effects of abiotic factors. *Journal of Biogeography*, *45*(3), 582–592.
Mishra, G. P., Stobdan, T., Thangaraj, P., Seth, T., & Singh, B. (2018). Prospects of local flora of trans-Himalayan region of Ladakh for various medicinal uses. In *Medicinal Plants: Promising Future for Health and New Drugs* (p. 239). CRC Press.
Mittermeier, R. A., Turner, W. R., Larsen, F. W., Brooks, T. M., & Gascon, C. (2011). Global biodiversity conservation: the critical role of hotspots. In *Biodiversity hotspots* (pp. 3–22). Berlin, Heidelberg: Springer.
Miyagi, A., Takahara, K., Kasajima, I., Takahashi, H., Kawai-Yamada, M., & Uchimiya, H. (2011). Fate of ^{13}C in metabolic pathways and effects of high CO_2 on the alteration of metabolites in *Rumex obtusifolius* L. *Metabolomics*, 7, 524–535.

Monastersky, R. (2013). Global carbon dioxide levels near worrisome milestone. *Nature, 497* (7447), 13.
Moore, C. M., Mills, M. M., Arrigo, K. R., Berman-Frank, I., Bopp, L., Boyd, P. W., et al. (2013). Processes and patterns of oceanic nutrient limitation. *Nature Geoscience, 6*(9), 701.
Moreira, X., Abdala-Roberts, L., Parra-Tabla, V., & Mooney, K. A. (2015). Latitudinal variation in herbivory: influences of climatic drivers, herbivore identity and natural enemies. *Oikos, 124*(11), 1444–1452.
Moss, B. (2012). Cogs in the endless machine: lakes, climate change and nutrient cycles: a review. *Science of the Total Environment, 434*, 130–142.
Moss, B., Kosten, S., Meerhoff, M., Battarbee, R. W., Jeppesen, E., Mazzeo, N., et al. (2011). Allied attack: climate change and eutrophication. *Inland Waters, 1*(2), 101–105.
Musavi, T., Migliavacca, M., Reichstein, M., Kattge, J., Wirth, C., Black, T. A., et al. (2017). Stand age and species richness dampen interannual variation of ecosystem-level photosynthetic capacity. *Nature Ecology & Evolution, 1*(2), 0048.
Ncube, B., Finnie, J. F., & Van Staden, J. (2012). Quality from the field: the impact of environmental factors as quality determinants in medicinal plants. *South African Journal of Botany, 82*, 11–20.
Nelson, G. C., Valin, H., Sands, R. D., Havlík, P., Ahammad, H., Deryng, D., et al. (2014). Climate change effects on agriculture: economic responses to biophysical shocks. *Proceedings of the National Academy of Sciences, 111*(9), 3274–3279.
Neugart, S., Fiol, M., Schreiner, M., Rohn, S., Zrenner, R., Kroh, L. W., et al. (2014). Interaction of moderate UV-B exposure and temperature on the formation of structurally different flavonol glycosides and hydroxycinnamic acid derivatives in kale (*Brassica oleracea* var. *sabellica*). *Journal of Agricultural and Food Chemistry, 62*(18), 4054–4062.
Newbold, T., Hudson, L. N., Hill, S. L., Contu, S., Lysenko, I., Senior, R. A., et al. (2015). Global effects of land use on local terrestrial biodiversity. *Nature, 520*(7545), 45.
Ngatia, L. W., Turner, B. L., Njoka, J. T., Young, T. P., & Reddy, K. R. (2015). The effects of herbivory and nutrients on plant biomass and carbon storage in Vertisols of an East African savanna. *Agriculture, Ecosystems & Environment, 208*, 55–63.
Nicotra, A. B., Atkin, O. K., Bonser, S. P., Davidson, A. M., Finnegan, E. J., Mathesius, U., et al. (2010). Plant phenotypic plasticity in a changing climate. *Trends in Plant Science, 15* (12), 684–692.
Nippert, J. B., Fay, P. A., Carlisle, J. D., Knapp, A. K., & Smith, M. D. (2009). Ecophysiological responses of two dominant grasses to altered temperature and precipitation regimes. *Acta Oecologica, 35*(3), 400–408.
Niu, S., Classen, A. T., Dukes, J. S., Kardol, P., Liu, L., Luo, Y., et al. (2016). Global patterns and substrate-based mechanisms of the terrestrial nitrogen cycle. *Ecology Letters, 19*(6), 697–709.
Noctor, G. (2006). Metabolic signalling in defence and stress: the central roles of soluble redox couples. *Plant, Cell & Environment, 29*(3), 409–425.
Noctor, G., & Mhamdi, A. (2017). Climate change, CO_2, and defense: the metabolic, redox, and signaling perspectives. *Trends in Plant Science, 22*(10), 857–870.
Nybakken, L., Sandvik, S. M., & Klanderud, K. (2011). Experimental warming had little effect on carbon-based secondary compounds, carbon and nitrogen in selected alpine plants and lichens. *Environmental and Experimental Botany, 72*(3), 368–376.
Okazaki, Y., & Saito, K. (2012). Recent advances of metabolomics in plant biotechnology. *Plant Biotechnology Reports, 6*(1), 1–15.
Oksanen, E., Keski-Saari, S., Kontunen-Soppela, S., & Keinänen, M. (2013). Metabolomics and transcriptomics increase our understanding about defence responses and genotypic differences of northern deciduous trees to elevating ozone, CO_2 and climate warming. *Climate Change, Air Pollution and Global Challenges*. Vol. 13. Elsevier Chapter 15.

Ossipov, V., Ossipova, S., Bykov, V., Oksanen, E., Koricheva, J., & Haukioja, E. (2008). Application of metabolomics to genotype and phenotype discrimination of birch trees grown in a long term open field experiment. *Metabolomics*, *4*, 39–51.
Paerl, H. W., & Paul, V. J. (2012). Climate change: links to global expansion of harmful cyanobacteria. *Water Research*, *46*(5), 1349–1363.
Pandey, P., Irulappan, V., Bagavathiannan, M. V., & Senthil-Kumar, M. (2017). Impact of combined abiotic and biotic stresses on plant growth and avenues for crop improvement by exploiting physio-morphological traits. *Frontiers in Plant Science*, *8*, 537.
Pandey, R., Kumar, B., Meena, B., Srivastava, M., Mishra, T., Tiwari, V., et al. (2017). Major bioactive phenolics in Bergenia species from the Indian Himalayan region: Method development, validation and quantitative estimation using UHPLC-QqQLIT-MS/MS. *PLoS One*, *12*(7)e0180950.
Parmesan, C., & Hanley, M. E. (2015). Plants and climate change: complexities and surprises. *Annals of Botany*, *116*(6), 849–864.
Patra, B., Schluttenhofer, C., Wu, Y., & Pattanaik, S. (2013). Transcriptional regulation of secondary metabolite biosynthesis in plants. *Biochimica et Biophysica Acta (BBA)-Gene Regulatory Mechanisms*, *1829*(11), 1236–1247.
Pauli, H., Gottfried, M., & Grabherr, G. (2014). Effects of climate change on the alpine and nival vegetation of the Alps. *Journal of Mountain Ecology*, 7.
Pauli, H., Gottfried, M., Dullinger, S., Abdaladze, O., Akhalkatsi, M., Alonso, J. L. B., et al. (2012). Recent plant diversity changes on Europe's mountain summits. *Science*, *336* (6079), 353–355.
Pedreschi, R., & Lurie, S. (2015). Advances and current challenges in understanding postharvest abiotic stresses in perishables. *Postharvest Biology and Technology*, *107*, 77–89.
Peng, H., Chen, Y., Yan, Z., & Han, W. (2016). Stage-dependent stoichiometric homeostasis and responses of nutrient resorption in *Amaranthus mangostanus* to nitrogen and phosphorus addition. *Scientific Reports*, *6*, 37219.
Peñuelas, J., & Sardans, J. (2009). Ecology: elementary factors. *Nature*, *460*(7257), 803.
Peñuelas, J., Sardans, J., Estiarte, M., Ogaya, R., Carnicer, J., Coll, M., et al. (2013). Evidence of current impact of climate change on life: a walk from genes to the biosphere. *Global Change Biology*, *19*(8), 2303–2338.
Peñuelas, J., Sardans, J., Filella, I., Estiarte, M., Llusià, J., Ogaya, R., et al. (2018). Assessment of the impacts of climate change on Mediterranean terrestrial ecosystems based on data from field experiments and long-term monitored field gradients in Catalonia. *Environmental and Experimental Botany*, *152*, 49–59.
Peñuelas, J., Sardans, J., Rivas-Ubach, A., & Janssens, I. A. (2012). The human-induced imbalance between C, N and P in Earth's life system. *Global Change Biology*, *18*(1), 3–6.
Pepin, N., Bradley, R. S., Diaz, H. F., Baraër, M., Caceres, E. B., Forsythe, N., et al. (2015). Elevation-dependent warming in mountain regions of the world. *Nature Climate Change*, *5*(5), 424.
Persson, J., Fink, P., Goto, A., Hood, J. M., Jonas, J., & Kato, S. (2010). To be or not to be what you eat: regulation of stoichiometric homeostasis among autotrophs and heterotrophs. *Oikos*, *119*(5), 741–751.
Pessarakli, M. (2016). *Handbook of plant and crop stress*. CRC Press.
Peterman, W., Waring, R. H., Seager, T., & Pollock, W. L. (2013). Soil properties affect pinyon pine—Juniper response to drought. *Ecohydrology*, *6*, 455–463.
Peters, K., Worrich, A., Weinhold, A., Alka, O., Balcke, G., Birkemeyer, C., et al. (2018). Current challenges in plant eco-metabolomics. *International Journal of Molecular Sciences*, *19*(5), 1385.
Petitpierre, B., McDougall, K., Seipel, T., Broennimann, O., Guisan, A., & Kueffer, C. (2016). Will climate change increase the risk of plant invasions into mountains? *Ecological Applications*, *26*(2), 530–544.

Philippot, L., Raaijmakers, J. M., Lemanceau, P., & Van Der Putten, W. H. (2013). Going back to the roots: the microbial ecology of the rhizosphere. *Nature Reviews Microbiology*, *11*(11), 789.

Poff, N. L., & Zimmerman, J. K. (2010). Ecological responses to altered flow regimes: a literature review to inform the science and management of environmental flows. *Freshwater Biology*, *55*(1), 194–205.

Poorter, H., Niklas, K. J., Reich, P. B., Oleksyn, J., Poot, P., & Mommer, L. (2012). Biomass allocation to leaves, stems and roots: meta-analyses of interspecific variation and environmental control. *New Phytologist*, *193*(1), 30–50.

Portmann, R. W., Daniel, J. S., & Ravishankara, A. R. (2012). Stratospheric ozone depletion due to nitrous oxide: influences of other gases. *Philosophical Transactions of the Royal Society of London B*, *367*(1593), 1256–1264.

Prater, C., Frost, P. C., Howell, E. T., Watson, S. B., Zastepa, A., King, S. S., et al. (2017). Variation in particulate C:N:P stoichiometry across the Lake Erie watershed from tributaries to its outflow. *Limnology and Oceanography*, *62*(S1).

Prieto, P., Peñuelas, J., Ogaya, R., & Estiarte, M. (2008). Precipitation-dependent flowering of *Globularia alypum* and *Erica multiflora* in Mediterranean shrubland under experimental drought and warming, and its inter-annual variability. *Annals of Botany*, *102*(2), 275–285.

Pujo-Pay, M., Conan, P., Oriol, L., Cornet-Barthaux, V., Falco, C., Ghiglione, J. F., et al. (2011). Integrated survey of elemental stoichiometry (C, N, P) from the western to eastern Mediterranean Sea. *Biogeosciences*, *8*(4), 883–899.

Pureswaran, D. S., De Grandpré, L., Paré, D., Taylor, A., Barrette, M., Morin, H., et al. (2015). Climate-induced changes in host tree–insect phenology may drive ecological state-shift in boreal forests. *Ecology*, *96*(6), 1480–1491.

Rahman, M. M., Tsukamoto, J., Tokumoto, Y., Shuvo, M., & Ashikur, R. (2013). The role of quantitative traits of leaf litter on decomposition and nutrient cycling of the forest ecosystems. *Journal of Forest and Environmental Science*, *29*(1), 38–48.

Raja, V., Majeed, U., Kang, H., Andrabi, K. I., & John, R. (2017). Abiotic stress: interplay between ROS, hormones and MAPKs. *Environmental and Experimental Botany*, *137*, 142–157.

Rajashekar, C. B. (2018). Elevated CO_2 levels affect phytochemicals and nutritional quality of food crops. *American Journal of Plant Sciences*, *9*(02), 150.

Rajkumar, M., Prasad, M. N. V., Swaminathan, S., & Freitas, H. (2013). Climate change driven plant–metal–microbe interactions. *Environment International*, *53*, 74–86.

Rasmann, S., Pellissier, L., Defossez, E., Jactel, H., & Kunstler, G. (2014). Climate-driven change in plant–insect interactions along elevation gradients. *Functional Ecology*, *28*(1), 46–54.

Reddy, A. R., Rasineni, G. K., & Raghavendra, A. S. (2010). The impact of global elevated CO_2 concentration on photosynthesis and plant productivity. *Current Science*, *99*(1), 46–57.

Rejeb, I. B., Pastor, V., & Mauch-Mani, B. (2014). Plant responses to simultaneous biotic and abiotic stress: molecular mechanisms. *Plants*, *3*(4), 458–475.

Richardson, A. E., Barea, J. M., McNeill, A. M., & Prigent-Combaret, C. (2009). Acquisition of phosphorus and nitrogen in the rhizosphere and plant growth promotion by microorganisms. *Plant and Soil*, *321*(1-2), 305–339.

Richardson, A. E., Lynch, J. P., Ryan, P. R., Delhaize, E., Smith, F. A., Smith, S. E., et al. (2011). Plant and microbial strategies to improve the phosphorus efficiency of agriculture. *Plant and Soil*, *349*(1-2), 121–156.

Rico, A., Jones, R., & Preston, G. M. (2009). Adaptation to the plant apoplast by plant pathogenic bacteria. In *Plant pathogenic bacteria: genomics and molecular biology* (pp. 63–89). Caister Academic Press.

Rivas-Ubach, A., Peñuelas i Reixach, J., & Sardans, J. (2014). *Stoichiometric and metabolomic shifts of organisms under environmental changes.* (Ph.D. thesis)Autonomous University of Barcelona.

Rivas-Ubach, A., Sardans, J., Pérez-Trujillo, M., Estiarte, M., & Peñuelas, J. (2012). Strong relationship between elemental stoichiometry and metabolome in plants. *Proceedings of the National Academy of Sciences, 109*(11), 4181–4186.

Robinson, A. R., Ukrainetz, N. K., Kang, K. Y., & Mansfield, S. D. (2007). Metabolite profiling of Douglas-fir (Pseudotsuga menziesii) field trials reveals strong environmental and weak genetic variation. *New Phytologist, 174*, 762–773.

Robinson, E. A., Ryan, G. D., & Newman, J. A. (2012). A meta-analytical review of the effects of elevated CO_2 on plant–arthropod interactions highlights the importance of interacting environmental and biological variables. *New Phytologist, 194*(2), 321–336.

Rogiers, S. Y., Hardie, W. J., & Smith, J. P. (2011). Stomatal density of grapevine leaves (*Vitis vinifera* L.) responds to soil temperature and atmospheric carbon dioxide. *Australian Journal of Grape and Wine Research, 17*(2), 147–152.

Rosenzweig, C., Elliott, J., Deryng, D., Ruane, A. C., Müller, C., Arneth, A., et al. (2014). Assessing agricultural risks of climate change in the 21st century in a global gridded crop model intercomparison. *Proceedings of the National Academy of Sciences, 111*(9), 3268–3273.

Rzanny, M., Kuu, A., & Voigt, W. (2013). Bottom–up and top–down forces structuring consumer communities in an experimental grassland. *Oikos, 122*(7), 967–976.

Sah, A. N. (2013). *Pharmacognostic phytochemical and pharmacological evaluation of Citrus medica Linn.* (Ph.D. thesis)Nainital, Uttarakhand: Department of Pharmaceutical Sciences Faculty of Technology, Kumaun University.

Sampaio, B. L., Edrada-Ebel, R., & Da Costa, F. B. (2016). Effect of the environment on the secondary metabolic profile of *Tithonia diversifolia*: a model for environmental metabolomics of plants. *Scientific Reports, 6*, 29265.

Sanchez, D. H., Schwabe, F., Erban, A., Udvardi, M. K., & Kopka, J. (2012). Comparative metabolomics of drought acclimation in model and forage legumes. *Plant, Cell & Environment, 35*(136), 149.

Sánchez-Rodríguez, E., Ruiz, J. M., Ferreres, F., & Moreno, D. A. (2012). Phenolic profiles of cherry tomatoes as influenced by hydric stress and rootstock technique. *Food Chemistry, 134*(2), 775–782.

Sandhu, H., & Sandhu, S. (2014). Linking ecosystem services with the constituents of human well-being for poverty alleviation in eastern Himalayas. *Ecological Economics, 107*, 65–75.

Sardans, J., Peñuelas, J., & Rivas-Ubach, A. (2011). Ecological metabolomics: overview of current developments and future challenges. *Chemoecology, 21*(4), 191–225.

Sardans, J., & Peñuelas, J. (2014). Climate and taxonomy underlie different elemental concentrations and stoichiometries of forest species: the optimum "biogeochemical niche" *Plant Ecology, 215*(4), 441–455.

Sardans, J., & Peñuelas, J. (2015). Potassium: a neglected nutrient in global change. *Global Ecology and Biogeography, 24*(3), 261–275.

Sardans, J., Peñuelas, J., Coll, M., Vayreda, J., & Rivas-Ubach, A. (2012). Stoichiometry of potassium is largely determined by water availability and growth in Catalonian forests. *Functional Ecology, 26*(5), 1077–1089.

Sardans, J., Peñuelas, J., Prieto, P., & Estiarte, M. (2008). Changes in Ca, Fe, Mg, Mo, Na, and S content in a Mediterranean shrubland under warming and drought. *Journal of Geophysical Research, Biogeosciences, 113*(G3).

Sardans, J., Rivas-Ubach, A., & Peñuelas, J. (2012a). The C:N:P stoichiometry of organisms and ecosystems in a changing world: a review and perspectives. *Perspectives in Plant Ecology, Evolution and Systematics, 14*(1), 33–47.

Sardans, J., Rivas-Ubach, A., & Peñuelas, J. (2012b). The elemental stoichiometry of aquatic and terrestrial ecosystems and its relationships with organismic lifestyle and ecosystem structure and function: a review and perspectives. *Biogeochemistry*, *111*(1-3), 1–39.
Sardans, J., Rivas-Ubach, A., Estiarte, M., Ogaya, R., & Peñuelas, J. (2013). Field-simulated droughts affect elemental leaf stoichiometry in Mediterranean forests and shrublands. *Acta Oecologica*, *50*, 20–31.
Sasaki, T., & Lauenroth, W. K. (2011). Dominant species, rather than diversity, regulates temporal stability of plant communities. *Oecologia*, *166*(3), 761–768.
Savage, J. A., Clearwater, M. J., Haines, D. F., Klein, T., Mencuccini, M., Sevanto, S., et al. (2016). Allocation, stress tolerance and carbon transport in plants: how does phloem physiology affect plant ecology? *Plant, Cell & Environment*, *39*(4), 709–725.
Saxena, R. K., Cui, X., Thakur, V., Walter, B., Close, T. J., & Varshney, R. K. (2011). Single feature polymosphisms (SFPs) for drought tolerance in pigeonpea. *Functional & Integrative Genomics*, *11*, 651–657.
Scharler, U. M., Ulanowicz, R. E., Fogel, M. L., Wooller, M. J., Jacobson-Meyers, M. E., Lovelock, C. E., et al. (2015). Variable nutrient stoichiometry (carbon: nitrogen: phosphorus) across trophic levels determines community and ecosystem properties in an oligotrophic mangrove system. *Oecologia*, *179*(3), 863–876.
Scheffer, M., Bascompte, J., Brock, W. A., Brovkin, V., Carpenter, S. R., Dakos, V., et al. (2009). Early-warning signals for critical transitions. *Nature*, *461*, 53–59.
Scherber, C., Eisenhauer, N., Weisser, W. W., Schmid, B., Voigt, W., Fischer, M., et al. (2010). Bottom-up effects of plant diversity on multitrophic interactions in a biodiversity experiment. *Nature*, *468*(7323), 553.
Schwahn, K., Beleggia, R., Omranian, N., & Nikoloski, Z. (2017). Stoichiometric correlation analysis: principles of metabolic functionality from metabolomics data. *Frontiers in Plant Science*, *8*, 2152.
Sehring, J. (2010). Framework conditions and challenges. In *Integrative and sustainability-oriented water management: cooperation potential between Germany and Central Asia* (p. 69).
Seinfeld, J. H., & Pandis, S. N. (2016). *Atmospheric chemistry and physics: from air pollution to climate change*. John Wiley & Sons.
Sharma, E., Chettri, N., & Oli, K. P. (2010). Mountain biodiversity conservation and management: a paradigm shift in policies and practices in the Hindu Kush-Himalayas. *Ecological Research*, *25*(5), 909–923.
Shen, J., Schoepfer, S. D., Feng, Q., Zhou, L., Yu, J., Song, H., et al. (2015). Marine productivity changes during the end-Permian crisis and Early Triassic recovery. *Earth-Science Reviews*, *149*, 136–162.
Shibata, H., Branquinho, C., McDowell, W. H., Mitchell, M. J., Monteith, D. T., Tang, J., et al. (2015). Consequence of altered nitrogen cycles in the coupled human and ecological system under changing climate: The need for long-term and site-based research. *Ambio*, *44*(3), 178–193.
Shinozaki, K., & Yamaguchi-Shinozaki, K. (2007). Gene networks involved in drought stress response and tolerance. *Journal of Experimental Botany*, *58*(2), 221–227.
Shrestha, U. B., Gautam, S., & Bawa, K. S. (2012). Widespread climate change in the Himalayas and associated changes in local ecosystems. *PLoS One*, 7(5)e36741.
Sierla, M., Waszczak, C., Vahisalu, T., & Kangasjärvi, J. (2016). Reactive oxygen species in the regulation of stomatal movements. *Plant Physiology*, *171*(3), 1569–1580.
Singh, A., & Chauhan, P. S. (2017). Ecological significance of soil-associated plant growth–promoting biofilm-forming microbes for stress management. In *Biofilms in plant and soil health* (p. 291). Wiley.
Singh, A., Baer, S. E., Riebesell, U., Martiny, A. C., & Lomas, M. W. (2015). C:N:P stoichiometry at the Bermuda Atlantic Time-series Study station in the North Atlantic Ocean. *Biogeosciences*, *12*(21), 6389–6403.

Singh, A., Jethva, M., Singla-Pareek, S. L., Pareek, A., & Kushwaha, H. R. (2016). Analyses of old "prokaryotic" proteins indicate functional diversification in *Arabidopsis* and *Oryza sativa*. *Frontiers in Plant Science*, 7, 304.

Singh, S. P., Bassignana-Khadka, I., Singh Karky, B., & Sharma, E. (2011). *Climate change in the Hindu Kush-Himalayas: the state of current knowledge*. International Centre for Integrated Mountain Development (ICIMOD).

Sinha, A. K., Jaggi, M., Raghuram, B., & Tuteja, N. (2011). Mitogen-activated protein kinase signaling in plants under abiotic stress. *Plant Signaling & Behavior*, *6*(2), 196–203.

Sistla, S. A., Appling, A. P., Lewandowska, A. M., Taylor, B. N., & Wolf, A. A. (2015). Stoichiometric flexibility in response to fertilization along gradients of environmental and organismal nutrient richness. *Oikos*, *124*(7), 949–959.

Sistla, S. A., & Schimel, J. P. (2012). Stoichiometric flexibility as a regulator of carbon and nutrient cycling in terrestrial ecosystems under change. *New Phytologist*, *196*(1), 68–78.

Sitters, J., Atkinson, C. L., Guelzow, N., Kelly, P., & Sullivan, L. L. (2015). Spatial stoichiometry: cross-ecosystem material flows and their impact on recipient ecosystems and organisms. *Oikos*, *124*(7), 920–930.

Slama, I., Abdelly, C., Bouchereau, A., Flowers, T., & Savouré, A. (2015). Diversity, distribution and roles of osmoprotective compounds accumulated in halophytes under abiotic stress. *Annals of Botany*, *115*(3), 433–447.

Smith, B., Warlind, D., Arneth, A., Hickler, T., Leadley, P., Siltberg, J., et al. (2014). Implications of incorporating N cycling and N limitations on primary production in an individual-based dynamic vegetation model. *Biogeosciences*, *11*, 2027–2054.

Smith, N. G. (2018). Plant respiration responses to elevated CO_2: an overview from cellular processes to global impacts. *Plant respiration: metabolic fluxes and carbon balance* (p. 69). Vol. 43. (p. 69). Springer.

Snart, C. J., Hardy, I. C., & Barrett, D. A. (2015). Entometabolomics: applications of modern analytical techniques to insect studies. *Entomologia Experimentalis et Applicata*, *155*(1), 1–17.

Somero, G. N. (2012). The physiology of global change: linking patterns to mechanisms. *Annual Review of Marine Science*, *4*, 39–61.

Sommaruga, R. (2015). When glaciers and ice sheets melt: consequences for planktonic organisms. *Journal of Plankton Research*, *37*(3), 509–518.

Song, G., Li, Y., Zhang, J., Li, M., Hou, J., & He, N. (2016). Significant phylogenetic signal and climate-related trends in leaf caloric value from tropical to cold-temperate forests. *Scientific Reports*, *6*, 36674.

Song, Y., Yao, Q., Wang, G., Yang, X., Pan, C., Johnston, E., et al. (2016). Integrating "omics" data into a biogeochemical model: a new model scheme to predict climate feedbacks from microbial function in tropical ecosystems. *AGU Fall Meeting Abstracts*.

Sperfeld, E., Wagner, N. D., Halvorson, H. M., Malishev, M., & Raubenheimer, D. (2017). Bridging ecological stoichiometry and nutritional geometry with homeostasis concepts and integrative models of organism nutrition. *Functional Ecology*, *31*(2), 286–296.

Spieb, N., Oufir, M., Matusikova, I., et al. (2012). Ecophysiological and transcriptomic responses of oak (*Quercus robur*) to long-term drought exposure and rewatering. *Environmental and Experimental Botany*, 77, 117–126.

Sproull, G. J., Quigley, M. F., Sher, A., & González, E. (2015). Long-term changes in composition, diversity and distribution patterns in four herbaceous plant communities along an elevational gradient. *Journal of Vegetation Science*, *26*(3), 552–563.

Srinivasarao, C., Kundu, S., Shanker, A. K., Naik, R. P., Vanaja, M., Venkanna, K., et al. (2016). Continuous cropping under elevated CO_2: differential effects on C4 and C3 crops, soil properties and carbon dynamics in semi-arid alfisols. *Agriculture, Ecosystems & Environment*, *218*, 73–86.

Srivastava, A., Kowalski, G. M., Callahan, D. L., Meikle, P. J., & Creek, D. J. (2016). Strategies for extending metabolomics studies with stable isotope labelling and fluxomics. *Metabolites*, *6*(4), 32.

Steffen, W., Persson, Å., Deutsch, L., Zalasiewicz, J., Williams, M., Richardson, K., et al. (2011). The Anthropocene: from global change to planetary stewardship. *Ambio*, *40* (7), 739.

Steinfath, M., Repsilber, D., Scholz, M., Walther, D., & Selbig, J. (2007). Integrated data analysis for genome-wide research. In *Plant systems biology* (pp. 309–329). Basel: Birkhäuser.

Sterner, R. W., & Elser, J. J. (2002). *Ecological stoichiometry: the biology of elements from molecules to the biosphere*. Princeton University Press.

Stott, P. A., Gillett, N. P., Hegerl, G. C., Karoly, D. J., Stone, D. A., Zhang, X., et al. (2010). Detection and attribution of climate change: a regional perspective. *Wiley Interdisciplinary Reviews: Climate Change*, *1*(2), 192–211.

Su, B., Pahlow, M., Wagner, H., & Oschlies, A. (2015). What prevents nitrogen depletion in the oxygen minimum zone of the eastern tropical South Pacific? *Biogeosciences (BG)*, *12*, 1113–1130.

Sun, C., Gao, X., Fu, J., Zhou, J., & Wu, X. (2015). Metabolic response of maize (Zea mays L.) plants to combined drought and salt stress. *Plant and Soil*, *388*(1-2), 99–117.

Suseela, V., & Tharayil, N. (2018). Decoupling the direct and indirect effects of climate on plant litter decomposition: accounting for stress-induced modifications in plant chemistry. *Global Change Biology*, *24*(4), 1428–1451.

Talarmin, A., Lomas, M. W., Bozec, Y., Savoye, N., Frigstad, H., Karl, D. M., et al. (2016). Seasonal and long-term changes in elemental concentrations and ratios of marine particulate organic matter. *Global Biogeochemical Cycles*, *30*(11), 1699–1711.

Tatarko, A. R., & Knops, J. M. (2018). Nitrogen addition and ecosystem functioning: both species abundances and traits alter community structure and function. *Ecosphere*, *9*(1).

Taub, D. R., & Wang, X. (2008). Why are nitrogen concentrations in plant tissues lower under elevated CO_2? A critical examination of the hypotheses. *Journal of Integrative Plant Biology*, *50*(11), 1365–1374.

Taylor, P. G., & Townsend, A. R. (2012). Stoichiometric control of organic carbon–nitrate relationships from soils to the sea. *Nature*, *464*(7292), 1178.

Telwala, Y., Brook, B. W., Manish, K., & Pandit, M. K. (2013). Climate-induced elevational range shifts and increase in plant species richness in a Himalayan biodiversity epicentre. *PLoS One*, *8*(2)e57103.

Temme, A. A., Liu, J. C., Cornwell, W. K., Cornelissen, J. H., & Aerts, R. (2015). Winners always win: growth of a wide range of plant species from low to future high CO_2. *Ecology and Evolution*, *5*(21), 4949–4961.

Thakur, M. P., Milcu, A., Manning, P., Niklaus, P. A., Roscher, C., Power, S., et al. (2015). Plant diversity drives soil microbial biomass carbon in grasslands irrespective of global environmental change factors. *Global Change Biology*, *21*(11), 4076–4085.

Theocharis, A., Clément, C., & Barka, E. A. (2012). Physiological and molecular changes in plants grown at low temperatures. *Planta*, *235*(6), 1091–1105.

Thomson, L. J., Macfadyen, S., & Hoffmann, A. A. (2010). Predicting the effects of climate change on natural enemies of agricultural pests. *Biological Control*, *52*(3), 296–306.

Tipping, E., Somerville, C. J., & Luster, J. (2016). The C:N:P:S stoichiometry of soil organic matter. *Biogeochemistry*, *130*(1-2), 117–131.

Tiwari, S. P., Kumar, P., Yadav, D., & Chauhan, D. K. (2013). Comparative morphological, epidermal, and anatomical studies of *Pinus roxburghii* needles at different altitudes in the North-West Indian Himalayas. *Turkish Journal of Botany*, *37*(1), 65–73.

Tobita, H., Yazaki, K., Harayama, H., & Kitao, M. (2016). Responses of symbiotic N_2 fixation in Alnus species to the projected elevated CO_2 environment. *Trees*, *30*(2), 523–537.

Torossian, J. L., Kordas, R. L., & Helmuth, B. (2016). Cross-scale approaches to forecasting biogeographic responses to climate change. *Advances in Ecological Research* (pp. 371–433). Vol. 55 (pp. 371–433). Academic Press.
Tranquillini, W. (2012). *Physiological ecology of the alpine timberline: tree existence at high altitudes with special reference to the European Alps.* (Vol. 31). Springer Science & Business Media.
Tripathi, S. N., & Raghubanshi, A. S. (2013). Seedling growth of five tropical dry forest tree species in relation to light and nitrogen gradients. *Journal of Plant Ecology*, 7(3), 250–263.
Trivedi, P., Anderson, I. C., & Singh, B. K. (2013). Microbial modulators of soil carbon storage: integrating genomic and metabolic knowledge for global prediction. *Trends in Microbiology*, *21*(12), 641–651.
Trolle, D., Hamilton, D. P., Pilditch, C. A., Duggan, I. C., & Jeppesen, E. (2011). Predicting the effects of climate change on trophic status of three morphologically varying lakes: implications for lake restoration and management. *Environmental Modelling & Software*, *26*(4), 354–370.
Trowbridge, A. M., Asensio, D., Eller, A. S., Way, D. A., Wilkinson, M. J., Schnitzler, J. P., et al. (2012). Contribution of various carbon sources toward isoprene biosynthesis in poplar leaves mediated by altered atmospheric CO_2 concentrations. *PLoS One*, 7(2) e32387.
Turkington, R. (2009). Top-down and bottom-up forces in mammalian herbivore–vegetation systems: an essay review. *Botany*, *87*(8), 723–739.
Tylianakis, J. M., Didham, R. K., Bascompte, J., & Wardle, D. A. (2008). Global change and species interactions in terrestrial ecosystems. *Ecology Letters*, *11*(12), 1351–1363.
Upchurch, R. G. (2008). Fatty acid unsaturation, mobilization, and regulation in the response of plants to stress. *Biotechnology Letters*, *30*(6), 967–977.
Urabe, J., Shimizu, Y., & Yamaguchi, T. (2018). Understanding the stoichiometric limitation of herbivore growth: the importance of feeding and assimilation flexibilities. *Ecology Letters*, *21*(2), 197–206.
Urban, M. C. (2015). Accelerating extinction risk from climate change. *Science*, *348*(6234), 571–573.
Uzuner, U., Sağlam, A., & Kadıoğlu, A. (2017). Salicylic acid: molecular basis of stress resistance in plants. In *Salicylic acid: a multifaceted hormone* (pp. 163–199). Singapore: Springer.
Valares Masa, C., Alias Gallego, J. C., Chaves Lobon, N., & Sosa Diaz, T. (2016). Intra-population variation of secondary metabolites in Cistus ladanifer L. *Molecules*, *21*(7), 945.
Valegård, K., Hasse, D., Andersson, I., & Gunn, L. H. (2018). Structure of Rubisco from *Arabidopsis thaliana* in complex with 2-carboxyarabinitol-1, 5-bisphosphate. *Acta Crystallographica Section D: Structural Biology*, 74(1).
van Dam, N. M., & Bouwmeester, H. J. (2016). Metabolomics in the rhizosphere: tapping into belowground chemical communication. *Trends in Plant Science*, *21*(3), 256–265.
Van Der Werff, J. (2011). *Spatial and temporal dynamics of phytoplankton nutrient status in the deep chlorophyll layer (DCL) of Lake Superior.* Master's ThesesUniversity of Minnesota Twin Cities.
Veneklaas, E. J., Lambers, H., Bragg, J., Finnegan, P. M., Lovelock, C. E., Plaxton, W. C., et al. (2012). Opportunities for improving phosphorus-use efficiency in crop plants. *New Phytologist*, *195*(2), 306–320.
Verdaguer, D., Díaz-Guerra, L., Font, J., González, J. A., & Llorens, L. (2018). Contrasting seasonal morphological and physio-biochemical responses to UV radiation and reduced rainfall of two mature naturally growing Mediterranean shrubs in the context of climate change. *Environmental and Experimental Botany*, *147*, 189–201.
Verdaguer, D., Jansen, M. A., Llorens, L., Morales, L. O., & Neugart, S. (2017). UV-A radiation effects on higher plants: exploring the known unknown. *Plant Science*, *255*, 72–81.

Verma, A., Kumar, S., Kumar, G., Saini, J. K., Agrawal, R., Satlewal, A., et al. (2018). Rhizosphere metabolite profiling: an opportunity to understand plant-microbe interactions for crop improvement. In *Crop improvement through microbial biotechnology* (pp. 343–361).
Verschoor, A. M., Van Dijk, M. A., Huisman, J. E. F., & Van Donk, E. (2013). Elevated CO2 concentrations affect the elemental stoichiometry and species composition of an experimental phytoplankton community. *Freshwater Biology*, *58*(3), 597–611.
Verspagen, J. M., Van de Waal, D. B., Finke, J. F., Visser, P. M., & Huisman, J. (2014). Contrasting effects of rising CO_2 on primary production and ecological stoichiometry at different nutrient levels. *Ecology Letters*, *17*(8), 951–960.
Vogel, A., Fester, T., Eisenhauer, N., Scherer-Lorenzen, M., Schmid, B., Weisser, W. W., et al. (2013). Separating drought effects from roof artefacts on ecosystem processes in a grassland drought experiment. *PLoS One*, *8*, e70997.
Wakeel, A., Gul, M., & Zörb, C. (2016). Potassium for sustainable agriculture. In *Soil science: agricultural and environmental prospectives* (pp. 159–182). Cham: Springer.
Wallace, S., Fleming, A., Wellman, C. H., & Beerling, D. J. (2011). Evolutionary development of the plant spore and pollen wall. *AoB Plants*, *2011*.
Wallenstein, M. D., Hess, A. M., Lewis, M. R., Steltzer, H., & Ayers, E. (2010). Decomposition of aspen leaf litter results in unique metabolomes when decomposed under different trees species. *Soil Biology and Biochemistry*, *42*, 484–490.
Wang, K., Jiang, H., Zhang, X., & Zhou, G. (2011). Analysis of spatial and temporal variations of carbon dioxide over China using SCIAMACHY satellite observations during 2003–2005. *International Journal of Remote Sensing*, *32*(3), 815–832.
Wang, M., Xie, B., Fu, Y., Dong, C., Hui, L., Guanghui, L., et al. (2015). Effects of different elevated CO_2 concentrations on chlorophyll contents, gas exchange, water use efficiency, and PSII activity on C3 and C4 cereal crops in a closed artificial ecosystem. *Photosynthesis Research*, *126*(2-3), 351–362.
Wang, W. Q., Sardans, J., Wang, C., Zeng, C. S., Tong, C., Asensio, D., et al. (2015). Ecological stoichiometry of C, N, and P of invasive *Phragmites australis* and native *Cyperus malaccensis* species in the Minjiang River tidal estuarine wetlands of China. *Plant Ecology*, *216*(6), 809–822.
Wargent, J. J., Nelson, B. C. W., McGhie, T. K., & Barnes, P. W. (2015). Acclimation to UV-B radiation and visible light in *Lactuca sativa* involves up-regulation of photosynthetic performance and orchestration of metabolome-wide responses. *Plant, Cell & Environment*, *38*(5), 929–940.
Warren, C. R., Aranda, I., & Cano, F. J. (2012). Metabolomic demonstrates divergent responses of two Eucalyptus species to water stress. *Metabolomics*, *8*, 186–200.
Webb, K. M., Freeman, C., & Broeckling, C. D. (2016). Metabolome profiling to understand the defense response of sugar beet (Beta vulgaris) to Rhizoctonia solani AG 2-2 IIIB. *Physiological and Molecular Plant Pathology*, *94*, 108–117.
Weber, E. (2017). *Invasive plant species of the world: a reference guide to environmental weeds*. CABI.
Wei, Y. D., & Ye, X. (2014). Urbanization, urban land expansion and environmental change in China. *Stochastic Environmental Research and Risk Assessment*, *28*(4), 757–765.
Weisany, W., Raei, Y., & Allahverdipoor, K. H. (2013). Role of some of mineral nutrients in biological nitrogen fixation. *Bulletin of Environment, Pharmacology and Life Sciences*, *2*(4), 77–84.
Werner, C., & Gessler, A. (2011). Diel variations in the carbon isotope composition of respired CO(2) and associated carbon sources: a review of dynamics and mechanisms. *Biogeosciences*, *8*(9).
Whipple, K. X. (2009). The influence of climate on the tectonic evolution of mountain belts. *Nature Geoscience*, *2*(2), 97.
Wirtz, K. W., & Kerimoglu, O. (2016). Autotrophic stoichiometry emerging from optimality and variable co-limitation. *Frontiers in Ecology and Evolution*, *4*, 131.

Wolfender, J. L., Marti, G., Thomas, A., & Bertrand, S. (2015). Current approaches and challenges for the metabolite profiling of complex natural extracts. *Journal of Chromatography A, 1382*, 136–164.

Xia, X. J., Zhou, Y. H., Shi, K., Zhou, J., Foyer, C. H., & Yu, J. Q. (2015). Interplay between reactive oxygen species and hormones in the control of plant development and stress tolerance. *Journal of Experimental Botany, 66*(10), 2839–2856.

Xiong, Y., Zeng, H., Xia, H., & Guo, D. (2014). Interactions between leaf litter and soil organic matter on carbon and nitrogen mineralization in six forest litter-soil systems. *Plant and Soil, 379*(1-2), 217–229.

Xu, J., & Zhang, S. (2015). Mitogen-activated protein kinase cascades in signaling plant growth and development. *Trends in Plant Science, 20*(1), 56–64.

Xu, J., Grumbine, R. E., Shrestha, A., Eriksson, M., Yang, X., Wang, Y. U. N., et al. (2009). The melting Himalayas: cascading effects of climate change on water, biodiversity, and livelihoods. *Conservation Biology, 23*(3), 520–530.

Xu, X., Shi, Z., Li, D., Zhou, X., Sherry, R. A., & Luo, Y. (2015). Plant community structure regulates responses of prairie soil respiration to decadal experimental warming. *Global Change Biology, 21*(10), 3846–3853.

Xu, Z., Jiang, Y., Jia, B., & Zhou, G. (2016). Elevated-CO_2 response of stomata and its dependence on environmental factors. *Frontiers in Plant Science, 7*, 657.

Yang, C. Q., Fang, X., Wu, X. M., Mao, Y. B., Wang, I. J., & Chen, X. Y. (2012). Transcriptional regulation of plant secondary metabolism. *Journal of Integrative Plant Biology, 54*, 703–712.

Yergeau, E., Bokhorst, S., Kang, S., Zhou, J., Greer, C. W., Aerts, R., et al. (2012). Shifts in soil microorganisms in response to warming are consistent across a range of Antarctic environments. *The ISME Journal, 6*(3), 692.

Yu, B., Liu, Y., Pan, Y., Liu, J., Wang, H., & Tang, Z. (2018). Light enhanced the biosynthesis of terpenoid indole alkaloids to meet the opening of cotyledons in process of photomorphogenesis of *Catharanthus roseus*. *Plant Growth Regulation, 84*(3), 617–626.

Yu, J., Chen, L., Xu, M., & Huang, B. (2012). Effects of elevated CO 2 on physiological responses of tall fescue to elevated temperature, drought stress, and the combined stresses. *Crop Science, 52*(4), 1848–1858.

Yu, Q., Chen, Q., Elser, J. J., He, N., Wu, H., Zhang, G., et al. (2010). Linking stoichiometric homoeostasis with ecosystem structure, functioning and stability. *Ecology Letters, 13*(11), 1390–1399.

Yuan, Z. Y., & Chen, H. Y. (2015). Decoupling of nitrogen and phosphorus in terrestrial plants associated with global changes. *Nature Climate Change, 5*(5), 465.

Yue, K., Peng, Y., Peng, C., Yang, W., Peng, X., & Wu, F. (2016). Stimulation of terrestrial ecosystem carbon storage by nitrogen addition: a meta-analysis. *Scientific Reports, 6*, 19895.

Zamboni, N., Fendt, S. M., Rühl, M., & Sauer, U. (2009). 13 C-based metabolic flux analysis. *Nature Protocols, 4*(6), 878.

Zechmeister-Boltenstern, S., Keiblinger, K. M., Mooshammer, M., Peñuelas, J., Richter, A., Sardans, J., et al. (2015). The application of ecological stoichiometry to plant–microbial–soil organic matter transformations. *Ecological Monographs, 85*(2), 133–155.

Zepp, R. G., Erickson Iii, D. J., Paul, N. D., & Sulzberger, B. (2011). Effects of solar UV radiation and climate change on biogeochemical cycling: interactions and feedbacks. *Photochemical & Photobiological Sciences, 10*(2), 261–279.

Zhang, H., Wu, H., Yu, Q., Wang, Z., Wei, C., Long, M., et al. (2013). Sampling date, leaf age and root size: implications for the study of plant C:N:P stoichiometry. *PLoS One, 8* (4)e60360.

Zhang, J., & Elser, J. J. (2017). Carbon:nitrogen:phosphorus stoichiometry in fungi: a meta-analysis. *Frontiers in Microbiology, 8*, 1281.

Zhang, J., Zhao, N., Liu, C., Yang, H., Li, M., Yu, G., et al. (2018). C:N:P stoichiometry in China's forests: from organs to ecosystems. *Functional Ecology*, *32*(1), 50–60.
Zhang, Z., Xue, Z., Lyu, X., Tong, S., & Jiang, M. (2017). Scaling of soil carbon, nitrogen, phosphorus and C:N:P ratio patterns in peatlands of China. *Chinese Geographical Science*, *27*(4), 507–515.
Zhao, Y., Zhao, C., Lu, X., Zhou, H., Li, Y., Zhou, J., et al. (2013). Investigation of the relationship between the metabolic profile of tobacco leaves in different planting regions and climate factors using a pseudotargeted method based on gas chromatography/mass spectrometry. *Journal of Proteome Research*, *12*, 5072–5083.
Zhao, M., & Running, S. W. (2010). Drought-induced reduction in global terrestrial net primary production from 2000 through 2009. *Science*, *329*(5994), 940–943.
Zhao, N., He, N., Wang, Q., Zhang, X., Wang, R., Xu, Z., et al. (2014). The altitudinal patterns of leaf C : N : P stoichiometry are regulated by plant growth form, climate and soil on Changbai Mountain, China. *PLoS One*, *9*(4)e95196.
Züst, T., & Agrawal, A. A. (2017). Trade-offs between plant growth and defense against insect herbivory: an emerging mechanistic synthesis. *Annual Review of Plant Biology*, *68*, 513–534.

FURTHER READING

Brodersen, K. E., Siboni, N., Nielsen, D. A., Pernice, M., Ralph, P. J., Seymour, J., et al. (2018). Seagrass rhizosphere microenvironment alters plant associated microbial community composition. *Environmental Microbiology*.
Cleveland, C. C., & Liptzin, D. (2007). C:N:P stoichiometry in soil: is there a "Redfield ratio" for the microbial biomass? *Biogeochemistry*, *85*(3), 235–252.
Grytnes, J. A., Kapfer, J., Jurasinski, G., Birks, H. H., Henriksen, H., Klanderud, K., et al. (2014). Identifying the driving factors behind observed elevational range shifts on European mountains. *Global Ecology and Biogeography*, *23*(8), 876–884.
Heuck, C., & Spohn, M. (2016). Carbon, nitrogen and phosphorus net mineralization in organic horizons of temperate forests: stoichiometry and relations to organic matter quality. *Biogeochemistry*, *131*(1-2), 229–242.
Scherling, C., Roscher, C., Giavalisco, P., Schulze, E. D., & Weckwerth, W. (2010). Metabolomics unravel contrasting effects of biodiversity on the performance of individual plant species. *PLoS One*, *5*, e12569.
Smith, V. H., & Schindler, D. W. (2009). Eutrophication science: where do we go from here? *Trends in Ecology & Evolution*, *24*(4), 201–207.
Williams, A. P., Allen, C. D., Macalady, A. K., Griffin, D., Woodhouse, C. A., Meko, D. M., et al. (2013). Temperature as a potent driver of regional forest drought stress and tree mortality. *Nature Climate Change*, *3*(3), 292.
Zhu, L., Xiao, H., Wang, Y., Jian, X., Zhang, Z., Zhang, H., et al. (2015). Physiological responses of macroalga *Gracilaria lemaneiformis* (Rhodophyta) to UV-B radiation exposure. *Chinese Journal of Oceanology and Limnology*, *33*(2), 389–399.
Zu Castell, W., & Ernst, D. (2012). Experimental 'omics' data in tree research: facing complexity. *Trees*, *26*(6), 1723–1735.

CHAPTER 5

Instrumentation Applied to Metabolomic Analysis

5.1 INTRODUCTION

Metabolomics is a discipline that investigates metabolites bearing a molecular weight of 50–2000 Da (Amantonico, Urban, & Zenobi, 2010; Lindon, Nicholson, & Holmes, 2011). It is an emerging field that evaluates diverse metabolic processes by analyzing the metabolomic interrogation of individual cells, tissues, organs, and biofluids from organisms or ecological backgrounds (Zenobi, 2013; Zhang, Sun, Wang, Han, & Wang, 2012). This discipline has led to a comprehension of not only organisms at the individual levels, but has offered a significant link for interpreting interactions between organisms and their environment (Bundy, Davey, & Viant, 2009; Fiehn, 2001). Metabolomics has emerged as an advanced method for deciphering perspectives pertaining to metabolism and functions of systems biology, as demonstrated in diverse fields (Putri et al., 2013; Wishart, 2016). Metabolomics has further broadened its scope in genomics, epigenomics, proteomics, and transcriptomics, which has offered us a more lucid and explicit explanation underlying organism function and their reactions to their changing environment (Horgan & Kenny, 2011; Yang et al., 2014). Further, these results can be used for disease diagnostics, improvement in the quality of agricultural produce, and the discovery of new and advanced therapeutics and pharmaceuticals (Nadella, Marla, & Kumar, 2012; Puchades-Carrasco & Pineda-Lucena, 2015). In a nutshell, metabolomics has encompassed sectors spanning from healthcare to ecology. Metabolomic research carried out during 2017 included the unraveling of mechanisms underlying pain (Patti et al., 2012), reproduction (Deepinder, Chowdary, & Agarwal, 2007), autophagy regulation (Masclaux-Daubresse et al., 2014), breath and sweat analysis in humans (Calderón-Santiago et al., 2015), and coronary atherosclerosis (Griffin, Atherton, Shockcor, & Atzori, 2011). Metabolomics encompasses analysis from single cells to the whole planet (Rubakhin, Lanni, & Sweedler, 2013), plant metabolomics (Hall & de Maagd, 2014; Hong, Yang, Zhang, & Shi, 2016; Schauer & Fernie, 2006), and

Ecometabolomics
https://doi.org/10.1016/B978-0-12-814872-3.00005-9

exposome-scale metabolomics, among others (Athersuch, 2016; Warth et al., 2017). With the advent of advanced mass spectrometry and nuclear magnetic resonance (NMR), there has been an increasing expansion of metabolomic studies each year. Novel high-throughput platforms, hyphenated technologies, competitive efficacy, and methods for data acquisition in metabolomics have resulted in more analysis of metabolomics data, preprocessing data, and interpretation and integration (Marshall & Powers, 2017; Pan & Raftery, 2007). Further, existing tools can be complemented with informatics, statistics, and computational biology, which provides innovative methods for comprehending metabolomics in a more lucid manner. This chapter intends to offer a concise review of metabolomic tools, software, and databases that were advanced during 2016–2017, thereby allowing readers and researchers to comprehend the systematic inventory of assets. Recently, a new tool was developed by Galaxy workflow, which integrates diverse platforms and data types (Boekel et al., 2015; Giacomoni, Fillatre, et al., 2014a; Giacomoni, Le Corguillé, et al., 2014b). Along with this, came the discovery of next-generation technology, which is comprised of capillary electrophoresis mass spectrometry (CE-MS), which looks promising as well. Numerous research scientists have also provided an explicit set of protocols for metabolomics and lipidomics via traveling-wave ion mobility mass spectrometry (IMS) (Paglia & Astarita, 2017). Various broad and comprehensive reviews have been detailed to include the prospects of metabolomic research, and evaluate its potential as a prime tool for chemical analysis (Wishart, 2008a, 2008b). Spicer et al. recently discovered freely accessible software and GitHub resource comprised of varied resources for metabolomic analysis (Grapov, Wanichthanarak, & Fiehn, 2015). More advanced databases, for example, MINE DB, were used for epimetabolite annotation (Lai, Kind, & Fiehn, 2017). Developments in analytical and hypothetic techniques are rationalized in numerous online resources, for example, metabolomic tools catalogues, and the following: OMIC tools http://omictools.com/ (Henry, Bandrowski, Pepin, Gonzalez, & Desfeux, 2014), Fiehn Lab resources http://fiehnlab.ucdavis.edu/, and http://metabolomics.ucdavis.edu/Downloads. Additional resources include: the Metabolomics Society's resource pages: http://metabolomicssociety.org/resources/metabolomics-databases, metabomatch: http://www.metabomatch.com/softwares/ (catalog), and software repositories such as the Comprehensive R Archives Network (CRAN): https://cran.rproject.org/web/packages/available_packages_by_name.html, Bioconductor https://www.bioconductor.org/, and GitHub https://github.com/. In addition to the classification of generalized contrivances, an outline of their delineated characteristics and main uses is also valuable. Based on the existing literature, we have summarized principal contrivances and data information in metabolomics into two main segments (1) spectroscopic approaches, and (2) mass spectroscopic approaches. These approaches are further classified as (3) GC-MS, liquid chromatography mass spectrometry (LC-MS), imaging mass spectrometry, (4) NMR tools, (5) untargeted data analysis, (6) ion mobility MS, (7) tools for isotope techniques, (8) data preprocessing tools, (9) annotation tools, (10) multifunctional tools, and (11) miscellaneous tools of interest.

5.2 SPECTROSCOPIC APPROACHES

Spectroscopic approaches are based on the absorbance or reorientation of electromagnetic radiation via chemical bonds. This can lead to the emission of radiation, or transition of the samples to higher energy states. Spectroscopic methods provide inclusive metabolic fingerprints of samples with instant spectral acquisition, thereby allowing high throughput evaluation and the spatial estimation of metabolites.

5.2.1 FT-IR Spectroscopy

FT-IR spectroscopy is a well-recognized, and constantly advancing analytical approach that facilitates the rapid (seconds per sample), high-throughput (1000s of samples per day), and non-exhaustive evaluation of a unique and extensive range of sample types. The main step in this method includes exposing the sample to an infrared beam, which allows the absorption of infrared radiation by functional groups that vibrate, when bent, stretched, or deformed (Aharoni et al., 2002; Hollywood, Brison, & Goodacre, 2006; Nicolaou, Xu, & Goodacre, 2010). These vibrations are associated with corresponding biochemical species, and a consequential spectrum can thus be illustrated as an infrared or metabolic fingerprint of chemical substance (Hameed, Ibraheam, & Kadhim, 2015). Diverse classes of metabolites respond to particular IR wavelengths, such as fatty acids, which can relate to IR bands from 2800 to 3050 m, amides to 1600–1800 cm, polysaccharides from 1000 to 1150 cm, and the rest fall into a mixed region, ranging from 1250 to 1450 cm (Clemens, Hands, Dorling, & Baker, 2014). FT-IR is comparatively more economical than mass spectrometry and other spectroscopic methods, and it offers instant and first-hand screening of metabolites (Corte et al., 2015; Derenne, Van Hemelryck, Lamoral-Theys, Kiss, &

Goormaghtigh, 2013). In addition, there are numerous sampling techniques; for example, high-throughput manifold sample analyzers, which can run 96–384 samples per run; and focal plane detector microscopes, in which artificial chemical image maps of samples may be constructed and attenuated by total reflectance methods (ATR) (Correa-García, Bermúdez-Moretti, Travo, Déléris, & Forfar, 2014). Apart from high reproducibility, accuracy, and speed of FT-IR, it neither requires exhaustive sample preparation, nor an elevated instrumental precision (Bacsik, Mink, & Keresztury, 2004; Siesler, Ozaki, Kawata, & Heise, 2008). In this method, the sample is directly injected on the sample plate composed of an IR transparent substance (e.g., ZnSe, CaF, or Si), or it may be dissolved in water earlier, before dehydration prior to measurement. However, FT-IR has limitations, in that it leads to high absorption of water in mid-IR, but this can be averted by dehydrating samples so to deduct water signals or effects of ATR (Baker et al., 2014; Yadav, Rani, Singh, & Murari, 2015). ATR can provide the estimation of samples that are acquiescent in nature; for example, leaves, stems, roots, and whole plants, which require no sample preparation, and are known to curtail any spectral variation between sample operators (Jaiswal et al., 2015; Jha, Jaiswal, Borah, Gautam, & Srivastava, 2015). In addition, the specific relevance of this technology to plant sciences and optical technology is constantly growing, with devices previously only accessible commercially, as portable, hand-held, and microdevices. These techniques would be more appropriate for assembling measurements; not only in a laboratory environment, but also in glass houses and in situ in the field with growing specimens of either natural or agricultural origins.

5.2.2 NMR Spectroscopy

NMR spectroscopy, in addition to having high precision and expansive throughput, requires only limited sample preparation. This procedure employs only nuclei with odd atomic or mass numbers that function as magnets, and can interact with external magnetic fields via nuclear spin mechanisms (Bharti & Roy, 2012; Montuschi et al., 2011). In ^{1}H-NMR, an external magnetic field is utilized to align the nuclear spin of counter-proton nuclei. These nuclei were bombarded with EM radiation at exact frequencies. The different nuclei are known to be in resonance where they shift from a low to high-energy spin state. The consequential emission of radiation during the nuclei shift to a ground state is identified as the absorption frequency (Larive, Barding, & Dinges, 2014; Wu, Southam,

Hines, & Viant, 2008). Protons exhibit varied frequencies, depending on the number of electrons revolving around the nucleus. The electrons ascertain magnetic fields against external magnetic fields of instruments in a process known as shielding, which leads to the absorption of EM radiation (Kim, Choi, & Verpoorte, 2011; Wishart, 2008a, 2008b). ^{1}H-NMR can supposedly offer an exclusive signal for each explicit hydrogen nucleus, thereby permitting the operator to assemble the compound's structure (Savorani, Rasmussen, Mikkelsen, & Engelsen, 2013; Tikunov, Johnson, Lee, Stoskopf, & Macdonald, 2010). As a majority of biomolecules contain hydrogen atoms, ^{1}H-NMR can be measured as nonbiased, in contrast to other metabolomics methods (Günther, 2013; Lanza et al., 2010). NMR is the most valuable and standardized metabolic profiling platform. Instead of offering only half of the required accuracy, as MS-based approaches do; NMR sensitivity varies depending on metabolites, with resolution and spectral stuffing also being significant (Arges & Ramani, 2013). Research reports of whole tissues and cellular material are presently emerging as more renowned because of the advancement of magic angle spinning (MAS). During MAS, the sample is heated via rotation at 54.7 degrees (magic angle), which confers high uniformity and spectral resolution (Renault et al., 2012). From convenient sample preparation to high precision in quantification, NMR-based metabolomics maintain delivery of high metabolite flux, which facilitates explicit metabolite detection and subsequent quantification. The contrivances and resources accessible for metabolomics assembly are detailed as follows:

NMR Proc Flow is chiefly employed for metabolic fingerprinting and targeted analysis that encompasses all spectra processing steps, including baseline correction, chemical shift calibration, and alignment (Bart et al., 2009; Duarte, Lanz, & Gruetter, 2011).

NMR Spec is mainly used for preprocessing and evaluation of NMR data, which is comprised of four main functional steps, which are spectrum alignment, baseline correction, peak extraction, and peak integration (Tulpan, Léger, Belliveau, Culf, & Čuperlović-Culf, 2011; Zhang, Gowda, et al., 2008; Zhang, Li, & Yan, 2008a).

nmrML is an accessible XML-based data exchange and storage format for NMR data (Rocca-Serra et al., 2016; Schober et al., 2017). This nmrML format is anticipated to be completely compatible with existing NMR data for chemical, biochemical, and metabolomic experiments. It acquires NMR data, and spectral parameters are validated for data acquisition, and spectral metadata, for example, chemical structures.

Moreover, it is compatible with standard compounds in NMR data for reference spectral libraries, and NMR data from complex biomixtures from metabolomics experiments.

MWAS tools are an exclusive R package fabricated for supplying an integrated pipeline to identify metabolomic data in expansive research studies. It enables quality control analysis, metabolome-wide association analysis via an array of models, statistical visualization, metabolite assignment via statistical total correlation spectroscopy (STOCCY) workflows, and elucidation of metabolome-wide association study (MWAS) data (Sugimoto, Kawakami, Robert, Soga, & Tomita, 2012; Xia, Mandal, Sinelnikov, Broadhurst, & Wishart, 2012).

speaq 2.0 is an another R package that facilitates NMR data analysis and provides spectral processing, for example, raw spectral alignment and processing together with peak-based processing steps (Costa, Maraschin, & Rocha, 2016). Furthermore, it provides analysis of complementary LC-MS workflows.

PepsNMR is also an R-based package intended for carrying out preprocessing of ^{1}H-NMR-based metabolomics data. This instrument executes solvent signal suppression; internal calibration; phase, baseline, and misalignment corrections; bucketing; and normalization. It further facilitates instant workflows, reproducibility, and resilient documenting, with other characteristics (Misra, 2018).

Plasmodesma is a Python-based package that analyzes independent, unsubstantiated processing of an expansive mass of 1D, 2D, and diffusion ordered spectroscopy (DOSY) experiments from a cycle of samples procured in diverse sample conditions (Day, 2011).

GISSMO (guided ideographic spin system model optimization) is a GUI-based web server for channelized ideographic spin system model optimization, which facilitates the proficient calculation and enhancement of spin system matrices aligned with experimental ^{1}D-^{1}H-NMR spectra of small molecules (Misra, 2018).

5.3 MASS SPECTROMETRIC APPROACHES

Metabolic profiling is expansively employed for procuring MS and chromatography-MS data, particularly in the plant sciences field (Ivanisevic et al., 2013; Theodoridis, Gika, & Wilson, 2011). Mass spectrometric methods generate ions that are separated in the analyzer region by their mass to charge (m/z) ratio (Dunn et al., 2013). The ions are then identified, and the signal is processed to the data system, where *m/z* ratios are acquired with their relative abundance for arrangement as a mass spectrum (Want, Nordström, Morita, & Siuzdak, 2007). In order to avoid problems with the ion abundance while it travels from one end of the instrument to the other, the MS analyzer, detector, and ionization source are executed under pressure (Neumann & Böcker, 2010). In order to detect metabolites, the linear time of flight (TOF)-MS is frequently used, wherein all ions are spun at constant kinetic energy, and the mass of ions is directly controlled by flight velocity, in addition to the flight tube. This validates the proportionate equation between arrival time in the detector, and analyte mass, demonstrating that the lighter analyte will reach the detector first. However, sometimes ions of similar molecular weights and different flight velocities are identified, leading to mass ion broadening; which results in mass ions reaching the detector first, and in a shorter flight time than the other (Khamis, Adamko, & El-Aneed, 2017). This limitation can be overcome by using reflectron (Re) TOF-MS, which is generally comprised of sequences of rings that function to produce an impeding field (Ágoston et al., 2014). The mass ions infiltrate, and are held within the field; mass ions with better flight velocities are preserved longer. Hence, the detection of mass ions is equalized, and peak broadening is prevented, leading to high resolution. Mass ions are evaluated by single-quadrupole MS, which executes on the principal of electric fields produced between arrays of four rods, through which mass ions pass (Erickson et al., 2015). Quadrupolar fields are produced due to voltages applied on the rods, which allows the mass ions to flow within a specified range; while ions beyond the range hit the rods and are discharged (Maeda et al., 2016). Negative rods attract positive mass ions, and the reverse is reported for negative mass ions, which are thereby represented by quadrupoles as mass filters (Sudakov, Mamontov, Xu, Xu, & Ding, 2016). Tandem MS can be employed for metabolite identification, and MS-MS allows mass fragmentation through collision with noble gases such as nitrogen or argon, which leads to collision-induced dissociation (Marshall et al., 2015). MS-MS methods are generally employed in single-quadrupole mass filters connected to a TOF flight tube and detector. Conversely, there is also di-quadrupole MS-MS, and now, triple quadrupole MS-MS, providing high mass ion specificity (Hoff, Pizzolato, Peralba, Díaz-Cruz, & Barceló, 2015). Tandem MS analysis can be carried out in sequence via an ion trap in order to select the ions that are to be fragmented; this MS approach leads to high structural analysis, and therefore more efficient metabolite detection. The resolution of MS strategies is generally enhanced with a pro-sample fractionation stage via varied chromatographic methods (Amais, Virgilio, Schiavo, & Nóbrega, 2015; Orrego-Lagarón, Vallverdú-Queralt, Martínez-Huélamo, Lamuela-Raventos, & Escribano-Ferrer, 2016). These diverse hyphenated MS methods have various applications, as well as limitations, but all signify sensitive and specific methods for qualitative or quantitative evaluation.

5.3.1 Gas Chromatography-Mass Spectrometry

This is an extensively used and well-established method for metabolic profiling. It is comparatively more economical than other spectrometric methods, and can analyze even expansive sample mixtures (Tranchida, Franchina, Dugo, & Mondello, 2016). GC-MS is generally inclined against non-volatile high molecular weight metabolites; whereas metabolites that volatize up to 250°C are readily analyzed. However, heat sensitive metabolites cannot be detected (Al-Tameme, Hameed, Idan, & Hadi, 2015; Hadi, Mohammed, & Hameed, 2016). GC-MS is employed for analyzing polar non-volatile metabolites, such as amino acids, sugars, and organic acids, via derivatization (Patel et al., 2017; Xiao et al., 2014). This chemical derivatization is a two-step process, especially in plant samples, wherein derivatization agents such as *O*-alkylhydroxylamine changes sample carbonyl groups to oximes for thermal stabilization (Jorge et al., 2016; Tseng, Chen, & Huang, 2014). Subsequently, treatment with a silylating agent such as *N*-methyl-*N*-(trimethylsilyl) trifluoroacetamide converts metabolites to volatile trimethylsilyl esters (Marcos & Pozo, 2015; Raynie, 2018). The electron impact (EI) leads to ion formation in GC-MS. The EI also implicates autonomous exclusion of sample solvents prior to the vaporized sample being carried into the ionization source, where it collides with the stable electron flow, with optimum energy to ionize the vaporized metabolite (Benton et al., 2014).

Diverse detectors are employed for mass identification such as single quadrupoles, TOF, or ion-trap-based mass analyzers. Detectors such as single quadrupoles exhibit a complete separation of metabolites, even in complex mixtures, within an hour of the chromatographic run (Benito et al., 2015). The time duration can further be reduced to 75% by using a high-speed TOF mass analyzer, coupled with deconvolution software (Kotte et al., 2016). The identification of metabolites is carried out via comparing the retention time and retention indices to those of internal standards. Generally, the mass spectral comparisons can be obtained through easily accessible databases; for example, the National Institute of Standards and Technology library, and others commonly used in house libraries. Advanced mass spectrometric instruments with improved efficiency, specificity, and resolution power have been developed in recent years (Delgado et al., 2011; McLafferty, 2008). The diverse multidimensional separation approaches are comprised of GC × GC-TOF-MS, where there is incomplete separation via a nonpolar GC column trailed by complete separation, via a second, shorter polar GC column, thereby providing lucid peaks, together with enhanced instrument assortment competence (Das et al., 2013; Gogus, Ozel, Kocak, Hamilton, & Lewis, 2011).

Sample Preparation

Sample preparation requires more generalized and broad extraction methodology; but this procedure should not hamper the stability of metabolites, or induce sample degradation or modification (Gullberg, Jonsson, Nordström, Sjöström, & Moritz, 2004; Lisec, Schauer, Kopka, Willmitzer, & Fernie, 2006). Moreover, it is necessary to separate fractions in order to analyze trace metabolites exhibiting a small number, but concentrated level of metabolites (Kim & Lee, 2002).

Derivatization

Derivatization is an imperative step that converts non-volatile metabolites to a volatile state, thereby rendering it more suitable for GC-MS. Derivatization is carried out through diverse derivatization agents such as alkylating, acylating, and preferably, silylating reagents (e.g., trimethylsilylated reagents most suitable for non- targeted metabolite profiling) (Fiehn & Major, 2005; Villas-Bôas, Smart, Sivakumaran, & Lane, 2011).

Separation

For effective separation by GC, most validated conditions are requisites for the separation of metabolites by GC, because small variations in gas flow conditions, temperature, and the kind of capillary column also modulate chromatographic retention, thereby affecting the elution order of metabolites (Shim et al., 2012).

Ionization

GC-MS carries out the ionization of samples through EI, which is the most exact and explicit procedure for averting ion suppression effects, which leads to more effective and enhanced levels of metabolite identification.

The EI transmits a constant amount of energy of −70 eV to metabolites, which are instantly transferred as molecular ions from GC outlets into a vacuum (Kraft, Weber, Grützke, Winter, & Nowak, 2015; Li, Gan, Bronja, & Schmitz, 2015a; Li, Zhang, Ma, Li, & Guo, 2015; Portolés, Mol, Sancho, & Hernández, 2014). This amount of energy is greater than the first ionization energy of all metabolites, resulting in resourceful production of molecular ions (Awad, Khamis, & El-Aneed, 2015). The remaining energy is dispersed through extremely competent, concentration-independent fragmentation of the molecular ions, which has two significant effects. First, all molecular ions bear one positive charge, which condenses spectral analysis. Second, high metabolite-specific mass spectral fragmentation patterns facilitate metabolite detection (Fiehn, 2016; Hübschmann, 2015). Apart from electron ionization, ionization can be carried out by chemical ionization (CI); by photons, through atmospheric pressure photoionization (APPI). EIs impart a great deal of energy, which surpasses even the first ionization energy, leading to an amplified level of molecular fragmentation (Qi & Volmer, 2017; van der Burgt, Finnegan, & Eden, 2015). On the contrary, soft ionization methods; for example, electrospray ionization (ESI), CI, and APPI, impart comparatively low energy, leading to less molecular fragmentation, and hence, all metabolites are not ionized (Hejazi, Guilhaus, Hibbert, & Ebrahimi, 2015; Meher & Chen, 2017). For the ionization of samples dispersed in liquids, ESI is the prime method employed in LC (Andrade et al., 2015). Here, the ionization of a solvent results in the generation of charged liquid droplets, which are sprayed into a vacuum (Banerjee & Mazumdar, 2012; Pagnotti, Chubatyi, & McEwen, 2011). As the charge concentrates at the surface, the solvent evaporates, resulting in the minimization of droplets. The charge is then transferred to the solutes, and ions are discharged from small droplets via electrostatic repulsion (Saleh et al., 2008). The resultant ions are consequently evaluated by mass spectrometric detectors. In contrast to EIs, ESI imparts low ionization energy to molecules; hence resulting in ion suppression, as all molecules are not ionized (Gabelica & Pauw, 2005). An alternative for ESI is advanced technology APPI, which employs a hot, inert gas flow, as opposed to a vacuum, in order to dehydrate the solvent and generate ions (Short, Hanold, Cai, & Syage, 2007). It also uses photons energized to about 10 eV, as opposed to electron or chemical reactions for molecular ionization (Karas, Glückmann, & Schäfer, 2000). Photoionization also induces ionization of less polar molecules, which are either not ionized, or generate unstable ions when exposed to other techniques (Giri et al., 2017; Heuser et al., 2015).

Detection

Three types of mass identification devices are employed in GC-MS coupling: the single-quadruple detector (QUAD), ion trap technology (TRAP), and TOF. The GC-QUAD-MS system is similar to high performance liquid chromatography (HPLC) in terms of its sample throughput, analyzing 10–20 samples per day (Kumar & Nagar, 2016; Lorenzo & Pico, 2017). Although GC-TRAP-MS technology has the same sample throughput, it is comprised of reaction monitoring competence wherein the predetermined sample is subjected to secondary fragmentation in order to produce daughter masses/fragments (Chauhan, Goyal, & Chauhan, 2014; Peterson et al., 2014). This further amplifies specificity, and reduces chemical noise, rendering an extra benefit to trace metabolite evaluation in complex samples (Mendes, Pereira, Ferreira, Chaves, & Vaz, 2017; Wong, Chin, Perlmutter, & Marriott, 2015). It also facilitates the detection of low-molecular-weight metabolites. The GC-TOF-MS method also provides high throughput, generating 10–50 scans; thereby facilitating the analysis of 30–40 samples per day (Rudnicka, Kowalkowski, Ligor, & Buszewski, 2011; Wiklund et al., 2008). The quadrupole detector is comprised of four metal rods that are electronically executed as a mass specific ion filter. Quadrupole detectors produce mass spectra via counting the ions that passes through the filter at each of the successively monitored masses (Driver et al., 2018; Li, Cao, Li, Wang, & He, 2016). Triple-quadrupole detectors are comprised of three quadrupole devices that are coupled in a linear array (Ye, Gao, Tian, Su, & Zhang, 2018; Zhao et al., 2015). Ionization generates ions that are passed through the first quadrupole, which functions as a mass-selective filter, allowing only specific parent ions of known mass. This identified parent ion then passes through the second quadrupole, which acts as a collision cell meant for colliding parent ions with gas molecules leading to the production of daughter ions from parent ions. These daughter ions then pass through a third quadrupole device, which functions as a second mass-selective filter that permits mass determination of daughter ions and their relative abundance. Triple quadrupole detectors are comparatively more effective due to their double filtration methods than single quadrupoles for simplifying even the most complex mixtures by excluding metabolites and passing target metabolites of relevance. Conversely, ion trap technology first assembles ions and creates ion reserves by spinning them into stable orbits, which is opposed to quadrupole technology, wherein ions permit linear movement (Eltony et al., 2016). In ion trap technology, as ions are generated from the device, they are simultaneously counted to provide ion abundance (Cinelli, Avino, Notardonato, Centola, & Russo, 2014;

Tian et al., 2018). Ion trap technology includes two steps that permit the production of mass spectra by trapping and generating ions of successive masses. Further, assembled primary ions can be fragmented again, thereby producing secondary fragments. This method has an immense utility in the determination of molecular structure, or in the identification of known metabolites in complex mixtures (Brownnutt, Kumph, Rabl, & Blatt, 2015).

TOF technology is a third mass-spectrometric technology for identification and ion estimation of macromolecules such as proteins, and polysaccharides via matrix-assisted laser desorption (MALDI-TOF). New interventions in this technology permit online coupling to GC. Ions are assembled into packages that are instantly accelerated along an evacuated flight tube, and identified at a definite distance (Granzotto & Sutherland, 2017; Perez-Gregorio, Mateus, & De Freitas, 2014). TOF differentiates ions on the basis of mass, as molecules of high mass travel faster than molecules with low mass. Each of these assembling, accelerating, and identification steps takes only a few milliseconds. Hence, GC-TOF-MS is preferably appropriate for instant screening of small volatiles; whereas matrix-assisted laser desorption/ionization (MALDI)-TOF MS is the most appropriate technology in polymer analysis and proteomics (Ni, Su, Qiu, Jia, & Du, 2016; Zhang, Andersson, Räder, & Müllen, 2015a; Zhang, Lei, Huhman, Sumner, & Zhao, 2015). Under standard conditions, each metabolite should be standardized via employing stable isotope-labeled isotopomers that are distinctly recognized by mass spectrometry, or a xenobiotic stereoisomer that is discernible by means of its chromatographic features.

Evaluation

Metabolites are detected, and then identified, by comparing their retention time, indices, and mass spectral fragmentation patterns to internal standards, or already identified metabolites in a database (Allen, Greiner, & Wishart, 2015; Vaniya & Fiehn, 2015). Generally, GC-QUAD-MS require precise information about the identification of fragment masses, and retention times of metabolites to minimize the possibility of misidentification (Vallarino & Osorio, 2016). Because it is cumbersome to assemble all metabolite information, it can be overcome by generating mass spectral and retention time index reference libraries for inclusive reports on metabolites (Giacomoni, Fillatre, et al., 2014a; Giacomoni, Le Corguillé, et al., 2014b; Merlet et al., 2016). However, this procedure requires at least 2 min per metabolite, and provides approximately 20 chromatogram files to be analyzed per day; but with massive metabolites, it leads to a rate limiting step (Wasylenko &

Stephanopoulos, 2015). GC-TOF-MS is a highly advanced technique that provides automatic and inclusive mass chromatogram extraction, integral mass spectral correction for coeluting compounds, retention time/indices calculation, and automatic selection of appropriate fragment mass for specific quantification (Chou, Kenig, Murray, Fritsen, & Doran, 2016; Obersteiner, Boenisch, & Engel, 2015). Use of GC-TOF-MS has increased in non-targeted and completely automatic metabolite detection; but there is still a need to improvise for the rectification of improper assignments (Genuit & Chaabani, 2017). Generally, almost 2 days are needed for carrying out the analysis of 50 samples via extraction and derivatization steps. The generalized steps include derivatization, separation, and ionization; which require 60–75 min via GC-QUAD-MS, or 35–45 min by GC-TOF-MS (Hoker, Obersteiner, Bönisch, & Engel, 2014). Sample throughput is further amplified via fingerprinting approaches, or targeted evaluation of metabolites. The main bottleneck is the identification and labor-intensive verification of misidentified compounds. GC-MS allows explicit and precise quantification of the metabolite up to a four-order magnitude under suitable validation of internal and external conditions (Long & Antoniewicz, 2014).

GC has the main disadvantage that it has no or little utility in thermally labile metabolites (Kirchner, Matisová, Hrouzková, & de Zeeuw, 2005; McNair & Miller, 2011). GC can be employed for the detection of volatile metabolites, or for metabolites that need to be derivatized (Koek, Jellema, van der Greef, Tas, & Hankemeier, 2011; Silva, Passos, & Câmara, 2011). At present, GC can identify macromolecules such as trisaccharides, steroids, diglycerides, and monophosphorylated metabolites (e.g., glycerol 3-phosphate and glucose 6-phosphate), but the majority of polyphosphorylated and activated metabolic intermediates are currently inaccessible to GC-MS analyses (Lai et al., 2017; Lerner et al., 2017; Tang et al., 2018).

GC-MS lacks effective peak resolution, and due to its amplified efficacy, it extracts even minor peaks, which clutter the whole peak spectrum, along with the resolution of capillaries, which creates an additional nuisance for the biologist (Farag, Porzel, & Wessjohann, 2012; Hübschmann, 2015). During extraction and sample preparation, some unidentifiable analytes can be produced, or mass fragmentation of parent ions may yield new metabolites (Ruiz-Matute, Hernandez-Hernandez, Rodriguez-Sanchez, Sanz, & Martinez-Castro, 2011; Xiao, Zhou, & Ressom, 2012). The identification of these metabolites is an imperative task that mutually further amplifies the competence of the GC-MS platform (Hummel, Strehmel, Selbig, Walther, & Kopka, 2010; Zhang et al., 2012). The precise and explicit advance implicates an addition to the battery of validated standard metabolites that are relevant to the biologist. GC-MS is quite ambiguous in peak comprehension, as it employs complete sample degradation, and hence renders samples incongruous for structural elucidation (Tsugawa, Tsujimoto, Arita, Bamba, & Fukusaki, 2011; Wishart, 2011). Moreover, GC generates below minimum levels of metabolites that could be subjected to structural explication, as NMR requires at least an upper-microgram to milligram-sample size (Dunn et al., 2013; Vinaixa et al., 2016).

5.3.2 Liquid Chromatography-Mass Spectrometry

LC-MS is emerging as the most suitable method for analyzing even thermally labile metabolites (Boundy, Selwood, Harwood, McNabb, & Turner, 2015; Gika, Theodoridis, Plumb, & Wilson, 2014). It is an efficient analytical method that integrates the resolution power of LC with the detection selectivity of mass spectrometry (de Villiers, Venter, & Pasch, 2016; Hird, Lau, Schuhmacher, & Krska, 2014). Integration of LC with MS is a significant landmark in the history of chromatography, which commenced in late 1980s. LC-MS is titled for its precision and regular flow for generating expansive metabolomic data owing to its high-resolution tandem mass spectrometry (Liu et al., 2017; Pérez-Victoria, Martín, & Reyes, 2016). LC first separates the sample components, and then launches them to the mass spectrometer, wherein they are identified, and subsequently charged (Cattaneo et al., 2016; Hubert, Houari, Rozet, Lebrun, & Hubert, 2015). The main relevance of LC-MS is in the generation of data pertaining to molecular weight, structure, identity, and magnitude of particular sample metabolites (Arapitsas et al., 2016). LC-MS offers the coherent evaluation of metabolites, even in the complex analyte (Arrivault et al., 2015). In contrast to GC-MS, LC-MS offers analytical utility to a more extensive range of metabolites (Fragkaki, Angelis, Kiousi, Georgakopoulos, & Lyris, 2015; Lu et al., 2016). GC-MS can be applied to a limited number of metabolites, ranging up to an order of 9 million, which are registered in online and in-house libraries (Naz, Moreira dos Santos, García, & Barbas, 2014). LC-MS generates very low heat, and hence, can be employed by the majority of metabolites, and cover organic compounds from a wide range of molecular weights (Trivedi, Panda, Sethi, & Jana, 2017; Zhang, Andersson, et al., 2015a; Zhang, Lei, et al., 2015). Therefore, LC-MS has relatively greater relevance than GC-MS in evaluating high, polar, ionic, thermally labile, and non-volatile metabolites (Steinborn et al., 2016; Vaiano et al., 2015).

Although some non-volatile metabolites can be subjected to GC-MS via chemical derivatization, LC-MS lessens this impediment, and allows the analysis of even non-volatile and thermally labile metabolites (Marcos & Pozo, 2015). LC-MS is comprised of HPLC-MS, capillary electrochromatography (CEC)-MS, and CE-MS. The main application of LC-MS is for use in the determination of the elemental composition and structural elucidation of the sample (Masiá, Suarez-Varela, Llopis-Gonzalez, & Pico, 2016).

Principle of LC-MS

Generally, LC-MS is an integration of HPLC, with MS via an ionization source. In this method, the sample is separated by LC, and then the separated samples are sprayed over an atmospheric pressure ion source, where they are ionized in a gaseous phase (Campillo, Viñas, Férez-Melgarejo, & Hernández-Córdoba, 2015; Naseri et al., 2014). Followed by this is the mass analyzer, which arranges ions as per their mass-to-charge ratio, and the detector then enumerates the ions that pass through the mass analyzer. It can also magnify signals generated from ions (Li, Gan, et al., 2015a; Li, Zhang, et al., 2015). Consequently, the mass spectrum reveals the scatter plot between the ion signals as a function of the mass-to-charge ratio, which is employed for determination of elements, or isotopes in the analytes, and for explicating the chemical structure of the metabolite (Kruppa & Zhong, 2016).

Tools of LC

Generally, LC employed in LC-MS is HPLC. The principal of separation in HPLC is either a normal or reverse phase of adsorption (Brown, 2012; Fekete, Kohler, Rudaz, & Guillarme, 2014). In normal-phase HPLC, the stationary phase is polar, and the mobile phase is nonpolar. The reverse phase contains a nonpolar stationary phase with a polar mobile phase (Yan et al., 2015; Zhao et al., 2014). Normal-phase chromatography is used for metabolite analysis in fruits and vegetables, and the reverse phase is extensively utilized in drug analysis (Contrepois, Jiang, & Snyder, 2015; Hutchings, Shields, Bianchi, & Schuur, 2018). The reverse phase employs columns such as octadecylsilane (ODS), C18, C8, C4, and so forth. LC-MS has been radically enhanced in recent years. It has been transformed from an expensive and intricate method, to an economical and simplified method (Danezis, Anagnostopoulos, Liapis, & Koupparis, 2016; Núñez, Gallart-Ayala, Martins, & Lucci, 2012). The genre of detectors was advanced further so as to confer expansion, and which adds to the relevance of this instrument

(Ignat, Volf, & Popa, 2011; Swartz, 2010; Zhang, Li, & Yan, 2008a; Zhang, Gowda, et al., 2008). The upgraded version of LC-MS is comprised of an assemblage of varied tools, such as an ionization source, a mass analyzer, and a tool for mass spectrometric data acquisition (Mu et al., 2012; Stafsnes, Dybwad, Brunsvik, & Bruheim, 2013).

Ionization Source

Generally used ionization sources in LC-MS are ESI, atmospheric pressure chemical ionization (APCI), and matrix-assisted laser desorption/ionization (MALDI). Similar to GC-MS, the EI and CI are also utilized as ionization sources in MS (Awad et al., 2015; Farré & Barceló, 2015; Ruddy, Hendrickson, Rodgers, & Marshall, 2018).

Electrospray Ionization

This kind of ionization can be generated by applying an intense electric field to samples flowing via capillary tubes with weak flux. This generates charged solvent molecules that are exposed to desolvation. Desolvation is parallel to the rise in charge density, leading to columbic repulsion to surmount the liquid surface tension, and it generates ion from droplets (Brown, Chen, & Zare, 2015; Cech & Enke, 2001; Wiseman et al., 2008), which forms the basis of this technique. This technique is highly sensitive, but is restricted to a range of 10–8 μL, and requires massive volumes to enhance sensitivity of the detection (Ifa, Wu, Ouyang, & Cooks, 2010; Konermann, Ahadi, Rodriguez, & Vahidi, 2013).

Atmospheric Pressure Chemical Ionization

This technique implicates mobile phase nebulization with nitrogen gas and vaporization via heating it to a comparatively elevated temperature of about 400°C. The resultant vapor is then passed through a corona discharge electrode to generate ions (Andrade et al., 2008; Chan et al., 2011). This ionization source is mainly used in LC-MS analysis of drugs, chemicals, and industrial effluents (Baker & Kasprzyk-Hordern, 2011; Zhao & Metcalfe, 2008).

Matrix-Assisted Laser Desorption/Ionization

This ionization technique is particularly applicable for massive and thermally labile metabolites such as proteins, polymers, and so forth. This technique implicated the implanting of the analyte in the matrix, which absorbs energy at a laser wavelength of 3337 nm (Bizzini, Durussel, Bille, Greub, &

Prod'hom, 2010; Gross, 2011). This laser energy once imparted to the matrix vacuum produces ions from analytes.

Quadrupole Mass Analyzer

Quadrupole and triple quadrupole are extensively utilized analyzers because of their ease of use, and the fact that they will encompass an extensive mass range of 10–4000 amu (Dawson, 2013; Michalski et al., 2011). The quadrupole analyzer offers substantial linearity for metabolite estimation and enhanced resolution up to 4000, with coherent mass spectra, and a scanning speed of 5000 amu per second and mass precision of 0.1–0.2 (Bourmaud, Gallien, & Domon, 2016).

Ion Trap Analyzer

This analyzer has already been discussed in detail in the previous section. This analyzer is more extensively used in GC-MS than LC-MS. As discussed earlier, an ion trap analyzer first stores ions in the ion trap, and then the ions are manipulated by employing DC and RF fields (Huerta-Fontela, Galceran, & Ventura, 2010; Zheng, Baker, & Hancock, 2006). The voltage intensity enables the analyzer to trap ions of a particular mass-to-charge ratio within the analyzing device, while non-specific ions exit from the trap by a trajectory induced by electrostatic forces (Bijlsma, Emke, Hernández, & De Voogt, 2013; Senko et al., 2013).

TOF Analyzer

This analyzer is employed in a single MS instrument and in MS/MS setup; the TOF is linked to a quadrupole (QTOF), or to an ion trap (QIT/TOF). The basic principle of the TOF analyzer is generation of ions from an ion source, and their subsequent acceleration to high velocity via an electric field established in the drift tube of an instrument. The accelerated ions are identified by a detector, via a linearized or reflective mode (Clark, Kaleta, Arora, & Wolk, 2013; Ogawa et al., 2012).

Magnetic Sector Mass Analyzer

These analyzers accelerate ions via magnetic fields as they pass through the flight tube and are separated by charge-to-mass ratios. Ions in the magnetic field are deflected at the same angle by both magnetic and centripetal forces exhibiting the same trajectory and centripetal force (Cappellin et al., 2012; Patel, 2013). Other ions collide with either side of the flight tube wall, and will not flow via slit to the detector (Grimalt, Sancho, Pozo, & Hernández, 2010; Passarelli & Winograd, 2011).

Fourier Transform-Ion Cyclotron Resonance

Electric and magnetic fields trap ions in circular orbits. These ions can produce a time-dependent current if they are excited by radio frequency in the electric field. This current is converted to an orbital frequency by Fourier transform, which corresponds to their mass-to-charge ratio (Han et al., 2008; Headley et al., 2011). Fourier transform-ion cyclotron resonance (FT-ICR) mass analyzers can execute multiple steps of mass spectrometry without extra analyzers. They also possess extensive mass ranges, and appreciable mass resolution (Kaiser et al., 2011; Wagner, Dittmar, & Jaffé, 2015).

Detectors

Mass spectrometry employs different types of detectors, for example, electron multipliers, dynolyte photomultipliers, and microchannel plates (MCPs) (Blakeman, Cavanaugh, Gilliland, & Ramsey, 2017; Kruppa & Zhong, 2016).

Electron multiplier dynodes are used to convert either –ve,+ve ions into electrons that will be augmented and identified. This will be extensively employed in quadrupole and ion trap instruments (Chen et al., 2015; McAlister & Coon, 2016).

Dynolyte photomultipliers convert charged ions into electrons. These electrons adsorb to phosphor and emit photons. These photons then hit the photomultiplier to acquire multiplied signals for recording (Iram, Iram, Arif, Siddiqui, & Husain, 2016; Kumar & Vijayan, 2014).

The MCP is the most generally employed spectrometer, exhibiting a low time response and an elevated level of precision, detecting <1 ns and single ion signals >50 mV, respectively.

Mass Spectrometric Data Acquisition System

The data acquired is preformatted to digitalize electrical signals from the detector, and pass them to the data system in the attuned format.

Scan and Selected Ion Monitoring

Mass spectrometers can be executed in either a scan mode, or a selected ion monitoring (SIM) mode.

Scan Mode

The instrument identifies a signal greater than a mass range from 50 to 2000 *m/z* during a short period of time (e.g., 2 s). In the scan period, the MS electronics successively detect the signals within constricted mass intervals until the full mass range is covered (Andrews, Simons, Young, Hawkridge, & Muddiman, 2011; Michalski et al., 2012). The spectra that are preserved symbolize the identifiable signal for the full mass range. Because full mass spectra are recorded, this execution mode is generally specified for qualitative evaluation, or for estimation when all analyte masses are unidentifiable (Kalli, Smith, Sweredoski, & Hess, 2013; Van der Heeft et al., 2009). Samples are passed through into the mass spectrometer via injection through HPLC. Then, it is important to compare the peak width and scan range (Gillet et al., 2012). As the peaks get narrower, the shorter the total scans time must be, so as to acquire suitable peak clarity (Law & Lim, 2013; Rajski, del Mar Gómez-Ramos, & Fernández-Alba, 2014). To obtain instant scan time, it is important to decline the scan range (Antignac et al., 2011; Schuhmann et al., 2012).

SIM Mode

Mass spectrometers can also execute in the SIM mode. Instead of continuous scanning, the SIM mode can be set only to detect mass-to-charge ratios (Yang et al., 2018; Zhang et al., 2014). The quadrupole takes an appreciable amount of time to sample each of the *m/z* values in parallel, and with extensive accuracy. Furthermore, the cycle time between data points is generally shorter than it is in scan mode; quantitative accuracy and precision are enhanced via moderate peak-shape profiling (Steingass, Langen, Carle, & Schmarr, 2015; Streng et al., 2016). Samples that are to be analyzed must have stored their *m/z* values as SIM, and are mainly utilized for target metabolite analysis (Hauler & Vetter, 2015). Target analysis via SIM mode can be time programmed to compare with the metabolite elution time window, in case of multiple target metabolomics (Ronsein et al., 2015). However, SIM cannot be utilized for quantitative analysis of non-specific metabolites (Davtian, Bard, Ménot, & Fagault, 2018; Lo Turco et al., 2016).

Applications of LS-MS/MS

LC-MS can be employed for the determination of molecular weight in the case of pharmaceuticals, proteins, or other macromolecules. Molecular weight determination is mostly carried out through API electrospray and APCI LC-MS systems. The data in the mass spectrum facilitates separation of compounds that are otherwise unresolved by chromatography; for example, benzodiazepines can be more effectively identified on MS detectors than ultraviolet (UV). LC-MS can also be employed for the detection of glycosylation in tyrptic proteins (Dong, Yan, Liang, & Stein, 2016; Jedrychowski et al., 2015). LC-MS can be used for the identification of very complex pharmaceutical drugs, such as trimipramine and thioridazine, which are used as antidepressants and tranquilizers, respectively. These drugs can be identified in urine samples by adopting SIM modes (Montenarh et al., 2015; Shipkova & Svinarov, 2016). The separation and detection of chiral impurities in pharmaceuticals such as drugs, and drug metabolites can be also be carried out through LC-MS.

The detection of food adulteration, as well as antinutritional factors, can be detected through LC-MS. Aflatoxins are lethal metabolites biosynthesized by some fungi in foods which could be exclusively identified by their mass spectra.

Tandem mass spectrometry can be employed for the determination of the structure revealing the information pertaining to mass spectral fragmentation (Bauer, Luetjohann, Rohn, Kuballa, & Jantzen, 2018; Jedrychowski et al., 2015).

MALDI-TOF MS is employed for SNP genotyping, DNA quantification, gene expression analysis, and DNA and RNA sequencing (Evans, Gallagher, Ratcliffe, & Williams, 2016; Reiding, Blank, Kuijper, Deelder, & Wuhrer, 2014).

LC-MS, coupled with APCI, has an extensive scope in environmental chemistry, as it can be employed for the detection of carbamates, monuron and diuron, which display explicit spectra that can be identified by LC-MS (Ojanperä, Mesihää, Rasanen, Pelander, & Ketola, 2016; Olmo-García, Bajoub, Fernández-Gutiérrez, & Carrasco-Pancorbo, 2016).

Fourier Transform Ion Cyclotron Resonance Mass Spectrometry

FT-ICR MS is a rapidly emerging technology that originates from ICR spectrometry. Its idea originated during the early 1930s, immediately following the discovery of the theory of ion cyclotron motion, which illustrates atomic nuclei by accelerating protons in the first-ever cyclotron accelerator (Han et al., 2008; Headley et al., 2011). In 1970, the first Fourier transform ICR mass spectrometer was developed by Melvin B. Comisarow and Alan G. Marshall (Comisarow & Marshall, 1996). Thereafter, FT-ICR MS has emerged as an authentic and validated method; in contrast to other methods of mass spectrometry, it has greater resolving power, mass resolution, and mass measurement precision, providing precise evaluation of an extensive range of biomolecules comprised of proteins and nucleotides (McKenna, Purcell, Rodgers, & Marshall, 2009; Scigelova, Hornshaw, Giannakopulos, & Makarov, 2011). The high precision in mass determination is due to the fact that FT-ICR MS estimates frequency to determine m/z (Chen et al., 2012; Nikolaev, Kostyukevich, & Vladimirov, 2016). FT-ICR was generally coupled with tandem MS spectrometry, comprising electron capture dissociation, continuous off-resonance irradiation collision activated dissociation (SORI-CAD)51, and infrared multiple photon dissociation, which make it convenient for illustrating non-covalent compounds and biological systems (Lobodin et al., 2015; Savory et al., 2011).

FT-ICR MS is based on the estimation of rotational frequency of ions in the magnetic field called "cyclotron frequency." According to principle, the cyclotron frequency is inversely proportional to the mass-to-charge ratio; and hence, it is likely to verify the m/q ratio, if this frequency and the magnetic field strength are known (Hendrickson et al., 2015; Nikolaev et al., 2016). This instrument is comprised of an ion trap, used as a magnet that generates a magnetic field within the trap, a method of ion generation and ion flow into the trap, an electronic system for excitation of ion motion, and signal detection induced by rotating magnetic field ions for the detection of electrodes in an ion trap. Further, it is comprised of data systems for signal recording, storage, and its conversion to mass spectra. The spectrometer is composed of an ion trap with varied modifications; hence creating the Penning trap (Atanasov, 2015; Block, 2016). Strong magnetic and electric fields are employed for trapping ions. The ions with low kinetic energy cannot flow perpendicular to the magnetic field, and to avert motion along the magnetic field in the Penning trap, the electric field is applied in the same direction as the trap's center via applying the appropriate symbol to the trapping electrodes (Orford et al., 2017; Ringle et al., 2016). Trapped ions strike against the magnetic field and spin with arbitrary phases in the plane perpendicular to the magnetic field, with frequencies close to cyclotron frequencies. The frequency analysis of any signal is generally evaluated via Fourier transformation, and this is FT-ICR MS. This method is generally repeated several times to collect the signal and acquire a spectrum with a suitable signal-to-noise ratio; hence, FT-ICR is pulsed technology. In advanced FT-ICR, spectrometer superconducting magnets are employed to generate highly strong magnetic fields (Buck et al., 2015; Smith, Podgorski, Rodgers, Blakney, & Hendrickson, 2018). The prime features of FT-ICR include resolving power, mass precision, mass assortment, and a pulsating range, depending heavily on magnetic field intensity. Advances in the FT-ICR technique during the past decade have been associated directly with advances in superconducting magnet technology. FT-ICR, or ion traps, are first-generation instruments that are reasonably easy devices to use (Ghaste, Mistrik, & Shulaev, 2016; Tfaily et al., 2015).

The FT-ICR mass spectrometer provides maximum resolving power and mass precision among all kinds of mass spectrometers. Its explicitly analytical features make FT-ICR a significant method for proteomics, metabolomics, and the evaluation of multifarious mixtures (Barry, Groseclose, Robichaud, Castellino, & Muddiman, 2015; Benigni, Porter, Ridgeway, Park, & Fernandez-Lima, 2018; Dungan et al., 2014). FT-ICR MS requires prolonged duration for signal acquisition, up to minutes. This ion-ion interaction significantly alters ion motion, and leads to a decline in resolving power. To comprehend these effects, there was an immediate need for intricate theory and supercomputer simulations; but they ultimately ended in the discovery of the ion trap, with dynamic harmonization, which illustrates the highest resolving power ever acquired (Anderson et al., 2016; Shaw, Robinson, & Paša-Tolić, 2016).

5.4 CAPILLARY ELECTROPHORESIS

CE is a type of technology that uses the small bore (20–200 μm) capillaries to carry out highly efficient separation of both large and small molecules (Klepárník, 2015; Stobaugh & Rabel, 2017). These separations are assisted by employing high voltage, which might build up the electroosmotic and electrophoretic flow of buffered solutions and ionic species, respectively, within the capillary (Redman, Mellors, Starkey, & Ramsey, 2016; Zhang, Hankemeier, & Ramautar, 2017). The features of separation and the resultant electropherogram have characteristics that resemble a cross between conventional polyacrylamide electrophoresis, and advanced HPLC.

CE is an explicit type of technology employing an electric field to carry out separation in complex mixtures (Grossman & Colburn, 2012; Kuhn & Hoffstetter-Kuhn, 2013). CE differs from generalized electrophoresis, as it is carried out within the boundaries of a narrow tube (Wiktorowicz & Colburn, 2012). In order to comprehend the performance of molecules under the influence of an electrical field inside the capillary; it is necessary to identify the mechanism that arises from the geometry of a capillary.

5.4.1 Principle

CE places samples to be analyzed between buffer solutions in a tube, and imparts an electric field, and the sample molecules are migrated in the proper direction at a rate determined by their charge and mobility. For separation, the sample is dispersed in an appropriate solvent, for example, water; but in the absence of an electrical field, ionic motion becomes random. When the electric field is applied, charged species start flowing to carry out separation (Beckers & Boček, 2000; Ostergaard & Heegaard, 2003). Conventionally, electrophoresis was carried out in anticonvective media; for example, polyacrylamide or agarose gels. Size-dependent separation of biological molecules; for example, nucleic acid and proteins, are carried out in the slab or tube format. Although this is an extensively used separation method, it takes a prolonged amount of time, performs separation, identification, and automation ineffectively (Hommerson, Khan, de Jong, & Somsen, 2011; Moser & Hage, 2008). In order to overcome these limitations of the slab format, electrophoretic separation is carried out in the capillaries. Capillaries are an anticonvective and heat controlling innovation, because of wide tube display band mixing and poor resolution due to low thermal gradients (Elbashir & Aboul-Enein, 2015; Simpson & Smith, 2005; Zhao & Chen, 2014). The large volume-to-area ratio of the capillary efficiently dispersed the heat generated. It is helpful to use high electric field leads to shorten the time required for analysis, and increase peak accuracy (Kitagawa & Otsuka, 2014).

5.4.2 Basic Instrumentation

CE is composed of capillaries with a narrow bore end, in which silica are fused and placed in buffer reservoirs. The composition of the reservoirs is similar to that of a capillary (Benhabib, Chiesl, Stockton, Scherer, & Mathies, 2010). Electrodes are embedded in the reservoirs in order to generate a power supply between high voltages (Felhofer, Blanes, & Garcia, 2010). By replacing one of the reservoirs, a sample is loaded onto the capillary, and the electric field is applied. Following the replacement of the buffer reservoir, and the application of an intense electric field, separation is carried out. Optical identification can be carried out at the opposite end, directly via a capillary wall (Stobaugh & Rabel, 2017).

5.4.3 Characteristics of CE

- Electrophoresis is carried out in narrow-bore fused silica capillaries (25–75-μm id).

- High voltages (10–30 kV) and high electric fields (100–500 V/cm) are imparted across the capillary.
- Capillaries are anticonvective, which restricts current supply and internal heating.
- The high electrical resistance of the capillary limits current generation and internal heating.
- Capillaries offer high efficiency ($N >$ 105–106) and rapid analysis.
- Identification can be carried out on the capillary (DAD, CCD, LIF), with an external detection cell (DAD) or online MS.
- Only a small sample volume is required (1–50 nL injected).
- Several methods are used to vary specificity, with an extensive relevance range.
- Capillaries executes in aqueous and nonaqueous media.
- They offer convenient method development and automated instrumentation.

5.4.4 Applications of CE

CE offers a novel format for LC and electrophoresis that:

- employs capillary tubing, within which the electrophoretic separation occurs;
- utilizes very high electric field strengths, often higher than 500 V/cm;
- uses modern detector technology, such as that of the electropherogram, which often resembles a chromatogram;
- has efficiencies on the order of capillary GC, or even greater;
- requires minute amounts of sample;
- is easily automated for precise quantitative analysis and ease of use;
- consumes limited quantities of reagents;
- is applicable to a wider selection of analytes compared with other analytical separation techniques.

The basic instrumental configuration for CE is relatively simple. All that is required is a fused-silica capillary with an optical viewing window, a controllable high voltage power supply, two electrode assemblies, two buffer reservoirs, and an UV detector. The ends of the capillary are placed in the buffer reservoirs, and the optical viewing window is aligned with the detector. After filling the capillary with buffer, the sample can be introduced by dipping the end of the capillary into the sample solution and elevating the immersed capillary a foot or so above the detector-side buffer reservoir. Virtually all of the pre-1988 work in CE was carried out on homemade devices following this basic configuration. While relatively easy to use for experimentation, these early systems were inconvenient for routine analysis, and too imprecise for quantitative analysis.

5.5 RAW DATA PROCESSING

The data procured from samples using varied analytical sources are the ultimate data resource, which can be used for understanding the metabolic changes in organisms against diverse environmental factors (Kašička, 2016; Vasa, Guex, Wilkinson, Weeks, & Giddings, 2008). But in order to keep pace with massive data generation from one step to the next, there is a need for significant data processing and statistical analysis, together with an appropriate data storage format (Ibáñez et al., 2012; Pluskal, Castillo, Villar-Briones, & Orešič, 2010). Metabolomic experiments can be easily carried out, but only if data mining competence and efficient contrivances of comparative analysis are sufficient (Ibáñez, Simó, García-Cañas, Cifuentes, & Castro-Puyana, 2013; Sugimoto, Wong, Hirayama, Soga, & Tomita, 2010). Data analysis in metabolomics is comprised of four significant steps:

- Raw data processing, which facilitates comparative analysis of diverse data sets.
- Identification of functional metabolites via data mining.
- Data production in a one-dimensional layout.
- Effective data storage.
- Database assembly.

However, metabolomic technologies are inefficient due to a lack of precision, which results in a chromatogram shift and mass drift (Kuehnbaum, Kormendi, & Britz-McKibbin, 2013; Soga et al., 2009). An expansive multivariate data set is assembled over varied MS runs, or between diverse instruments, and the deconvolution of overlapping chromatogram peaks and chromatogram arrangements to surmount drift are both needed (Arayne, Sultana, Zuberi, & Siddiqui, 2009; Sârbu et al., 2012). This can be achieved by using varied software methods, including metabolomics spectral configuring and conversion tools (Duran, Yang, Wang, & Sumner, 2003); associative optimized distorting for chromatogram comparison (Wang, Zhang, & Harrington, 2014); METALIGN for correction of baseline, reducing noise, chromatogram alignment; and MS peak selection by employing sample alignment algorithms for deconvolution (Lommen, 2009; Lommen & Kools, 2012). Data alignment can be ensured by exerting baseline correction and noise reduction, but this can lead to the loss of important data.

Data mining can be accomplished via employing varied statistical tools that facilitate high throughput multivariate analysis of diverse components within the data set, leading to an assortment of variables that exhibit a substantial response (Blat et al., 2016; Dumancas et al., 2018; Wu, 2013). Computer-based statistical tools are competent enough to identify the samples, which are similar or diverse, and emphasize the functional variables (Gowda et al., 2014; Sugimoto et al., 2012). Generally, metabolic profiles are rich in metabolites such as GC-MS, and are comprised of about 300–500 variables; whereas an FT-IR spectra includes numerous variables and different wave numbers, with prospective of imaging FT-IR, which may be comprised of about 64 spectral dimensions (De Iorio, Ebbels, & Stephens, 2008; Worley & Powers, 2013). Expansive data generation leads to considerable constraints while analyzing data, so in order to overcome these limitations, multivariate statistical methods are used. Principal component and independent analysis are insignificant approaches that can execute a decline in dimensionality, which summarizes the metabolite profile using limited factors with negligible data loss (Jonsson et al., 2005; Ramadan, Jacobs, Grigorov, & Kochhar, 2006). Principal component analysis demonstrates the variability among the multivariate data set as per the set of variables termed "principal components." Principal component analysis implicates the distension of an (X) matrix drafted from a specified integer of samples (N), with a particular number of variables; for example, a peak area for MS chromatogram, or intensity for the FT-IR wavelength against multidimensional space (Bro & Smilde, 2014; Nyamundanda, Brennan, & Gormley, 2010). Principal components (PCs) are linear arrangements of original variables termed "PC loadings," which are employed in the projection of the X matrix. Principal component analysis (PCA) representation might be known as both linear and additive, with a generalized limited set of 10 PCs demonstrating >90% of the total estimated variance (Gu et al., 2011; Zwanenburg, Hoefsloot, Westerhuis, Jansen, & Smilde, 2011). PCA is an invalid technique that is unable to discern data set structures, as well as models, on the total variance within the data set. High variance was demonstrated by a scatter plot representing a variation of each PC. Hence, PCA allows instantaneous methods of representing and evaluating a data set, with actual variance between sample classes being signified by differential clustering (Worley & Powers, 2013; Xu & Goodacre, 2012). The differentiated variables among original data can be specified from ranking the PC loading vectors from which the model is derived. Discriminant function analysis (DFA) was employed for evaluating high variability of data sets, such as in plant samples (Brereton & Lloyd, 2014; Gromski et al., 2015). DFA can be employed where algorithms are tailored with validated information on the data structure, including the number of analytical classes and replicates within them (Miyagi et al., 2010). DFA implies the data of the experimental structure to generate estimates between and within class disparity. DFA uses this data to amplify the between-class discrepancy while decreasing the class disparity, as represented in the Fisher ratio, as per the validated experimental class structure (Misra & der Hooft, 2016; Ren, Hinzman, Kang, Szczesniak, & Lu, 2015).

During the late 1990s, the relevance of genetic algorithms and genetic programming to metabolomics was analyzed (Hageman, van den Berg, Westerhuis, van der Werf, & Smilde, 2008; Trevino & Falciani, 2006). Genetic algorithms symbolize the class of algorithms that replicate the processes implicated in Darwinian natural selection and evolutionary genetics. To authenticate PCA, PC-DFA, and genetic algorithms (GA) methods for analysis, the experimental, or biological, replicates were inflated into PCA or PC-DFA ordinate space (Graham & Castner, 2012; Gromski et al., 2014). This implicates each class among the data set that is indiscriminately divided by creating a test sample and training set. PCA, PC-DFA, or GA models are produced separately for the training set, and the test set is then consequently distended into a similar ordinate space. The close association between training and test clusters signifies the authenticity of the PCA, PC-DFA, and GA model. This is important when a number of variables is significantly amplified greater than the sample number, and when it's necessary to check the accuracy of the hypothesis produced (Green, Gilmore, & Seah, 2014; Ropodi, Panagou, & Nychas, 2016).

The estimation of metabolites gives the fundamental data pertaining to biological responses to physiological or ecological variations. Metabolite profiling facilitates the shift from hypothesis-driven research to the analysis of system-wide responses, particularly when it is integrated with other profiling procedures (Ufer, Juif, Boof, Muehlan, & Dingemanse, 2017; Wolfender, Marti, Thomas, & Bertrand, 2015). Following the characterization, the next step is demonstration of regulatory processes. Methodical investigation of the entire metabolome provides a potent and equitable approach for detecting the site at which functional evaluation of all metabolites within an organism allows imperative methods to function as to modify fluxes. Datasets from metabolite profiling could be chemometrically evaluated to unravel the association between metabolites and the expression levels of particular genes or proteins (Cheng, Lan, Zheng, & Gao, 2018; Wu et al., 2017). During the postgenomic era, metabolite profiling will be increasingly used to phenotype mutants and transgenic organisms in order to characterize gene function. The major constraint in functional genomics is allocating functions to the inadequately annotated genes; metabolite profiling will allow an explanation for those that encode proteins implicated in metabolism. The queries pertaining to molecular composition and its involvement in nutrition and health are advancing to the forefront. The advent of new innovations is rapidly emerging, resulting in new varieties and the discovery of new breeding programs; either via transgenic technology, or by employing molecular markers integrated with extensive crosses. Metabolite profiling can be employed to identify this diversity phenotypically, related to the metabolite composition, facilitating a potent tool for directing breeding programs, and to prepare the researchers at an initial stage for constructive or undesirable traits. The efficiency of this method is expansively amplified if it is integrated with an authenticated metabolite profile of the commercial plant produce. Metabolite profiling can provide risk assessment, together with baseline information. Metabolite profiling can be employed to detect the new metabolites, which can be evaluated with the assortment of metabolites in produce that were previously unobtainable. Furthermore, metabolite profiling will assist in revealing significant information pertaining to nutraceutical value, which is becoming more acceptable in the plant produce market (Arbona, Iglesias, & Gómez-Cadenas, 2015; Strehmel, Kopka, Scheel, & Böttcher, 2014). The major constraints for life sciences in the coming decades is to advance beyond the comprehensive information pertaining to organisms and their responses in a controlled or laboratory environment in order to comprehend the intricate associations in an innate ecosystem, and during evolution. The integration of high throughput and expansiveness will enable metabolite profiling to be combined with environmental studies that need extensive sampling approaches, but have generally gone through limitations of presumed research. Metabolite profiling can also be employed for assessment of genotypic diversity, without the expansion of molecular contrivances for specialized species (Venkatesh et al., 2016). In the coming years, metabolomics will be the prime method for deciphering communication between the environment and organisms through metabolic fluxes.

5.6 CHALLENGE OF COMPARTMENTALIZED METABOLOMICS

Numerous primary and secondary metabolites have been detected in plant cells using diverse, expansive, and high-throughput metabolic technologies. But there is some major lacuna with respect to the perception of the mechanism underlying the plant metabolome, especially in metabolites synthesized within the specified compartmental boundaries. Metabolomics has thus emerged as an inclusive and in-depth picture of the plant metabolome, describing metabolite composition, and thereby indicating its significant role in functional genomics and system biology (Nägele, 2014; Töpfer, Kleessen, & Nikoloski, 2015). Metabolomics is extensively utilized for evaluating metabolic responses to varying gene expressions (Kaever et al., 2015), biotic and abiotic stresses (Farag, Al-Mahdy, Meyer, Westphal, & Wessjohann, 2017), to identify genetic and metabolic diversity (Beatty et al., 2016), and it has been integrated with other omics technologies in systems biology-driven research (Bock, Farlik, & Sheffield, 2016; Karahalil, 2016). However, certain unanticipated observations have generated advanced pathways, together with advances in their regulation and evolution. It has been ascertained that cell metabolism is highly intricate, linking the genotype and resulting phenotype to record cellular responses (Anwar, Lakra, Singla-Pareek, & Pareek, 2016). Although previous studies have outlined milestones in demonstrating biological systems, there is still a dearth of reports pertaining to temporal and spatial regulation of the metabolome (Saviano et al., 2016; Todaka et al., 2017). The compartmentalization of metabolites is imperative for metabolism and functionality in the eukaryotic cells; consequently, the association of metabolic pathways within and between compartments requires decoding. While subcellular localization of enzymes is anticipated at the computational and experimental level, the evaluation of the subcellular localization of metabolites, products, and substrates is more complex because of superfluous pathways, transport, and storage (Bragança, Oliveira, & dos Santos Isaias, 2016; Kistler & Broz, 2015). Moreover, the complexities for authentic metabolite estimations in subcellular compartments reveal instant yields and the remarkably rapid translocation of metabolites between compartments (Gabaldón & Pittis, 2015; Heise et al., 2014). Due to this, an approach allowing precise data procurement on the subcellular allocation of varied metabolites is still limited.

Storage techniques result in metabolomic variations that need to be taken into account for categorization of metabolic fingerprints into different classes. The varied factors such as long storage, temperature, and antibacterial preservatives have been evaluated so that experiments are executed consequently. Preservatives such as borate affect ^{1}H-NMR peaks of some metabolites such as citrate, mannitol, α-hydroxyisobutyric acid, and methylmalonic acid; however, these effects are negligible at the interindividual level (Kuroda, Kunimura, Fukaya, Nakamura, & Ohno, 2014; Zheng, Zhao, Wu, Dong, & Feng, 2016). On the contrary, biological samples can be stored up to 9 months at −40°C without any significant change to the ^{1}H-NMR spectra (Mahrous & Farag, 2015; Markley et al., 2017). For authenticated metabolomic experiments, metabolites change with respect to time duration, age, sex, strain, and diet, which needs to be estimated to ensure that these sources of variations are not misunderstood (Dona et al., 2016; Xia, Sinelnikov, Han, & Wishart, 2015). For example, metabolite variation in urine is highly significant; hence, several research reports have determined this variation more often in urine samples than in any biofluid (Khamis et al., 2017; Thévenot, Roux, Xu, Ezan, & Junot, 2015). Metabolite fingerprints also vary with respect to age, as several studies demonstrated significantly high variations in urine samples from adults and children (Ivey et al., 2017; Petrakis, Cagliani, Polissiou, & Consonni, 2015). Some research studies resulted in highly precise biomarkers by detecting metabolites that continually vary against particular conditions, irrespective of other factors (Mastrangelo, Ferrarini, Rey-Stolle, García, & Barbas, 2015). It is a generalized fact that genes are encoded for proteins that are biosynthesized when the gene is expressed, and these proteins are further processed into metabolites that function as regulatory switches for executing cellular metabolism. The integrative approach that combines these three steps using genomics, proteomics, and metabolomics can immensely advance our perception underlying biological processes, along with sequential execution of the events at the three levels. Metabolomics enables a more precise metabolic estimation, even at lower levels, and its integration with varied fields such as omics presents an inclusive snapshot of biological pathways and their associations (Sergushichev et al., 2016; Vijayakumar, Conway, Lió, & Angione, 2018). It is immensely important that variation in gene expression be analyzed against metabolic activity at the level of the whole organism to completely comprehend its biological functions. Further, integrative omics-metabolic analysis is a constraint-based method that originated as a process to combine proteomics and metabonomic data with genomic-scale metabolic models (Boccio et al., 2016; Tayade et al., 2018). Integration of omics with metabolic analysis is becoming a quadratic programming quandary that focuses on identifying the steady-state flux equation that tracks mass balance and enzyme directionality limitations, and is reliable for analyzing fluxes determined by Michaelis-Menten kinetics (Eberwein, Shen, & Jenerette, 2017; Lervik, Kjelstrup, & Qian, 2015). Metabolomic data is associated with gene expression, and traced via KEGG (Kyoto Encyclopedia of Genes and Genomes) to allow further illustration of the pathways impacted by orotic acid-induced fatty livers. The association between gene expression and metabolomic data can be demonstrated via partial least square (PLS) (Dumancas, Ramasahayam, Bello, Hughes, & Kramer, 2015). In order to investigate phenotypic and metabolomic changes, it is necessary to investigate genomic expression instead of relying on the artificial experimental time scale (Rajasundaram & Selbig, 2016; Wen et al., 2014). Even to date, few research reports have integrated the omics fields because of the lack of metabolic data comparable with other fields; the extent of metabolomic data is constantly expanding as rapid and conventional methods, for example, NMR spectroscopy, become more prevalent (Ashbrook & McKay, 2016; Foroozandeh et al., 2014).

5.7 CONCLUSIONS

Metabolomics has immense potential for illustrating inclusive information pertaining to deciphering pathways in systems biology; however, it has certain limitations, as with any other discipline. Biological variability in metabolomic data is useful for disease diagnostics. The concentration of metabolites varies with respect to circadian rhythms, diet, age, sex, and weight, which is too complex as compared with control groups, particularly in human studies where there are numerous ethical and financial restrictions. The majority of these factors have been revealed to vary significantly, so their effects can be mitigated via the use of automated filters to decrease redundant disparities. In case samples are to be evaluated for an extended period of time, the superfluous variability, for example, metabolite degradation, can be decreased using control samples comprised of aliquots from each sample. This quality control set can be evaluated throughout the analytical run to display different patterns in the metabolic composition over time. ^{1}H-NMR spectroscopy is relatively insensitive to metabolites present in low concentrations. However, metabolites in high concentrations exist in varied pathways, and are not, therefore, exclusive to specific responses. Even though highly concentrated metabolites can vary immensely in response to stress or environmental conditions, their ubiquity reduces their ability to be used as robust biomarkers, because tracking the precise pathways is quite cumbersome. However, the high-throughput nature of metabolomics has the ability to compensate for this; many different metabolites can be calculated, and the slight variations can be efficiently elucidated. The field of metabolomics has evolved during recent decades, but it is still developing. The field of metabolomics has enormous potential for the advancement of healthcare and biomedical science. Metabolic phenotyping has, and will, facilitate drug discovery, disease diagnosis, and personalization of healthcare and identification of drug toxicity. Metabolomics integrated with genomics and proteomics represents an unknown element in the comprehension of biological systems, which opens the door for enormous possibilities.

REFERENCES

Ágoston, K., Gyémánt, G., Kalmár, L., Kerékgyártó, J., Szurmai, Z., Döncző, B., et al. (2014). Synthesis and MALDI-TOF MS analysis of protected oligosaccharide components of N-glycoproteins. *Journal of Carbohydrate Chemistry*, *33*(6), 326–343.

Aharoni, A., Ric de Vos, C. H., Verhoeven, H. A., Maliepaard, C. A., Kruppa, G., Bino, R., et al. (2002). Nontargeted metabolome analysis by use of Fourier transform ion cyclotron mass spectrometry. *OMICS*, *6*(3), 217–234.

Allen, F., Greiner, R., & Wishart, D. (2015). Competitive fragmentation modeling of ESI-MS/MS spectra for putative metabolite identification. *Metabolomics*, *11*(1), 98–110.

Al-Tameme, H. J., Hameed, I. H., Idan, S. A., & Hadi, M. Y. (2015). Biochemical analysis of *Origanum vulgare* seeds by Fourier-transform infrared (FT-IR) spectroscopy and gas chromatography-mass spectrometry (GC-MS). *Journal of Pharmacognosy and Phytotherapy*, *7*(9), 221–237.

Amais, R. S., Virgilio, A., Schiavo, D., & Nóbrega, J. A. (2015). Tandem mass spectrometry (ICP-MS/MS) for overcoming molybdenum oxide interferences on Cd determination in milk. *Microchemical Journal*, *120*, 64–68.

Amantonico, A., Urban, P. L., & Zenobi, R. (2010). Analytical techniques for single-cell metabolomics: state of the art and trends. *Analytical and Bioanalytical Chemistry*, *398*(6), 2493–2504.

Anderson, L. C., DeHart, C. J., Kaiser, N. K., Fellers, R. T., Smith, D. F., Greer, J. B., et al. (2016). Identification and characterization of human proteoforms by top-down LC-21 tesla FT-ICR mass spectrometry. *Journal of Proteome Research*, *16*(2), 1087–1096.

Andrade, F. J., Shelley, J. T., Wetzel, W. C., Webb, M. R., Gamez, G., Ray, S. J., et al. (2008). Atmospheric pressure chemical ionization source. 1. Ionization of compounds in the gas phase. *Analytical Chemistry*, *80*(8), 2646–2653.

Andrade, G. C. R. M., Monteiro, S. H., Francisco, J. G., Figueiredo, L. A., Botelho, R. G., & Tornisielo, V. L. (2015). Liquid chromatography–electrospray ionization tandem mass spectrometry and dynamic multiple reaction monitoring method for determining multiple pesticide residues in tomato. *Food Chemistry*, *175*, 57–65.

Andrews, G. L., Simons, B. L., Young, J. B., Hawkridge, A. M., & Muddiman, D. C. (2011). Performance characteristics of a new hybrid quadrupole time-of-flight tandem mass spectrometer (TripleTOF 5600). *Analytical Chemistry*, *83*(13), 5442–5446.

Antignac, J. P., Courant, F., Pinel, G., Bichon, E., Monteau, F., Elliott, C., et al. (2011). Mass spectrometry-based metabolomics applied to the chemical safety of food. *TrAC Trends in Analytical Chemistry*, *30*(2), 292–301.

Anwar, K., Lakra, N., Singla-Pareek, S. L., & Pareek, A. (2016). Investigating abiotic stress response machinery in plants: the metabolomic approach. In *Innovative saline agriculture* (pp. 303–319). New Delhi: Springer.

Arapitsas, P., Della Corte, A., Gika, H., Narduzzi, L., Mattivi, F., & Theodoridis, G. (2016). Studying the effect of storage conditions on the metabolite content of red wine using HILIC LC–MS based metabolomics. *Food Chemistry, 197*, 1331–1340.

Arayne, M. S., Sultana, N., Zuberi, M. V., & Siddiqui, F. A. (2009). Spectrophotometric quantitation of metformin in bulk drug and pharmaceutical formulations using multivariate technique. *Indian Journal of Pharmaceutical Sciences, 71*(3), 331.

Arbona, V., Iglesias, D. J., & Gómez-Cadenas, A. (2015). Non-targeted metabolite profiling of citrus juices as a tool for variety discrimination and metabolite flow analysis. *BMC Plant Biology, 15*(1), 38.

Arges, C. G., & Ramani, V. (2013). Two-dimensional NMR spectroscopy reveals cation-triggered backbone degradation in polysulfone-based anion exchange membranes. *Proceedings of the National Academy of Sciences, 110*(7), 2490–2495.

Arrivault, S., Guenther, M., Fry, S. C., Fuenfgeld, M. M., Veyel, D., Mettler-Altmann, T., et al. (2015). Synthesis and use of stable-isotope-labeled internal standards for quantification of phosphorylated metabolites by LC–MS/MS. *Analytical Chemistry, 87*(13), 6896–6904.

Ashbrook, S. E., & McKay, D. (2016). Combining solid-state NMR spectroscopy with first-principles calculations—a guide to NMR crystallography. *Chemical Communications, 52* (45), 7186–7204.

Atanasov, D. (2015). Recent results from the Penning-trap mass spectrometer ISOLTRAP. *Verhandlungen der Deutschen Physikalischen Gesellschaft, 48*(2), 1.

Athersuch, T. (2016). Metabolome analyses in exposome studies: profiling methods for a vast chemical space. *Archives of Biochemistry and Biophysics, 589*, 177–186.

Awad, H., Khamis, M. M., & El-Aneed, A. (2015). Mass spectrometry, review of the basics: ionization. *Applied Spectroscopy Reviews, 50*(2), 158–175.

Bacsik, Z., Mink, J., & Keresztury, G. (2004). FTIR spectroscopy of the atmosphere. I. Principles and methods. *Applied Spectroscopy Reviews, 39*(3), 295–363.

Baker, D. R., & Kasprzyk-Hordern, B. (2011). Multi-residue analysis of drugs of abuse in wastewater and surface water by solid-phase extraction and liquid chromatography–positive electrospray ionisation tandem mass spectrometry. *Journal of Chromatography A, 1218*(12), 1620–1631.

Baker, M. J., Trevisan, J., Bassan, P., Bhargava, R., Butler, H. J., Dorling, K. M., et al. (2014). Using Fourier transform IR spectroscopy to analyze biological materials. *Nature Protocols, 9*(8), 1771.

Banerjee, S., & Mazumdar, S. (2012). Electrospray ionization mass spectrometry: a technique to access the information beyond the molecular weight of the analyte. *International Journal of Analytical Chemistry, 2012*.

Barry, J. A., Groseclose, M. R., Robichaud, G., Castellino, S., & Muddiman, D. C. (2015). Assessing drug and metabolite detection in liver tissue by UV-MALDI and IR-MALDESI mass spectrometry imaging coupled to FT-ICR MS. *International Journal of Mass Spectrometry, 377*, 448–455.

Bart, J., Kolkman, A. J., Oosthoek-de Vries, A. J., Koch, K., Nieuwland, P. J., Janssen, H., et al. (2009). A microfluidic high-resolution NMR flow probe. *Journal of the American Chemical Society, 131*(14), 5014–5015.

Bauer, A., Luetjohann, J., Rohn, S., Kuballa, J., & Jantzen, E. (2018). Ion chromatography tandem mass spectrometry (IC-MS/MS) multimethod for the determination of highly polar pesticides in plant-derived commodities. *Food Control, 86*, 71–76.

Beatty, P. H., Klein, M. S., Fischer, J. J., Lewis, I. A., Muench, D. G., & Good, A. G. (2016). Understanding plant nitrogen metabolism through metabolomics and computational approaches. *Plants, 5*(4), 39.

Beckers, J. L., & Boček, P. (2000). Sample stacking in capillary zone electrophoresis: principles, advantages and limitations. *Electrophoresis*, *21*(14), 2747–2767.

Benhabib, M., Chiesl, T. N., Stockton, A. M., Scherer, J. R., & Mathies, R. A. (2010). Multichannel capillary electrophoresis microdevice and instrumentation for in situ planetary analysis of organic molecules and biomarkers. *Analytical Chemistry*, *82*(6), 2372–2379.

Benigni, P., Porter, J., Ridgeway, M. E., Park, M. A., & Fernandez-Lima, F. (2018). Increasing analytical separation and duty cycle with nonlinear analytical mobility scan functions in TIMS-FT-ICR MS. *Analytical Chemistry*, *90*(4), 2446–2450.

Benito, S., Abrego, Z., Sánchez, A., Unceta, N., Goicolea, M. A., & Barrio, R. J. (2015). Characterization of organic gunshot residues in lead-free ammunition using a new sample collection device for liquid chromatography–quadrupole time-of-flight mass spectrometry. *Forensic Science International*, *246*, 79–85.

Benton, H. P., Ivanisevic, J., Mahieu, N. G., Kurczy, M. E., Johnson, C. H., Franco, L., et al. (2014). Autonomous metabolomics for rapid metabolite identification in global profiling. *Analytical Chemistry*, *87*(2), 884–891.

Bharti, S. K., & Roy, R. (2012). Quantitative 1H NMR spectroscopy. *TrAC Trends in Analytical Chemistry*, *35*, 5–26.

Bijlsma, L., Emke, E., Hernández, F., & De Voogt, P. (2013). Performance of the linear ion trap Orbitrap mass analyzer for qualitative and quantitative analysis of drugs of abuse and relevant metabolites in sewage water. *Analytica Chimica Acta*, *768*, 102–110.

Bizzini, A., Durussel, C., Bille, J., Greub, G., & Prod'hom, G. (2010). Performance of matrix-assisted laser desorption ionization-time of flight mass spectrometry for identification of bacterial strains routinely isolated in a clinical microbiology laboratory. *Journal of Clinical Microbiology*, *48*(5), 1549–1554.

Blakeman, K. H., Cavanaugh, C. A., Gilliland, W. M., & Ramsey, J. M. (2017). High pressure mass spectrometry of volatile organic compounds with ambient air buffer gas. *Rapid Communications in Mass Spectrometry*, *31*(1), 27–32.

Blat, J., Evans, A., Kim, H., Imre, E., Polok, L., Ila, V., et al. (2016). Big data analysis for media production. *Proceedings of the IEEE*, *104*(11), 2085–2113.

Block, M. (2016). Recent developments in Penning-trap mass spectrometry. *Nuclear Instruments and Methods in Physics Research Section B: Beam Interactions with Materials and Atoms*, *376*, 265–269.

Boccio, P., Rossi, C., Ioia, M., Cicalini, I., Sacchetta, P., & Pieragostino, D. (2016). Integration of metabolomics and proteomics in multiple sclerosis: From biomarkers discovery to personalized medicine. *Proteomics - Clinical Applications*, *10*(4), 470–484.

Bock, C., Farlik, M., & Sheffield, N. C. (2016). Multi-omics of single cells: strategies and applications. *Trends in Biotechnology*, *34*(8), 605–608.

Boekel, J., Chilton, J. M., Cooke, I. R., Horvatovich, P. L., Jagtap, P. D., Käll, L., et al. (2015). Multi-omic data analysis using Galaxy. *Nature Biotechnology*, *33*(2), 137.

Boundy, M. J., Selwood, A. I., Harwood, D. T., McNabb, P. S., & Turner, A. D. (2015). Development of a sensitive and selective liquid chromatography–mass spectrometry method for high throughput analysis of paralytic shellfish toxins using graphitised carbon solid phase extraction. *Journal of Chromatography A*, *1387*, 1–12.

Bourmaud, A., Gallien, S., & Domon, B. (2016). Parallel reaction monitoring using quadrupole-Orbitrap mass spectrometer: principle and applications. *Proteomics*, *16*(15-16), 2146–2159.

Bragança, G. P., Oliveira, D. C., & dos Santos Isaias, R. M. (2016). Compartmentalization of metabolites and enzymatic mediation in nutritive cells of Cecidomyiidae galls on *Piper arboreum* Aubl. (Piperaceae). *Journal of Plant Studies*, *6*(1), 11.

Brereton, R. G., & Lloyd, G. R. (2014). Partial least squares discriminant analysis: taking the magic away. *Journal of Chemometrics*, *28*(4), 213–225.

Bro, R., & Smilde, A. K. (2014). Principal component analysis. *Analytical Methods*, *6*(9), 2812–2831.
Brown, P. (2012). *High pressure liquid chromatography: biochemical and biomedical applications.* Elsevier.
Brown, T. A., Chen, H., & Zare, R. N. (2015). Identification of fleeting electrochemical reaction intermediates using desorption electrospray ionization mass spectrometry. *Journal of the American Chemical Society*, *137*(23), 7274–7277.
Brownnutt, M., Kumph, M., Rabl, P., & Blatt, R. (2015). Ion-trap measurements of electric-field noise near surfaces. *Reviews of Modern Physics*, *87*(4), 1419.
Buck, A., Ly, A., Balluff, B., Sun, N., Gorzolka, K., Feuchtinger, A., et al. (2015). High-resolution MALDI-FT-ICR MS imaging for the analysis of metabolites from formalin-fixed, paraffin-embedded clinical tissue samples. *The Journal of Pathology*, *237*(1), 123–132.
Bundy, J. G., Davey, M. P., & Viant, M. R. (2009). Environmental metabolomics: a critical review and future perspectives. *Metabolomics*, *5*(1), 3.
Calderón-Santiago, M., Priego-Capote, F., Turck, N., Robin, X., Jurado-Gámez, B., Sanchez, J. C., et al. (2015). Human sweat metabolomics for lung cancer screening. *Analytical and Bioanalytical Chemistry*, *407*(18), 5381–5392.
Campillo, N., Viñas, P., Férez-Melgarejo, G., & Hernández-Córdoba, M. (2015). Dispersive liquid–liquid microextraction for the determination of flavonoid aglycone compounds in honey using liquid chromatography with diode array detection and time-of-flight mass spectrometry. *Talanta*, *131*, 185–191.
Cappellin, L., Karl, T., Probst, M., Ismailova, O., Winkler, P. M., Soukoulis, C., et al. (2012). On quantitative determination of volatile organic compound concentrations using proton transfer reaction time-of-flight mass spectrometry. *Environmental Science & Technology*, *46*(4), 2283–2290.
Cattaneo, G., Ubiali, D., Calleri, E., Rabuffetti, M., Höfner, G. C., Wanner, K. T., et al. (2016). Development, validation and application of a 96-well enzymatic assay based on LC-ESI-MS/MS quantification for the screening of selective inhibitors against *Mycobacterium tuberculosis* purine nucleoside phosphorylase. *Analytica Chimica Acta*, *943*, 89–97.
Cech, N. B., & Enke, C. G. (2001). Practical implications of some recent studies in electrospray ionization fundamentals. *Mass Spectrometry Reviews*, *20*(6), 362–387.
Chan, G. C. Y., Shelley, J. T., Wiley, J. S., Engelhard, C., Jackson, A. U., Cooks, R. G., et al. (2011). Elucidation of reaction mechanisms responsible for afterglow and reagent-ion formation in the low-temperature plasma probe ambient ionization source. *Analytical Chemistry*, *83*(10), 3675–3686.
Chauhan, A., Goyal, M. K., & Chauhan, P. (2014). GC-MS technique and its analytical applications in science and technology. *Journal of Analytical and Bioanalytical Techniques*, *5*(6), 222.
Chen, C. H., Chen, T. C., Zhou, X., Kline-Schoder, R., Sorensen, P., Cooks, R. G., et al. (2015). Design of portable mass spectrometers with handheld probes: aspects of the sampling and miniature pumping systems. *Journal of the American Society for Mass Spectrometry*, *26*(2), 240–247.
Chen, X., Shen, B., Sun, J., Wang, C., Shan, H., Yang, C., et al. (2012). Characterization and comparison of nitrogen compounds in hydrotreated and untreated shale oil by electrospray ionization (ESI) Fourier transform ion cyclotron resonance mass spectrometry (FT-ICR MS). *Energy & Fuels*, *26*(3), 1707–1714.
Cheng, J., Lan, W., Zheng, G., & Gao, X. (2018). Metabolomics: a high-throughput platform for metabolite profile exploration. In *Computational systems biology* (pp. 265–292). New York, NY: Humana Press.
Chou, L., Kenig, F. P. H., Murray, A. E., Fritsen, C. H., & Doran, P. T. (2016, February). GC × GC-TOF-MS of metabolites of Lake Vida Brine (Antarctica): evidence for past and current biogeochemical processes. In *AGU fall meeting abstracts*.

Cinelli, G., Avino, P., Notardonato, I., Centola, A., & Russo, M. V. (2014). Study of XAD-2 adsorbent for the enrichment of trace levels of phthalate esters in hydroalcoholic food beverages and analysis by gas chromatography coupled with flame ionization and ion-trap mass spectrometry detectors. *Food Chemistry*, *146*, 181–187.
Clark, A. E., Kaleta, E. J., Arora, A., & Wolk, D. M. (2013). Matrix-assisted laser desorption ionization–time of flight mass spectrometry: a fundamental shift in the routine practice of clinical microbiology. *Clinical Microbiology Reviews*, *26*(3), 547–603.
Clemens, G., Hands, J. R., Dorling, K. M., & Baker, M. J. (2014). Vibrational spectroscopic methods for cytology and cellular research. *Analyst*, *139*(18), 4411–4444.
Comisarow, M. B., & Marshall, A. G. (1996). The early development of Fourier transform ion cyclotron resonance (FT-ICR) spectroscopy. *Journal of Mass Spectrometry*, *31*(6), 581–585.
Contrepois, K., Jiang, L., & Snyder, M. (2015). Optimized analytical procedures for the untargeted metabolomic profiling of human urine and plasma by combining hydrophilic interaction (HILIC) and reverse-phase liquid chromatography (RPLC)–mass spectrometry. *Molecular & Cellular Proteomics*, *14*(6), 1684–1695.
Correa-García, S., Bermúdez-Moretti, M., Travo, A., Déléris, G., & Forfar, I. (2014). FTIR spectroscopic metabolome analysis of lyophilized and fresh Saccharomyces cerevisiae yeast cells. *Analytical Methods*, *6*(6), 1855–1861.
Corte, L., Tiecco, M., Roscini, L., De Vincenzi, S., Colabella, C., Germani, R., et al. (2015). FTIR metabolomic fingerprint reveals different modes of action exerted by structural variants of *n*-alkyltropinium bromide surfactants on *Escherichia coli* and *Listeria innocua* cells. *PLoS One*, *10*(1), e0115275.
Costa, C., Maraschin, M., & Rocha, M. (2016). An R package for the integrated analysis of metabolomics and spectral data. *Computer Methods and Programs in Biomedicine*, *129*, 117–124.
Danezis, G. P., Anagnostopoulos, C. J., Liapis, K., & Koupparis, M. A. (2016). Multi-residue analysis of pesticides, plant hormones, veterinary drugs and mycotoxins using HILIC chromatography–MS/MS in various food matrices. *Analytica Chimica Acta*, *942*, 121–138.
Das, M. K., Bishwal, S. C., Das, A., Dabral, D., Varshney, A., Badireddy, V. K., et al. (2013). Investigation of gender-specific exhaled breath volatome in humans by GCxGC-TOF-MS. *Analytical Chemistry*, *86*(2), 1229–1237.
Davtian, N., Bard, E., Ménot, G., & Fagault, Y. (2018). The importance of mass accuracy in selected ion monitoring analysis of branched and isoprenoid tetraethers. *Organic Geochemistry*, *118*, 58–62.
Dawson, P. H. (Ed.), (2013). *Quadrupole mass spectrometry and its applications*: Elsevier.
Day, I. J. (2011). On the inversion of diffusion NMR data: Tikhonov regularization and optimal choice of the regularization parameter. *Journal of Magnetic Resonance*, *211*(2), 178–185.
De Iorio, M., Ebbels, T. M., & Stephens, D. A. (2008). Statistical techniques in metabolic profiling. In *Handbook of statistical genetics* (4th ed., pp. 347–373): Wiley.
de Villiers, A., Venter, P., & Pasch, H. (2016). Recent advances and trends in the liquid-chromatography–mass spectrometry analysis of flavonoids. *Journal of Chromatography A*, *1430*, 16–78.
Deepinder, F., Chowdary, H. T., & Agarwal, A. (2007). Role of metabolomic analysis of biomarkers in the management of male infertility. *Expert Review of Molecular Diagnostics*, 7(4), 351–358.
Delgado, J. L., Filippone, S., Martín-Domenech, A., Altable, M., Maroto, E., Langa, F., et al. (2011). Mass spectrometry studies of the retro-cycloaddition reaction of pyrrolidino and 2-pyrazolinofullerene derivatives under negative ESI conditions. *Journal of the American Society for Mass Spectrometry*, *22*(3), 557–567.

Derenne, A., Van Hemelryck, V., Lamoral-Theys, D., Kiss, R., & Goormaghtigh, E. (2013). FTIR spectroscopy: a new valuable tool to classify the effects of polyphenolic compounds on cancer cells. *Biochimica et Biophysica Acta (BBA) - Molecular Basis of Disease, 1832*(1), 46–56.

Dona, A. C., Kyriakides, M., Scott, F., Shephard, E. A., Varshavi, D., Veselkov, K., et al. (2016). A guide to the identification of metabolites in NMR-based metabonomics/metabolomics experiments. *Computational and Structural Biotechnology Journal, 14*, 135–153.

Dong, Q., Yan, X., Liang, Y., & Stein, S. E. (2016). In-depth characterization and spectral library building of glycopeptides in the tryptic digest of a monoclonal antibody using 1D and 2D LC–MS/MS. *Journal of Proteome Research, 15*(5), 1472–1486.

Driver, J. A., Nagornov, K. O., Kozhinov, A. N., Tsybin, Y. O., Kharchenko, A., & Amster, I. J. (2018). Multiparticle simulations of quadrupolar ion detection in an ion cyclotron resonance cell with four narrow aperture detection electrodes. *Journal of the American Society for Mass Spectrometry, 29*(1), 51–62.

Duarte, J., Lanz, B., & Gruetter, R. (2011). Compartmentalized cerebral metabolism of [1, 6-13C] Glucose determined by in vivo 13C NMR Spectroscopy at 14.1 T. Frontiers in Neuroenergetics, 3, 3.

Dumancas, G. G., Bello, G. A., Hughes, J., Murimi, R., Viswanath, L. C. K., Orndorff, C. O. N., et al. (2018). Visualization tools for big data analytics in quantitative chemical analysis: a tutorial in chemometrics. In *Handbook of research on big data storage and visualization techniques* (pp. 873–917): IGI Global.

Dumancas, G. G., Ramasahayam, S., Bello, G., Hughes, J., & Kramer, R. (2015). Chemometric regression techniques as emerging, powerful tools in genetic association studies. *TrAC Trends in Analytical Chemistry, 74*, 79–88.

Dunn, W. B., Erban, A., Weber, R. J., Creek, D. J., Brown, M., Breitling, R., et al. (2013). Mass appeal: metabolite identification in mass spectrometry-focused untargeted metabolomics. *Metabolomics, 9*(1), 44–66.

Duran, A. L., Yang, J., Wang, L., & Sumner, L. W. (2003). Metabolomics spectral formatting, alignment and conversion tools (MSFACTs). *Bioinformatics, 19*(17), 2283–2293.

Eberwein, J., Shen, W., & Jenerette, G. D. (2017). Michaelis-Menten kinetics of soil respiration feedbacks to nitrogen deposition and climate change in subtropical forests. *Scientific Reports*, 7(1), 1752.

Elbashir, A. A., & Aboul-Enein, H. Y. (2015). Separation and analysis of triazine herbcide residues by capillary electrophoresis. *Biomedical Chromatography, 29*(6), 835–842.

Eltony, A. M., Gangloff, D., Shi, M., Bylinskii, A., Vuletić, V., & Chuang, I. L. (2016). Technologies for trapped-ion quantum information systems. *Quantum Information Processing, 15*(12), 5351–5383.

Erickson, B. K., Jedrychowski, M. P., McAlister, G. C., Everley, R. A., Kunz, R., & Gygi, S. P. (2015). Evaluating multiplexed quantitative phosphopeptide analysis on a hybrid quadrupole mass filter/linear ion trap/orbitrap mass spectrometer. *Analytical Chemistry, 87*(2), 1241–1249.

Evans, M., Gallagher, J. A., Ratcliffe, I., & Williams, P. A. (2016). Determination of the degree of polymerisation of fructans from ryegrass and chicory using MALDI-TOF Mass Spectrometry and Gel Permeation Chromatography coupled to multiangle laser light scattering. *Food Hydrocolloids, 53*, 155–162.

Farag, M. A., Al-Mahdy, D. A., Meyer, A., Westphal, H., & Wessjohann, L. A. (2017). Metabolomics reveals biotic and abiotic elicitor effects on the soft coral *Sarcophyton ehrenbergi* terpenoid content. *Scientific Reports*, 7(1), 648.

Farag, M. A., Porzel, A., & Wessjohann, L. A. (2012). Comparative metabolite profiling and fingerprinting of medicinal licorice roots using a multiplex approach of GC–MS, LC–MS and 1D NMR techniques. *Phytochemistry, 76*, 60–72.

Farré, M., & Barceló, D. (2015). Ambient ionization techniques. *In Vol. 68. Comprehensive Analytical Chemistry* (pp. 245–273): Elsevier.
Fekete, S., Kohler, I., Rudaz, S., & Guillarme, D. (2014). Importance of instrumentation for fast liquid chromatography in pharmaceutical analysis. *Journal of Pharmaceutical and Biomedical Analysis, 87*, 105–119.
Felhofer, J. L., Blanes, L., & Garcia, C. D. (2010). Recent developments in instrumentation for capillary electrophoresis and microchip-capillary electrophoresis. *Electrophoresis, 31* (15), 2469–2486.
Fiehn, O. (2001). Combining genomics, metabolome analysis, and biochemical modelling to understand metabolic networks. *Comparative and Functional Genomics, 2* (3), 155–168.
Fiehn, O. (2016). Metabolomics by gas chromatography–mass spectrometry: combined targeted and untargeted profiling. *Current Protocols in Molecular Biology, 114*, 30.4.1–30.4.32.
Fiehn, O., & Major, H. (2005). Exact molecular mass determination of polar plant metabolites using GCT with chemical ionization. *Waters Application Library Number 720001260EN*.
Foroozandeh, M., Adams, R. W., Meharry, N. J., Jeannerat, D., Nilsson, M., & Morris, G. A. (2014). Ultrahigh-resolution NMR spectroscopy. *Angewandte Chemie International Edition, 53*(27), 6990–6992.
Fragkaki, A. G., Angelis, Y. S., Kiousi, P., Georgakopoulos, C. G., & Lyris, E. (2015). Comparison of sulfo-conjugated and gluco-conjugated urinary metabolites for detection of methenolone misuse in doping control by LC-HRMS, GC-MS and GC-HRMS. *Journal of Mass Spectrometry, 50*(5), 740–748.
Gabaldón, T., & Pittis, A. A. (2015). Origin and evolution of metabolic sub-cellular compartmentalization in eukaryotes. *Biochimie, 119*, 262–268.
Gabelica, V., & Pauw, E. D. (2005). Internal energy and fragmentation of ions produced in electrospray sources. *Mass Spectrometry Reviews, 24*(4), 566–587.
Genuit, W., & Chaabani, H. (2017). Comprehensive two-dimensional gas chromatography-field ionization time-of-flight mass spectrometry (GCxGC-FI-TOFMS) for detailed hydrocarbon middle distillate analysis. *International Journal of Mass Spectrometry, 413*, 27–32.
Ghaste, M., Mistrik, R., & Shulaev, V. (2016). Applications of Fourier transform ion cyclotron resonance (FT-ICR) and orbitrap based high resolution mass spectrometry in metabolomics and lipidomics. *International Journal of Molecular Sciences, 17*(6), 816.
Giacomoni, F., Fillatre, Y., Rothwell, J., Eisner, R., Cesaire, D., Pujos-Guillot, E., et al. (2014a). PhytoHub version 1.0: a food metabolome database dedicated to dietary phytochemicals. *11th NuGOweek nutrigenomics of foods. 2014; Castellammare di Stabia, ITA, 2014-09-08-2014-09-11*95.
Giacomoni, F., Le Corguillé, G., Monsoor, M., Landi, M., Pericard, P., Pétéra, M., et al. (2014b). Workflow4Metabolomics: a collaborative research infrastructure for computational metabolomics. *Bioinformatics, 31*(9), 1493–1495.
Gika, H. G., Theodoridis, G. A., Plumb, R. S., & Wilson, I. D. (2014). Current practice of liquid chromatography–mass spectrometry in metabolomics and metabonomics. *Journal of Pharmaceutical and Biomedical Analysis, 87*, 12–25.
Gillet, L. C., Navarro, P., Tate, S., Röst, H., Selevsek, N., Reiter, L., et al. (2012). Targeted data extraction of the MS/MS spectra generated by data-independent acquisition: a new concept for consistent and accurate proteome analysis. *Molecular & Cellular Proteomics, 11* (6)O111.016717.
Giri, A., Coutriade, M., Racaud, A., Okuda, K., Dane, J., Cody, R. B., et al. (2017). Molecular characterization of volatiles and petrochemical base oils by photo-ionization GC × GC-TOF-MS. *Analytical Chemistry, 89*(10), 5395–5403.

Gogus, F., Ozel, M. Z., Kocak, D., Hamilton, J. F., & Lewis, A. C. (2011). Analysis of roasted and unroasted *Pistacia terebinthus* volatiles using direct thermal desorption-GCxGC–TOF/MS. *Food Chemistry*, *129*(3), 1258–1264.

Gowda, H., Ivanisevic, J., Johnson, C. H., Kurczy, M. E., Benton, H. P., Rinehart, D., et al. (2014). Interactive XCMS online: simplifying advanced metabolomic data processing and subsequent statistical analyses. *Analytical Chemistry*, *86*(14), 6931–6939.

Graham, D. J., & Castner, D. G. (2012). Multivariate analysis of ToF-SIMS data from multicomponent systems: the why, when, and how. *Biointerphases*, 7(1-4), 49.

Granzotto, C., & Sutherland, K. (2017). Matrix assisted laser desorption ionization mass fingerprinting for identification of acacia gum in microsamples from works of art. *Analytical Chemistry*, *89*(5), 3059–3068.

Grapov, D., Wanichthanarak, K., & Fiehn, O. (2015). MetaMapR: pathway independent metabolomic network analysis incorporating unknowns. *Bioinformatics*, *31*(16), 2757–2760.

Green, F. M., Gilmore, I. S., & Seah, M. P. (2014). Molecular structure and identification through G-SIMS and SMILES. In *Surface analysis and techniques in biology* (pp. 141–170). Cham: Springer.

Griffin, J. L., Atherton, H., Shockcor, J., & Atzori, L. (2011). Metabolomics as a tool for cardiac research. *Nature Reviews Cardiology*, *8*(11), 630.

Grimalt, S., Sancho, J. V., Pozo, Ó. J., & Hernández, F. (2010). Quantification, confirmation and screening capability of UHPLC coupled to triple quadrupole and hybrid quadrupole time-of-flight mass spectrometry in pesticide residue analysis. *Journal of Mass Spectrometry*, *45*(4), 421–436.

Gromski, P. S., Muhamadali, H., Ellis, D. I., Xu, Y., Correa, E., Turner, M. L., et al. (2015). A tutorial review: metabolomics and partial least squares-discriminant analysis–a marriage of convenience or a shotgun wedding. *Analytica Chimica Acta*, *879*, 10–23.

Gromski, P. S., Xu, Y., Kotze, H. L., Correa, E., Ellis, D. I., Armitage, E. G., et al. (2014). Influence of missing values substitutes on multivariate analysis of metabolomics data. *Metabolites*, *4*(2), 433–452.

Gross, J. H. (2011). Matrix-assisted laser desorption/ionization. In *Mass spectrometry* (pp. 507–559). Berlin, Heidelberg: Springer.

Grossman, P. D., & Colburn, J. C. (Eds.), (2012). *Capillary electrophoresis: theory and practice*. Academic Press.

Gu, H., Pan, Z., Xi, B., Asiago, V., Musselman, B., & Raftery, D. (2011). Principal component directed partial least squares analysis for combining nuclear magnetic resonance and mass spectrometry data in metabolomics: application to the detection of breast cancer. *Analytica Chimica Acta*, *686*(1-2), 57–63.

Gullberg, J., Jonsson, P., Nordström, A., Sjöström, M., & Moritz, T. (2004). Design of experiments: an efficient strategy to identify factors influencing extraction and derivatization of *Arabidopsis thaliana* samples in metabolomic studies with gas chromatography/mass spectrometry. *Analytical Biochemistry*, *331*(2), 283–295.

Günther, H. (2013). *NMR spectroscopy: basic principles, concepts and applications in chemistry*. John Wiley & Sons.

Hadi, M. Y., Mohammed, G. J., & Hameed, I. H. (2016). Analysis of bioactive chemical compounds of *Nigella sativa* using gas chromatography-mass spectrometry. *Journal of Pharmacognosy and Phytotherapy*, *8*(2), 8–24.

Hageman, J. A., van den Berg, R. A., Westerhuis, J. A., van der Werf, M. J., & Smilde, A. K. (2008). Genetic algorithm based two-mode clustering of metabolomics data. *Metabolomics*, *4*(2), 141–149.

Hall, R. D., & de Maagd, R. A. (2014). Plant metabolomics is not ripe for environmental risk assessment. *Trends in Biotechnology*, *32*(8), 391–392.

Hameed, I. H., Ibraheam, I. A., & Kadhim, H. J. (2015). Gas chromatography mass spectrum and Fourier-transform infrared spectroscopy analysis of methanolic extract of *Rosmarinus oficinalis* leaves. *Journal of Pharmacognosy and Phytotherapy*, 7(6), 90–106.
Han, J., Danell, R. M., Patel, J. R., Gumerov, D. R., Scarlett, C. O., Speir, J. P., et al. (2008). Towards high-throughput metabolomics using ultrahigh-field Fourier transform ion cyclotron resonance mass spectrometry. *Metabolomics*, *4*(2), 128–140.
Hauler, C., & Vetter, W. (2015). A non-targeted gas chromatography/electron capture negative ionization mass spectrometry selected ion monitoring screening method for polyhalogenated compounds in environmental samples. *Rapid Communications in Mass Spectrometry*, *29*(7), 619–628.
Headley, J. V., Barrow, M. P., Peru, K. M., Fahlman, B., Frank, R. A., Bickerton, G., et al. (2011). Preliminary fingerprinting of Athabasca oil sands polar organics in environmental samples using electrospray ionization Fourier transform ion cyclotron resonance mass spectrometry. *Rapid Communications in Mass Spectrometry*, *25*(13), 1899–1909.
Heise, R., Arrivault, S., Szecowka, M., Tohge, T., Nunes-Nesi, A., Stitt, M., et al. (2014). Flux profiling of photosynthetic carbon metabolism in intact plants. *Nature Protocols*, *9*(8), 1803.
Hejazi, L., Guilhaus, M., Hibbert, D. B., & Ebrahimi, D. (2015). Gas chromatography with parallel hard and soft ionization mass spectrometry. *Rapid Communications in Mass Spectrometry*, *29*(1), 91–99.
Hendrickson, C. L., Quinn, J. P., Kaiser, N. K., Smith, D. F., Blakney, G. T., Chen, T., et al. (2015). 21 Tesla Fourier transform ion cyclotron resonance mass spectrometer: a national resource for ultrahigh resolution mass analysis. *Journal of the American Society for Mass Spectrometry*, *26*(9), 1626–1632.
Henry, V. J., Bandrowski, A. E., Pepin, A. S., Gonzalez, B. J., & Desfeux, A. (2014). OMICtools: an informative directory for multi-omic data analysis. *Database*, *2014*.
Heuser, S., Sabbar, M., Boge, R., Lucchini, M., Gallmann, L., Cirelli, C., et al. (2015). Photo ionization time delay in molecular hydrogen. *Ultrafast phenomena* (pp. 36–39). Vol. XIX (pp. 36–39). Cham: Springer.
Hird, S. J., Lau, B. P. Y., Schuhmacher, R., & Krska, R. (2014). Liquid chromatography-mass spectrometry for the determination of chemical contaminants in food. *TrAC Trends in Analytical Chemistry*, *59*, 59–72.
Hoff, R. B., Pizzolato, T. M., Peralba, M. D. C. R., Díaz-Cruz, M. S., & Barceló, D. (2015). Determination of sulfonamide antibiotics and metabolites in liver, muscle and kidney samples by pressurized liquid extraction or ultrasound-assisted extraction followed by liquid chromatography–quadrupole linear ion trap-tandem mass spectrometry (HPLC–QqLIT-MS/MS). *Talanta*, *134*, 768–778.
Hoker, J., Obersteiner, F., Bönisch, H., & Engel, A. (2014). Initializing a digital chromatography data archive for tropospheric air samples on Taunus Observatory Frankfurt by GC-TOF-MS. *EGU general assembly conference abstracts*. Vol. 16.
Hollywood, K., Brison, D. R., & Goodacre, R. (2006). Metabolomics: current technologies and future trends. *Proteomics*, *6*(17), 4716–4723.
Hommerson, P., Khan, A. M., de Jong, G. J., & Somsen, G. W. (2011). Ionization techniques in capillary electrophoresis-mass spectrometry: principles, design, and application. *Mass Spectrometry Reviews*, *30*(6), 1096–1120.
Hong, J., Yang, L., Zhang, D., & Shi, J. (2016). Plant metabolomics: an indispensable system biology tool for plant science. *International Journal of Molecular Sciences*, *17*(6), 767.
Horgan, R. P., & Kenny, L. C. (2011). 'Omic' technologies: genomics, transcriptomics, proteomics and metabolomics. *The Obstetrician & Gynaecologist*, *13*(3), 189–195.
Hubert, C., Houari, S., Rozet, E., Lebrun, P., & Hubert, P. (2015). Towards a full integration of optimization and validation phases: an analytical-quality-by-design approach. *Journal of Chromatography A*, *1395*, 88–98.

Hübschmann, H. J. (2015). *Handbook of GC-MS: fundamentals and applications*. John Wiley & Sons.

Huerta-Fontela, M., Galceran, M. T., & Ventura, F. (2010). Fast liquid chromatography–quadrupole-linear ion trap mass spectrometry for the analysis of pharmaceuticals and hormones in water resources. *Journal of Chromatography A*, *1217*(25), 4212–4222.

Hummel, J., Strehmel, N., Selbig, J., Walther, D., & Kopka, J. (2010). Decision tree supported substructure prediction of metabolites from GC-MS profiles. *Metabolomics*, *6*(2), 322–333.

Hutchings, J. A., Shields, M. R., Bianchi, T. S., & Schuur, E. A. (2018). A rapid and precise method for the analysis of underivatized amino acids in natural samples using volatile-ion-pairing reverse-phase liquid chromatography–electrospray ionization tandem mass spectrometry. *Organic Geochemistry*, *115*, 46–56.

Ibáñez, C., Simó, C., García-Cañas, V., Cifuentes, A., & Castro-Puyana, M. (2013). Metabolomics, peptidomics and proteomics applications of capillary electrophoresis-mass spectrometry in foodomics: a review. *Analytica Chimica Acta*, *802*, 1–13.

Ibáñez, C., Simó, C., García-Cañas, V., Gómez-Martínez, Á., Ferragut, J. A., & Cifuentes, A. (2012). CE/LC-MS multiplatform for broad metabolomic analysis of dietary polyphenols effect on colon cancer cells proliferation. *Electrophoresis*, *33*(15), 2328–2336.

Ifa, D. R., Wu, C., Ouyang, Z., & Cooks, R. G. (2010). Desorption electrospray ionization and other ambient ionization methods: current progress and preview. *Analyst*, *135*(4), 669–681.

Ignat, I., Volf, I., & Popa, V. I. (2011). A critical review of methods for characterisation of polyphenolic compounds in fruits and vegetables. *Food Chemistry*, *126*(4), 1821–1835.

Iram, F., Iram, H., Arif, M., Siddiqui, A. A., & Husain, A. (2016). Bioanalytical method development and validation by hyphenated technique (LC-MS/MS). *Journal of Pharmaceutical and Medicinal Chemistry*, *2*(1).

Ivanisevic, J., Zhu, Z. J., Plate, L., Tautenhahn, R., Chen, S., O'Brien, P. J., et al. (2013). Toward 'omic scale metabolite profiling: a dual separation–mass spectrometry approach for coverage of lipid and central carbon metabolism. *Analytical Chemistry*, *85*(14), 6876–6884.

Ivey, K. L., Rimm, E. B., Kraft, P., Clish, C. B., Cassidy, A., Hodgson, J., et al. (2017). Identifying the metabolomic fingerprint of high and low flavonoid consumers. *Journal of Nutritional Science*, *6*.

Jaiswal, P., Jha, S. N., Borah, A., Gautam, A., Grewal, M. K., & Jindal, G. (2015). Detection and quantification of soymilk in cow–buffalo milk using attenuated total reflectance Fourier transform infrared spectroscopy (ATR–FTIR). *Food Chemistry*, *168*, 41–47.

Jedrychowski, M. P., Wrann, C. D., Paulo, J. A., Gerber, K. K., Szpyt, J., Robinson, M. M., et al. (2015). Detection and quantitation of circulating human irisin by tandem mass spectrometry. *Cell Metabolism*, *22*(4), 734–740.

Jha, S. N., Jaiswal, P., Borah, A., Gautam, A. K., & Srivastava, N. (2015). Detection and quantification of urea in milk using attenuated total reflectance-Fourier transform infrared spectroscopy. *Food and Bioprocess Technology*, *8*(4), 926–933.

Jonsson, P., Johansson, A. I., Gullberg, J., Trygg, J., Grung, B., Marklund, S., et al. (2005). High-throughput data analysis for detecting and identifying differences between samples in GC/MS-based metabolomic analyses. *Analytical Chemistry*, 77(17), 5635–5642.

Jorge, T. F., Rodrigues, J. A., Caldana, C., Schmidt, R., van Dongen, J. T., Thomas-Oates, J., et al. (2016). Mass spectrometry-based plant metabolomics: metabolite responses to abiotic stress. *Mass Spectrometry Reviews*, *35*(5), 620–649.

Kaever, A., Landesfeind, M., Feussner, K., Mosblech, A., Heilmann, I., Morgenstern, B., et al. (2015). MarVis-pathway: integrative and exploratory pathway analysis of non-targeted metabolomics data. *Metabolomics*, *11*(3), 764–777.

Kaiser, N. K., Savory, J. J., McKenna, A. M., Quinn, J. P., Hendrickson, C. L., & Marshall, A. G. (2011). Electrically compensated Fourier transform ion cyclotron resonance cell for complex mixture mass analysis. *Analytical Chemistry*, *83*(17), 6907–6910.

Kalli, A., Smith, G. T., Sweredoski, M. J., & Hess, S. (2013). Evaluation and optimization of mass spectrometric settings during data-dependent acquisition mode: focus on LTQ-Orbitrap mass analyzers. *Journal of Proteome Research*, *12*(7), 3071–3086.

Karahalil, B. (2016). Overview of systems biology and omics technologies. *Current Medicinal Chemistry*, *23*(37), 4221–4230.

Karas, M., Glückmann, M., & Schäfer, J. (2000). Ionization in matrix-assisted laser desorption/ionization: singly charged molecular ions are the lucky survivors. *Journal of Mass Spectrometry*, *35*(1), 1–12.

Kašička, V. (2016). Recent developments in capillary and microchip electroseparations of peptides (2013–middle 2015). *Electrophoresis*, *37*(1), 162–188.

Khamis, M. M., Adamko, D. J., & El-Aneed, A. (2017). Mass spectrometric based approaches in urine metabolomics and biomarker discovery. *Mass Spectrometry Reviews*, *36*(2), 115–134.

Kim, H. K., Choi, Y. H., & Verpoorte, R. (2011). NMR-based plant metabolomics: where do we stand, where do we go? *Trends in Biotechnology*, *29*(6), 267–275.

Kim, N. S., & Lee, D. S. (2002). Comparison of different extraction methods for the analysis of fragrances from *Lavandula* species by gas chromatography–mass spectrometry. *Journal of Chromatography A*, *982*(1), 31–47.

Kirchner, M., Matisová, E., Hrouzková, S., & de Zeeuw, J. (2005). Possibilities and limitations of quadrupole mass spectrometric detector in fast gas chromatography. *Journal of Chromatography A*, *1090*(1-2), 126–132.

Kistler, H. C., & Broz, K. (2015). Cellular compartmentalization of secondary metabolism. *Frontiers in Microbiology*, *6*, 68.

Kitagawa, F., & Otsuka, K. (2014). Recent applications of on-line sample preconcentration techniques in capillary electrophoresis. *Journal of Chromatography A*, *1335*, 43–60.

Klepárník, K. (2015). Recent advances in combination of capillary electrophoresis with mass spectrometry: methodology and theory. *Electrophoresis*, *36*(1), 159–178.

Koek, M. M., Jellema, R. H., van der Greef, J., Tas, A. C., & Hankemeier, T. (2011). Quantitative metabolomics based on gas chromatography mass spectrometry: status and perspectives. *Metabolomics*, 7(3), 307–328.

Konermann, L., Ahadi, E., Rodriguez, A. D., & Vahidi, S. (2013). Electrospray ionization (ESI) generates intact gas-phase ions from analytes in solution for mass spectrometric investigations. ESI can proceed via different mechanisms. Low molecular weight analytes follow the ion evaporation model (IEM), whereas the charged residue model (CRM) applies to large globular species. A chain ejection model (CEM) has been proposed for disordered polymers. *Analytical Chemistry*, *85*(1), 2–9.

Kotte, S. C. B., Ch, R. R., Tulam, V. K., Raghavan, S., Dubey, P. K., & Murali, P. M. (2016). Qualitative analysis of gefitinib by quadrupole–time of flight (LCMS) coupled with PDA and ELSD detectors, using dual electrospray ionization. *International Journal of Research in Pharmaceutical Sciences*, *3*(1), 113–119.

Kraft, V., Weber, W., Grützke, M., Winter, M., & Nowak, S. (2015). Study of decomposition products by gas chromatography-mass spectrometry and ion chromatography-electrospray ionization-mass spectrometry in thermally decomposed lithium hexafluorophosphate-based lithium ion battery electrolytes. *RSC Advances*, *5*(98), 80150–80157.

Kruppa, G., & Zhong, W. (2016). Applications of high-resolving power, high-accuracy mass spectrometry for the structural elucidation of natural products. *Data acquisition and applications to compound classes* (pp. 177–198). *Modern NMR approaches to the structure elucidation of natural products: Vol. 2* (pp. 177–198). Royal Society of Chemistry.

Kuehnbaum, N. L., Kormendi, A., & Britz-McKibbin, P. (2013). Multisegment injection-capillary electrophoresis-mass spectrometry: a high-throughput platform for metabolomics with high data fidelity. *Analytical Chemistry*, *85*(22), 10664–10669.

Kuhn, R., & Hoffstetter-Kuhn, S. (2013). *Capillary electrophoresis: principles and practice.* Springer Science & Business Media.

Kumar, K. J., & Vijayan, V. (2014). An overview of liquid chromatography-mass spectroscopy instrumentation. *Pharmaceutical Methods*, *5*(2), 47–55.

Kumar, P. R., & Nagar, G. (2016). Identification of metabolites through GC/LC–MS processed data using different reference libraries and their comparison. *Journal of Pharmaceutical and Biomedical Sciences*, *6*(6), 363–368.

Kuroda, K., Kunimura, H., Fukaya, Y., Nakamura, N., & Ohno, H. (2014). 1H NMR evaluation of polar and nondeuterated ionic liquids for selective extraction of cellulose and xylan from wheat bran. *ACS Sustainable Chemistry & Engineering*, *2*(9), 2204–2210.

Lai, Z., Kind, T., & Fiehn, O. (2017). Using accurate mass gas chromatography–mass spectrometry with the MINE database for epimetabolite annotation. *Analytical Chemistry*, *89* (19), 10171–10180.

Lanza, I. R., Zhang, S., Ward, L. E., Karakelides, H., Raftery, D., & Nair, K. S. (2010). Quantitative metabolomics by 1H-NMR and LC-MS/MS confirms altered metabolic pathways in diabetes. *PLoS One*, *5*(5)e10538.

Larive, C. K., Barding, G. A., Jr., & Dinges, M. M. (2014). NMR spectroscopy for metabolomics and metabolic profiling. *Analytical Chemistry*, *87*(1), 133–146.

Law, K. P., & Lim, Y. P. (2013). Recent advances in mass spectrometry: data independent analysis and hyper reaction monitoring. *Expert Review of Proteomics*, *10*(6), 551–566.

Lerner, B. M., Gilman, J. B., Aikin, K. C., Atlas, E. L., Goldan, P. D., Graus, M., et al. (2017). An improved, automated whole air sampler and gas chromatography mass spectrometry analysis system for volatile organic compounds in the atmosphere. *Atmospheric Measurement Techniques*, *10*(1), 291.

Lervik, A., Kjelstrup, S., & Qian, H. (2015). Michaelis–Menten kinetics under non-isothermal conditions. *Physical Chemistry Chemical Physics*, *17*(2), 1317–1324.

Li, D. X., Gan, L., Bronja, A., & Schmitz, O. J. (2015a). Gas chromatography coupled to atmospheric pressure ionization mass spectrometry (GC-API-MS). *Analytica Chimica Acta*, *891*, 43–61.

Li, T., Cao, J., Li, Z., Wang, X., & He, P. (2016). Broad screening and identification of β-agonists in feed and animal body fluid and tissues using ultra-high performance liquid chromatography-quadrupole-orbitrap high resolution mass spectrometry combined with spectra library search. *Food Chemistry*, *192*, 188–196.

Li, X. Q., Zhang, Q. H., Ma, K., Li, H. M., & Guo, Z. (2015b). Identification and determination of 34 water-soluble synthetic dyes in foodstuff by high performance liquid chromatography–diode array detection–ion trap time-of-flight tandem mass spectrometry. *Food Chemistry*, *182*, 316–326.

Lindon, J. C., Nicholson, J. K., & Holmes, E. (Eds.), (2011). *The handbook of metabonomics and metabolomics.* Elsevier.

Lisec, J., Schauer, N., Kopka, J., Willmitzer, L., & Fernie, A. R. (2006). Gas chromatography mass spectrometry–based metabolite profiling in plants. *Nature Protocols*, *1*(1), 387.

Liu, J., Liu, Y., Wang, Y., Abozeid, A., Zu, Y. G., & Tang, Z. H. (2017). The integration of GC–MS and LC–MS to assay the metabolomics profiling in *Panax ginseng* and *Panax quinquefolius* reveals a tissue-and species-specific connectivity of primary metabolites and ginsenosides accumulation. *Journal of Pharmaceutical and Biomedical Analysis*, *135*, 176–185.

Lo Turco, V., Di Bella, G., Potortì, A. G., Tropea, A., Casale, E. K., Fede, M. R., et al. (2016). Determination of plasticisers and BPA in Sicilian and Calabrian nectar honeys by selected ion monitoring GC/MS. *Food Additives & Contaminants: Part A*, *33*(11), 1693–1699.

Lobodin, V. V., Nyadong, L., Ruddy, B. M., Curtis, M., Jones, P. R., Rodgers, R. P., et al. (2015). DART Fourier transform ion cyclotron resonance mass spectrometry for analysis of complex organic mixtures. *International Journal of Mass Spectrometry, 378*, 186–192.

Lommen, A. (2009). MetAlign: interface-driven, versatile metabolomics tool for hyphenated full-scan mass spectrometry data preprocessing. *Analytical Chemistry, 81*(8), 3079–3086.

Lommen, A., & Kools, H. J. (2012). MetAlign 3.0: performance enhancement by efficient use of advances in computer hardware. *Metabolomics, 8*(4), 719–726.

Long, C. P., & Antoniewicz, M. R. (2014). Quantifying biomass composition by gas chromatography/mass spectrometry. *Analytical Chemistry, 86*(19), 9423–9427.

Lorenzo, M., & Pico, Y. (2017). Gas chromatography and mass spectroscopy techniques for the detection of chemical contaminants and residues in foods. In *Chemical contaminants and residues in food* (2nd ed., pp. 15–50). (2nd ed.). Elsevier.

Lu, Y., Wang, Y., Ong, C. N., Subramaniam, T., Choi, H. W., Yuan, J. M., et al. (2016). Metabolic signatures and risk of type 2 diabetes in a Chinese population: an untargeted metabolomics study using both LC-MS and GC-MS. *Diabetologia, 59*(11), 2349–2359.

Maeda, H., Yamazaki, T., Nishiyama, Y., Hamada, M., Hashi, K., Shimizu, T., et al. (2016). Development of super-high-field NMR operated beyond 1 GHz using high-temperature superconducting coils. *eMagRes*. https://doi.org/10.1002/9780470034590.emrstm1492.

Mahrous, E. A., & Farag, M. A. (2015). Two dimensional NMR spectroscopic approaches for exploring plant metabolome: a review. *Journal of Advanced Research, 6*(1), 3–15.

Marcos, J., & Pozo, O. J. (2015). Derivatization of steroids in biological samples for GC–MS and LC–MS analyses. *Bioanalysis*, 7(19), 2515–2536.

Markley, J. L., Brüschweiler, R., Edison, A. S., Eghbalnia, H. R., Powers, R., Raftery, D., et al. (2017). The future of NMR-based metabolomics. *Current Opinion in Biotechnology, 43*, 34–40.

Marshall, D. D., & Powers, R. (2017). Beyond the paradigm: combining mass spectrometry and nuclear magnetic resonance for metabolomics. *Progress in Nuclear Magnetic Resonance Spectroscopy, 100*, 1–16.

Marshall, D. D., Lei, S., Worley, B., Huang, Y., Garcia-Garcia, A., Franco, R., et al. (2015). Combining DI-ESI–MS and NMR datasets for metabolic profiling. *Metabolomics, 11*(2), 391–402.

Masclaux-Daubresse, C., Clément, G., Anne, P., Routaboul, J. M., Guiboileau, A., Soulay, F., et al. (2014). Stitching together the multiple dimensions of autophagy using metabolomics and transcriptomics reveals impacts on metabolism, development, and plant responses to the environment in *Arabidopsis*. *The Plant Cell, 26*(5), 1857–1877.

Masiá, A., Suarez-Varela, M. M., Llopis-Gonzalez, A., & Pico, Y. (2016). Determination of pesticides and veterinary drug residues in food by liquid chromatography-mass spectrometry: a review. *Analytica Chimica Acta, 936*, 40–61.

Mastrangelo, A., Ferrarini, A., Rey-Stolle, F., García, A., & Barbas, C. (2015). From sample treatment to biomarker discovery: a tutorial for untargeted metabolomics based on GC-(EI)-Q-MS. *Analytica Chimica Acta, 900*, 21–35.

McAlister, G. C., & Coon, J. J. (2016). Methods for multi-stage ion processing involving ion/ion chemistry in a quadrupole linear ion trap. *Applications of Ion Trapping Devices* (p. 59). *Practical Aspects of Trapped Ion Mass Spectrometry: Vol. V.* (p. 59). Taylor and Francis.

McKenna, A. M., Purcell, J. M., Rodgers, R. P., & Marshall, A. G. (2009). Identification of vanadyl porphyrins in a heavy crude oil and raw asphaltene by atmospheric pressure photoionization Fourier transform ion cyclotron resonance (FT-ICR) mass spectrometry. *Energy & Fuels, 23*(4), 2122–2128.

McLafferty, F. W. (2008). Mass spectrometry across the sciences. *Proceedings of the National Academy of Sciences, 105*(47), 18088–18089.

McNair, H. M., & Miller, J. M. (2011). *Basic gas chromatography*. John Wiley & Sons.
Meher, A. K., & Chen, Y. C. (2017). Electrospray modifications for advancing mass spectrometric analysis. *Mass Spectrometry*, *6*(2), S0057.
Mendes, T. P., Pereira, I., Ferreira, M. R., Chaves, A. R., & Vaz, B. G. (2017). Molecularly imprinted polymer-coated paper as a substrate for highly sensitive analysis using paper spray mass spectrometry: quantification of metabolites in urine. *Analytical Methods*, *9*(43), 6117–6123.
Merlet, B., Paulhe, N., Vinson, F., Frainay, C., Chazalviel, M., Poupin, N., et al. (2016). A computational solution to automatically map metabolite libraries in the context of genome scale metabolic networks. *Frontiers in Molecular Biosciences*, *3*, 2.
Michalski, A., Damoc, E., Hauschild, J. P., Lange, O., Wieghaus, A., Makarov, A., et al. (2011). Mass spectrometry-based proteomics using Q Exactive, a high-performance benchtop quadrupole Orbitrap mass spectrometer. *Molecular & Cellular Proteomics*, *10*(9), M111–011015.
Michalski, A., Damoc, E., Lange, O., Denisov, E., Nolting, D., Müller, M., et al. (2012). Ultra high resolution linear ion trap Orbitrap mass spectrometer (Orbitrap Elite) facilitates top down LC MS/MS and versatile peptide fragmentation modes. *Molecular & Cellular Proteomics*, *11*(3), O111–013698.
Misra, B. B. (2018). New tools and resources in metabolomics: 2016–2017. *Electrophoresis*, *39*(7), 909–923.
Misra, B. B., & der Hooft, J. J. (2016). Updates in metabolomics tools and resources: 2014–2015. *Electrophoresis*, *37*(1), 86–110.
Miyagi, A., Takahashi, H., Takahara, K., Hirabayashi, T., Nishimura, Y., Tezuka, T., et al. (2010). Principal component and hierarchical clustering analysis of metabolites in destructive weeds; polygonaceous plants. *Metabolomics*, *6*(1), 146–155.
Montenarh, D., Hopf, M., Warth, S., Maurer, H. H., Schmidt, P., & Ewald, A. H. (2015). A simple extraction and LC-MS/MS approach for the screening and identification of over 100 analytes in eight different matrices. *Drug Testing and Analysis*, 7(3), 214–240.
Montuschi, P., Paris, D., Melck, D., Lucidi, V., Ciabattoni, G., Raia, V., ... & Motta, A. (2011). NMR spectroscopy metabolomic profiling of exhaled breath condensate in patients with stable and unstable cystic fibrosis. Thorax, 67(3):222-8. doi: https://doi.org/10.1136/thoraxjnl-2011-200072
Moser, A. C., & Hage, D. S. (2008). Capillary electrophoresis-based immunoassays: Principles and quantitative applications. *Electrophoresis*, *29*(16), 3279–3295.
Mu, R., Shi, H., Yuan, Y., Karnjanapiboonwong, A., Burken, J. G., & Ma, Y. (2012). Fast separation and quantification method for nitroguanidine and 2, 4-dinitroanisole by ultrafast liquid chromatography–tandem mass spectrometry. *Analytical Chemistry*, *84*(7), 3427–3432.
Nadella, K. D., Marla, S. S., & Kumar, P. A. (2012). Metabolomics in agriculture. *OMICS*, *16*(4), 149–159.
Nägele, T. (2014). Linking metabolomics data to underlying metabolic regulation. *Frontiers in Molecular Biosciences*, *1*, 22.
Naseri, M. T., Shamsipur, M., Babri, M., Saeidian, H., Sarabadani, M., Ashrafi, D., et al. (2014). Determination of lewisite metabolite 2-chlorovinylarsonous acid in urine by use of dispersive derivatization liquid-liquid microextraction followed by gas chromatography–mass spectrometry. *Analytical and Bioanalytical Chemistry*, *406*(21), 5221–5230.
Naz, S., Moreira dos Santos, D. C., García, A., & Barbas, C. (2014). Analytical protocols based on LC–MS, GC–MS and CE–MS for nontargeted metabolomics of biological tissues. *Bioanalysis*, *6*(12), 1657–1677.
Neumann, S., & Böcker, S. (2010). Computational mass spectrometry for metabolomics: identification of metabolites and small molecules. *Analytical and Bioanalytical Chemistry*, *398*(7-8), 2779–2788.

Ni, Y., Su, M., Qiu, Y., Jia, W., & Du, X. (2016). ADAP-GC 3.0: improved peak detection and deconvolution of co-eluting metabolites from GC/TOF-MS data for metabolomics studies. *Analytical Chemistry, 88*(17), 8802–8811.

Nicolaou, N., Xu, Y., & Goodacre, R. (2010). Fourier transform infrared spectroscopy and multivariate analysis for the detection and quantification of different milk species. *Journal of Dairy Science, 93*(12), 5651–5660.

Nikolaev, E. N., Kostyukevich, Y. I., & Vladimirov, G. N. (2016). Fourier transform ion cyclotron resonance (FT ICR) mass spectrometry: theory and simulations. *Mass Spectrometry Reviews, 35*(2), 219–258.

Núñez, O., Gallart-Ayala, H., Martins, C. P., & Lucci, P. (2012). New trends in fast liquid chromatography for food and environmental analysis. *Journal of Chromatography A, 1228*, 298–323.

Nyamundanda, G., Brennan, L., & Gormley, I. C. (2010). Probabilistic principal component analysis for metabolomic data. *BMC Bioinformatics, 11*(1), 571.

Obersteiner, F., Boenisch, H., & Engel, A. (2015). *GC/TOF-MS as a new method for halocarbon observation in the atmosphere.* signal 250, 3.

Ogawa, M., Yamamoto, S., Kousa, Y., Nakamura, F., Yukawa, R., Fukushima, A., et al. (2012). Development of soft x-ray time-resolved photoemission spectroscopy system with a two-dimensional angle-resolved time-of-flight analyzer at SPring-8 BL07LSU. *Review of Scientific Instruments, 83*(2)023109.

Ojanperä, I., Mesihää, S., Rasanen, I., Pelander, A., & Ketola, R. A. (2016). Simultaneous identification and quantification of new psychoactive substances in blood by GC-APCI-QTOFMS coupled to nitrogen chemiluminescence detection without authentic reference standards. *Analytical and Bioanalytical Chemistry, 408*(13), 3395–3400.

Olmo-García, L., Bajoub, A., Fernández-Gutiérrez, A., & Carrasco-Pancorbo, A. (2016). Evaluating the potential of LC coupled to three alternative detection systems (ESI-IT, APCI-TOF and DAD) for the targeted determination of triterpenic acids and dialcohols in olive tissues. *Talanta, 150*, 355–366.

Orford, R., Clark, J. A., Nystrom, A., Savard, G., Aprahamian, A., Brodeur, M., et al. (2017). Phase-imaging mass measurements with the Canadian Penning trap mass spectrometer. *Proceedings of the 14th International Symposium on Nuclei in the Cosmos (NIC2016)*011102.

Orrego-Lagarón, N., Vallverdú-Queralt, A., Martínez-Huélamo, M., Lamuela-Raventos, R. M., & Escribano-Ferrer, E. (2016). Metabolic profile of naringenin in the stomach and colon using liquid chromatography/electrospray ionization linear ion trap quadrupole-Orbitrap-mass spectrometry (LC-ESI-LTQ-Orbitrap-MS) and LC-ESI-MS/MS. *Journal of Pharmaceutical and Biomedical Analysis, 120*, 38–45.

Ostergaard, J., & Heegaard, N. H. (2003). Capillary electrophoresis frontal analysis: Principles and applications for the study of drug-plasma protein binding. *Electrophoresis, 24*(17), 2903–2913.

Paglia, G., & Astarita, G. (2017). Metabolomics and lipidomics using traveling-wave ion mobility mass spectrometry. *Nature Protocols, 12*(4), 797.

Pagnotti, V. S., Chubatyi, N. D., & McEwen, C. N. (2011). Solvent assisted inlet ionization: an ultrasensitive new liquid introduction ionization method for mass spectrometry. *Analytical Chemistry, 83*(11), 3981–3985.

Pan, Z., & Raftery, D. (2007). Comparing and combining NMR spectroscopy and mass spectrometry in metabolomics. *Analytical and Bioanalytical Chemistry, 387*(2), 525–527.

Passarelli, M. K., & Winograd, N. (2011). Lipid imaging with time-of-flight secondary ion mass spectrometry (ToF-SIMS). *Biochimica et Biophysica Acta (BBA) - Molecular and Cell Biology of Lipids, 1811*(11), 976–990.

Patel, D. P., Krausz, K. W., Xie, C., Beyoğlu, D., Gonzalez, F. J., & Idle, J. R. (2017). Metabolic profiling by gas chromatography-mass spectrometry of energy metabolism in high-fat diet-fed obese mice. *PLoS One, 12*(5)e0177953.

Patel, R. (2013). Matrix-assisted laser desorption ionization–time of flight mass spectrometry in clinical microbiology. *Clinical Infectious Diseases*, *57*(4), 564–572.

Patti, G. J., Yanes, O., Shriver, L. P., Courade, J. P., Tautenhahn, R., Manchester, M., et al. (2012). Metabolomics implicates altered sphingolipids in chronic pain of neuropathic origin. *Nature Chemical Biology*, *8*(3), 232.

Perez-Gregorio, M. R., Mateus, N., & De Freitas, V. (2014). Rapid screening and identification of new soluble tannin–salivary protein aggregates in saliva by mass spectrometry (MALDI-TOF-TOF and FIA-ESI-MS). *Langmuir*, *30*(28), 8528–8537.

Pérez-Victoria, I., Martín, J., & Reyes, F. (2016). Combined LC/UV/MS and NMR strategies for the dereplication of marine natural products. *Planta Medica*, *82*(09/10), 857–871.

Peterson, A. C., Hauschild, J. P., Quarmby, S. T., Krumwiede, D., Lange, O., Lemke, R. A., et al. (2014). Development of a GC/Quadrupole-Orbitrap mass spectrometer, part I: design and characterization. *Analytical Chemistry*, *86*(20), 10036–10043.

Petrakis, E. A., Cagliani, L. R., Polissiou, M. G., & Consonni, R. (2015). Evaluation of saffron (*Crocus sativus* L.) adulteration with plant adulterants by 1H NMR metabolite fingerprinting. *Food Chemistry*, *173*, 890–896.

Pluskal, T., Castillo, S., Villar-Briones, A., & Orešič, M. (2010). MZmine 2: modular framework for processing, visualizing, and analyzing mass spectrometry-based molecular profile data. *BMC Bioinformatics*, *11*(1), 395.

Portolés, T., Mol, J. G., Sancho, J. V., & Hernández, F. (2014). Use of electron ionization and atmospheric pressure chemical ionization in gas chromatography coupled to time-of-flight mass spectrometry for screening and identification of organic pollutants in waters. *Journal of Chromatography A*, *1339*, 145–153.

Puchades-Carrasco, L., & Pineda-Lucena, A. (2015). Metabolomics in pharmaceutical research and development. *Current Opinion in Biotechnology*, *35*, 73–77.

Putri, S. P., Nakayama, Y., Matsuda, F., Uchikata, T., Kobayashi, S., Matsubara, A., et al. (2013). Current metabolomics: practical applications. *Journal of Bioscience and Bioengineering*, *115*(6), 579–589.

Qi, Y., & Volmer, D. A. (2017). Electron-based fragmentation methods in mass spectrometry: an overview. *Mass Spectrometry Reviews*, *36*(1), 4–15.

Rajasundaram, D., & Selbig, J. (2016). More effort—more results: recent advances in integrative 'omics' data analysis. *Current Opinion in Plant Biology*, *30*, 57–61.

Rajski, Ł., del Mar Gómez-Ramos, M., & Fernández-Alba, A. R. (2014). Large pesticide multiresidue screening method by liquid chromatography-Orbitrap mass spectrometry in full scan mode applied to fruit and vegetables. *Journal of Chromatography A*, *1360*, 119–127.

Ramadan, Z., Jacobs, D., Grigorov, M., & Kochhar, S. (2006). Metabolic profiling using principal component analysis, discriminant partial least squares, and genetic algorithms. *Talanta*, *68*(5), 1683–1691.

Raynie, D. E. (2018). Analyte derivatization as an important tool for sample preparation. *The LCGC 35th Anniversary Edition*, *36*(1), 14–17.

Redman, E. A., Mellors, J. S., Starkey, J. A., & Ramsey, J. M. (2016). Characterization of intact antibody drug conjugate variants using microfluidic capillary electrophoresis–mass spectrometry. *Analytical Chemistry*, *88*(4), 2220–2226.

Reiding, K. R., Blank, D., Kuijper, D. M., Deelder, A. M., & Wuhrer, M. (2014). High-throughput profiling of protein N-glycosylation by MALDI-TOF-MS employing linkage-specific sialic acid esterification. *Analytical Chemistry*, *86*(12), 5784–5793.

Ren, S., Hinzman, A. A., Kang, E. L., Szczesniak, R. D., & Lu, L. J. (2015). Computational and statistical analysis of metabolomics data. *Metabolomics*, *11*(6), 1492–1513.

Renault, M., Pawsey, S., Bos, M. P., Koers, E. J., Nand, D., Tommassen-van Boxtel, R., et al. (2012). Solid-state NMR spectroscopy on cellular preparations enhanced by

dynamic nuclear polarization. *Angewandte Chemie International Edition*, *51*(12), 2998–3001.
Ringle, R., Bollen, G., Brodeur, M., Cooper, K., Eibach, M., Gulyuz, K., et al. (2016, September). First direct determination of the superallowed beta-decay QEC-value for O-14 via Penning trap mass spectrometry at the LEBIT facility. In *APS Division of Nuclear Physics Meeting Abstracts*.
Rocca-Serra, P., Salek, R. M., Arita, M., Correa, E., Dayalan, S., Gonzalez-Beltran, A., et al. (2016). Data standards can boost metabolomics research, and if there is a will, there is a way. *Metabolomics*, *12*(1), 14.
Ronsein, G. E., Pamir, N., von Haller, P. D., Kim, D. S., Oda, M. N., Jarvik, G. P., et al. (2015). Parallel reaction monitoring (PRM) and selected reaction monitoring (SRM) exhibit comparable linearity, dynamic range and precision for targeted quantitative HDL proteomics. *Journal of Proteomics*, *113*, 388–399.
Ropodi, A. I., Panagou, E. Z., & Nychas, G. J. (2016). Data mining derived from food analyses using non-invasive/non-destructive analytical techniques; determination of food authenticity, quality & safety in tandem with computer science disciplines. *Trends in Food Science & Technology*, *50*, 11–25.
Rubakhin, S. S., Lanni, E. J., & Sweedler, J. V. (2013). Progress toward single cell metabolomics. *Current Opinion in Biotechnology*, *24*(1), 95–104.
Ruddy, B. M., Hendrickson, C. L., Rodgers, R. P., & Marshall, A. G. (2018). Positive ion electrospray ionization suppression in petroleum and complex mixtures. *Energy & Fuels*, *32*(3), 2901–2907.
Rudnicka, J., Kowalkowski, T., Ligor, T., & Buszewski, B. (2011). Determination of volatile organic compounds as biomarkers of lung cancer by SPME–GC–TOF/MS and chemometrics. *Journal of Chromatography B*, *879*(30), 3360–3366.
Ruiz-Matute, A. I., Hernandez-Hernandez, O., Rodriguez-Sanchez, S., Sanz, M. L., & Martinez-Castro, I. (2011). Derivatization of carbohydrates for GC and GC–MS analyses. *Journal of Chromatography B*, *879*(17-18), 1226–1240.
Saleh, N., Kim, H. J., Phenrat, T., Matyjaszewski, K., Tilton, R. D., & Lowry, G. V. (2008). Ionic strength and composition affect the mobility of surface-modified FeO nanoparticles in water-saturated sand columns. *Environmental Science & Technology*, *42*(9), 3349–3355.
Sârbu, C., Naşcu-Briciu, R. D., Kot-Wasik, A., Gorinstein, S., Wasik, A., & Namieśnik, J. (2012). Classification and fingerprinting of kiwi and pomelo fruits by multivariate analysis of chromatographic and spectroscopic data. *Food Chemistry*, *130*(4), 994–1002.
Saviano, G., Paris, D., Melck, D., Falasca, A., Trupiano, D., Iorizzi, M., et al. (2016). Monitoring spatial and temporal metabolic dynamics of woody poplar root under mechanical stress conditions by NMR-based metabolomics. *Metabolomics*, *12*(4), 65.
Savorani, F., Rasmussen, M. A., Mikkelsen, M. S., & Engelsen, S. B. (2013). A primer to nutritional metabolomics by NMR spectroscopy and chemometrics. *Food Research International*, *54*(1), 1131–1145.
Savory, J. J., Kaiser, N. K., McKenna, A. M., Xian, F., Blakney, G. T., Rodgers, R. P., et al. (2011). Parts-per-billion Fourier transform ion cyclotron resonance mass measurement accuracy with a "walking" calibration equation. *Analytical Chemistry*, *83*(5), 1732–1736.
Schauer, N., & Fernie, A. R. (2006). Plant metabolomics: towards biological function and mechanism. *Trends in Plant Science*, *11*(10), 508–516.
Schober, D., Jacob, D., Wilson, M., Cruz, J. A., Marcu, A., Grant, J. R., et al. (2017). nmrML: a community supported open data standard for the description, storage, and exchange of NMR data. *Analytical Chemistry*, *90*(1), 649–656.
Schuhmann, K., Almeida, R., Baumert, M., Herzog, R., Bornstein, S. R., & Shevchenko, A. (2012). Shotgun lipidomics on a LTQ Orbitrap mass spectrometer by successive switching between acquisition polarity modes. *Journal of Mass Spectrometry*, *47*(1), 96–104.

Scigelova, M., Hornshaw, M., Giannakopulos, A., & Makarov, A. (2011). Fourier transform mass spectrometry. *Molecular & Cellular Proteomics*, *10*(7), M111–009431.

Senko, M. W., Remes, P. M., Canterbury, J. D., Mathur, R., Song, Q., Eliuk, S. M., et al. (2013). Novel parallelized quadrupole/linear ion trap/Orbitrap tribrid mass spectrometer improving proteome coverage and peptide identification rates. *Analytical Chemistry*, *85*(24), 11710–11714.

Sergushichev, A. A., Loboda, A. A., Jha, A. K., Vincent, E. E., Driggers, E. M., Jones, R. G., et al. (2016). GAM: a web-service for integrated transcriptional and metabolic network analysis. *Nucleic Acids Research*, *44*(W1), W194–W200.

Shaw, J. B., Robinson, E. W., & Paša-Tolić, L. (2016). Vacuum ultraviolet photodissociation and Fourier transform–ion cyclotron resonance (FT-ICR) mass spectrometry: revisited. *Analytical Chemistry*, *88*(6), 3019–3023.

Shim, S. M., Kim, J. Y., Lee, S. M., Park, J. B., Oh, S. K., & Kim, Y. S. (2012). Profiling of fermentative metabolites in kimchi: Volatile and non-volatile organic acids. *Journal of Korean Society for Applied Biological Chemistry*, *55*(4), 463–469.

Shipkova, M., & Svinarov, D. (2016). LC–MS/MS as a tool for TDM services: where are we? *Clinical Biochemistry*, *49*(13–14), 1009–1023.

Short, L. C., Hanold, K. A., Cai, S. S., & Syage, J. A. (2007). Electrospray ionization/atmospheric pressure photoionization multimode source for low-flow liquid chromatography/mass spectrometric analysis. *Rapid Communications in Mass Spectrometry*, *21*(10), 1561–1566.

Siesler, H. W., Ozaki, Y., Kawata, S., & Heise, H. M. (Eds.), (2008). *Near-infrared spectroscopy: principles, instruments, applications*. John Wiley.

Silva, C. L., Passos, M., & Câmara, J. S. (2011). Investigation of urinary volatile organic metabolites as potential cancer biomarkers by solid-phase microextraction in combination with gas chromatography-mass spectrometry. *British Journal of Cancer*, *105*(12), 1894.

Simpson, D. C., & Smith, R. D. (2005). Combining capillary electrophoresis with mass spectrometry for applications in proteomics. *Electrophoresis*, *26*(7-8), 1291–1305.

Smith, D. F., Podgorski, D. C., Rodgers, R. P., Blakney, G. T., & Hendrickson, C. L. (2018). 21 tesla FT-ICR mass spectrometer for ultrahigh-resolution analysis of complex organic mixtures. *Analytical Chemistry*, *90*(3), 2041–2047.

Soga, T., Igarashi, K., Ito, C., Mizobuchi, K., Zimmermann, H. P., & Tomita, M. (2009). Metabolomic profiling of anionic metabolites by capillary electrophoresis mass spectrometry. *Analytical Chemistry*, *81*(15), 6165–6174.

Stafsnes, M. H., Dybwad, M., Brunsvik, A., & Bruheim, P. (2013). Large scale MALDI-TOF MS based taxa identification to identify novel pigment producers in a marine bacterial culture collection. *Antonie Van Leeuwenhoek*, *103*(3), 603–615.

Steinborn, A., Alder, L., Michalski, B., Zomer, P., Bendig, P., Martinez, S. A., et al. (2016). Determination of glyphosate levels in breast milk samples from Germany by LC-MS/MS and GC-MS/MS. *Journal of Agricultural and Food Chemistry*, *64*(6), 1414–1421.

Steingass, C. B., Langen, J., Carle, R., & Schmarr, H. G. (2015). Authentication of pineapple (*Ananas comosus* [L.] Merr.) fruit maturity stages by quantitative analysis of γ-and δ-lactones using headspace solid-phase microextraction and chirospecific gas chromatography–selected ion monitoring mass spectrometry (HS-SPME–GC–SIM-MS). *Food Chemistry*, *168*, 496–503.

Stobaugh, J. F., & Rabel, S. R. (2017). Capillary electrophoresis. In *Chromatographic analysis of pharmaceuticals* (2nd edition, pp. 209–238). (2nd edition). Routledge.

Strehmel, N., Kopka, J., Scheel, D., & Böttcher, C. (2014). Annotating unknown components from GC/EI-MS-based metabolite profiling experiments using GC/APCI (+)-QTOFMS. *Metabolomics*, *10*(2), 324–336.

Streng, A. S., de Boer, D., Bouwman, F. G., Mariman, E. C., Scholten, A., van Dieijen-Visser, M. P., et al. (2016). Validation, optimisation, and application data in support of the development of a targeted selected ion monitoring assay for degraded cardiac troponin T. *Data in Brief*, 7, 397–405.

Sudakov, M., Mamontov, E., Xu, F., Xu, C., & Ding, C. F. (2016). Possibility of operating quadrupole mass filter at high resolution. *International Journal of Mass Spectrometry*, *408*, 9–19.

Sudasinghe, N., Dungan, B., Lammers, P., Albrecht, K., Elliott, D., Hallen, R., et al. (2014). High resolution FT-ICR mass spectral analysis of bio-oil and residual water soluble organics produced by hydrothermal liquefaction of the marine microalga *Nannochloropsis salina*. *Fuel*, *119*, 47–56.

Sugimoto, M., Kawakami, M., Robert, M., Soga, T., & Tomita, M. (2012). Bioinformatics tools for mass spectroscopy-based metabolomic data processing and analysis. *Current Bioinformatics*, 7(1), 96–108.

Sugimoto, M., Wong, D. T., Hirayama, A., Soga, T., & Tomita, M. (2010). Capillary electrophoresis mass spectrometry-based saliva metabolomics identified oral, breast and pancreatic cancer-specific profiles. *Metabolomics*, *6*(1), 78–95.

Swartz, M. (2010). HPLC detectors: a brief review. *Journal of Liquid Chromatography & Related Technologies*, *33*(9-12), 1130–1150.

Tang, K., Liu, H., Li, X. Z., Hassan, Y. I., Shao, S., & Zhou, T. (2018). An efficient gas chromatography–mass spectrometry approach for the simultaneous analysis of deoxynivalenol and its bacterial metabolites 3-keto-DON and 3-epi-DON. *Journal of Food Protection*, *81*(2), 233–239.

Tayade, R., Nguyen, T., Oh, S. A., Hwang, Y. S., Yoon, I. S., Deshmuk, R., et al. (2018). Effective strategies for enhancing tolerance to high-temperature stress in rice during the reproductive and ripening stages. *Plant Breeding and Biotechnology*, *6*(1), 1–18.

Tfaily, M. M., Chu, R. K., Tolić, N., Roscioli, K. M., Anderton, C. R., Paša-Tolić, L., et al. (2015). Advanced solvent based methods for molecular characterization of soil organic matter by high-resolution mass spectrometry. *Analytical Chemistry*, *87*(10), 5206–5215.

Theodoridis, G., Gika, H. G., & Wilson, I. D. (2011). Mass spectrometry-based holistic analytical approaches for metabolite profiling in systems biology studies. *Mass Spectrometry Reviews*, *30*(5), 884–906.

Thévenot, E. A., Roux, A., Xu, Y., Ezan, E., & Junot, C. (2015). Analysis of the human adult urinary metabolome variations with age, body mass index, and gender by implementing a comprehensive workflow for univariate and OPLS statistical analyses. *Journal of Proteome Research*, *14*(8), 3322–3335.

Tian, Y., Decker, T. K., McClellan, J. S., Bennett, L., Li, A., De la Cruz, A., et al. (2018). Improved miniaturized linear ion trap mass spectrometer using lithographically patterned plates and tapered ejection slit. *Journal of the American Society for Mass Spectrometry*, *29*(2), 213–222.

Tikunov, A. P., Johnson, C. B., Lee, H., Stoskopf, M. K., & Macdonald, J. M. (2010). Metabolomic investigations of American oysters using 1H-NMR spectroscopy. *Marine Drugs*, *8*(10), 2578–2596.

Todaka, D., Zhao, Y., Yoshida, T., Kudo, M., Kidokoro, S., Mizoi, J., et al. (2017). Temporal and spatial changes in gene expression, metabolite accumulation and phytohormone content in rice seedlings grown under drought stress conditions. *The Plant Journal*, *90*(1), 61–78.

Töpfer, N., Kleessen, S., & Nikoloski, Z. (2015). Integration of metabolomics data into metabolic networks. *Frontiers in Plant Science*, *6*, 49.

Tranchida, P. Q., Franchina, F. A., Dugo, P., & Mondello, L. (2016). Comprehensive two-dimensional gas chromatography-mass spectrometry: recent evolution and current trends. *Mass Spectrometry Reviews*, *35*(4), 524–534.

Trevino, V., & Falciani, F. (2006). GALGO: an R package for multivariate variable selection using genetic algorithms. *Bioinformatics*, *22*(9), 1154–1156.

Trivedi, M. K., Panda, P., Sethi, K. K., & Jana, S. (2017). Metabolite profiling in *Withania somnifera* roots hydroalcoholic extract using LC/MS, GC/MS and NMR spectroscopy. *Chemistry & Biodiversity, 14*(3).

Tseng, W. C., Chen, P. S., & Huang, S. D. (2014). Optimization of two different dispersive liquid–liquid microextraction methods followed by gas chromatography–mass spectrometry determination for polycyclic aromatic hydrocarbons (PAHs) analysis in water. *Talanta, 120*, 425–432.

Tsugawa, H., Tsujimoto, Y., Arita, M., Bamba, T., & Fukusaki, E. (2011). GC/MS based metabolomics: development of a data mining system for metabolite identification by using soft independent modeling of class analogy (SIMCA). *BMC Bioinformatics, 12* (1), 131.

Tulpan, D., Léger, S., Belliveau, L., Culf, A., & Čuperlović-Culf, M. (2011). MetaboHunter: an automatic approach for identification of metabolites from 1 H-NMR spectra of complex mixtures. *BMC Bioinformatics, 12*(1), 400.

Ufer, M., Juif, P. E., Boof, M. L., Muehlan, C., & Dingemanse, J. (2017). Metabolite profiling in early clinical drug development: current status and future prospects. *Expert Opinion on Drug Metabolism & Toxicology, 13*(8).

Vaiano, F., Serpelloni, G., Focardi, M., Fioravanti, A., Mari, F., & Bertol, E. (2015). LC–MS/MS and GC–MS methods in propofol detection: evaluation of the two analytical procedures. *Forensic Science International, 256*(1-6).

Vallarino, J. G., & Osorio, S. (2016). Simultaneous determination of plant hormones by GC-TOF-MS. In *Plant signal transduction* (pp. 229–237). New York, NY: Humana Press.

van der Burgt, P. J., Finnegan, S., & Eden, S. (2015). Electron impact fragmentation of adenine: partial ionization cross sections for positive fragments. *The European Physical Journal D, 69*(7), 173.

Van der Heeft, E., Bolck, Y. J. C., Beumer, B., Nijrolder, A. W. J. M., Stolker, A. A. M., & Nielen, M. W. F. (2009). Full-scan accurate mass selectivity of ultra-performance liquid chromatography combined with time-of-flight and orbitrap mass spectrometry in hormone and veterinary drug residue analysis. *Journal of the American Society for Mass Spectrometry, 20*(3), 451–463.

Vaniya, A., & Fiehn, O. (2015). Using fragmentation trees and mass spectral trees for identifying unknown compounds in metabolomics. *TrAC Trends in Analytical Chemistry, 69*, 52–61.

Vasa, S. M., Guex, N., Wilkinson, K. A., Weeks, K. M., & Giddings, M. C. (2008). ShapeFinder: a software system for high-throughput quantitative analysis of nucleic acid reactivity information resolved by capillary electrophoresis. *RNA, 14*(10), 1979–1990.

Venkatesh, T. V., Chassy, A. W., Fiehn, O., Flint-Garcia, S., Zeng, Q., Skogerson, K., et al. (2016). Metabolomic assessment of key maize resources: GC-MS and NMR profiling of grain from B73 hybrids of the nested association mapping (NAM) founders and of geographically diverse landraces. *Journal of Agricultural and Food Chemistry, 64*(10), 2162–2172.

Vijayakumar, S., Conway, M., Lió, P., & Angione, C. (2018). Optimization of multi-omic genome-scale models: methodologies, hands-on tutorial, and perspectives. In *Metabolic network reconstruction and modeling* (pp. 389–408). New York, NY: Humana Press.

Villas-Bôas, S. G., Smart, K. F., Sivakumaran, S., & Lane, G. A. (2011). Alkylation or silylation for analysis of amino and non-amino organic acids by GC-MS? *Metabolites, 1*(1), 3–20.

Vinaixa, M., Schymanski, E. L., Neumann, S., Navarro, M., Salek, R. M., & Yanes, O. (2016). Mass spectral databases for LC/MS-and GC/MS-based metabolomics: state of the field and future prospects. *TrAC Trends in Analytical Chemistry, 78*, 23–35.

Wagner, S., Dittmar, T., & Jaffé, R. (2015). Molecular characterization of dissolved black nitrogen via electrospray ionization Fourier transform ion cyclotron resonance mass spectrometry. *Organic Geochemistry, 79*, 21–30.

Wang, Z., Zhang, M., & Harrington, P. D. B. (2014). Comparison of three algorithms for the baseline correction of hyphenated data objects. *Analytical Chemistry*, *86*(18), 9050–9057.
Want, E. J., Nordström, A., Morita, H., & Siuzdak, G. (2007). From exogenous to endogenous: the inevitable imprint of mass spectrometry in metabolomics. *Journal of Proteome Research*, *6*(2), 459–468.
Warth, B., Spangler, S., Fang, M., Johnson, C. H., Forsberg, E. M., Granados, A., et al. (2017). Exposome-scale investigations guided by global metabolomics, pathway analysis, and cognitive computing. *Analytical Chemistry*, *89*(21), 11505–11513.
Wasylenko, T. M., & Stephanopoulos, G. (2015). Metabolomic and 13C-metabolic flux analysis of a xylose-consuming *Saccharomyces cerevisiae* strain expressing xylose isomerase. *Biotechnology and Bioengineering*, *112*(3), 470–483.
Wen, W., Li, D., Li, X., Gao, Y., Li, W., Li, H., et al. (2014). Metabolome-based genome-wide association study of maize kernel leads to novel biochemical insights. *Nature Communications*, *5*, 3438.
Wiklund, S., Johansson, E., Sjöström, L., Mellerowicz, E. J., Edlund, U., Shockcor, J. P., et al. (2008). Visualization of GC/TOF-MS-based metabolomics data for identification of biochemically interesting compounds using OPLS class models. *Analytical Chemistry*, *80*(1), 115–122.
Wiktorowicz, J. E., & Colburn, J. C. (2012). Protein analyses by capillary electrophoresis. In *Capillary electrophoresis: theory and practice* (p. 273). Elsevier.
Wiseman, J. M., Ifa, D. R., Zhu, Y., Kissinger, C. B., Manicke, N. E., Kissinger, P. T., et al. (2008). Desorption electrospray ionization mass spectrometry: imaging drugs and metabolites in tissues. *Proceedings of the National Academy of Sciences*, *105*(47), 18120–18125.
Wishart, D. S. (2008a). Metabolomics: applications to food science and nutrition research. *Trends in Food Science & Technology*, *19*(9), 482–493.
Wishart, D. S. (2008b). Quantitative metabolomics using NMR. *TrAC Trends in Analytical Chemistry*, *27*(3), 228–237.
Wishart, D. S. (2011). Advances in metabolite identification. *Bioanalysis*, *3*(15), 1769–1782.
Wishart, D. S. (2016). Emerging applications of metabolomics in drug discovery and precision medicine. *Nature Reviews Drug Discovery*, *15*(7), 473.
Wolfender, J. L., Marti, G., Thomas, A., & Bertrand, S. (2015). Current approaches and challenges for the metabolite profiling of complex natural extracts. *Journal of Chromatography A*, *1382*, 136–164.
Wong, Y. F., Chin, S. T., Perlmutter, P., & Marriott, P. J. (2015). Evaluation of comprehensive two-dimensional gas chromatography with accurate mass time-of-flight mass spectrometry for the metabolic profiling of plant–fungus interaction in *Aquilaria malaccensis*. *Journal of Chromatography A*, *1387*, 104–115.
Worley, B., & Powers, R. (2013). Multivariate analysis in metabolomics. *Current Metabolomics*, *1*(1), 92–107.
Wu, G. (2013). *Statistical methods for the integrative analysis of high-throughput genomic data*. The Johns Hopkins University.
Wu, H., Southam, A. D., Hines, A., & Viant, M. R. (2008). High-throughput tissue extraction protocol for NMR-and MS-based metabolomics. *Analytical Biochemistry*, *372*(2), 204–212.
Wu, Z. Y., Zeng, Z. D., Xiao, Z. D., Mok, D. K. W., Liang, Y. Z., Chau, F. T., et al. (2017). Semiautomated alignment of high-throughput metabolite profiles with chemometric tools. *Journal of Analytical Methods in Chemistry*, *2017*.
Xia, J., Mandal, R., Sinelnikov, I. V., Broadhurst, D., & Wishart, D. S. (2012). MetaboAnalyst 2.0—a comprehensive server for metabolomic data analysis. *Nucleic Acids Research*, *40*(W1), W127–W133.
Xia, J., Sinelnikov, I. V., Han, B., & Wishart, D. S. (2015). MetaboAnalyst 3.0—making metabolomics more meaningful. *Nucleic Acids Research*, *43*(W1), W251–W257.

Xiao, J. F., Zhou, B., & Ressom, H. W. (2012). Metabolite identification and quantitation in LC-MS/MS-based metabolomics. *TrAC Trends in Analytical Chemistry*, *32*, 1–14.
Xiao, Z., Yu, D., Niu, Y., Chen, F., Song, S., Zhu, J., et al. (2014). Characterization of aroma compounds of Chinese famous liquors by gas chromatography–mass spectrometry and flash GC electronic-nose. *Journal of Chromatography B*, *945*, 92–100.
Xu, Y., & Goodacre, R. (2012). Multiblock principal component analysis: an efficient tool for analyzing metabolomics data which contain two influential factors. *Metabolomics*, *8*(1), 37–51.
Yadav, J., Rani, A., Singh, V., & Murari, B. M. (2015). Prospects and limitations of non-invasive blood glucose monitoring using near-infrared spectroscopy. *Biomedical Signal Processing and Control*, *18*, 214–227.
Yan, Z., Zhang, W., Gao, J., Lin, Y., Li, J., Lin, Z., et al. (2015). Reverse-phase high performance liquid chromatography separation of positional isomers on a MIL-53 (Fe) packed column. *RSC Advances*, *5*(50), 40094–40102.
Yang, D., Du, X., Yang, Z., Liang, Z., Guo, Z., & Liu, Y. (2014). Transcriptomics, proteomics, and metabolomics to reveal mechanisms underlying plant secondary metabolism. *Engineering in Life Sciences*, *14*(5), 456–466.
Yang, M., Luo, J., Li, K., Hu, S., Ding, T., Liu, H., et al. (2018). Determination and pharmacokinetic study of guaiol in rat plasma by gas chromatography–mass spectrometry with selected ion monitoring. *Journal of Chromatography B*, *1085*, 30–35.
Ye, J., Gao, Y., Tian, S., Su, J., & Zhang, W. (2018). A novel and effective mode-switching triple quadrupole mass spectrometric approach for simultaneous quantification of fifteen ginsenosides in *Panax ginseng*. *Phytomedicine*, *44*, 164–172.
Zenobi, R. (2013). Single-cell metabolomics: analytical and biological perspectives. *Science*, *342*(6163), 1243259.
Zhang, A., Sun, H., Wang, P., Han, Y., & Wang, X. (2012). Modern analytical techniques in metabolomics analysis. *Analyst*, *137*(2), 293–300.
Zhang, B., Li, X., & Yan, B. (2008a). Advances in HPLC detection—towards universal detection. *Analytical and Bioanalytical Chemistry*, *390*(1), 299–301.
Zhang, L., Li, P., Sun, X., Hu, W., Wang, X., Zhang, Q., et al. (2014). Untargeted fatty acid profiles based on the selected ion monitoring mode. *Analytica Chimica Acta*, *839*, 44–50.
Zhang, S., Gowda, G. N., Asiago, V., Shanaiah, N., Barbas, C., & Raftery, D. (2008b). Correlative and quantitative 1H NMR-based metabolomics reveals specific metabolic pathway disturbances in diabetic rats. *Analytical Biochemistry*, *383*(1), 76–84.
Zhang, W., Andersson, J. T., Räder, H. J., & Müllen, K. (2015a). Molecular characterization of large polycyclic aromatic hydrocarbons in solid petroleum pitch and coal tar pitch by high resolution MALDI ToF MS and insights from ion mobility separation. *Carbon*, *95*, 672–680.
Zhang, W., Hankemeier, T., & Ramautar, R. (2017). Next-generation capillary electrophoresis–mass spectrometry approaches in metabolomics. *Current Opinion in Biotechnology*, *43*, 1–7.
Zhang, W., Lei, Z., Huhman, D., Sumner, L. W., & Zhao, P. X. (2015b). MET-XAlign: a metabolite cross-alignment tool for LC/MS-based comparative metabolomics. *Analytical Chemistry*, *87*(18), 9114–9119.
Zhao, S. S., & Chen, D. D. (2014). Applications of capillary electrophoresis in characterizing recombinant protein therapeutics. *Electrophoresis*, *35*(1), 96–108.
Zhao, W. W., Zhang, C. Y., Yan, Z. G., Bai, L. P., Wang, X., Huang, H., et al. (2014). Separations of substituted benzenes and polycyclic aromatic hydrocarbons using normal-and reverse-phase high performance liquid chromatography with UiO-66 as the stationary phase. *Journal of Chromatography A*, *1370*, 121–128.
Zhao, W., Huang, X., Li, X., Zhang, F., Chen, S., Ye, M., et al. (2015). Qualitative and quantitative analysis of major triterpenoids in *Alismatis rhizoma* by high performance

liquid chromatography/diode-array detector/quadrupole-time-of-flight mass spectrometry and ultra-performance liquid chromatography/triple quadrupole mass spectrometry. *Molecules*, *20*(8), 13958–13981.

Zhao, X., & Metcalfe, C. D. (2008). Characterizing and compensating for matrix effects using atmospheric pressure chemical ionization liquid chromatography – tandem mass spectrometry: analysis of neutral pharmaceuticals in municipal wastewater. *Analytical Chemistry*, *80*(6), 2010–2017.

Zheng, X., Baker, H., & Hancock, W. S. (2006). Analysis of the low molecular weight serum peptidome using ultrafiltration and a hybrid ion trap-Fourier transform mass spectrometer. *Journal of Chromatography A*, *1120*(1-2), 173–184.

Zheng, X., Zhao, Y., Wu, H., Dong, J., & Feng, J. (2016). Origin identification and quantitative analysis of honeys by nuclear magnetic resonance and chemometric techniques. *Food Analytical Methods*, *9*(6), 1470–1479.

Zwanenburg, G., Hoefsloot, H. C., Westerhuis, J. A., Jansen, J. J., & Smilde, A. K. (2011). ANOVA–principal component analysis and ANOVA–simultaneous component analysis: a comparison. *Journal of Chemometrics*, *25*(10), 561–567.

FURTHER READING

Altieri, K. E., Seitzinger, S. P., Carlton, A. G., Turpin, B. J., Klein, G. C., & Marshall, A. G. (2008). Oligomers formed through in-cloud methylglyoxal reactions: chemical composition, properties, and mechanisms investigated by ultra-high resolution FT-ICR mass spectrometry. *Atmospheric Environment*, *42*(7), 1476–1490.

Cajka, T., & Fiehn, O. (2015). Toward merging untargeted and targeted methods in mass spectrometry-based metabolomics and lipidomics. *Analytical Chemistry*, *88*(1), 524–545.

Coutinho, I. D., Baker, J. M., Ward, J. L., Beale, M. H., Creste, S., & Cavalheiro, A. J. (2016). Metabolite profiling of sugarcane genotypes and identification of flavonoid glycosides and phenolic acids. *Journal of Agricultural and Food Chemistry*, *64*(21), 4198–4206.

Fiehn, O. (2008). Extending the breadth of metabolite profiling by gas chromatography coupled to mass spectrometry. *TrAC Trends in Analytical Chemistry*, *27*(3), 261–269.

Kim, D. H., Achcar, F., Breitling, R., Burgess, K. E., & Barrett, M. P. (2015). LC–MS-based absolute metabolite quantification: application to metabolic flux measurement in trypanosomes. *Metabolomics*, *11*(6), 1721–1732.

Kok, W. (2013). *Capillary electrophoresis: instrumentation and operation.* (Vol. 4). Springer Science & Business Media.

Paglia, G., Williams, J. P., Menikarachchi, L., Thompson, J. W., Tyldesley-Worster, R., Halldórsson, S., et al. (2014). Ion mobility derived collision cross sections to support metabolomics applications. *Analytical Chemistry*, *86*(8), 3985–3993.

Korytár, P., Parera, J., Leonards, P. E. G., De Boer, J., & Brinkman, U. T. (2005). Quadrupole mass spectrometer operating in the electron-capture negative ion mode as detector for comprehensive two-dimensional gas chromatography. *Journal of Chromatography A*, *1067*(1-2), 255–264.

Shi, Q., Hou, D., Chung, K. H., Xu, C., Zhao, S., & Zhang, Y. (2010). Characterization of heteroatom compounds in a crude oil and its saturates, aromatics, resins, and asphaltenes (SARA) and non-basic nitrogen fractions analyzed by negative-ion electrospray ionization Fourier transform ion cyclotron resonance mass spectrometry. *Energy & Fuels*, *24*(4), 2545–2553.

Sleighter, R. L., & Hatcher, P. G. (2008). Molecular characterization of dissolved organic matter (DOM) along a river to ocean transect of the lower Chesapeake Bay by ultrahigh resolution electrospray ionization Fourier transform ion cyclotron resonance mass spectrometry. *Marine Chemistry*, *110*(3-4), 140–152.

CHAPTER 6

Ecological Metabolomics: Challenges and Perspectives

6.1 INTRODUCTION

The advent of advanced metabolomic technologies has paved a way toward the understanding of mechanisms underlying physiological and ecological studies, thereby simplifying the most intricate relationships between organisms and the environment (Peñuelas & Sardans, 2009a, 2009b; Sardans, Penuelas, & Rivas-Ubach, 2011). Metabolomics reveals information about metabolites in organisms by providingus with a snapshot of the metabolome (Kusari, Hertweck, & Spiteller, 2012; Macel, Van, Nicole, & Keurentjes, 2010). Metabolomics provides phenotypical expression at the metabolic level in specific ecological conditions. Furthermore, it is an imperative tool for assessing the phenotypic variation of one genotype against varied ecological conditions, particularly in abiotic and biotic stresses (Mittler & Shulaev, 2013). Metabolomic studies enable instantaneous determination of primary and secondary metabolites that function as a defensive screen (Jones et al., 2013). Metabolomics offers more efficient evaluation methods of diverse competencies presented by phenotypic plasticity in plants, thereby displaying the actual metabolic pathway implicated in the phenotype response (Rivas-Ubach, Sardans, Pérez-Trujillo, Estiarte, & Peñuelas, 2012). Metabolomics integrated with genomics can facilitate the instant determination of genes implicated in adaptive processes (Moshelion & Altman, 2015; Van Aggelen et al., 2010). This could assist in revealing gene functions and ascertaining data networks, thereby advancing our knowledge pertaining to the advancement of phenotypic plasticity (Kuhlisch & Pohnert, 2015) and its evolutionary outlay (Hollert, Seiler, Blaha, & Young, 2007). Metabolomics has comparatively more benefits than previously established analytical approaches used in chemoecology, including sample fractionation, purification, and bioassays (Noecker et al., 2016; Prince & Pohnert, 2010). Metabolomics can estimate metabolites in a single step, instead of taking exhaustive steps for analyzing and identifying even unstable metabolites (Brunetti, George, Tattini, Field, & Davey, 2013; Rivas-Ubach et al., 2013). Metabolomics

Ecometabolomics
https://doi.org/10.1016/B978-0-12-814872-3.00006-0

can also be used for initial screening of metabolomic responses, which includes instant and resultant utilization of target chemical determinations (Kind et al., 2009). In recent years, there has been advancement of analytical methods and progression of computer hardware and software to decode and record expansive data sets (Kouzes, Anderson, Elbert, Gorton, & Gracio, 2009). This has increased the prospective of detecting and instantly estimating a high number of primary and secondary metabolites (Jonsson, Bruce, et al., 2005; Jonsson, Johansson, et al., 2005). These developments will provide us not only a snapshot of the metabolome, but also a dynamic picture of metabolites in organisms (Xia, Psychogios, Young, & Wishart, 2009; Zhang, Sun, Wang, Han, & Wang, 2012). Ecological metabolomics can function as an imperative signal for determining organism lifestyle. Hence, using metabolomics, we can have a lucid and inclusive layout of metabolism, which provides us with the health status of organisms, populations, and ecosystems; and in this way, we can pave the way toward innovative improvements in the overall ecology (Muller et al., 2018; Ouborg & Vriezen, 2007; Peters et al., 2018). This chapter will provide the state of the art of ecological relevance of the metabolomic technique, providing an outline of the present observations acquired by its utilization in ecophysiological and ecological research studies (Bundy, Davey, & Viant, 2009). To acquire such advancement, we emphasize the significance of integrating metabolomic studies with other omics research studies; for example, genomics, transcriptomics, and proteomics. The objective is to assess an outline of an organism's response against ecological variations at varied time scales, and from genotype to phenotype (Orgogozo, Morizot, & Martin, 2015; Seger & Sturm, 2007). We also demonstrate the environmental themes where metabolomics could be effectively employed. Among these themes, we emphasize interaction between organisms and the environment, and the utilization of imperative statistical tools in data mining. The word "ecometabolomics" has been derived from words adopted from Fiehn (2002) and Schripsema (2010) describing the integration of metabolomic techniques into ecological studies. Hence, ecometabolomics can be defined as the discipline that integrates metabolomic profiling into ecophysiological or ecological studies to demonstrate the effects of biotic or abiotic factors on particular inclusive pathways. Ecometabolomics also investigates the pathways used for detecting the metabolite or set of metabolites implicated in organism's response against ecological stress using the usual qualitative and quantitative analytical approaches, for example, nuclear magnetic resonance (NMR) or gas chromatography (GC)-mass spectrometry (MS), and data mining of statistical analysis (Simmler, Napolitano, McAlpine, Chen, & Pauli, 2014; Wolfender, Marti, Thomas, & Bertrand, 2015). Hence, metabolomic studies using only target techniques, for example, HPLC focused on a limited number of metabolites, are not known ecometabolomic studies (Tenenboim & Brotman, 2016). When specific metabolites, and not every metabolite, are identified and estimated, it becomes convenient to classify samples as per their origin or their ecological or ecophysiological significance. This is referred to as metabolic fingerprinting; which, in the ecological context, is referred to as ecometabolic fingerprinting. This method provides impartial investigation and assessment of sample molecular biochemistry, and, via an appropriate assessment, can be used to explain plant responses to ecological changes (Van Veld et al., 2018). If we need to know inclusive information about the whole metabolome, we have to analyze samples using NMR, GC-MS, which provides an overall estimation of all metabolites (Bingol et al., 2015; Vinaixa et al., 2016). These analytical approaches can demonstrate maximum data pertaining to the metabolome of the biological system referred to as metabolomics (Cai, Weng, Guo, Peng, & Zhu, 2015; Gowda et al., 2014). When we intend to differentiate metabolomic response of an organism against varied environmental changes, we define it as ecometabolomics. This discipline is most suitable for demonstrating inclusive responses of an organism to its environment.

6.2 METABOLOMIC DATABASES

Metabolomics is a data-based subject that is primarily based on characterization via an expansive utilization of databases. The relevance of the database ranges from compound characterizations, including actual mass, GC-EI spectrum, or MS/MS spectrum, in comparison with data integration and visualization represented as reactions, pathways, and cell compartments. The prime objective of metabolomics is to detect and estimate all the metabolites in biological samples (Fiehn, Kind, & Barupal, 2018; Vinaixa et al., 2016). Hence, it encompasses both metabolic fingerprinting and other metabolomic methods that use high-dimensional, nonannotated variables and multivariate statistics to explore biomarkers that have the possibility to be structurally identified in the following steps (Welzenbach et al., 2016). Hence, a significant step in metabolomics will be the detection or structural annotation of compounds in a high-throughput way. MS is a prime method for unbiased evaluation of small metabolites in life sciences. Using appropriate sample preparation and mass spectrometric methods, we can identify hundreds to thousands of metabolites (Cajka & Fiehn, 2015; Gika, Theodoridis, Plumb, & Wilson, 2014). However, the majority of metabolites in an expansive biological sample, such as plant tissues, are nonannotated and unknown metabolites because of the unavailability of experimental databases, chemical complexity, and the dynamic nature of an organism's metabolome (Ahmed, 2016; Sangwan et al., 2015). The prime step in metabolomics is structural analysis, commencing from the determination of elemental composition, as metabolites cannot be sequenced like proteins or polynucleotides (Kind & Fiehn, 2006). In addition to identification and estimation of metabolites, mass spectra can also be manipulated for structural explication of metabolites to decrease the integer of de novo explication for metabolomic studies. This is a practical approach that could instigate the instant annotation of metabolomic mass spectra with a list of metabolites that match the elemental composition of metabolites present in easily accessible databases (Smith et al., 2005). The significant metabolites vary with size, as well as quality of investigated databases; but even structures with different chemical confrontations, such as ladderanes, could not be debarred as a priori, on the basis of numerical databases (Dunn et al., 2013; Wolf, Schmidt, Müller-Hannemann, & Neumann, 2010). The provisional list of annotations could be further limited in the following steps by inculcating more physiochemical or biological data, for example, comparing anticipated outcomes against identified MS/MS fragmentation patterns (Gehlenborg et al., 2010), or similarity evaluation from manipulating genomic data pertaining to biochemical pathways of an organism (Hou, Acharya, Zhu, & Cheng, 2015; Oberhardt, Palsson, & Papin, 2009). There is the possibility of impropriety in determining actual mass spectrometric methods because of integrated combinatorial detonation without using reference standards (Green, Łatuszyński, Pereyra, & Robert, 2015). MS is ineffective in differentiating between stereoisomers (Hofmann, Hahm, Seeberger, & Pagel, 2015; Wolf et al., 2010). The whole molecular mass of each element is a rational number derived from different numbers of neutrons, protons, and electrons (Todeschini & Consonni, 2009). Hence, the actual mass of the metabolite is different from the nominal mass. This characteristic can be manipulated for recursively calculating the elemental composition from the unidentified metabolite mass spectrum in the ranges of the measurement of error (Schrimpe-Rutledge, Sherrod, & McLean, 2018; Vaniya & Fiehn, 2015).

Mass spectrometers can determine mass/charge ratios with high <5 ppm error; parts per million, or very high mass accuracy (<1 ppm) (Eiler et al., 2017), and can be acquired using software algorithms that originate a list of possible elemental compositions from the estimated monoisotopic mass (del Mar Gómez-Ramos, Lozano, Rajski, & Fernández-Alba, 2017). For this method, we not only require precision in mass determination, but also high resolving power of the mass spectrometer (Groessl, Graf, & Knochenmuss, 2015; Silverstein, Webster, Kiemle, & Bryce, 2014). The output mass spectrum is characterized as a Gaussian peak shape, and some peaks overlap with each other due to low resolving power (Marcoux et al., 2015; Marty et al., 2015). Resolving power can be determined at particular *m/z* with a full-width, half-height maximum of the peak. Quadrupole mass spectrometers generally can access 3000 (Loboda, Krutchinsky, Bromirski, Ens, & Standing, 2000); TOF analyzers up to 10,000, and Fourier transform ion cyclotron resonance (FT-ICR) mass spectrometers can have resolving power up to 1,000,000 or larger (Gavard, Rossell, Spencer, & Barrow, 2017). On the contrary, isobaric masses cannot be determined by MS only, for this we need chromatography for separating overlapping metabolites (Rowland, Robbins, Corilo, Marshall, & Rodgers, 2014). Further information is needed that can be easily collected from mass spectra such as the abundance of natural isotopes in metabolites that display the fraction in which isotopes of an element are present in natural sources on earth. The pattern of isotopic abundance of a metabolite's mass spectrum can function as an imperative limitation for eliminating elemental composition (Schuhmann et al., 2017). Determination of isotope abundance can be determined by an isotope ratio mass spectrometer, but it has relatively fewer applications for determining the elemental composition of unknown metabolites (Chahrour, Cobice, & Malone, 2015). The isotopic abundance of a particular molecular formula can be determined by solving polynomial equations, or via Fourier transformation (Jang, Chen, & Rabinowitz, 2018). Isotopic patterns can be determined from very root mean square by using an isotope abundance filter, particularly if the implication of further metabolites can be overlooked by integrating compound isolation into mass spectrometric detection via liquid (LC) or gas chromatography (GC) (Olson & Yergey, 2009; Rockwood & Haimi, 2006). Using chromatography or high resolution MS, we can reveal lucid and clear mathematical interpretations of pseudo-molecular ions (Bhardwaj & Hanley, 2014; Henke & Kelleher, 2016). Mass spectra resulting from these instruments can produce fragmentation, rearrangement, and adducts (Li, Gan, Bronja, & Schmitz, 2015; Silverstein et al., 2014). The unique elemental composition could be determined from mass spectrum, and this molecular interpretation can be provided for metabolite annotation via two different approaches:

(a) Exhaustive computation of isomeric structures
(b) Query of databases of known metabolites

An exhaustive method employs either the deterministic method or the stochastic molecular isomer generator. Exhaustive methods for a particular molecular formula, numerous hundreds to billions of isomers, can be built depending on the number and nature of elements given by the chemical composition. The number of molecular formulas for the 11 most common elements at 1000u is known to be more than 350 million (Kind & Fiehn, 2006). Using the deterministic method, we can analyze small molecules by electron impact MS via MOLGEN-MS (Hufsky, Rempt, Rasche, Pohnert, & Böcker, 2012; Kind & Fiehn, 2010). On the contrary, the deterministic method can be employed for determination of high molecular weight compounds; however, this method is confronted by computing power limits due to the combinatorial explosion of isomeric structures, which may provide a stochastic isomer generator that is more promising for the future (Saunders, 2004). Facilitating the automatic annotation of metabolites in metabolomic screens appears to be practical for exploring the existing chemical structures, or even restricting the search for known natural product databases (Dias et al., 2016; Wolfender et al., 2015). There are various repositories, such as the life science oriented PubChem database of the U.S. National Institutes of Health, and its sub-DB Chemi ID plus, the Kyoto ligand biochemical pathway database (KEGG), the CRC dictionary of natural products (DNP), and an expansive compendium of organic chemical structures (Beilstein), range of commercially available chemicals employed for authenticating any given hit (MDL), mass spectrum library utilized in GC-MS (NIST5.0), and list of complementary small molecules as represented in the Chemical Abstracts Service (CAS) database. Comprehensive repositories such as CAS or DNP acquire a more expansive range of hits than a database procured from KEGG (Go, 2010; Kind et al., 2009). Hence, it is imperative to limit our automatic annotation of mass spectra to KEGG exploration (Cheng, Lan, Zheng, & Gao, 2018; Kanehisa, 2008, 2016). However, any metabolite procured from a KEGG pathway exhibits a greater resemblance to known metabolites because of the focus on conserved biochemical pathways signified in KEGG (Kanehisa, Furumichi, Tanabe, Sato, & Morishima, 2016; Kotera, Moriya, Tokimatsu, Kanehisa, & Goto, 2015). Unlike the KEGG database, CAS signifies expansive databases that are accessible for small molecules containing $\sim$ 20 million organic chemicals (Karp et al., 2015); but CAS cannot function as an appropriate option for usual metabolite exploration. CAS encompasses synthetically synthesized compounds; hence, it is improbable that it will represent natural compounds. On the contrary, the CAS SciFinder front end facilitates restricted and deliberate formula searches, providing an exploration of one formula at a time, but not a series or batches of queries (Swoger & Helms, 2015). For these two reasons, CAS queries can be debarred from automated annotation efforts of complex metabolomic surveys. However, for identification relevance of specified unidentified metabolites in biomarker studies, the CAS database still offers the main inclusive outline. Though PubChem is a rapidly growing and freely available database, it retrieves only 19 hits for $C_{15}H_{12}O_7$ as compared with DNP, retrieving 129 diverse isomeric structures for the same compound. In recent years, PubChem has reached its limit of 5 million entries, which has grown from its initial record of 800,000 entries. Because PubChem is an easily accessible database, and encompasses KEGG, ChemIDplus, NCBI, and numerous other databases, it therefore should be included in automatic metabolite annotations (Kim, Dreher et al., 2015; Kim, Thiessen et al., 2015; Wang & Bryant, 2014). A thoroughly molecular diversity calculation could exhibit overlap and for an automated method; in a DNP database, we could employ the SD file format (*.sdf), while semi-automatic methods would be promising for Beilstein database. Therefore, for comprehensive exploration of unidentified metabolites commencing from elemental composition, DNP and PubChem searches should be integrated.

Following is a summary of the numerous metabolomic databases:

6.2.1 Comprehensive Metabolomic Databases

Human Metabolome Database (HMDB)

This database is freely accessible, and encompasses inclusive data pertaining to small metabolites in the human body. This database is comprised of three types of data a) Chemical data b) Clinical data c) Biochemistry/Biology data. HMDB is comprised of data on more than 6500 metabolites, along with 1500 proteins and some DNA sequences related to these metabolites. In this database, information is stored in MetaboCard, which contains more than 100 data fields, with 2/3 of data pertaining to chemical/clinical data, and 1/3 related to enzymatic or biochemical data. Numerous data fields are hyperlinked to other databases, such as KEGG, PubChem, MetaCyc, ChEBI, PDB, Swiss-Prot, and GenBank, and diverse structures and pathways presentation applets.

BiGG

The BiGG database is a metabolic reconstruction of human metabolism designed for systems biology simulations and metabolic flux balance modeling. It is a comprehensive, literature-based, genome-scale metabolic reconstruction that accounts for the functions of 1,496 ORFs, 2,004 proteins,

2,766 metabolites, and 3,311 metabolic and transport reactions. It was assembled from Build 35 of the human genome.

SetupX

The XML-compatible web-based metabolomics LIMS developed SetupX under the supervision of the Fiehn Laboratory at UC Davis, and was constructed around a relational database management core. This metabolomic database is particularly specified for GC-MS data via its metabolic database called BinBase. BinBase is a GC-TOF metabolomic database built for data queries.

SYSTOMONAS (SYSTems biology of PseudOMONAS)

This database is based on the biological research on the Pseudomonas species, which encompasses widespread transcriptomic, proteomic, and metabolomic data; together with the metabolic reconstruction of this pathogen. Comparative genomics was employed to reconstruct metabolic networks in SYSTOMONAS. Extensive integration with conventional databases BRENDA, KEGG, and PRODORIC is also regulated. Numerous contrivances for the evaluation of accumulated data and for the identification of the analogous results are provided, facilitating an instant comprehension of metabolic pathways, genomic measures, and significant promoter structures.

MetaboLights Database

This is a cross-species, cross-technique database exclusively developed for metabolomic experiments and resultant data covering all metabolites and their known spectra, plus their biological functions, sites, accumulation levels, and investigational records from diverse research studies. MetaboLights provides user-submission contrivances, and has immense documenting aptitude. We would employ and advance effectively referenced formats where diverse compounds are summarized; for example, the programmed spectral and chromatographic information and related data pertaining to the chemical configuration, together with metadata illustrating assays and the overall research.

6.2.2 Metabolic Pathway Databases

KEGG

KEGG (Kyoto Encyclopedia of Genes and Genomes) is the inclusive and extensively utilized database comprised of metabolic pathways from diverse organisms. These pathways are attached to metabolites and proteins, as well as enzyme data. Currently, KEGG has more than fifteen thousand metabolites procured from diverse sources such as plants, animals, and bacteria. There are about 7,742 entries for drugs comprised of various salt forms and drug carriers, and almost 11,000 glycan structures.

MetaCyc

This is a unique database that experimentally demonstrates metabolic networks. MetaCyc includes more than 1,100 metabolic networks derived from 1,500 varied organisms. MetaCyc is premeditated from systematic investigational literature, and includes pathways involved in both primary and secondary metabolism associated with metabolites, enzymes, and genes.

HumanCyc

This database is based on bioinformatics that illustrate human metabolic pathways and the human genome. The current format of HumanCyc was built via Build 31 of the human genome. The resulting pathway/genome database (PGDB) includes data on 28,783 genes, their products, and metabolic networks, and the pathways they catalyze.

BioCyc

BioCyc includes more than 371 genome databases derived from metabolic networks and genomes from a whole organism. This database includes a BioCyc assortment arranged into layers based on a manual review and update of the content they have received.

Reactome

This database is built by combining a peer-reviewed database of biological pathways with protein trafficking and signalling pathways. Reactome includes different types of reactions in its schematic representation assortment, consisting of experimentally validated, manually interpreted, and digitally surmised reactions. Reactome includes pathway data from different organisms, more than 20 of which a primary source homosapiens. Reactome includes pathway diagrams from more than 2,700 proteins, 2,800 reactions, and 860 pathways for humans.

WikiPathways

WikiPathways is an open, shared platform for execution and allocation of models of biological pathways for data visualization and investigation. The human assembly includes more than 800 pathways, including more than 7,500 genes. WikiPathways also includes pathways with more than 1,000 metabolites.

6.2.3 Compound or Compound-Specific Databases

PubChem

PubChem is a common and easily accessible database of chemical structures of low-molecular-weight metabolites, and information on their biological relevance. It collects information on structure, nomenclature, and evaluated physio-chemical data, and is associated with NIH PubMed/Entrez data. PubChem was built from three associated databases among the NCBI's Entrez data retrieval system; PubChem chemical structures, PubChem metabolites, and PubChem bioassays. PubChem also contributes an instantaneous similarity exploratory tool. PubChem has more than 19 million discreet chemical structures.

ChEBI (Chemical Entities of Biological Interest)

This database is easily accessible assemblage of molecular compounds targeted on low-molecular-weight metabolites. The chemical structures in ChEBI are generated through organic routes via metabolites or synthetic products used to mediate the process of living organisms. ChEBI includes structures and nomenclature data, with together with hyperlinks to several well-established databases. The database has over 15,500 chemical entities.

ChemSpider

ChemSpider is an exclusive collection of organic compounds, including more than 20 million metabolites from different sources. At present, the database is comprised of information from diverse sources, including a marine natural product database, ACD-Labs chemical databases, the EPA's databases, and those from a series of chemical retailers. It has extensive investigative relevance, and the majority of metabolites have a wide range of determined physic-chemical property values.

KEGG Glycan

This database is an assortment of established glycan structures. It includes all distinct structures obtained from CarbBank, structures entered from recent publications, and structures present in KEGG pathways. KEGG glycan has more than 11,000 glycan structures from a wide range of eukaryotic and prokaryotic sources.

IIMDB

This database produces in vivo/in silico metabolite assemblages, including both documented and digitally resulting metabolites. This database can be accessed at http://metabolomics.pharm.uconn.edu/iimdb/, and includes more than 23,000 detected metabolites; for example, human metabolites, secondary metabolites, and glycerol–phospholipids acquired from available biochemical databases, together with more than 400,000 digitally produced human phase-I and phase II metabolites of these compounds. The IIMDB database identifies an accessible web boundary and programmer-friendly soporific web service.

6.2.4 Drug Databases

DrugBank

This database is a combination of bioinformatics and chemo-informatics sources that combine with elaborated drug (i.e., pharmaceutical, pharmacological, and chemical) data with demonstrated drug targets, comprised of sequence, structure, and pathway data. This database includes more than 4,800 drugs, including >1,350 FDA-authenticated small molecule drugs, 123 FDA-approved biotech (protein/peptide) drugs, 71 nutraceuticals, and >3,243 experimental drugs. DrugBank also includes an extensive range of SNP-drug data that is used for pharmacogenomics studies.

Therapeutic Target Database

This database contributes information regarding the remedial protein and nucleic acid targets demonstrated in the documented reports, the targeted disease conditions, the pathway data, and analogue drugs/ligands directed at each of these targets. The database presently includes 1,535 targets and 2,107 drugs/ligands.

PharmGKB

This database is the main repository for genetic, molecular, and cellular phenotypes, and clinical data regarding people who are involved in pharmacogenomics research studies. This database assists researchers in understanding genetic variations among individuals, providing their divergence in regard to reactions against drugs. PharmGKB includes investigative data on genes

(>20,000), diseases (>3000), drugs (>2500), and pathways (53). It is comprised of data on 470 genetic variants (SNP data) influencing drug metabolism.

STITCH (Search tool for interactions of chemicals)

This is an investigative database that integrates data linked to associations from metabolic pathways, crystal structures, binding experiments, and drug-target connections. This database contributes text mining and chemical structure similarity used for calculating the involvement between chemicals. This suggested relationship can be traced back to the primary data sources. The database includes association data for 68,000 varied chemicals, including 2,200 drugs, and relates them to 1.5 million genes, and diagonally to 373 genomes.

Super Target

This database includes a main data set of about 7,300 drug target connections, of which 4,900 relations have been subjected to more extensive manual annotation investigation. Super Target contributes tools for 2D drug screening and sequence assessment of the targets. The database includes more than 2,500 target proteins, which are annotated with 7,300 links to 1,500 drugs.

6.2.5 Spectral Databases

BMRB (BioMagResBank) is the main repository for investigative NMR spectral data, primarily for macromolecules. The BMRB also includes a recently identified subsection for metabolite data. The current metabolomics database includes structures on behalf of applets, nomenclature data, extensive 1D and 2D spectral peak lists (from 1D, TOCSY, DEPT, and HSQC experiments), raw spectra, and FIDs appropriate for 500 molecules. This data is available for download.

MMCD (Madison Metabolomic Consortium Database)

This database is built of biologically imperative small metabolites procured from a digital database and scientific data. It includes approximately 10,000 metabolite entries and spectral data on more than 500 metabolites. In this database, each metabolite is registered with an average of 50 separate data fields that contribute the chemical formulas, names, synonyms, physical and chemical features, and NMR and MS data on actual metabolites under exclusive conditions, where available. NMR chemical shifts are determined by empirical or theoretical approaches, data on the incidence of the metabolites in varied biological species, and an extensive range of images, references, and linked public databases.

Mass Bank

This mass spectral database is comprised of experimentally acquired, high-resolution MS spectra metabolites. It is regulated and funded by the JST-BIRD project; it contributes varied inquiries for standard spectra acquired from Keio University, RIKENPSC, and other Japanese research institutions. It is legally authenticated by the Mass Spectrometry Society of Japan. This database has MS data and an outstanding spectral/structure investigating relevance. There are more than 13,000 spectra from 1,900 varied accessible metabolites.

6.2.6 Golm Metabolome Database

This database offers easy access to conventional GC/MS libraries, which are preserved as mass spectral (MS) and retention time index (RTI) libraries (MSRI). The software NIST/AMDIS can be used to preserve the spectral and retention time indices for identification of metabolites according to spectral tags and RI's. The libraries are both explicable and downloadable, and have been carefully collected under specific conditions on diverse kinds of GC-MS tools, such as quadrupole and TOF.

6.2.7 Metlin

This metabolite database is a research repository for mass spectral data. Metlin is organized by Siuzdak and Abagyan groups under the regulation of the Center for Mass Spectrometry at The Scripps Research Institute. METLIN investigates the metabolites by name, mass, formula, or structure. It includes almost 15,000 structures, including those of greater than 8,000 di and tripeptides. METLIN includes MS/MS, LC/MS, and FT/MS data that can be investigated by peak lists, mass ranges, biological origins, or disease.

6.2.8 BML-NMR (Birmingham Metabolite Library Nuclear Magnetic Resonance)

This is a public database that includes 3,328 NMR spectra of 208 generalized metabolite standards. This database includes both 2-D 1H J- determined spectra, and 1D 1H spectra acquired at 500 MHz via diverse water repression approaches and authenticated features for solution at pH values of 6.6, 7.0, and 7.4. The unprocessed and processed data, including connected metadata, are preserved in a rationale build MySQL database that is adaptable, with metabolomic standards initiative (MSI)-certified reporting necessities, with several imperative alterations. Library data can be investigated effortlessly and analyzed through custom-written web boundaries. FIDs, NMR spectra, and associated metadata can be downloaded to newly implicated, MSI well-matched XML schema.

6.2.9 mzCloud

This metabolomic database is an open and free access platform for characterizing the assemblage of high resolution/exact mass spectral trees via a third-generation spectra connection algorithm. mzCloud analyzes an elaborate dataset that describes the work of public corporate dedicated research and scientific groups focusing on validating a detailed library of high-quality spectral trees to promote the structure elucidation of unknown metabolites. mzCloud focuses on elaborate characterization bottlenecks by permitting all mass spectrometrically relevant characteristics, identifying several experimental and computational characteristics, and in a few cases, characterization of unknowns, even if they are not yet revealed in the library.

6.2.10 Disease and Physiology Databases

OMIM (Online Mendelian Inheritance in Man)

This is an elaborate collection of human genes and genetic phenotypes. This metabolomic database encloses full text, referenced in OMIM, which includes data on all identified Mendelian disorders and more than 12,000 genes. OMIM focuses on the correlation between phenotypes and genotypes. It is reconstructed frequently; the entrants include several links to other genetics resources. OMIM includes 379 diseases with associated gene sequence data, along with 2,385 conditions with disease phenotypes and identified genetic causes.

METAGENE

METAGENE is a database of inborn errors of metabolism, contributing data pertaining to diseases, genetic reasons, treatments, and identifiable metabolite levels or clinical tests that might be employed to identify or assess conditions. It encloses data on 431 genetic diseases.

OMMBID (On-Line Metabolic and Molecular Basis to Inherited Disease)

This is an online HTML-format index organized by Charles Scriver at McGill University elaborating on the genetics, metabolism, diagnosis, and treatments of hundreds of metabolic disorders, based on data contributed by a number of professionals. It also includes elaborate reviews, explained pathways, chemical entities, and physiological records and tables that are often used by clinical biochemists. Most of the university libraries contributed to preserving this source.

6.3 METABOLOME ANALYSIS AS AN IMPERATIVE TOOL FOR ECOLOGICAL STUDIES

The relevance of ecometabolomics has been expansively advanced further in field studies, wherein free living organisms are assessed for their inter-individual metabolic flux resulting from ecological interactions with their immediate and unrestrained environments (Kuhlisch & Pohnert, 2015; Macel et al., 2010; Sardans et al., 2011). Numerous research studies have been carried out to explain the organism's metabolic response in environments for deciphering environmental bio-monitoring, such as in marine mussels (*Mytilus galloprovincialis*). This mussel was used as a sentinel to analyze stress-induced phenotypes produced as a result of hypoxia, as specified in animals samples taken directly from the coastline that are exposed to diverse environmental conditions; in comparison with the same stress response in animals grown under controlled laboratory conditions prior to analysis (Jones et al., 2013). In contrast to our presumptions, there was an enhancement in metabolic variability against 60 h stabilization in the laboratory in adductor muscle tissue; thus covering metabolic changes related to hypoxia. Even after following the highly efficient sampling strategy, an ecological assessment of free living wildlife is expected to divulge expansive inter-individual metabolic diversity (Brunetti et al., 2013; van Dam & van der Meijden, 2018; Viant, 2009). This validates the requirement for the phenotypic anchoring of the molecular level estimations, so as to enhance the lucidity of the metabolomic data. Some research reports established the imperative for individual species, and sex is requisite to identify more delicate inter-individual metabolic changes that result from an environmental stressor. The majority of metabolic diversity in the mantle tissue of marine mussels was due to sexual disparity (Atasaral-Şahin et al., 2015). This provides a platform for sex identification in mussels via their metabolic fingerprinting. Various researchers documented the competence of NMR metabolomics as a prime diagnostic contrivance for detection of disparity in liver metabolomes of European flounder (*Platichthys flesus*); thereby facilitating identification differences among clean versus polluted sites (Williams et al., 2014). Several research organizations from developed countries in Europe, North America, and Australia analyze and validate the competence of ^{1}H NMR metabolomics to produce analogous datasets from environmentally originated samples (Gallo et al., 2015; Kikuchi, Ito, & Date, 2018). In addition to the relevance of metabolomics in aquatic ecotoxicology research, there have been metabolomic studies carried in earthworms in metal-polluted sites. Bundy et al. (2007) procured data from autochthonous *L. rubellus* and *L. terrestris,* and allochthonous *E. andrei* populations thriving around a zinc-contaminated site. The responses observed by *E. andrei* were uncertain due to a high mortality rate at the contamination site. Moreover, *L. terrestris* exhibited a decline in both histidine and *N,N*-α-dimethylhistidine, or "methylhistidine," at the most contaminated sites. In spite of diverse geographic sites and varied physiochemical characteristics between sites, researchers were still able to detect zinc as the prime pollutant significantly affecting metabolism in organisms. The study showed that it is also possible to identify biomarker metabolites associated with zinc across all diverse sites (Bundy et al., 2009). It has been established now that plant populations are significantly affected by diverse biotic interactions and metabolites, particularly secondary metabolites that exert a functional role on varied ecological processes. Research on such metabolites will probably increase as the study of metabolomics becomes used more frequently for tackling ecological issues (Yi et al., 2016). Herbivory is another exemplar for illustrating ecological issues. Metabolomic strategies have begun to be used in analyzing the metabolic variations in plants exposed to either generalized or specific herbivores. Arany et al. (2008) reported an inland and coastal natural population of *A. thaliana,* and demonstrated disparity in the metabolome of each species. Further, they found these variations primarily in glucosinolate concentrations, which affect the growth of specialist or generalist herbivores, entailing chemical adaptation to herbivory types at the population level. Riipi et al. (2004) demonstrated disparity in the leaf chemistry of the mountain birch (*Betula pubescens*) among different phenological stages. This study evaluated numerous temporal changes in leaf metabolites that can be employed for anti-herbivory purposes, for example, hydrolysable tannins and proanthocyanidins. This highlights the significance of evaluating temporal scales when analyzing the metabolomic variations in the field, particularly for ecological studies. Hjältén et al. (2007) demonstrated genetically modified aspen trees exhibiting overexpression of sucrose phosphate synthetase so as to increase cell sucrose concentrations and biomass; thereby changing the metabolic composition of secondary metabolites related to anti-herbivores. Widarto et al. (2006) employed 1D and 2D NMR for monitoring the diverse insect pests at varied instar stages to analyze the metabolic response of *Brassica rapa*. They evaluated variation in glucosinolates and phenylpropanoids, together with essential carbon metabolites, for example, sugars and amino acids, via a non-targeted approach. Peiris et al. (2008) differentiated three species of wood-decomposing basidiomycetes by evaluating the alterations in 120 metabolic peaks procured via GC-MS. They have also evaluated metabolomic changes via analyzing organic acids, carbohydrates, and phenolics where species are grown together and compete for the same resources. Symbiosis is a generalized subject in both biotic and abiotic interactions. Cao et al. (2017) illustrated the metabolomic changes in symbiotic relationships existing between an endophytic fungus and its host plant (*Lolium perenne*). This study observed variations in the metabolome of infected plants, chiefly comprised of metabolites such as mannitol, peramine, and perloline. Numerous research reports have been documented that illustrate the metabolomic changes in different plant parts via employing diverse chemometric tests and a multitude of hyphenated chromatographic techniques. Previous research studies in animal models have also demonstrated the relevance of microprobe NMR spectroscopy for profiling the haemlymph of solitary- and crowd-reared forms of the locust *Schistocerca gregaria* (Rahman et al., 2002). They have demonstrated numerous metabolites that exhibited varied forms, such as increased levels of putrescine, trehalose in the solitary phase, acetate, and ethanol; and it was conjectured that polyamine metabolism might be associated with hormone production, which could then induce swarming behavior (Ladero, Calles-Enríquez, Fernández, & Alvarez, 2010). Miller, Islam, Claridge, Dodgson, and Simpson (2008) employed bio-assay mediated method that established that an alkylated DOPA analogue was accountable for transmission of epigenetic information influencing swarming behavior in *S. gregaria*. All the ecometabolomic research studies carried out so far have included diverse steps from sample collections, and laboratory evaluation, followed by the application of a multitude of statistical tools that could facilitate the identification of differences in metabolic fingerprints of both control and stressed organisms. However, only a limited number of metabolic characterizations has been carried out, due to a restrained level of rationalization in molecular pathways; thus, this approach is referred to as "fishing" (Eckert & Bajorath, 2007). Metabolomics has thus opened up an exciting new field of study that has provided the model for many new researchers to initiate research programs in the environmental metabolomics. For some specific research studies, there is only a need to collect, grind, measure, and analyze data in order to tackle questions of interest. However, a more significant outlook can be comprehended if it is explored beyond the secure boundaries of our environmental omics community, and asking what significant approaches and new biological interventions have been discovered using metabolomics, from the perspective of most scientists. Therefore, numerous steps are taken to advance environmental metabolomics, including the subjects associated with advanced knowledge of baseline metabolism, advanced metabolite detection, and improved utilization of genetic resources and data. Furthermore, from the technical standpoint, researchers need to be more precise and exact in their selection of analytical and bioinformatics tools to tackle their specific questions of interest. For example, if a researcher is interested in explaining the comprehensive perspective of the pathways implicated in response to a toxicant, a technique with adequate sensitivity for identifying an extensive range of metabolites, most likely based on MS, would be a suitable choice. In the case of identifying the differences between metabolic phenotypes, especially when high sample throughput is required, optical methods, for example, FT-IR would be suitable. Similarly, in cases of oxidative stress, we can employ techniques, such as electrochemical detection, that estimate redox-active metabolites with exquisite sensitivity. While for comparatively instant, quantitative, and non-targeted analysis of mass metabolite data, we can rely on NMR spectroscopy (Smolinska, Blanchet, Buydens, & Wijmenga, 2012). Likewise, we can select a technique for analyzing data, such as principal component analysis (PCA) and partial least squares discriminant analysis (PLS-DA), that should not be used for all types of research data, unless the data is expansive and entails multiple variables (Brereton & Lloyd, 2014; Rezzi et al., 2005). In a similar way, we need to address which methods, whether univariate or multivariate, and unsupervised or supervised, should be selected to answer particular questions. The experimental design of various ecological metabolomic studies should also change. The information of varying metabolite levels, independently, can hardly lead to the ultimate perspective on biological process. In well-established model organisms, different controlled experiments can be carried out to describe the alterations in biological systems; but this cannot be applied to ecologically relevant organisms. Therefore, substitutive metabolomic studies could be carried out, along with other techniques comprising molecular, physiological, and even population-level measurements (Johnson, Ivanisevic, & Siuzdak, 2016). In recent years, there has been an advent of novel concepts such as "Phenotypic anchoring," which is the main concept utilized by omics communities, wherein research studies are drafted to link particular molecular modifications commencing from gene expression profiling, proteomics, and metabolomics to particular harmful effects determined by established parameters (Daston, 2008; Paules, 2003). This strategy has been termed "molecular estimation," with a specific perspective on fields other than omics. Whatever the perspective, future efforts should be inclined toward the integration of metabolomics evaluation with an enhanced level of biological data, so as to amplify the interpretability and understanding acquired from the molecular data (Rochfort, 2005). One of the major advantages of metabolomics for ecological research studies is that strategy is relevant to all species, excluding any prior data on the genome sequence (Lankadurai, Nagato, & Simpson, 2013; Lin, Viant, & Tjeerdema, 2006). For example, lactate is the same essential entity in fish as it is in plants and animals. Therefore, from the technological viewpoint, metabolomics can be relevant to all species; but this extensive range of applicability has been counteracted by significant problems. In 2007–2008, ecological metabolomic research was documented in more than 20 species comprised of eight freshwater, marine fish; and nine species of aquatic and terrestrial invertebrates, together with plants and microbes (Iacono et al., 2010; Miller, 2007; Viant, 2008). Conversely, it has been confirmed now that in order to provide inclusive inferences on ecological metabolomics, the intricate knowledge of basic biology is requisite (Keurentjes, Molenaar, & Zwaan, 2013). For example, parameters associated with the life cycle, developmental stages, and reproductive methods are significant contrivances to instantly detect or ascertain the species, sex, and age. Metabolomic studies of natural and anthropogenic stress-inducing agents will be milder without this baseline data. However, we are not researching only the "sentinel" species to be used as model organisms, and excluding other organisms (Basu et al., 2007; Regoli et al., 2006). The relevance of metabolomics in extreme environments, such as that of extremophiles, will be exciting. Metabolomics can, however, deliver more robust and easily comprehensible findings in a more timely manner if energy, time, and resources are dedicated to a specific number of sentinel species (Rivas-Ubach et al., 2012; Sardans, Peñuelas, Coll, Vayreda, & Rivas-Ubach, 2012). Conversely, ecologists are employing transcriptomics limited to the interpretation of specific species for entirely technical reasons. Although these investigations have an advantage, in that they are focused on a comparatively inadequate number of species; these scientists have been limited by the improper documentation of genome resources on conventional sentinel species within the ecological field. However, with the discovery of high throughput DNA sequencing, this is about to change. So, we consider this an indispensable era for ecological omics research, in the sense that if metabolomics and transcriptomics communities can resolve the issue of determining most suitable sentinel species to study for biological reasons, then advancement could be instantaneous (Beierkuhnlein, Thiel, Jentsch, Willner, & Kreyling, 2011; Sardans & Peñuelas, 2013). Research reports have documented the collaborative efforts in extoxicogenomics, which should be applied to a restricted to number of organisms. These are comprised of microbes such as *Saccharomyces cerevisiae*, aquatic and terrestrial invertebrates such as *Daphnia magna*, *Caenorhabditis elegans*, and plants such as *Lemna minor*, *Arabidopsis thaliana*), an amphibian (e.g., *Xenopus tropicalis*) and a fish (e.g., *Danio rerio*). Until now, most ecological studies have failed to characterize numerous metabolites. It is pertinent that advancements in metabolite characterization will immensely augment the intervention of ecological metabolomics in such areas as biomarker discovery in ecotoxicogenomics, detecting metabolic pathways that are modified against biological perturbation, or determining new signaling compounds. Both analytical and informatics strategies related to metabolite characterization continue to advance, and design considerations for contributing explicit metabolite characterizations via the Metabolomics Standards Initiative are now available (Alla, Khedr, Serag, Abu-Alnaga, & Nada, 2012). The issue of metabolite characterization has been renowned for numerous years, and has facilitated the validation of the requirement for expansive metabolite databases. Schemes, for example, the Human Metabolome database, are now assisting the extraction and procuring of peaks and their inference (Wanichthanarak, Fan, Grapov, Barupal, & Fiehn, 2017). However, the metabolites that are characterized are inclined toward mammalians; in particular, human biology, and so ecological metabolomic studies that are specifically required for metabolite characterization will be particularly useful for comparison between studies. Ecological metabolomics could be helped significantly from the acquisition of metabolite libraries dedicated to nonmammalian organisms. Even though the variety of plant and animal kingdoms is immense, focusing on the judicious utilization of numerous sentinel species can offer a potential solution. Until now, a majority of metabolomic research studies have focused on soluble, polar metabolites; and there has not been an immense agreement of highlighting on lipid analysis. Certainly, there are unknown lipid metabolites yet to be identified from invertebrates, without additional plant species. It is expected that lipid analysis will play a significant role in ecological metabolomics, if only as a fraction of an initiative to comprehend alterations linked with environmental stresses.

6.4 METABOLOMIC ANALYSIS OF PLANT—ENVIRONMENT INTERACTIONS

Variations of plant metabolism are of prime importance in plant developmental processes that describe keystone varied means by which plants respond to the environment. The relevance of metabolomics is significant in the identification of phenotypic effects of varied abiotic and biotic stresses on plants. However, this observation, in accordance with analyzing samples that are applicable to ecologically based hypotheses; is known as "environmental metabolomics" (Lankadurai et al., 2013). The expansion of environmental metabolomics is one of the novel approaches in omics technologies that has been one of the primary advancements in plant physiology (Tenenboim & Brotman, 2016). It is relevant to the whole landscape of plant ecology, from the perspective of plant plasticity and adjustment, to community composition, and even genetic alteration in crops (Ahanger, Gul, Ahmad, & Akram, 2018; Hong, Yang, Zhang, & Shi, 2016). An expansive range of new research studies uses diverse metabolomics approaches commencing from the first stages of tissue sampling, through to sample preservation, transportation, and analysis (Bonneau, Tetreault, Robitaille, Boucher, & De Guire, 2016). Hence, it can conjectured that the idea and relevance of plant ecological metabolomics is an ecologically significant analytical contrivance. It investigates the prime techniques employed in situ within field sites, with a specific focus on sampling and processing, and those more suitable for utilization in a laboratory-based setting, with an emphasis on secondary metabolite investigation (Kirwan et al., 2018; Zampieri, Sekar, Zamboni, & Sauer, 2017).

The principal step for metabolomic research relevance is in the perspective of how much of the variability for metabolomes is affected by inherent genetic factors comprised of maternal effects on seed quality, and by the external ecological factors (Das, Kim, Khadka, Rakwal, & Rohila, 2017; Rosental et al., 2016). Until now, the studies have signified that environment influences the metabolome far more than genetic variation. Robinson, Dauwe, and Mansfield (2018) described the metabolome of Douglas fir trees, and observed that metabolite-like carbohydrates, lignin biosynthetic metabolites, and phenotypic (tree morphology, fiber, and wood chemistry) qualities were strongly associated with the site; while a similar relation linked to genetic structure was weaker. Frank, Röhlig, Davies, Barros, and Engel (2012) and Matsuda et al. (2012) demonstrated significant variation in metabolomes of maize and rice, respectively, related to changes in phenology and seasons in comparison with genetic variations induced by disparity due to strains or genetic alterations. The increased level of disparity of the natural environment presents immense constraints for plants in terms of their competence with surrounding plants, and responds suitably to diverse abiotic and biotic threats. Abiotic stress is comprised of changes in availability in nutrient resources, irradiance, and temperature, which can change immensely on spatial scales sorting from mere square meters, to total landscapes (Pianka, 2011; Waring & Running, 2010). Besides naturally prevailing heterogeneity, anthropogenic effects on the environment, for example, atmospheric CO_2 enrichment and nitrogen deposition, generate further ecological disparity (Alados et al., 2017). However, phenotypic plasticity sustainability under ecological stress across both short and long time scales is regulated by genes that not only control the traits of an organism, but also its level of response against varied environmental stimuli (Rivas-Ubach et al., 2016; Veley et al., 2017). The phenotypic plasticity is determined by the ability of plants to modify morphological traits and physiological traits. There is an extensive range of disparity in the extent to which these metabolic traits respond to ecological pressure, ranging from insignificant shifts in metabolite turnover, for example, that observed against variation in C/N/P/K within the local environment (Kravitz et al., 2016; Rivas-Ubach et al., 2012), through to activation or upregulation of exceedingly particular metabolic pathways (Cañas et al., 2015; Field & Lake, 2011). Finally, characters have been favored via evolutionary processes of natural selection that are applicable to the range of conditions within species habitat and distribution. The comprehension of the alteration in these metabolic traits is necessary to perceive how the environment affects plant growth, and how it is recognized environmentally. So, to acquire this perspective on biochemical alteration, plant physiologists have taken the lead in recent improvements in analytical chemistry and bioinformatics, which is progress that has provided an inclusive evaluation of the biochemical composition of plant tissue. This strategy is termed "metabolomics," and when applied to evaluating samples that are applicable to ecologically based hypotheses, it is termed "ecological metabolomics." Its relevance includes the complete report of plant ecological physiology from the perspective of plant plasticity and adaptation via community composition, and even genetic alterations in crops. Recently, numerous key reviews and viewpoints on this issue have contributed a sketch of how metabolomics can to be employed to facilitate the detection of natural ecological pressures and pollution effects on ecosystems (Sardans et al., 2011; Simpson, Simpson, & Soong, 2012; Weckwerth, 2011a, 2011b). The accessibility of tissues should be measured at the commencement of the experiment. This will depend on the analysis pipeline, even though generally 10–100 mg of fresh tissue is requisite for most analytical applications, namely GC, MS, and NMR (Simpson et al., 2012; Szecowka et al., 2013). The numerous samples required for correct biological interpretation must be also considered, even though this number could be established by the number of samples the researcher has access to, particularly if exceptional or hard-to-obtain plants are being analyzed. Broadhurst and Kell (2006) illustrated an exceptional synopsis of the statistical considerations comprised sample size, which is requisite for metabolomics experiments. It is also indispensable to consider the purity of the samples being harvested for metabolite analysis of plant tissue that is polluted with bacteria, fungi, or pests, for example, aphids, which may impede the results, particularly if the samples are originated from natural habitats and not from controlled growth room facilities (Tietenberg & Lewis, 2016). The relevance of chemicals to control pests, for example, aphids and sciarid flies, in growth room facilities should be taken into account when investigating the metabolome of plants subjected to such chemicals (Cloyd, 2018). In order to carry out the metabolomic analysis, it is necessary to avert the persistence of enzymatic and oxidative processes within cells, as some metabolite's turnover rates are <1 s. This guarantees the metabolome remains unaffected, and hence, symbolizes a lucid "snapshot" outline of the plant's metabolism at the time of sampling. There are varied methods employed for sampling in the field of the environment; for example, the method of freezing leaf samples in liquid nitrogen prior to laboratory analysis (Ossipov et al., 2008). There are some alternative methods for freezing the samples directly, via liquid nitrogen enclosed in a dry shipper, dry ice, or by putting the tissue inside cold methanol. The samples procured from a marine environment have a high salt concentration, so in order to remove excess salt, specific methods are employed for preventing the hurdles associated with metabolite extraction (Macintyre et al., 2014). Moreover, the transportation of plant samples from the field to the laboratory can be cumbersome with respect to cost and logistics, so partial metabolite analysis can be carried out on site, such as pigment estimation via solvent extraction of cut leaf discs using spectrophotometry (Giannarelli et al., 2010).

6.4.1 Extracting Metabolites

There are diverse extraction methods that vary with respect to polarity of the metabolites and the solvent system (Kim & Verpoorte, 2010). The most generalized extraction methods employed varied solvents such as methanol, chloroform, and water for homogenization of tissue samples (Lin, Wu, Tjeerdema, & Viant, 2007; Sasidharan, Chen, Saravanan, Sundram, & Latha, 2011). The resultant bi-phasic solution is employed for evaluation of polar metabolites.

6.4.2 Analyzing Metabolites

There is an extensive range of technologies used for the evaluation of metabolite evaluation in the plant solvent extracts. The selection of evaluation will depend on the accessibility of equipment; but more significantly, it will be determined by the research hypothesis being tested (Kothari, 2004). The two chief strategies are generally bifurcated between metabolite fingerprinting and metabolite profiling. These methods are completely different, as metabolite fingerprinting is a semi-quantitative approach with expansively implicit metabolite detection, while metabolite profiling implicates precise chemical identification and estimation (Carlson & Cravatt, 2007; Roessner et al., 2001). The method of metabolite fingerprinting is a proficient method for detecting broad patterns in the metabolite response to stress, or in population studies (Field & Lake, 2011; Simpson et al., 2012). These fingerprinting methods primarily focus on investigating metabolites in complex mixtures by either directly injecting a crude extract into a mass spectrometer, termed "direct injection MS," (Walker, Budhathoki-Uprety, Novak, & Muddiman, 2011) or by procuring NMR spectra of the total mixture (Forseth & Schroeder, 2011). NMR fingerprinting is more efficient than MS due to its expansive range of metabolite characterizations, and ability to decipher the structural information of unknown metabolites without requiring metabolite detection and ionization. The metabolite fingerprinting procured from NMR needs to be validated by quantitative metabolic profiling techniques (Farag, Porzel, & Wessjohann, 2012; van der Kooy, Maltese, Choi, Kim, & Verpoorte, 2009). These profiling methods can isolate metabolites before metabolite detection, either by GC, GC × GC techniques where manifold GC columns are pooled to amplify the separation, resolution, and data acquisition time per sample (Almstetter, Oefner, & Dettmer, 2012); or, LC (Theodoridis, Gika, Want, & Wilson, 2012). Choi and Jung (2017) described different assays for sample preparation, derivatization, and analysis by HPLC and GC-MS. These can profile animal or plant samples that can be detected either by their molecular mass, fragmentation pattern via MS, or chemical shift using NMR. Viant and Sommer (2013), Bowne, Bacic, Tester, and Roessner (2011) and Viant (2009) contributed reviews on how these detection methods can be employed to allocate function to the metabolites inside an environmental studies framework.

6.4.3 Data Interpretation and Bioinformatics

The expansive datasets that are originated from metabolomic research reports are generally evaluated, annotated, and envisioned via curated online portals such as www.plantcyc.org (Zhang et al., 2010) and www.plantmetabolomics.org, where ecological and genetic information linked to the experiments can also be considered as covariables (Bais et al., 2010; Bais, Moon-Quanbeck, Nikolau, & Dickerson, 2011). There are numerous sites, for example, the ECOMICS portal for metabolomics studies associated with biomass production (https://database.riken.jp/ecomics/), and the metabolic network exchange (www.metnetdb.org). The expansive datasets can also be evaluated via univariate and multivariate statistics, including PCA using online servers such as Metaboanalyst (http://www.metaboanalyst.ca/; Xia & Wishart, 2011). These portals, and how metabolomic data can be integrated into other omics, is reviewed by Mochida and Shinozaki (2011) and Bassel et al. (2012).

Using advanced metabolomic methods offers discreet prospects to investigate the functional phenotype of individuals; thereby deciphering potent impending actions in plant plastic responses against varied abiotic and biotic stresses (Nicotra et al., 2010) at the individual (Reed, Schindler, & Waples, 2011), population (Murren et al., 2015), and even community levels (Field & Lake, 2011). This can be procured using the metabolome characterization, followed by multivariate statistical analyses of species within an entire genus distributed across numerous populations and diverse geographical locations. Numerous research reports have illustrated population dynamics wherein identification of different origins of varied crops was traced, such as the origin of coffee beans, identification of tobacco plants from China and Zimbabwe, and identification of populations of the arctic-alpine plant *Arabidopsis lyrata* ssp. *petraea* from isolated regions across

Europe (Davey, Woodward, & Quick, 2009; Li et al., 2011; Zhang et al., 2013). It was established by different research studies that there is no significant association between the metabolomic and genetic variation within a region. Such population studies are restricted to higher plants, as confirmed via detection of metabolomic shifts of macro-microalgal species linked with day and time, sampling period, and geographical site (Riedel, Biester, & Dittmar, 2012). Scherling, Roscher, Giavalisco, Schulze, and Weckwerth (2010) demonstrated that metabolomic variation is a functional indicator for competence and consequent biodiversity of plants within experimental plant communities. This study confirmed higher metabolic diversity in small herbaceous species in contrast with taller, more dominant plants within their experimental field plots. Metabolic profiles signified that carbon and nitrogen was inadequate in smaller plants when subjected to higher diversity, indicating that varieties in resources availability impels biodiversity. Further studies confirmed that metabolic diversity is directly associated with genotypic abundance within populations of wild plants (Moore, Andrew, Külheim, & Foley, 2014). The most exciting relevance of metabolomics is the evaluation of biochemical traits at the landscape scale (Toubiana et al., 2012). The advent of airborne hyperspectral analysis, whereby variation in absorbance and reflectance patterns of particular metabolite within leaves is observed over an entire landscape via light aircraft and remote-sensing techniques, provides an assessment of biochemical phenotypes transversely comparatively significant (Lewinsohn & Gijzen, 2009; Santos, Corrêa, & Costa, 2011).

Metabolomics can also offer considerably to ecological questions and hypotheses on gross carbon, nitrogen, and phosphate distribution in plants, particularly when the habitat is under environmental stress (Peñuelas & Sardans, 2009a, 2009b; Rivas-Ubach et al., 2012). There are several research studies on the metabolic responses in plants to drought (Kusano et al., 2011), cold temperatures (Miyagi, Takahara, Takahashi, Kawai-Yamada, & Uchimiya, 2010), and atmospheric nitrogen deposition in lichens (Freitag et al., 2012). Metabolomic techniques have facilitated the estimation of metabolic responses to exposure to pesticides, for example, glyphosate (Petersen et al., 2011), and have been employed in diverse studies to incline toward drought tolerance in model plants and crops (Bechtold, 2018; Bowne et al., 2012). Besides abiotic stresses, plants are subjected to a wide array of other stressors for biotic association within their natural habitat. Numerous chemical ecologists have developed several exciting research studies that have been developed to investigate the metabolic responses and interaction of herbivores using different techniques such as ultra-performance LC MS and NMR (Forseth & Schroeder, 2011). This method is not just limited to higher plants, as it has also been used to investigate chemical defense responses in red algae (Nylund, Weinberger, Rempt, & Pohnert, 2011). Based on the relevance and methods for the metabolite profiling of plants, it can be concluded that metabolomics has significant and functional prospects for ecological plant sciences. In addition to the equipment and techniques that are relevant to an extensive range of scientists, the data procured in these studies will facilitate a step-change on the perspective of how an extensive range of plants link with each other, with other species, and with the environment (Blackburn et al., 2014; Flood et al., 2016).

6.5 LINKING METABOLOMICS DATA TO UNDERLYING METABOLIC REGULATION

Numerous research studies have reported the impact of varying environments on the homeostasis of cellular metabolism via employing diverse metabolomic approaches (Nägele, 2014; Nägele, Fürtauer, Nagler, Weiszmann, & Weckwerth, 2016). The illustration of experimental analysis of an organism's metabolomes plays a functional role in the strategies of basic system biology. Several methods have been established that integrate of varied hyphenated chromatographic techniques, such as LC-MS, GC-MS, CE-MS, and H NMR, which offer an illustrative snapshot of specific metabolic homeostasis (Rahman, 2018; Gorrochategui, Jaumot, Lacorte, & Tauler, 2016). To investigate the regulatory pathways implicated in cellular metabolism, the prime step will be identification of metabolites and the estimation of metabolite levels (Draisma et al., 2015; Stahl et al., 2016). This advances our perspective on the molecular organization and regulation of cellular metabolism, as well as the whole organism (Smith, Sindhu, & Meissner, 2016; Van Veld et al., 2018). However, it is challenging to extract the relevant data, which is contained in metabolomic data sets. Based on the prime concept of molecular biology, metabolite concentration and its fluctuation is due to the intended data flux resulting from activation of genes via transcription to translation, followed by post-translational alterations (Fan, Krautkramer, Feldman, & Denu, 2015; Nägele et al., 2014). Thus, metabolomic data signify the overall output of a metabolic system consisting of varied levels of molecular organization. Therefore, the converse task of metabolomic reports to basic metabolic development should capitulate data, which, if decoded, appropriately provides an illustrative perspective into the metabolic system. However, the derivatization of regulatory pathways is highly intricate, not only due to an expansive number of metabolites, but also a higher level of cellular compartmentalization and differentiation. There are diverse hyphenated chromatographic techniques, such as illustrative GC × GC—*time of flight MS* (GC × GC-TOF-MS), in which two columns with orthogonal segregation features are pooled to generate greater peak intensity (Almstetter et al., 2012; Koek, Jellema, van der Greef, Tas, & Hankemeier, 2011), and may also facilitate higher detection confidence by resolving coeluting metabolites (Castro-Perez, 2007; Xiao, Zhou, & Ressom, 2012). However, on the level of MS, such coelution might be determined, such as by employing approaches of tandem-MS or MS^n (Andrews, Simons, Young, Hawkridge, & Muddiman, 2011; El-Aneed, Cohen, & Banoub, 2009). The consequential data matrix, which generally is comprised of metabolites of prime carbon/nitrogen metabolism; for example, primary macromolecules such as sugars, amino acids, organic acids, and lipids, characterizes a metabolic homeostasis, or its stimulation via mediated dynamics (Aldridge & Rhee, 2014; Leon, Garcia-Canaveras, Donato, & Lahoz, 2013). Chemical derivatization extends the metabolite spectrum, which can be evaluated by GC techniques, rendering them more volatile and thermally stable (Orata, 2012). The generalized method is a two-step derivatization process that consists of oximation, which is comprised of stabilized open ring conformation; and silylation steps, which stabilize molecules via the replacement of hydrogen in functional polar groups (e.g., the hydroxyl group, by trimethylsilyl group [$-Si(CH_3)_3$]) (Hill & Roessner, 2013). GC-MS specifically provides the estimation of volatile and uncharged metabolites, and liquid chromatography coupled with MS (LC-MS) is the technique of choice to analyze semi- or non-volatile and thermally unstable metabolites (Hopfgartner, Lesur, & Varesio, 2013). Both GC-MS and LC-MS could be employed to investigate isomeric metabolites, but high chromatographic resolution of GC, and specifically GC × GC, make it the appropriate analytical technique to determine structurally strongly linked metabolites (Meinert & Meierhenrich, 2012). In LC-MS, structural data pertaining to metabolites can be procured via collision-induced dissociation (Ma & Chowdhury, 2013; Wu et al., 2015). Diverse methods in MS, as well as their characteristic features, are extensively reviewed (Lössl, van de Waterbeemd, & Heck, 2016; Nemkov, D'Alessandro, & Hansen, 2015; Sali et al., 2018); but it can be generalized that the integration of these techniques is supposed to amplify the metabolome cover considerably (Sumner, Lei, Nikolau, & Saito, 2015; Yi et al., 2016). Though the identification confidence is obviously the limiting factor, the output of the GC and LC-MS platform provides an illustrative snapshot of metabolic homeostasis and its rejoinder against diverse genomics or ecological perturbations (Kim et al., 2017). Moreover, there are varied analytical methods, such as UV, IR, FT-IR, and FT-ICR spectroscopy (Van Agthoven, Delsuc, Bodenhausen, & Rolando, 2013), NMR (Simmler et al., 2014), or capillary electrophoresis (Kuehnbaum & Britz-Mckibbin, 2013), which can still expand this metabolic data space and augment its assertion. The organism and sample type analyzed determines the metabolic coverage contributed by metabolomic approaches. The metabolome of a diverse organism yields a varied mass of metabolites, ranging from hundreds of metabolites, such as 750 from *E. coli* (Sévin, Fuhrer, Zamboni, & Sauer, 2016), to more than a thousand metabolites from a eukaryotic metabolome, such as yeast (Reed, Baer, & Edison, 2017), and over more than one hundred thousand in plants (Hagel et al., 2015). The metabolomic data interpretation of humans is highly intricate, not only due to an expansive range of metabolites; but also enhanced levels of cellular compartmentalization and differentiation. Numerous research reports have focused on metabolite dynamics and underlying fluxes (Arrivault et al., 2014; Masakapalli, Kruger, & Ratcliffe, 2013). Because of an expansive subcellular level of resolution, it is possible to comprehend the astonishing and exciting characteristics of metabolic regulation. An example is the subcellular resolution of plastidial and cytosolic pathways implicated in carbohydrate oxidation, for example, glycolysis or the oxidative pentose phosphate pathway (PPP). Masakapalli et al. (2010) reported that multiple data sets can be fixed completely to the models with modified subcellular compartmentation of the PPP in a heterotrphic *Arabidopsis* cell suspension using steady state isotope labeling in a subcellular flux analysis. It is indispensable for the significance of experimental data procured on the subcellular level to decrease the ambiguity pertaining to biochemical regulatory pathways. The elaborate determination of leaf metabolism in *Arabidopsis* was presented by employing methodology of $^{13}CO_2$-labeling to determine a time-dependent outline and kinetics in the metabolome (Szecowka et al., 2013). Hyphenated chromatographic approaches such as LC and GC-MS were employed for evaluating the variations in isotope patterns of about 40 primary metabolites; including prime carbohydrates as well as organics and amino acids, plus phosphorylated intermediates procured from the Calvin-Benson cycle. Though numerous experimental observations were included, as per earlier reports, numerous unknown characteristics of labeling kinetics could be revealed because of the subcellular resolution (Szecowka et al., 2013). Ultimately, the authors determine that data pertaining to metabolite compartmentation is a requirement for representing photosynthetic or allied metabolic processes in multicellular eukaryotic tissues (Rossi, Kalde, Srisakvarakul, Kruger, & Ratcliffe, 2017).

These research studies offer rational and detailed strategies to investigate intricate systems in a comparatively systematic manner. Though there exists an obvious incongruity between the number of metabolites that are estimated (~10–100), or detected and (comparatively) estimated (~100–1000), and that are anticipated in a metabolome based on genome-wide predictions (~10^3–10^5) (Danchin, 2017; Watson, Yilmaz, & Walhout, 2015). Further, the aim of expanding the data output and confidence of an analytical metabolomic platform is its specific challenge to deduce the resultant experimental information in a biochemically significant method, which is a requirement of a testable hypothesis. As highlighted herein, this is primarily due to high data output on the underlying molecular and structural organization being assembled in a metabolomics data set.

6.6 DEVELOPING REGULATORY STRATEGIES FROM METABOLIC FLASH

The complexity of decoding all-inclusive metabolomics information arrays can simply be continued by a simple illustration wherein a trial including samples from two discreet groups, a wild-type (control group) and a *knock out* mutant, in general, leads to m autonomous natural repeats of each, with n methodological repeats for each group. Assume "p" as metabolite level to be estimated in the metabolomic analysis. Following this, the exact overall investigational figures implicated in this trial terminate in two data assortments, each with $p \times n \times m$ dimensions. Assuming the biological variance of the procedure is much greater than technical variance ($var_{tech} \gg var_{biol}$), the scientific mean value can be generated for every biological repeat and data set to deduce data matrices exhibiting $p \times m$ magnitude (Strogatz, 1994). Even though fundamental univariate statistical experiments and evaluations instantaneously allow for a direct assessment of different data sets, and contribute immediate information of the underlying natural procedure, the produced supposition is ambiguous because of varied metabolic illustrations for each metabolomic feature. The divergence can be deduced from the covariance matrixes of such metabolomics data. This can provide us with an actual estimate of the correlation between two metabolites with respect to each other.

$$\mathrm{Cov}(x, y) = \frac{1}{m-1}\sum_{i=1}^{m}(x_i - \overline{x})(y_i - \overline{y})$$

The symbols x and y represent variables having sample size m, whereas $\overline{x}$ and $\overline{y}$ are mean values. For that reason, the pxm matrix deduces a $p \times p$ covariance matrix. This quadratic association between the numerous metabolites, and probably metabolic associations, discreetly illustrates why metabolomic data arrays (e.g., $p < 10$) are still cumbersome to decode without mathematical formulas. Expanding the metabolomic coverage of investigational methods is instantly associated with an exponential increase in illustration, which must be known. Therefore, in order to explore the main rationale and inclusive biochemical illustration for all determined investigational data, the instantaneous relevance of varied statistical tools that validates, or even matches with each other, has emerged as a generalized strategy. There are several tools and software programs to thoroughly validate the fundamental statistical approaches, and several podiums and graphical user interfaces have emerged to enable and assist the application of such methods (Sheth & Thaker, 2014; Sugimoto, Kawakami, Robert, Soga, & Tomita, 2012). Most of the statistical contrivances are proficient enough to investigate, determine, and actually visualize multivariate problems, via methods such as PCA, independent component analysis (ICA) (Steuer, Gross, Selbig, & Blasius, 2006), or independent PCA (Yao, Coquery, & Lê Cao, 2012). Depending on the covariance data, these multivariate approaches decrease the dimensionality of the dataset, while preserving maximum data, i.e., variance. This yields abstract variables, that is, components that entail the most prominent and characteristic features of multivariate data sets, thus contributing and helping with the building of biological hypotheses. Along with other supervised and unsupervised statistical ways, including hierarchical clustering, k-means clustering, and PLS-DA; molecular compounds contribute to the sample separation to an extent, and can be detected (Le Cao, Boitard, & Besse, 2011; Sun & Weckwerth, 2012; Bellaire et al., 2014). The multivariate data analysis is a significant part of systems biology approaches consisting not only of metabolomics, but also other data sets originated from varied omics approaches, for example, transcriptomics and proteomics (Weckwerth, 2008). The combined evaluation of varied experimental high throughput data sets employs different specific software packages for complete regression analysis, correlation, and visualization. An outstanding idea for software contrivances, platforms, and workflows of computational tasks in systems biology was offered earlier (Ghosh, Matsuoka, Asai, Hsin, & Kitano, 2011). Even though these platforms contribute diverse statistical and mathematical approaches, and the ultimate result that systems biologist are focused on most times may be comparable, the detection of particular metabolic clusters and patterns, that is, samples with similar statistical features assort the control or pathway from treatment group are vital. This ultimately contributes to an outline pertaining to metabolic steps or pathways that are affected by external alterations, and helps us resolve the hypothesis underlying biochemical and molecular biological mechanisms.

Several research studies from diverse biological research fields have confirmed the relevance and utility of such computational workflows (Altaf-Ul-Amin, Afendi, Kiboi, & Kanaya, 2014). However, parallel to this finding, the derived data on the regulatory association between molecular organization levels has been known to be available only to a certain extent. This is primarily because of the intricate and nonlinear association that exists among the levels of transcripts, proteins, and metabolites. Therefore, any metabolite variation in organisms is due to diverse genomic or ecological alterations, which can be due to varied reasons, including:

1. Forward/feedback regulation of enzymes exhibiting nonlinear kinetics.
2. Post-translational alteration of enzymes/proteins implicated in the metabolite production, and inter-conversion or transport.
3. Variation in the rate of protein production and degradation.

Therefore, it is not possible to present explicit conclusions from varied metabolite levels on regulatory methods in a metabolic network, excluding the following specific examples:

(a) All metabolites of the metabolic network are recognized and biochemically identified.

(b) All associations between levels of molecular organization are due to linear generalizations.

Even though during recent years, expansive data pertaining to genome-wide metabolic network structure and principles has not been reported, it is necessary to mention the molecular systems of prokaryote (Carrera et al., 2014), yeast (Sanchez, Perez-Correa, & Agosin, 2014), algae (Chang et al., 2011), higher plants (Mintz-Oron et al., 2012; Hill et al., 2013), human metabolism (Mardinoglu, Gatto, & Nielsen, 2013), and disease and medicine (Vandamme, Fitzmaurice, Kholodenko, & Kolch, 2013). On the contrary, the approaches of mathematical modeling facilitate the method of linearization for these examples

6.7 LINEARIZATION OF METABOLIC FUNCTIONS

Estimating the precise solution of intricate and disorganized systems (Colzani et al., 2009) is a generalized strategy in diverse scientific fields, such as mathematics and physics, to produce linearization of nonlinear functions. In a general symbolic framework, one could consider polynomial function $y(x)$, comprised of numerous minima and maxima. Until now, by supposing and evaluating the function $y(x)$ for all x, we have only been concerned with the performance of $y(x)$ in one certain point $x(0)$. In this example, the function is described by the coordinates ($x0$, $y(x0)$). We can approximate the function $y(x)$ by drawing the tangent in ($x0$, $y(x0)$). With this, we have approximated the original function by a linear function. In a metabolic context, $x0$ would represent a certain steady-state metabolite concentration, while $y(x)$ symbolizes the so-called metabolite function, taking the steady-state value $y(x0)$. In the subsequent steps, the time-dependent metabolic function will be redrafted as $fM(t)$. When the original function is compared with the linearized estimate, it becomes evident that these solutions are applicable for limited and predefined intervals of the fundamental function (Pang & Maslov, 2011). But they contribute exceedingly helpful information, and can facilitate background knowledge pertaining to the original function, specifically if this function is exceedingly intricate. Finally, they may permit replication and calculation of the systems performance, consequently expanding the existing knowledge about the intricacies of the background information of a system (Morel & Levy, 2009). To transfer this approach to a systems biology method, we need a time-dependent alteration in metabolite concentration (M). Mathematically, this can be estimated by an ordinary differential equation. While the left side of the equation demonstrates the alteration in M with changes in time t, the right side of the equation describes all rates influencing the metabolite concentration (M). This can be recapitulated by the metabolic function $fM(t)$. The reaction rates are determined by diverse parameters' variables and functions; for example, inhibitor/activator concentrations, thermodynamic limitations, post-translational modifications, protein levels, and so on. This mechanically links the regulation of metabolite levels to all other levels of molecular organization (Vallino, 2010). Every function is comprised of varied nonlinear elements, for example, enzyme kinetics (Michaelis-Menten/Hill) or thermodynamic equations (Arrhenius), leading to a highly nonlinear narrative of the biological system. The utilization of the principle of linearization contributes to detection of a nonlinear system around an experimentally evaluated metabolic steady state, by substituting the nonlinear with the linear function (Meysman & Bruers, 2010). The most convenient two-dimensional example can be demonstrated by the tangent in one point of the original function. This can be evaluated by discriminating the metabolic function, with reference to reaction variables, for example, metabolites. Such mathematical equations can be applied to all measured system variables at the same metabolic steady state, and the fractional derivatives are integrated in so-called Jacobian matrices (Sun & Weckwerth, 2012). By activating perturbations and evaluating the deflections of the metabolic system, distinctive characteristics, such as stability of stead states or flux sensitivities, can be measured and recapitulated in the Jacobian matrix (Reznik & Segre, 2010). Although there are well-established mathematical theories that primarily regulate this method (Doerfler et al., 2013), and they are generally applicable in engineering sciences (Föllinger & Konigorski, 2013), the direct incorporation of experimental (metabolomics) data is still demanding. Currently, we have focused on the advancement of approaches for a direct combination of metabolomics data, integrating the covariance data with genome-wide metabolic data to approaches of subcellular compartmentization (Nägele, 2014) and control of enzyme activity (Nägele et al., 2014). Summarizing all the listed strategies together, with numerous others from diverse research fields of biomathematics theoretical biology, bioinformatics and cybernetics will produce an exciting variety of inclusive models and approaches, which have been advanced during recent decades. The prime constriction of the next decade's metabolomics research will be the combination and relevance of numerous platforms, to manipulate high-throughput data as efficiently as possible. Concluding the aforementioned findings, open questions, and strategies, metabolomics plays a prime role in currently systems biological research. The prospects on the combination of diverse experimental metabolomic approaches will extend the coverage of the metabolome. The explanation of resultant multidimensional data arrays, with reference to metabolic network data at the genome scale, will extensively advance the perspective of intricate metabolic networks (Moseley, 2013). Integration and blending of diverse statistical methods and approaches of mathematical modeling is a prospective strategy to advance our expertise in interpreting illustrative metabolomic data, with reference to biochemical regulation in multilayered and highly compartmentalized biological systems. This will fundamentally supply exceptionally reflective information to all realms of life, and offer us with novel ideas and perspectives to resolve impending issues of universal concern.

6.8 IMPLEMENTATION OF STATISTICAL TOOLS TO DECIPHER POPULATION METABOLOMES

Metabolomics data offers an exceptional prospect for decoding the metabolic mechanisms by investigating expansive magnitude of metabolites (Chadeau-Hyam et al., 2013). Statistical analysis is employed for comprehensively interpreting the data quality and intricate batch effects (Misra, Fahrmann, & Grapov, 2017; Wehrens et al., 2016). Metabolomics is a comparatively novel high-throughput innovation that focuses on the estimation of entire endogenous metabolites within biological samples in an impartial way (Misra, 2018; Misra et al., 2017; Sarpe & Schriemer, 2017). Metabolomics is a relatively new high-throughput technology that aims at measuring entire endogenous metabolomes within a biological sample in an impartial manner. The metabolic profiles procured may act as a physiological state, and have been known to encompass effects of genetic regulation, along with ecological parameters (Lüttge, 2007; Spicer & Gaston, 2009). This prospect to associate genotypic to phenotypic data guarantees novel impending approaches and biomarkers for diverse research fields, including biomedical and pharmaceutical research. Several methods from other omics fields can be reutilized in the statistical analysis of metabolomic data. However, numerous specific contrivances for metabolomic data have been developed during recent years (Misra, 2018; Misra et al., 2017). We will discuss the recent progressive technological interventions in the analysis of metabolomic data, particularly by employing Gaussian graphical models (GGMs) and ICA (Bartel, Krumsiek, & Theis, 2013; Hendriks et al., 2011). There are diverse methodologies employed for decoding the high-throughput metabolomic data adapted from previous emerged omics technologies, mainly developed for transcriptomics analysis (Boekel et al., 2015; Joyce & Palsson, 2006; Tan, Ipcho, Trengove, Oliver, & Solomon, 2009). Conventional analytical techniques focus on investigating group-wise disparity, either in a parameter-by-parameter manner, called the univariate method (e.g., *t*-test), analysis of variance (ANOVA), or employing multivariate methodologies (e.g., MANOVA, ASCA, PCA, PLS) (Hervé, Nicolè, & Lê Cao, 2018; Ren, Hinzman, Kang, Szczesniak, & Lu, 2015). Univariate methodologies are generally utilized for decreasing the expansive metabolite data to provide coherent and compact responses under investigated conditions (Borland, Brickhouse, Thomas, & Fountain, 2010). The univariate methods, such as two-way ANOVA, are employed to assess medication-induced levels of changes in individual metabolites (Saccenti, Hoefsloot, Smilde,

Westerhuis, & Hendriks, 2014). Also, a Wilcoxon rank-sum test integrated with ANOVA can describe diverse cancer development stages, ranging from benign prostate to metastatic disease (Klein et al., 2017). However, univariate methods fail to discriminate between groups if there are only minor differences on single molecule level, even if multi-molecule combinations would delineate them on a systems level. Hence, the multivariate analysis methods aim to not only identify variations of single metabolites between diverse groups, but also employ the variable structures between the individual metabolites (Chatfield, 2018; Mertler & Reinhart, 2016). Possibly the most generalized multivariate analysis strategies that focus on illustration in the metabolomic field are PCA, cluster analysis, and PLS regression (PLSR), including derivative methods (Melaku, Gill, Taylor, Adams, & Shi, 2017; Varmuza & Filzmoser, 2016). PCA symbolizes an *unsubstantiated* linear mixture model that focuses on demonstrating the variance within a dataset by a small number of mutually decorrelated principal components (PCs) (Granato, Santos, Escher, Ferreira, & Maggio, 2018). In the case of metabolomics data, PCs are vectors of metabolite contributions that are constructed such that they are pairwise orthogonal to each other, and arrayed by the extent of variability they exhibit (Alonso-Gutierrez et al., 2015; Bro & Smilde, 2014). PCA can be decoded as a linear mixture model, where the data matrix is factorized into two matrices; a score matrix, which is comprised of the positions of the observations in a novel, rotated coordinate system; and a loading matrix, which encompasses the weights for the original variables to convert them into the scores. Due to its relevance in dimensionality reduction, data visualization, clustering, and sample group discrimination, PCA is generally employed as an initial point for data analysis, particularly in a premise-free, investigative experimental system (Kuligowski et al., 2014; Song et al., 2017). Some applications among various others in the metabolomic field include metabolomic analysis of urine in the kidneys for cancer diagnostics (Emwas et al., 2015), or urine and serum metabolites in the detection of Parkinson's disease (Luan et al., 2015) and diabetes (Guasch-Ferré et al., 2016). Szymańska, Saccenti, Smilde, and Westerhuis (2012) developed a multivariate ANOVA method called component analysis (ASCA), which is a blend of ANOVA and PCA methodologies. ASCA is specifically suitable for the evaluation of datasets with an intricate underlying experimental design, comprised of varied concurrently calculated covariates. It thus allocates in a straight line, and recounts disparity in the data to the diverse plan parameters. Moreover, hypothetical characteristics regarding ASCA, in addition to relevant metabolomics data; for example, the outcome of an oral rinse on human saliva metabolic profiles can be observed.

Cluster analysis corresponds to another unsubstantiated multivariate method appropriate for the investigation of metabolomics data with a self-organizing map, hierarchical cluster analysis (HCA) (Patterson et al., 2008), with *k*-means clustering (Xia et al., 2009) being the most important mode. In general, clustering methods group and visualize samples, along with intrinsic resemblance in their dimensions, irrespective of sample groupings. There are some research reports that document common queries of clustering strategies such as error propagation, complicated interpretability, and deprived reproducibility of the recognized clusters (Alyass, Turcotte, & Meyre, 2015; Cressie, Calder, Clark, Hoef, & Wikle, 2009).

6.8.1 Gaussian Graphical Models

Metabolites are cellular entities that are members of highly tangled biological pathways, and hence, exhibit a high level of interactivity (Karp et al., 2009; Kelder et al., 2011). A method to methodically model and instinctively understand such interdependencies is the representation as a graph or network. This method has become accepted and extensively utilized over the last decade. Networks are generally comprised of nodes, typically representing molecules (genes, proteins, metabolites); while associates between the nodes describe their interactions (van Iersel et al., 2008). The node symbolizes metabolites, and relates to a metabolic interaction in a metabolic network. At present, numerous publicly accessible metabolic databases exist that focus on global reconstruction of metabolic pathways, such as KEGG (Kanehisa et al., 2013), human Recon 1 (Sigurdsson, Jamshidi, Steingrimsson, Thiele, & Palsson, 2010), and EHMN (Ma et al., 2007; Stobbe et al., 2013). The biological networks included in such databases can be employed for direct statistical analysis in a functional manner (Hecker, Lambeck, Toepfer, Van Someren, & Guthke, 2009). The idea of a network-directed investigation was employed to, for example, identify diverse cancer subtypes by the detection of surrounding particular patterns of protein-protein interactions (PPI) networks or metabolic pathways (Wang et al., 2010), or for the deduction of an individualized therapy. Apparently, metabolomic databases still include incomplete information, which introduces partiality to a pathway-based evaluation influencing the requirement for novel complementary approaches (Cascante & Marin, 2008; Prosser, Larrouy-Maumus, & de Carvalho, 2014). One substitute, unparalleled strategy is to design pathway networks directly from the data. While this strategy was inadequate in the documentation of regulatory networks in specific in mammalian cells (Llaneras & Picó, 2008), the biochemical nature of metabolite pathways has exhibited improved suitability for such a restoration (Feist, Herrgård, Thiele, Reed, & Palsson, 2009). To facilitate reconstruction of metabolic pathways directly from the data, statistical methods develop the obviously occurring biological variation in the metabolite affluence between biological replicates (Duarte et al., 2007; Thiele et al., 2013). Such disparity in metabolite levels might occur either due to intrinsic fluctuations in temperature or pH (Woodward et al., 2016), or due to extrinsic factors, for example, variations in enzyme levels induced by diverse regulatory states) in the system (Dollo, Yi, & Lee, 2010; Wolfender et al., 2015). It is significant to document that these disparities in the metabolite concentration frequently occur in cumulative way, as such reflecting the wiring of the underlying metabolic network. The upgrading of pathways from high throughput is entirely dependent on Bayesian networks (Mitra, Müller, & Ji, 2016), or correlation-based measures. Bayesian networks are probabilistic graphical models that describe arbitrary variables, for example, gene expression levels or metabolite concentrations, (Aguilera, Fernández, Fernández, Rumí, & Salmerón, 2011; Zhu et al., 2012). The significant examples are gene networks deduced from gene expression data. However, Bayesian networks can only rebuild acyclic graphs, though authentic biological networks are recognized to enclose cycles and feedback loops (Landuyt et al., 2013; Lo et al., 2012). The easiest prospect for avoiding this inadequacy is pair-wise correlation methods (Bartel et al., 2013, where two nodes are associated if their individual correlation lies above a certain threshold. Even though their relevance has been reported in numerous applications (Sun & Weckwerth, 2012; Weckwerth, 2011a, 2011b), authentic correlation-based techniques lack the competence to distinguish between direct and indirect correlation; a highly significant correlation between two metabolites could be interceded by one or more unsolvable variables that are the definite origin for the practical correlation. GGMs, which are dependent on supposed partial correlation coefficients, deduce circumlocutory associations via training each pair wise correlation between two variables adjacent to all residual variables. These graphical models are unidirectional, where each node corresponds to arbitrary variables, and an edge between two nodes can be drawn if variables are conditionally dependent (Liu, Han, Yuan, Lafferty, & Wasserman, 2012). GGMs have a wide scope in the field of transcriptomics, as well as in metabolomics data interpretation (Rahman, 2018; Vinciotti, Augugliaro,

Abbruzzo, & Wit, 2016). Although the estimation of full order partial correlations generally needs a higher number of samples than variables, the requirement is not currently being met by prevailing experimental designs. Numerous techniques focus on this subject, recommending substitutive quantitative algorithms employing low-order partial correlation (Liu, Ren, Zhao, Yang, & Gao, 2016), bootstrap resampling (Gromski et al., 2015; Gromski, Xu, Hollywood, Turner, & Goodacre, 2015), or shrinkage quantification (Zhou et al., 2016). However, the chemical detection of yet unidentified metabolites in MS remains a significant subject (Dunn et al., 2013). Generally, these new metabolites cannot be allocated because present metabolic libraries lack entries, or minimum information on a nonsignificant number of metabolites. Different research reports have identified numerous metabolites using GGMs in combination with genomic data, providing clues at an extensive range of prospective relevance of GGMs for functional metabolomics (Krumsiek, Suhre, Illig, Adamski, & Theis, 2011). Although these methods, including PLS-DA, PCA, and GGMs are efficient in the evaluation of multivariate data, their application is limited to second-order statistical dependencies between the variables (Grootveld, 2014). However, these higher levels of dependencies originated from nonlinear metabolic processes, and are intrinsically insufficient for traditional statistical methods (Tufféry, 2011). The linearity of the association between estimated entities is an estimation that is only exact for normally distributed data (Bell & Jones, 2015). The most convenient method for dealing with discreetly distributed data is to employ a rank spearman correlation test (Fukushima, Kusano, Redestig, Arita, & Saito, 2011). ICA is a strategy that is used to define higher-order dependencies by expanding the idea of regular correlation to statistical dependence (Choi, 2015; Du & Swamy, 2014). ICA has proved to have immense relevance in biomedical research fields, such as the neurobiological field, including electroencephalographics (Brookes et al., 2011), and functional magnetic resonance imaging (Song et al., 2011). Further, ICA exhibits significant applicability in the field of molecular biology for the classification of cancer types, and in the investigation of cell proliferation processes (Yao et al., 2012) from transcriptomics data. Furthermore, ICA has demonstrated potential for metabolomics with reference to the evaluation of toxicological effects on *Arabidopsis thaliana* thriving in toxins, and for the evaluation of starch metabolism (Yonekura-Sakakibara et al., 2012), and the advancement of colitis in mice (Chen et al., 2008). Although ICA is quite similar to PCA, the main mathematical disparity lies between associations of determined variables. The ICA expands the decorrelation concept from PCA to statistical independence, which is a significant condition if the data is non-Gaussian. In this context, ICA petrifies the data matrix of an estimated metabolomics profile into *k*-statistically independent components (ICs). Therefore, metabolomic estimation symbolizes a blend of these discreet metabolic processes that we intend to unravel (Bathe & Farshidfar, 2014). As the quality and resolution of metabolomics methods advances, the blend of diverse, high-throughput datasets on the single-cell level becomes more of a possibility (Kueger, Steinhauser, Willmitzer, & Giavalisco, 2012; Scheltema et al., 2009). We assume that expanding high-quality datasets will advance not only the development of novel statistical tools, but also proceed to a higher competence of current techniques; ultimately contributing to an in-depth perspective of an organism's biology.

6.9 IDENTIFYING MEGA-METABOLOMES THROUGH MULTIVARIATE DATA MINING

With advancements in the omics field, for example, genomics, transcriptomics, proteomics, and metabolomics; there has been significant progress in the interpretation of molecular biological data (Agarwal & Dhar, 2014; Provost & Fawcett, 2013). This has also increased the efficacy of storage, data recovery, integration, and data evaluation (McCartney, 2015). There is an immediate prerequisite for an extensive data pool encompassing the life sciences for developing data-intensive or data-driven discoveries in biology (Altaf-Ul-Amin et al., 2014). The evaluation of this data pool will provide novel logic approaches and contrivances for unravelling novel trends, associations, discontinuities, and exceptions that uncover features of basic biology (Afendi et al., 2013). The emergence of big data biology, which is a data-intensive subject, is supposed to contribute to the exceedingly growing omics data generated by genomics, transcriptomics, proteomics, and metabolomics (Swan, 2013). The utilization of KNApSAcK Family DB in metabolomics demonstrates mining approaches, for example, PCA, PLSR, and the multiway model, and exhibits their relevance on the identification of the mega-metabolome. Metabolomic data is essentially multivariate; even for a single study, the initial point for data analysis is a single matrix $N \cdot D$ where, N is the number of samples and D is the number of variables (such as metabolites, binned chemical shifts, wave numbers, masses, retention times etc.). Each independent variable may be considered a single geometric dimension, such that each sample may considered to exist as a single point in an abstract concept known as D-dimensional hyperspace. The function of the analyst is to comprehend the basic characteristics of the data points dispersed in this hyperspace. This might be carried out in the form of unsupervised dimensionality reduction, or clustering or by correlation analysis, or by supervised pattern recognition encompassing multivariate regression and discriminant analysis (Afendi et al., 2011). This might be an imperative task given the existing number and the speed in which novel approaches are being developed.

6.9.1 KNApSAcK Family Database

Similar to other scientific subjects, omics biology is an age of instant increases of data, which has led to the emergence of big data biology (Nakamura et al., 2014; Shinbo et al., 2006). An extensive range of sequencing centers, high-throughput analytical facilities, and different laboratories have produced expansive amounts of data, for example, nucleotides and protein sequences, gene expression estimations, protein and genetic associations, mass spectra of metabolites, and phenotype studies (Nakamura et al., 2014; Wijaya et al., 2014). In KNApSAcK Core DB, there has been the development of species-metabolite associations to enable accessibility to metabolite data procured from analytical approaches. So far, nine databases of the KNApSAcK family (except DietDish) have been developed, which are linked with the KNApSAcK Core DB that simply procures candidates of secondary metabolites in species employed in numerous applications (Ikeda et al., 2013; Nakamura et al., 2013). The KNApSAcK Core DB was employed in extremely varied contrivances of metabolomic studies encompassing metabolite detection, the construction of integrated databases ("DB"), and bioinformatics and systems biology ("Bioinfo") (Kumar et al., 2016; Liu et al., 2015). Further, it is applicable in species spanning from microbes to plants and animals; thus far including twenty eight species of plants, for example, *Angelica acutiloba*, *Arabidopsis lyrata* ssp. *petraea*, *Arabidopsis thaliana*, *Atriplex halimus*, *Bacillus subtilis*, *Brassica oleraceae* var *capitata*, *Brufelsia calycina*, *Capsicum* sp., *Citrus sinensis*, *Curcuma longa*, *Ephedra* sp., *Escherichia coli*, *Fragaria x ananassa*, *Fragaria vesca*, *Glycine max*, *Glycyrrhiza uralensis*, *Hordeum vulgare*, *Jatropha curcas*, *Malus x domestica*, *Oryza sativa*, *Papaver somniferum*, *Solanum lycopersicum*, *Solanum tuberosum*, and *Zea mays* (Okada et al., 2013; Tung, 2014; Wijaya et al., 2014). The principal mining studies of blended herbal medicines for systematically comprehending the actual symphony of plants to efficacies on humans are based on PCA, which makes it feasible to classify the components in character combination systems, and PLS that can link the constituents of plants to the competence and N-PLS that can associate multi-factors to the efficacies.

6.9.2 Principal Component Analysis

This kind of statistical analysis is discussed earlier in Section 6.8, wherein we have demonstrated the implementation of statistical tools in the metabolomic data interpretation. PCA is a linear transformation of an expansive range of interconnected variables into a novel set of variables, called PCs, which are unrelated and patterned so that only a few preserve the maximum variation present in all the original variables (Bro and Smilde, 2014; Jolliffe and Cadima, 2016; Xanthopoulos et al., 2013)

6.9.3 Partial Least Squares

PLSR is a regression method that supposes basic parameters among the predictors that account for most of the response variation (Brereton and Lloyd, 2014; Gromski, Muhamadali, et al., 2015; Gromski, Xu, et al., 2015). These basic factors of X-variate

$$T = XW$$

are procured by amplifying their covariance with the equivalent fundamental factors of Y-variate, where X is an $n \times m$ matrix of predictors, Y is an $n \times p$ matrix of responses, T is an $n \times c$ matrix of X-score factors, and W is $m \times c$ matrix of weight. Note that n is the number of observations, m is the number of predictors, p is the number of responses, and c is the number of components.

6.9.4 Multiway Model

An elaborate PLSR associated with multidimensional data, multiway PLS, has been developed by Bro (Allegrini and Olivieri, 2014), and is also referred to as N-PLS. In this model, the same standard of PLSR for two-dimensional data is employed, that is, both analysts and response blocks are decayed consecutively into multilinear models, such that the pair-wise scores have maximal covariance (Nguyen et al., 2014; Stein et al., 2015). The attainment of the analyst is then regressed to the response variable. This illustrates the disintegration of the N-PLS model. Moreover, the N-PLS model can also be used for differentiation purposes, which is called N-PLS-DA, which is the multiway version of PLS-DA, by employing the dummy matrix of group membership as the response variable.

Multivariate analysis can provide the following data, at a minimum:

- Explanation of the workflow.
- Information about the algorithm employed, which must be clearly given and include the software packages and description integers or date used

for computation. State-of-the-art, radical or unconventional algorithms should, in general, be averted if they have not been separately assessed and accepted and published in a methodology paper.

- Information about how statistics are alienated into training and external corroboration sets, including, as much as possible, verification that both sets are uniformly representative of the sample space/data distribution.
- Descriptions of how internal validation/meta-parameter optimization is evaluated.
- Descriptions about the selected metric for evaluating the analytical skill of the model.
- Calculation scores for both training and external test sets.
- Particulars on data interpretation and visualization.

6.10 ECOMETABOLOMICS AS A CONDUIT BETWEEN BIOCHEMISTRY AND ECOLOGY

The fields of biochemistry and ecology function at entirely different spatio-temporal levels, with distinct biochemical resolutions. The interactions between organisms were primarily resolved by investigating the diversity of natural products, with an objective of characterizing the explicit metabolites (Raguso et al., 2015). However, this approach has faced enormous hurdles due to multi-functionality of metabolites across spatial and temporal scales, as implicated in various organism interactions. On the contrary, eco-metabolomics is an imperative interdisciplinary subject that has developed to integrate these different levels. For example, climate change invokes different changes at varied levels. The varied temperatures and altered moisture can alter gene expression and the metabolism of plants that function at diverse scales among species (Peñuelas et al., 2013). At these intrinsic levels, biochemical responses can be identified by estimating variations in metabolite levels (Brunetti et al., 2013). Meanwhile, the diverse individual responses of plant species in the community alter species composition functioning at the population and community levels (Leibold et al., 2004). This leads to species migration, which is evident at spatiotemporally coarse levels, ranging from a few, up to numerous kilometers. Ultimately, all these diverse types of variations effect ecosystem functions. A metabolite profile can be employed to estimate plant responses with diverse biochemical resolutions (Peters et al., 2014). They can also be employed to detect biochemical characteristics that function as markers, for example, for phenotypic plasticity, biochemical associations with other organisms, or determining the insidious potential of exotic plants (Bhat et al., 2016; Tewes et al., 2018). Hence, ecometabolomics is a subject that enables researchers to illustrate associations between processes functioning at discreet spatiotemporal scales. Because it employs metabolites for mechanistically recounting these mechanisms, it is assumed as the mediator between diverse research methods (Jones et al., 2013). For the complete data processing and data analysis, bioinformatics and biostatistical tools are imperative (Wiggins et al., 2016). Ecometabolomics employs two different approaches for inferring the interactions between organisms, as well as with the environment. These two approaches include a bottom-up approach, which encompasses spatiotemporally fine scales within plants, including cellular processes, physiological characteristics, and growth parameters to spatiotemporally coarse scales such as population fitness and crop productivity (Bowne et al., 2011). In comparison, a top-down strategy is generally adopted for ecological studies that conjecture from spatiotemporally coarse levels, including association processes at the population and community levels to spatiotemporally finer scales distinguishing physio-biochemical traits and morphological parameters of plants (Sun et al., 2016).

The bottom-up strategy encompasses biochemistry that targets central metabolic processes in individual organisms. The central function of metabolites is illustrated in biological pathways implicated at the cellular level, or physiological processes on the organ level of common model species. Ecometabolomics focuses on both systemic and fundamental responses to environmental parameters. In order to comprehend the application of the metabolic responses, it is imperative to characterize metabolites that are altered under limited conditions, or that differentiate organisms or species that interact at the population and community levels (Rivas-Ubach et al., 2016, 2017). Such metabolites can illustrate methods at spatiotemporally larger levels; for example, variations in crop productivity, or identify alterations in species interaction at the population and community levels (Rivas-Ubach et al., 2016). Numerous research reports have illustrated the metabolomic changes that induced plasticity against varied stresses in parallel with an increase in metabolite pools. For example, a differential response of tomato plants to varied salinity levels demonstrates a pronounced accumulation of carotenoids (Van Meulebroek et al., 2016). Similar studies were reported in maize subjected to varied temperature levels that induced accumulation of metabolites that are in concomitant with tolerance against heat and cold. Similar results were reported in *Arabidopsis thaliana* subjected to drought stress (Frolov et al., 2017). The species-specific metabolic flux in response to symbiotic interactions between numerous plants and arbuscular mycorrhizal fungi was reported (Nacoulma et al., 2013). Furthermore, plant metabolome alterations subjected to biotic stress such as nematode and aphids interferes with roots and shoots (Kutyniok and Müller, 2012), which has been studied by researchers. At lower-scale levels, the intricacy of the biological phenomenon is expansive, and turnover of processes, for example, translation into molecules, enzymatic activity, biochemical pathways, and cell cycles occurs within seconds, to a few hours (Kutyniok and Müller, 2013). In order to reduce the intricacy in data assemblage, biochemical research is generally carried out with Arabidopsis, Medicago, tobacco, or tomato plants. The expansive data on the function of metabolites in these model plants facilitates the investigation of non-model plants, which are generally used in ecometabolomics (Gutbrodt et al., 2012). This is generally more feasible, along with intricate experimental designs, for example, three factorial analyses were employed for estimating responses of *Brassica oleraceato* to herbivory and drought stress. Skoneczny et al. (2017) revealed a significant relationship between the physiological levels of two *Echium* species, root-producing shikonin to that of the phenological stage at the population level. Most of the studies used untargeted metabolite profiling to determine the varied states of the organism. However, metabolomic techniques can also be competitively employed for the detection of actual mediators of interactions (Kuhlisch and Pohnert, 2015). In case there is data available on the chemical properties of metabolites, targeted profiling can be used to detect candidate metabolites. And their functions can be further validated by bioassays; for example, detection of the first sex pheromones of unicellular diatoms (Gillard et al., 2013).

In the top-down approach, including coarse to fine spatiotemporal levels; ecological experiments, environmental effects, and biotic interactions of organisms were illustrated. This is acquired at spatiotemporally coarse levels at which responses of diverse plant species in an ecosystem are primarily determined (Leibold et al., 2004). Processes such as species migration and community shifts can take only a short time, or can take years to defend against climatic change (Gutbrodt et al., 2012). At larger scales, this is likely because of the intricacy and expansive number of profiles that need to be procured. The intricacies that exist between organisms are assumed to be greater than intrinsic biological mechanisms at coarse scales (Dunn and Ellis, 2005). Owing to expansive complexity in metabolite profiles, there are very few metabolites that are analyzed at the community level (Legendre and Legendre, 2012). Kursar et al. (2009) analyzed a community of 37 species of Inga trees sharing herbivores and pathogens at local and regional spatial scales. The observations obtained from these interactions were also imperative for niche discrimination of diverse congeneric Inga species in the community. The ecological coexistence is substantiated by diverse biochemical assemblages, and the interspecific variations allow niche segregation among congeneric tree species originating from chemical defenses. Metabolite profiles are determined by the level of diversity, intensity of varied competition, and surrounding plants (Sedio et al., 2017). The metabolite profiles of microbial communities revealed the metabolic relationship between soil species richness, niche width, and distribution. It has been established that the rhizosphere consists of highly variable community of microbes that interact with root exudates (Scherling et al., 2010; Zuppinger-Dingley et al., 2015). We assume that comparable research studies investigating rhizosphere metabolomes interacting with plants will produce the same new insights, and might employ analogous practical strategies, as with community metabolomics of microbial colonies (Okie et al., 2015). Generally, numerous organisms from a single species were investigated at the population level, with respect to ecological variables (Field and Lake, 2011). Field and Lake (2011) demonstrated the discrepancy in foliar metabolite profiles that can be related to genetic variability, soil characteristics, and growth-associated traits. Gargallo-Garriga et al. (2015) demonstrated differential metabolic responses by the shoots and roots of the two grasses *Holcus lanatus* and *Alopecurus pratensis* in response to drought and warming. On the contrary, the shrub *Sambucus nigra* displayed specific interactions between plants and the microbial community in the phyllosphere (Gargallo-Garriga et al., 2016). Research reports on exotic species demonstrated discreet and explicit metabolite pools in comparison with native congeners (Macel et al., 2014). Hence, ecometabolomics revealed the poor herbivore species' performance on exotics, thereby forwarding evidence that advances the concept of the "Novel weapon hypothesis" (Callaway and Ridenour, 2004). In contradiction to biochemistry, ecology adopts a top-down research strategy wherein complex ecological processes are simplified to smaller sets, and illustrated autonomously to exhibit cryptic biochemical characteristics that function upon diverse species (Sedio et al., 2017). With this reference, extremely explicit metabolites, also known as markers, could function as an alternative for "eco-chemical" characteristics, and enable a significant classification primarily based on metabolites (Drenovsky et al., 2012; Ehlers et al., 2016; Hopkins et al., 2009). In biochemistry, metabolites have an imperative function in characterizing mechanistic metabolites of processes; but in ecology, this function is conventionally completed by characteristics such as morphological, physiological, or phenological traits of each plant. Ecometabolomics ascertained the biochemical characteristics that illustrate the relationship with plant developmental stages, and the allelo-chemicals of trees or plant defense traits in both indigenous and exotic populations (Skoneczny et al., 2017). Furthermore, ecometabolomics can be employed to explore eco-chemical characteristics that illustrate ecological processes, for example, species co-existence, multi-trophic interactions, and stress plasticity can be observed in the reviews (Schweiger et al., 2014; Valares Masa et al., 2016).

6.11 CURRENT CHALLENGES IN ECOMETABOLOMICS

In order to scale up from the biochemical to the ecological level, we need to characterize metabolites that influence ecology. There are some insect pheromones to which insects are sensitive, even over long distances (Blande, 2017). For example, fruit bats, along with frugivorous primates, employ fruit fragrances to identify ripe fruits (Hodgkison et al., 2007; Nevo et al., 2015, 2016). In a similar way, the generation of volatiles can have an immense effect on pollination (Dormont et al., 2014; Majetic et al., 2008) and dispersal (Rodríguez et al., 2013; Valenta et al., 2013). GC-MS can be used in the detection of volatiles. Due to the presence of numerous metabolites that are generated as a result of plant-animal interaction, it is still difficult to identify the metabolites or their blend that uses central ecological functions (Mithöfer and Boland, 2012; Pitts-Singer et al., 2017). In comparison with other analytical approaches, NMR exhibits more competence due to wider metabolite coverage, including both polar and nonpolar metabolites, and facilitates their characterization by evaluating their resonance frequencies and line shapes in spectra with spectral libraries (Pimenta et al., 2013; Rivas-Ubach et al., 2013). Data processing is highly intricate due to the need for preprocessing of raw data; for example, the production of bucketing tables. Several strategies to generate bucketing tables (De Meyer et al., 2008; Sousa et al., 2013) and an expansive range of diverse contrivances and methods accessible for these processing steps (Rocca-Serra et al., 2016; Spicer et al., 2017) are reported, such as NMR data processing and metabolite characterization, which remain remarkably colossal. Likewise, in ecometabolomics, there is also a constraint to "scale down" from ecological to biochemical scales, that is, to explore significant metabolites in diverse organisms (eco-chemical characteristics) that can be associated with explicit ecological interactions. Because organisms produce numerous metabolites, suitable experimental setups and varied bio-statistical methods are indispensable to screening the contenders that trigger variability/diversity, or can be ascribed to the particular interaction (Scherling et al., 2010). Ecometabolomics can also be employed on either side of the organism interaction; for example, between plants and herbivores. In order to identify coumaroylquinic acids, we can define the idea of the metabolic interface between plants and caterpillars. This class of metabolite classes was loaded in both jasmonic acid induced plants and caterpillars feeding on them. By inclining the trophic chains, they are expected to influence higher trophic organisms; for example, the larvae of endoparastic wasps (Qiu et al., 2009), which feeds on the caterpillar's fat, and has no direct contact with the plant.

For top-down approaches in ecometabolomics, characterization of single metabolites is generally not suitable. For this purpose, metabolic fingerprint analysis is performed, in which a set of characteristics is investigated, instead of a single feature (Gutbrodt et al., 2012; Leiss et al., 2013). In general, metabolic fingerprints are produced from the metabolites that are shared, or that are discreet for diverse species (Gao and Xu, 2015; Jansen et al., 2009; Sue et al., 2011). The collaborative effort made by participants at a three-day workshop at the German Centre for Integrative Biodiversity Research from 16 to 18 October 2017 summarized five major challenges in ecometabolomic studies:

6.11.1 Composite Experimental Setups and Variable Metabolite Profiles

The main constraint in carrying out ecometabolomics studies is the subsistence of the intricate experimental setups as they are generally investigated in field conditions, and with non-model species (Bowne et al., 2011; Roessner and Dias, 2013). Metabolite profiles are further altered due to varied genetic backgrounds, different developmental stages of the individuals, by short- and long-term environmental variations, for example, seasonal and weather patterns, and also by a changing milieu of biotic interactions. Hence, the diversity among metabolite profiles within the treatment classes is more expansive than in traditional metabolomic studies. Though pilot studies are rarely employed, they can be ideal for evaluating the variability level among treatment groups, as well as in large-scale experiments. But, pilot studies facilitate the quantification of the numerous samples required, and therefore the number of significant metabolite profiles to be procured to validate effect sizes in a statistically lucid manner (Trutschel et al., 2015). There are easily accessible R packages and templates to sustain scientists with corresponding statistics (Kelley, 2007). Numerous ecological investigations that target population and community levels are designed in such a way that the highest number of replicates is made at the level where the largest variability among samples is anticipated (Trogisch et al., 2017).

In order to carry out ecometabolomic studies at the community and population level, experiments are designed in such way so as to accommodate the highest replications in samples where higher- level variability is expected. Ecometabolomic investigations hence follow complex designs in a top-down approach covering spatiotemporally coarse levels. While in bottom-up approaches generally carried out in glasshouse or growth chambers, launching the suitable number of controls is also important, so as to ensure an effective size between control and treatments. Numerous top-down strategies that function at spatiotemporally coarse scales intend to cover the intricacy of intrinsic biological mechanisms by investigating methods of biological organization that are mediated by sets of eco-chemicals (Maier et al., 2010). The handling of the samples and usual metabolite acquisition approach is generally explicit to the underlying research queries (Dunn et al., 2012). Expansive sampling operations at the community and population levels generally need samples to be pooled and homogenized, which improves the reproducibility of estimations by reducing spatial heterogeneities. However, the pooling might hamper thorough insights at spatiotemporally fine scales (Calf et al., 2018; Kuzina et al., 2009).

Instruments run through batches lead to shifts in m/z or RT among the subject samples because of the matrix effects. Ecometabolomic studies have not yet involved strategies such as quality control of the instrumental setup and interspersal of quality control and mixed quality control, which are necessary to identify and authenticate these shifts (Bouhifd, 2015). The analytical reproducibility for metabolomic approaches is modulated by instrumental configurations and types of separation techniques (Arens et al., 2015; Kiontke et al., 2016; Rivas-Ubach et al., 2013). In order to avert the batch to batch effects involved in instrumental setup and variance during sampling, we need to implement normalization techniques, including a reasonable number of blanks (Birkemeyer and Kopka, 2007; Hutschenreuther et al., 2010).

6.11.2 Feature Extraction

Diverse bioinformatics tools were employed for the extraction of metabolite characteristics from raw metabolite profiles. Using XCMS and OpenMS, numerous studies have implicated other tools and algorithms that are accessible for processing and alignment of raw metabolite data (Röst et al., 2016; Smith et al., 2006). The model organism is employed for validating data generation from the designed bioinformastical operations, including raw data, peak detection, feature extraction, and retention time shift (Kumar et al., 2016; Spicer et al., 2017). Optimization of feature extraction and alignment of data to tackle the diverse organisms, multifactorial experimental designs, and resultant diversity of samples needs to be validated for ecometabolomic data. In a similar way, MS of raw data are generally processed by optimizing the instrumental settings; for example, the signal to noise threshold and maximal m/z deviations for peak detection for the specific experimental setup. Establishment of the optimized features will facilitate the peak identification algorithms to separate peaks from noise consistently, and arrange analogous characterization across samples in an authentic manner, and allocate distinct characteristic identifiers (Smith et al., 2006). Moreover, investigating the mathematical transformations; for example, log or sqrt on the characteristic matrix, could be indispensable to acquire a semi-normal distribution of values to a greater extent (Van den Berg et al., 2006). In comparison with primitive metabolomic approaches, ecometabolomic experiments with intricate designs might need diverse parameter settings for characteristic alignment known as "grouping" in XCMS, or feature linking in OpenMS to appropriately match extracted characteristics among diverse samples (Tautenhahn et al., 2008; Weisser et al., 2013).

Metabolite profiling via LC-MS and GC-MS characteristics is generally comprised of repeated data in the form of adduct isotopes and in-source fragments. These are significant for metabolite detection, although it leads to the co-linearity of the characteristic features of peaks. Co-linearity could result in consequent statistical investigation (Röst et al., 2016). Co-linearity can be lowered by assembling characteristics that were classified by the R package CAMERA into a similar pseudo-compound class. Peters et al. (2018) anticipated the function "get Reduced Peak-list" to the CAMERA (version 1.33.3 or later), which can be employed instead of the regular "get Peak list" function.

6.11.3 Metabolite Identification

The detection of metabolite characteristics in data acquisition is still a constraint. Furthermore, ecometabolomics can be employed for metabolite profiling of the unknown compounds in non-model organisms

(Wishart, 2009: Viant, 2009). The authenticated standard for metabolite characterization is the comparison with procured MS data of a reference standard. However, there are scarce reference standards available for non-model species. Further, it will be more cumbersome to purify the substantial amount of compound and then adequately purify it for structural identification via NMR. Efforts are being made in the field of computational methods to compensate for the lack of reference standards (Xiao et al., 2012). Diverse detection pipelines have been developed to be used for acquisition methods in ecometabolomics (Jandová et al., 2015; Moore et al., 2004; Lankadurai et al., 2013).

Metabolite profiling through GC-MS has yielded expected retention times and consistent spectral data, and the application of expansive libraries, for example, NIST, MoNA, and the Golm Metabolome Database (GMD) due to their vigorous capillary columns and specific ionization sources (Hernández et al., 2012; Kopka et al., 2005; Vinaixa et al., 2016). Detection of novel metabolites can be aided by employing source separation approaches that avert vigorous chromatographic segmentation (Domingo-Almenara et al., 2015, 2016). Consequently, metabolite characteristics are more consistent in comparison with LC-MS. Because LC-MS is loaded with numerous other adducts, isotopes, and ion source fragments, it is comprised of numerous repeated characteristics that belong to metabolites (Domingo-Almenara, Montenegro-Burke, Benton, & Siuzdak, 2017). Peak grouping and annotation, and numerous algorithms are determined by inputs of identified m/z distances among general adducts, fragments, and multiplied charged ions. In spite of expansive bioinformatics tools and spectral libraries to annotate and detect metabolites, the LC/MS-MS data processing pipeline is still intricate and time consuming to use, because it still requires expansive manual data assessment (Viant, 2009). The comparison of purified standard spectra in reference to libraries is generally a pre-requisite for metabolite identification. The utilization of various molecular networks, for example, the Global Natural Products Social (GNPS) Molecular Networking database and structural matching tools, facilitates the comparison of unknown compounds with fragments of the same metabolite that have similar substructures (Wang et al., 2016). The predicative retention times for metabolites can be employed as an extra orthogonal characteristic for filtering out the candidate metabolites (Cao et al., 2015; Creek et al., 2011; Jonsson, Bruce, et al., 2005; Jonsson, Johansson, et al., 2005; Krauss et al., 2010; McEachran et al., 2018).

Recent advances in computational annotation tools, for example, MetFrag and MetFamily, facilitate matching of estimated spectra with reference to spectra of metabolite classes. Numerous other bioinformatics tools that investigate in-silico prediction have been equipped with identified metabolites that originated from *Arabidopsis thaliana* and *Lycopersicum esculentum* (Domingo-Almenara et al., 2017; Shahaf et al., 2016). This enables additional unclear matches of characteristics with analogous spectra that allow more precise annotations of "unknown unknowns" (Ruttkies et al., 2015). In several cases, this unclear annotation might be adequate for ecologists to illustrate some biological interpretations, for example, disparity in herbivore resistance. Herbivore resistance can be further elucidated by target analysis to isolate and identify which of the metabolites in the metabolite family, such as particular flavonoids, use the underlying agent for the effect. Machine erudition strategies have remarkable abilities for in-silico prediction of "unknown unknowns" with tools such as CFM-ID (Allen et al., 2014) and CSI:Finger ID (Dührkop et al., 2015). Integrating targeted spectral libraries with computational de-replication techniques has been anticipated to detect metabolites in non-model versus model species (Allard et al., 2016; Chervin et al., 2017).

6.11.4 Statistical Analyses

Statistical analysis is an indispensable tool in ecometabolomics for dealing with expansive data matrices and multifactorial experimental set ups, plus untargeted approaches. There are diverse statistical methods employed in ecometabolomics, such as GLM, PCA, ANOVA, HCA, Kruskal-Wallis test, LDA, Canonical variate analysis, NMDS, Tukey HSD, DCA, Pearson correlation, discriminant analysis, dbRDA, Mantel test, *F*-test, RDA, MANOVA, PERMANOVA, DA, linear mixed model, REML, Tukey-Kramer test, PLS-DA, OPLS-DA, Diversity indices, a priori path models (PROC CALIS), upfield and downfield diversity, cluster heat map average linkage, Mann-Whitney *U* test, Dunn test, volcano plot, Friedman ANOVA, Wilcoxon signed-rank test, heat map, volcano plot, Chi2 test, the Venn-Euler diagram, and so forth (Caseys et al., 2012; Herden et al., 2016; Huseby et al., 2012; Jandová et al., 2015). The utility of all these statistical tests is determined by a number of predictors, factor levels, and the data type (independent vs. dependent data) and distribution of data such as normal vs. non-normal distribution. Ecometabolomic studies employ both parametric tests such as ANOVA, as well as non-parametric tests; that is, Kruskal–Wallis and Mann–Whitney U tests. The linear mixed effect models are employed for nested experimental setups used for facilitating the accountability of random factors and to figure out different errors in the diverse hierarchical levels. Various post-hoc tests such as Tukey's HSD can be used for estimating p values among the diverse groups. But while dealing with diverse metabolites, there is the possibility of false positives among all statistical analyses. To tackle these false positives, the Bonferroni correction that controls the family wise error rate, or the Benjamini–Hochberg method, which controls the false discovery rate, is recommended in metabolomics (Broadhurst and Kell, 2006; Vinaixa et al., 2012). Numerous ecometabolomic studies have employed Holm–Sidak the method, and the Levene's test for multi-comparison correction (Guo & Lankmayr, 2010; Newson and Team, 2003). Linear mixed effect models are also employed for investigating multivariate data to identify multiple predictors, plus controlling the imprecise parameters and replicate samples in single techniques (Jamil et al., 2013; Tewes et al., 2018). In addition to tests for statistical significance, quantitative variation can be estimated for individual metabolites to signify how positive their intensities differ between groups. The observations from statistical tests can be integrated with fold variation analyses to assess which metabolite is more interesting for further investigation. Owing to enormous dimensionality of data and numerous research queries, several feature matrices are added with untargeted approaches. The most commonly used statistical tools for multivariate analysis include Principal Coordinate Analysis (PCoA) and (Non-metric) Multidimensional Scaling (N) MDS) for comparing the metabolite profiles between samples, and to select sets of metabolite candidates (Smith et al., 2015; Weber et al., 2013). The metabolites that exhibit maximum variations among groups are often compounds of interest. These metabolites are determined by post-hoc tests, for example, non-parametric PERMANOVA (PERmutational Multivariate ANalysis Of VAriance) (Warton et al., 2012). Although in ecometabolomics there are generally two or more data matrices, as usual, the characteristic table is comprised of the biochemical data of the metabolite profiles and another matrix with numerous ecological parameters (Legendre and Legendre, 2012). In order to analyze tables conjointly, ordination methods such as Redundancy Analysis (RDA), distance-based RDA (dbRDA) plus Discriminant Analysis (DA) (esp. Linear DA), Orthogonal PLS (OPLS), Hierarchical Clustering (HCA), classification and machine learning such as Random Forests (RF), and Support Vector Machines (SVM) are used. To facilitate the association between ecosystems functions and metabolites, the inclusion of meta-data from a database, and data from elemental analyzers will be imperative for discerning metabolomic traits (Jesch et al., 2018).

6.11.5 Bioinformatics Software Tools and Workflows

Numerous bioinformatics software contrivances, data processing work flows, and databases employed in metabolomics were standardized for model species, or were validated for clinical purposes. Hence, they are indirectly relevant to ecometabolomics studies. However, it is cumbersome for bioinformaticians to choose the most appropriate software tools and data sources for the idiosyncratic ecometabolomic experimental setup. The prime constraint for ecometabolomics includes data repositories and libraries, because both ecological and biochemical information is generally fragmented. Most of these repositories are included under the control of institutes or commercial agencies. Moreover, primary data are not constantly shared with the scientific community, are limited to a few project members, or are mislaid after the paper is published (Wilkinson et al., 2016). Numerous biochemical databases, for example, PubChem or KEGG, primarily possess chemical structures and data about model species. For non-model species, species-metabolite association is deciphered through databases such as KNApSAcK or NPASS (Nakamura et al., 2014; Zeng et al., 2018). Databases such as MetaCyc, GMD, or BioCyc can be employed for retrieving the association between metabolites and biochemical traits (Caspi et al., 2016; Karp et al., 2015). However, presently they are not sufficient for leveling up to processes at ecological levels. Numerous ecometabolomic studies depend on ecological data sources that are scattered between countries or limited to local floras. The Plant Trait Database (TRY) agglomerates ecological characteristics of diverse organisms universally, and can also be employed as a source in ecometabolomics (Kattge et al., 2011). However, there still remains the requirement for coalesced characteristics that assemble the data from the numerous small local databases (Soranno et al., 2015). There are no exclusive repositories for preserving ecometabolomics' primary data; however, there are a few repositories, such as MetaboLights, Metabolomic Workbench, Metabolome express, or GNPS. But these repositories are not regularly used for preserving raw profiles and documenting metadata (Carroll et al., 2010; Haug et al., 2013; Sneddon et al., 2014; Wang et al., 2016). The reproducibility and quality assurance are imperative for scientific investigation, which is enumerated as firm rules for analytical validity (Bouhifd, 2015). But reproducibility for data processing and computational investigations is always hard to acquire (Russell, 2013; Wilkinson et al., 2016). This intricacy is partly due to the expansive biological and ecological systems (Goodacre et al., 2004), the variability of technological platforms employed in metabolomics (Eghbalnia et al., 2017; Zhang et al., 2012), and the varied number of accessible bioinformatics tools. For example, Spicer et al. (2017) enlisted more than 130 open bioinformatics contrivances and numerous labs, and also employed proprietary vendor software (Ioannidis et al., 2009; Peng, 2011). In order to make ecometabolomics studies easier, and to replicate experimental setups, it is suggested that data sets, metadata, and the equivalent bioinformatics data processing pipelines be shared with the scientific community (Kanani et al., 2008; Meier et al., 2017). Wilkinson et al. (2016) demonstrated FAIR, which stands for Findability, Accessibility, Interoperability, and Reusability guiding principles, which are an array of basic rules that contribute to suitable data management and prolonged data execution. During the recent past, bioinformatics workflow formats were established to include all inclusive data processing pipelines, starting from data downloads from public repository, to data quality control (Godzien et al., 2015), and various biostatistical analyses. It is imperative to reproduce the data processing pipeline to give the scientific community a consistent perspective at any level of study (Leipzig, 2016). Presently, the scientific community is eagerly tussling with the next steps because of the technical complexity of the software employed. During recent years, the galaxy workflow platform has become exceedingly accepted with numerous "omics" technologies (Afgan et al., 2016). At present, there are few metabolomic workflow platforms that can be employed for ecometabolomic studies, such as Galaxy workflow systems, Workflow4Metabolomics (Giacomoni et al., 2015), and Galaxy-M (Davidson et al., 2016) and the KNIME workflow system, integrated with OpenMS modules (Fillbrunn et al., 2017; Pfeuffer et al., 2017). But these platforms require improvisation so as to accommodate ecometabolomics modules that allow data processing for non-bioformaticians (Meier et al., 2017)

6.12 POTENTIAL CONSTRAINTS IN ECOMETABOLOMICS

Metabolomic approaches can be employed for the estimation of any deviation in normal metabolite levels in organisms that render organisms incapable of attaining the steady state level, also known as homeostasis. Modeling in systems biology is fundamentally determined by these deviations, and permits scale up from spatiotemporally fine to a coarser scale (Weckwerth, 2011a, 2011b). Although there was an extensive variation in the metabolite profiles of the same plant procured at different times, as observed in model plants such as *Arabidopsis*, *Silene*, and CAM plants (Abraham et al., 2016; Dötterl et al., 2012; Kim et al., 2017; Tarakhovskaya et al., 2017).

Hence, we can postulate that the metabolite profile is a static snapshot of the biochemistry of the plant at a particular point in time. Because the biotic ecosystem can alleviate the effects encompassing climate change, it is imperative that the ecometabolomics community focus on the whole, along with its surroundings, instead of just focusing on the individual (Fester, 2015). Endophytic microbes can colonize internal tissues of host plants, they can affect metabolite profiles of plants, and they might even yield exogenous metabolites (Ryan et al., 2008). Endophytes can form varied associations with their host plants that can be harmful, for example, pathogenic fungi, but this can also be advantageous to both partners (Liu et al., 2017). This, at times, renders it cumbersome to furnish concrete conclusions from spatiotemporally fine to coarser scales that are, however, imperative to determining effects on ecosystems' services (Kumar et al., 2016). Though the susceptibility of the analytical policy can be "consent" in bottom-up strategies in biochemistry, and has enabled numerous comprehensive insights into processes among organisms that would not have been likely to be identified if not, it might be "cursed" with top-down techniques (Jones et al., 2013). The sensitivity of metabolite profiles may encompass undesired short-term variations, including weather conditions, and it therefore generally requires acquisition of numerous metabolite profiles, and the estimation of extra non-biochemical characteristics of plants and ecological parameters, to exclude side effects or to be approved for shifts and variations (Jones et al., 2013). Various statistical methods are employed specifically for tackling expansive variabilities in ecology (Legendre and Legendre, 2012). It is also imperative to think that biochemical characteristics might not be implicated in every kind of ecological interaction (Jones et al., 2013). Keeping in view the rationale, it is significant for top-down ecometabolomics studies that quick research queries or hypotheses are determined even earlier than the samples are taken. There is no single analytical method that can encompass all metabolites immediately. Certainly, the metabolite coverage exceedingly depends on the extraction method, and the kind of instrumentation employed. It is beyond the scope of this chapter to enumerate all advantages and limitations of the varied kinds of acquisition methods. However, there are numerous reviews published so far in this context (Hernández et al., 2012; Sardans et al., 2011; van Dam & van der Meijden, 2011; Viant and Sommer, 2013).

6.13 NEW PERSPECTIVES IN ECOMETABOLOMICS

It is now well established that ecometabolomics contributes new strategies to retort basic ecological queries. For example, numerous methods in ecology are induced by interactions, for example, those between soil microbes and plants that are distant to the human eye (MacArthur, 1955). Structural equation modeling can design relationships between multivariate data, including cause-effect at diverse levels, so as to decipher the organismal interactions or linkages in ecology (Clough, 2012; Douma et al., 2012; Eisenhauer et al., 2012; Lamb, 2008; Lamb et al., 2009; Menéndez et al., 2007). Because metabolites are intermediates in these processes, they have the potential to be employed for SEM (Canarini et al., 2016; Cortois et al., 2017; Eisenhauer et al., 2015; Wei et al., 2015). SEM is analogous to techniques employed in systems biology and cellular processes (Weckwerth, 2011a, 2011b), and we anticipate that SEM will be used frequently in ecometabolomics in the future. This will permit estimation of the number and intensity of ecological linkages, and therefore help illustrate the varied kinds of biochemical interactions organisms comprehend in ecosystems. Moreover, inherent variables in SEMs might be employed to construct "eco-chemical" characteristics that demonstrate a portion of the functioning of ecosystems. We have illustrated with the top-down strategies generally utilized in ecometabolomics that it is generally suitable to characterize singular metabolites. To a certain extent, an array of metabolite characteristics, annotating metabolite classes, and associating them with ecological functions are some of the prime objectives. The rejoinders acquired by these kinds of studies can be employed to obtain and assemble novel questions and hypotheses that are of imperative interest for biochemistry. Hence, following characterizing fundamental ecological associations, the data can be employed for "zooming in" and investigating bottom-up strategies that contribute novel illustrated insights into ecological processes arbitrated by metabolites (Chervin et al., 2017; Poisot, Bever, Nemri, Thrall, & Hochberg, 2011). Even though it should be known that metabolomics, in principle, is a hypothesis-producing strategy. If the objective of the research is to identify explicit metabolites responsible for ecological interactions, top-down studies must be pursued by schematic experiments. Hence, we can summarize the utility of ecometabolomics under two approaches; bottom-up and top-down approaches that have their own exceptional benefits and limitations, and can contribute to science immensely. In the future, ecometabolomics is likely to become more intricate by incorporating more parameters into the experimental designs. Moreover, to amplify the metabolite coverage, an integration of varied extraction procedures and analytical approaches might even be employed in the future. The recent trend has also flipped toward the utilization of non-model species with both types of methods (Bhat et al., 2016; Iriti & Vitalini, 2018). The constraints we detected in this chapter are not restricted to the plant realm. Actually, other realms, for example, metabolomic investigations of wild animals, have the same constraints, such as tackling the "unknown unknowns," and dealing with non-model species. We validated that the term "ecometabolomics" is not employed extensively by the scientific community. The distribution of ecometabolomic primary data in repositories will immensely contribute to making ecometabolomics accessible to an extensive scientific audience, and will permit metadata studies that re-evaluate observations from varied ecometabolomic reports in the future (Fester, 2015). But, there is still the requirement for an advanced level of blending between spatiotemporal scales, strong cooperation between researchers to advance databases for both non-model species and spectral data, and hence, also among scientists' alliances around the world.

6.14 CONCLUSION

Like other scientific subjects, biology encompasses all instant data procurement techniques, and scientists determine the accessibility of the data, for example, nucleotides and protein sequences, protein and gene expressions, metabolite flux, and so forth. The underlying principle of the present systematic perspective of big biology pertaining to health, nutrition, and other societal queries has presently become the focal point of scholars in collective studies of science and data. The increase of community databases such as the KNApSAcK family DB has been powerfully linked with bolstering data-intensive science. The central question is whether scientists can deduce how systems and whole organisms work from this stream of molecular data. To advance this method, a data-intensive method is required for deciphering intra and inters associations in individual layers. The first method can be solved based on the kind of multivariate analysis, for example, cluster analysis and PCA. Even though the latter is more intricate, numerous methods, including PLS and N-PLS, make it possible to elucidate these associations. Big data biology has become an important part of biology, and the laws of nature could be elucidated based on comprehensive investigation of big data biology, the age of which has already commenced. For centuries, biological research primarily conducted on experiments, and for many years, computational evaluations have generally followed experimentation; but in the future, it might be the opposite, that is, computational analysis may be done primarily to direct experimental design assisted by multifaceted and freely accessible omics data in diverse databases.

REFERENCES

Abraham, P. E., Yin, H., Borland, A. M., Weighill, D., Lim, S. D., De Paoli, H. C., et al. (2016). Transcript, protein and metabolite temporal dynamics in the CAM plant Agave. *Nature Plants*, *2*, 16178.

Afendi, F. M., Okada, T., Yamazaki, M., Hirai-Morita, A., Nakamura, Y., Nakamura, K., et al. (2011). KNApSAcK family databases: integrated metabolite–plant species databases for multifaceted plant research. *Plant and Cell Physiology*, *53*(2), e1.

Afendi, F. M., Ono, N., Nakamura, Y., Nakamura, K., Darusman, L. K., Kibinge, N., et al. (2013). Data mining methods for omics and knowledge of crude medicinal plants toward big data biology. *Computational and Structural Biotechnology Journal*, *4*(5) e201301010.

Afgan, E., Baker, D., van den Beek, M., Blankenberg, D., Bouvier, D., Čech, M., et al. (2016). The Galaxy platform for accessible, reproducible and collaborative biomedical analyses: 2016 update. *Nucleic Acids Research*, *44*, W3–W10.

Agarwal, R., & Dhar, V. (2014). Editorial—Big data, data science, and analytics: the opportunity and challenge for IS research. *Information Systems Research*, *25*(3), 443–448.

Aguilera, P. A., Fernández, A., Fernández, R., Rumí, R., & Salmerón, A. (2011). Bayesian networks in environmental modelling. *Environmental Modelling & Software*, *26*(12), 1376–1388.

Ahanger, M. A., Gul, F., Ahmad, P., & Akram, N. A. (2018). Environmental stresses and metabolomics—deciphering the role of stress responsive metabolites. In *Plant Metabolites and Regulation Under Environmental Stress* (pp. 53–67).

Ahmed, M. F. Y. M. (2016). *Development of computational analysis tools for natural products research and metabolomics*. Ph.D. Dissertation Kyoto University.

Alados, C. L., Saiz, H., Gartzia, M., Nuche, P., Escós, J., Navarro, T., et al. (2017). Plant–plant interactions scale up to produce vegetation spatial patterns: the influence of long- and short-term process. *Ecosphere*, *8*(8).

Aldridge, B. B., & Rhee, K. Y. (2014). Microbial metabolomics: innovation, application, insight. *Current Opinion in Microbiology*, *19C*, 90–96. https://doi.org/10.1016/j.mib.2014.06.009.

Alla, M. M. N., Khedr, A. H. A., Serag, M. M., Abu-Alnaga, A. Z., & Nada, R. M. (2012). Regulation of metabolomics in *Atriplex halimus* growth under salt and drought stress. *Plant Growth Regulation*, *67*, 281–304.

Allard, P. -M., Péresse, T., Bisson, J., Gindro, K., Marcourt, L., Pham, V. C., et al. (2016). Integration of molecular networking and in-silico MS/MS fragmentation for natural products dereplication. *Analytical Chemistry*, *88*, 3317–3323.

Allegrini, F., & Olivieri, A. C. (2014). IUPAC-consistent approach to the limit of detection in partial least-squares calibration. *Analytical Chemistry*, *86*(15), 7858–7866.

Allen, F., Pon, A., Wilson, M., Greiner, R., & Wishart, D. (2014). CFM-ID: a web server for annotation, spectrum prediction and metabolite identification from tandem mass spectra. *Nucleic Acids Research*, *42*, W94–W99.

Almstetter, M. F., Oefner, P. J., & Dettmer, K. (2012). Comprehensive two-dimensional gas chromatography in metabolomics. *Analytical and Bioanalytical Chemistry*, *402*(6), 1993–2013.

Alonso-Gutierrez, J., Kim, E. M., Batth, T. S., Cho, N., Hu, Q., Chan, L. J. G., et al. (2015). Principal component analysis of proteomics (PCAP) as a tool to direct metabolic engineering. *Metabolic Engineering*, *28*, 123–133.

Altaf-Ul-Amin, M., Afendi, F. M., Kiboi, S. K., & Kanaya, S. (2014). Systems biology in the context of big data and networks. *BioMed Research International*, *2014*.

Alyass, A., Turcotte, M., & Meyre, D. (2015). From big data analysis to personalized medicine for all: challenges and opportunities. *BMC Medical Genomics*, *8*(1), 33.

Andrews, G. L., Simons, B. L., Young, J. B., Hawkridge, A. M., & Muddiman, D. C. (2011). Performance characteristics of a new hybrid quadrupole time-of-flight tandem mass spectrometer (TripleTOF 5600). *Analytical Chemistry*, *83*(13), 5442–5446.
Arany, A. M., De Jong, T. J., Kim, H. K., Van Dam, N. M., Choi, Y. H., Verpoorte, R., et al. (2008). Glucosinolates and other metabolites in the leaves of *Arabidopsis thaliana* from natural populations and their effects on a generalist and a specialist herbivore. *Chemoecology*, *18*(2), 65–71.
Arens, N., Döll, S., & Mock, H. -P. (2015). The reproducibility of liquid chromatography separation technology and its potential impact on large scale plant metabolomics experiments. *Journal of Chromatography B*, *991*, 41–45.
Arrivault, S., Guenther, M., Florian, A., Encke, B., Feil, R., Vosloh, D., et al. (2014). Dissecting the subcellular compartmentation of proteins and metabolites in *Arabidopsis* leaves using non-aqueous fractionation. *Molecular & Cellular Proteomics*, *13*, 2246–2259. https://doi.org/10.1074/mcp.M114.038190.
Atasaral-Şahin, Ş., Romero, M. R., Cueto, R., González-Lavín, N., Marcos, M., & Diz, A. P. (2015). Subtle tissue and sex-dependent proteome variation in mussel (*Mytilus galloprovincialis*) populations of the Galician coast (NW Spain) raised in a common environment. *Proteomics*, *15*(23-24), 3993–4006.
Bais, P., Moon, S. M., He, K., Leitao, R., Dreher, K., Walk, T., et al. (2010). PlantMetabolomics.org: a web portal for plant metabolomics experiments. *Plant Physiology*, *152*(4), 1807–1816.
Bais, P., Moon-Quanbeck, S. M., Nikolau, B. J., & Dickerson, J. A. (2011). Plantmetabolomics.org: mass spectrometry-based *Arabidopsis* metabolomics—database and tools update. *Nucleic Acids Research*, *40*(D1), D1216–D1220.
Bartel, J., Krumsiek, J., & Theis, F. J. (2013). Statistical methods for the analysis of high-throughput metabolomics data. *Computational and Structural Biotechnology Journal*, *4*(5) e201301009.
Bassel, G. W., Gaudinier, A., Brady, S. M., Hennig, L., Rhee, S. Y., & De Smet, I. (2012). Systems analysis of plant functional, transcriptional, physical interaction, and metabolic networks. *The Plant Cell*, *24*(10), 3859–3875.
Basu, N., Scheuhammer, A. M., Bursian, S. J., Elliott, J., Rouvinen-Watt, K., & Chan, H. M. (2007). Mink as a sentinel species in environmental health. *Environmental Research*, *103*(1), 130–144.
Bathe, O. F., & Farshidfar, F. (2014). From genotype to functional phenotype: unraveling the metabolomic features of colorectal cancer. *Genes*, *5*(3), 536–560.
Bechtold, U. (2018). Plant life in extreme environments: how do you improve drought tolerance? *Frontiers in Plant Science*, *9*.
Beierkuhnlein, C., Thiel, D., Jentsch, A., Willner, E., & Kreyling, J. (2011). Ecotypes of European grass species respond differently to warming and extreme drought. *Journal of Ecology*, *99*, 703–713.
Bell, A., & Jones, K. (2015). Explaining fixed effects: Random effects modeling of time-series cross-sectional and panel data. *Political Science Research and Methods*, *3*(1), 133–153.
Bellaire, A., Ischebeck, T., Staedler, Y., Weinhaeuser, I., Mair, A., Parameswaran, S., et al. (2014). Metabolism and development—integration of micro computed tomography data and metabolite profiling reveals metabolic reprogramming from floral initiation to silique development. *The New Phytologist*, *202*, 322–335. https://doi.org/10.1111/nph.12631.
Bhardwaj, C., & Hanley, L. (2014). Ion sources for mass spectrometric identification and imaging of molecular species. *Natural Product Reports*, *31*(6), 756–767.
Bhat, H. A., Kaur, T., Bhat, R., & Vyas, D. (2016). Physiological and biochemical plasticity of Lepidium latifolium as 'sleeper weed' in Western Himalayas. *Physiologia Plantarum*, *156*, 278–293.

Bingol, K., Bruschweiler-Li, L., Yu, C., Somogyi, A., Zhang, F., & Brüschweiler, R. (2015). Metabolomics beyond spectroscopic databases: a combined MS/NMR strategy for the rapid identification of new metabolites in complex mixtures. *Analytical Chemistry*, *87*(7), 3864–3870.

Birkemeyer, C., & Kopka, J. (2007). Design of metabolite recovery by variations of the metabolite profiling protocol. In *Concepts in plant metabolomics* (pp. 45–69). Dordrecht, The Netherlands: Springer.

Blackburn, T. M., Essl, F., Evans, T., Hulme, P. E., Jeschke, J. M., Kühn, I., et al. (2014). A unified classification of alien species based on the magnitude of their environmental impacts. *PLoS Biology*, *12*(5)e1001850.

Blande, J. D. (2017). Plant communication with herbivores. *Advances in botanical research* (pp. 281–304). Vol. 82 (pp. 281–304). New York, NY: Elsevier 978-0-12-801431-8.

Boekel, J., Chilton, J. M., Cooke, I. R., Horvatovich, P. L., Jagtap, P. D., Käll, L., et al. (2015). Multi-omic data analysis using Galaxy. *Nature Biotechnology*, *33*(2), 137.

Bonneau, E., Tetreault, N., Robitaille, R., Boucher, A., & De Guire, V. (2016). Metabolomics: perspectives on potential biomarkers in organ transplantation and immunosuppressant toxicity. *Clinical Biochemistry*, *49*(4-5), 377–384.

Borland, L., Brickhouse, M., Thomas, T., & Fountain, A. W. (2010). Review of chemical signature databases. *Analytical and Bioanalytical Chemistry*, *397*(3), 1019–1028.

Bouhifd, M. (2015). Quality assurance of metabolomics. *ALTEX*, *32*, 319–326.

Bowne, J., Bacic, A., Tester, M., & Roessner, U. (2011). Abiotic stress and metabolomics. R. D. Hall (Ed.), *Annual plant reviews* (pp. 61–85). Vol. 43 (pp. 61–85). Oxford, UK: Wiley-Blackwell. ISBN 978-1-4443-3995-6.

Bowne, J. B., Erwin, T. A., Juttner, J., Schnurbusch, T., Langridge, P., Bacic, A., et al. (2012). Drought responses of leaf tissues from wheat cultivars of differing drought tolerance at the metabolite level. *Molecular Plant*, *5*(2), 418–429.

Brereton, R. G., & Lloyd, G. R. (2014). Partial least squares discriminant analysis: taking the magic away. *Journal of Chemometrics*, *28*(4), 213–225.

Bro, R., & Smilde, A. K. (2014). Principal component analysis. *Analytical Methods*, *6*(9), 2812–2831.

Broadhurst, D. I., & Kell, D. B. (2006). Statistical strategies for avoiding false discoveries in metabolomics and related experiments. *Metabolomics*, *2*(4), 171–196.

Brookes, M. J., Woolrich, M., Luckhoo, H., Price, D., Hale, J. R., Stephenson, M. C., et al. (2011). Investigating the electrophysiological basis of resting state networks using magnetoencephalography. *Proceedings of the National Academy of Sciences*, *108*(40), 16783–16788. https://doi.org/10.1073/pnas.1112685108.

Brunetti, C., George, R. M., Tattini, M., Field, K., & Davey, M. P. (2013). Metabolomics in plant environmental physiology. *Journal of Experimental Botany*, *64*(13), 4011–4020.

Bundy, J. G., Davey, M. P., & Viant, M. R. (2009). Environmental metabolomics: a critical review and future perspectives. *Metabolomics*, *5*(1), 3.

Bundy, J. G., Keun, H. C., Sidhu, J. K., Spurgeon, D. J., Svendsen, C., Kille, P., et al. (2007). Metabolic profile biomarkers of metal contamination in a sentinel terrestrial species are applicable across multiple sites. *Environmental Science & Technology*, *41*(12), 4458–4464.

Cai, Y., Weng, K., Guo, Y., Peng, J., & Zhu, Z. J. (2015). An integrated targeted metabolomic platform for high-throughput metabolite profiling and automated data processing. *Metabolomics*, *11*(6), 1575–1586.

Cajka, T., & Fiehn, O. (2015). Toward merging untargeted and targeted methods in mass spectrometry-based metabolomics and lipidomics. *Analytical Chemistry*, *88*(1), 524–545.

Calf, O. W., Huber, H., Peters, J. L., Weinhold, A., & van Dam, N. M. (2018). Glycoalkaloid composition explains variation in slug resistance in Solanum dulcamara. *Oecologia*.

Callaway, R. M., & Ridenour, W. M. (2004). Novel weapons: invasive success and the evolution of increased competitive ability. *Frontiers in Ecology and the Environment*, *2*, 436–443.

Canarini, A., Carrillo, Y., Mariotte, P., Ingram, L., & Dijkstra, F. A. (2016). Soil microbial community resistance to drought and links to C stabilization in an Australian grassland. *Soil Biology and Biochemistry*, *103*, 171–180.
Cañas, R. A., Canales, J., Muñoz-Hernández, C., Granados, J. M., Ávila, C., García-Martín, M. L., et al. (2015). Understanding developmental and adaptive cues in pine through metabolite profiling and co-expression network analysis. *Journal of Experimental Botany*, *66*(11), 3113–3127.
Cao, M., Fraser, K., Huege, J., Featonby, T., Rasmussen, S., & Jones, C. (2015). Predicting retention time in hydrophilic interaction liquid chromatography mass spectrometry and its use for peak annotation in metabolomics. *Metabolomics*, *11*, 696–706.
Cao, M., Fraser, K., Jones, C., Stewart, A., Lyons, T., Faville, M., et al. (2017). Untargeted metabotyping *Lolium perenne* reveals population-level variation in plant flavonoids and alkaloids. *Frontiers in Plant Science*, *8*, 133.
Carlson, E. E., & Cravatt, B. F. (2007). Chemoselective probes for metabolite enrichment and profiling. *Nature Methods*, *4*(5), 429.
Carrera, J., Estrela, R., Luo, J., Rai, N., Tsoukalas, A., & Tagkopoulos, I. (2014). An integrative, multi-scale, genome-wide model reveals the phenotypic landscape of *Escherichia coli*. *Molecular Systems Biology*, *10*, 735. https://doi.org/10.15252/msb.20145108.
Carroll, A. J., Badger, M. R., & Millar, A. H. (2010). The MetabolomeExpress Project: enabling web-based processing, analysis and transparent dissemination of GC/MS metabolomics datasets. *BMC Bioinformatics*, *11*, 376.
Cascante, M., & Marin, S. (2008). Metabolomics and fluxomics approaches. *Essays in Biochemistry*, *45*, 67–82.
Caseys, C., Glauser, G., Stölting, K. N., Christe, C., Albrectsen, B. R., & Lexer, C. (2012). Effects of interspecific recombination on functional traits in trees revealed by metabolomics and genotyping-by-resequencing. *Plant Ecology and Diversity*, *5*, 457–471.
Caspi, R., Billington, R., Ferrer, L., Foerster, H., Fulcher, C. A., Keseler, I. M., et al. (2016). The MetaCyc database of metabolic pathways and enzymes and the BioCyc collection of pathway/genome databases. *Nucleic Acids Research*, *44*, D471–D480.
Castro-Perez, J. M. (2007). Current and future trends in the application of HPLC-MS to metabolite-identification studies. *Drug Discovery Today*, *12*(5-6), 249–256.
Chadeau-Hyam, M., Campanella, G., Jombart, T., Bottolo, L., Portengen, L., Vineis, P., et al. (2013). Deciphering the complex: methodological overview of statistical models to derive OMICS-based biomarkers. *Environmental and Molecular Mutagenesis*, *54*(7), 542–557.
Chahrour, O., Cobice, D., & Malone, J. (2015). Stable isotope labelling methods in mass spectrometry-based quantitative proteomics. *Journal of Pharmaceutical and Biomedical Analysis*, *113*, 2–20.
Chang, R. L., Ghamsari, L., Manichaikul, A., Hom, E. F., Balaji, S., Fu, W., et al. (2011). Metabolic network reconstruction of Chlamydomonas offers insight into light-driven algal metabolism. *Molecular Systems Biology*, *7*, 518. https://doi.org/10.1038/msb.2011.52.
Chatfield, C. (2018). *Introduction to multivariate analysis*. Routledge.
Chen, C., Shah, Y. M., Morimura, K., Krausz, K. W., Miyazaki, M., Richardson, T. A., et al. (2008). Metabolomics reveals that hepatic stearoyl-CoA desaturase 1 downregulation exacerbates inflammation and acute colitis. *Cell Metabolism*, *7*(2), 135–147.
Cheng, J., Lan, W., Zheng, G., & Gao, X. (2018). Metabolomics: a high-throughput platform for metabolite profile exploration. In *Computational systems biology* (pp. 265–292). New York, NY: Humana Press.
Chervin, J., Stierhof, M., Tong, M. H., Peace, D., Hansen, K. Ø., Urgast, D. S., et al. (2017). Targeted de-replication of microbial natural products by high-resolution MS and predicted LC retention time. *Journal of Natural Products*, *80*, 1370–1377.

Choi, S. (2015). Independent component analysis. In *Encyclopedia of biometrics* (pp. 917–924).
Choi, S. J., & Jung, M. Y. (2017). Simple and fast sample preparation followed by gas chromatography-tandem mass spectrometry (GC-MS/MS) for the analysis of 2-and 4-methylimidazole in cola and dark beer. *Journal of Food Science, 82*(4), 1044–1052.
Clough, Y. (2012). A generalized approach to modeling and estimating indirect effects in ecology. *Ecology, 93*, 1809–1815.
Cloyd, R. A. (2018). Insect management for disease control in florists' crops. In *Handbook of florists' crops diseases* (pp. 69–85). Springer.
Colzani, M., Waridel, P., Laurent, J., Faes, E., Ruegg, C., & Quadroni, M. (2009). Metabolic labeling and protein linearization technology allow the study of proteins secreted by cultured cells in serum-containing media. *Journal of Proteome Research, 8*(10), 4779–4788.
Cortois, R., Veen, G. F. C., Duyts, H., Abbas, M., Strecker, T., Kostenko, O., et al. (2017). Possible mechanisms underlying abundance and diversity responses of nematode communities to plant diversity. *Ecosphere, 8*, e01719.
Creek, D. J., Jankevics, A., Breitling, R., Watson, D. G., Barrett, M. P., & Burgess, K. E. V. (2011). Toward global metabolomics analysis with hydrophilic interaction liquid chromatography–mass spectrometry: improved metabolite identification by retention time prediction. *Analytical Chemistry, 83*, 8703–8710.
Cressie, N., Calder, C. A., Clark, J. S., Hoef, J. M. V., & Wikle, C. K. (2009). Accounting for uncertainty in ecological analysis: the strengths and limitations of hierarchical statistical modeling. *Ecological Applications, 19*(3), 553–570.
Danchin, A. (2017). From chemical metabolism to life: the origin of the genetic coding process. *Beilstein Journal of Organic Chemistry, 13*, 1119.
Das, A., Kim, D. W., Khadka, P., Rakwal, R., & Rohila, J. S. (2017). Unraveling key metabolomic alterations in wheat embryos derived from freshly harvested and water-imbibed seeds of two wheat cultivars with contrasting dormancy status. *Frontiers in Plant Science, 8*, 1203.
Daston, G. P. (2008). Gene expression, dose-response, and phenotypic anchoring: applications for toxicogenomics in risk assessment. *Toxicological Sciences, 105*(2), 233–234.
Davey, M. P., Woodward, F. I., & Quick, W. P. (2009). Intraspecfic variation in cold-temperature metabolic phenotypes of *Arabidopsis lyrata* ssp. *petraea*. *Metabolomics, 5*(1), 138–149.
Davidson, R. L., Weber, R. J. M., Liu, H., Sharma-Oates, A., & Viant, M. R. (2016). Galaxy-M: a Galaxy work flow for processing and analyzing direct infusion and liquid chromatography massspectrometry-based metabolomics data. *GigaScience, 5*.
De Meyer, T., Sinnaeve, D., Van Gasse, B., Tsiporkova, E., Rietzschel, E. R., De Buyzere, M. L., et al. (2008). NMR-based characterization of metabolic alterations in hypertension using an adaptive, intelligent binning algorithm. *Analytical Chemistry, 80*, 3783–3790.
del Mar Gómez-Ramos, M., Lozano, A., Rajski, Ł., & Fernández-Alba, A. R. (2017). Time-of-flight and Orbitrap mass spectrometry for target and nontarget analysis of pesticide residues in food. In *Multiresidue methods for the analysis of pesticide residues in food*. Boca Raton, FL: CRC Press.
Dias, D. A., Jones, O. A., Beale, D. J., Boughton, B. A., Benheim, D., Kouremenos, K. A., et al. (2016). Current and future perspectives on the structural identification of small molecules in biological systems. *Metabolites, 6*(4), 46.
Doerfler, H., Lyon, D., Nägele, T., Sun, X., Fragner, L., Hadacek, F., et al. (2013). Granger causality in integrated GC–MS and LC–MS metabolomics data reveals the interface of primary and secondary metabolism. *Metabolomics, 9*(3), 564–574.
Dollo, V. H., Yi, S. X., & Lee, R. E., Jr. (2010). High temperature pulses decrease indirect chilling injury and elevate ATP levels in the flesh fly, Sarcophaga crassipalpis. *Cryobiology, 60*(3), 351–353.

Domingo-Almenara, X., Montenegro-Burke, J. R., Benton, H. P., & Siuzdak, G. (2017). Annotation: a computational solution for streamlining metabolomics analysis. *Analytical Chemistry*, *90*, 480–489.
Domingo-Almenara, X., Perera, A., & Brezmes, J. (2016). Avoiding hard chromatographic segmentation: A moving window approach for the automated resolution of gas chromatography–mass spectrometry-based metabolomics signals by multivariate methods. *Journal of Chromatography A*, *1474*, 145–151.
Domingo-Almenara, X., Perera, A., Ramírez, N., Cañellas, N., Correig, X., & Brezmes, J. (2015). Compound identification in gas chromatography/mass spectrometry-based metabolomics by blind source separation. *Journal of Chromatography A*, *1409*, 226–233.
Dormont, L., Delle-Vedove, R., Bessière, J. -M., & Schatz, B. (2014). Floral scent emitted by white and coloured morphs in orchids. *Phytochemistry*, *100*, 51–59.
Dötterl, S., Jahreiß, K., Jhumur, U. S., & Jürgens, A. (2012). Temporal variation of flower scent in Silene otites (Caryophyllaceae): a species with a mixed pollination system: temporal variation of flower scent. *Botanical Journal of the Linnean Society*, *169*, 447–460.
Douma, J. C., Shipley, B., Witte, J. -P., Aerts, R., & Van Bodegom, P. M. (2012). Disturbance and resource availability act differently on the same suite of plant traits: revisiting assembly hypotheses. *Ecology*, *93*, 825–835.
Draisma, H. H., Pool, R., Kobl, M., Jansen, R., Petersen, A. K., Vaarhorst, A. A., et al. (2015). Genome-wide association study identifies novel genetic variants contributing to variation in blood metabolite levels. *Nature Communications*, *6*, 7208.
Drenovsky, R. E., Grewell, B. J., D'Antonio, C. M., Funk, J. L., James, J. J., Molinari, N., et al. (2012). A functional trait perspective on plant invasion. *Annals of Botany*, *110*, 141–153.
Du, K. L., & Swamy, M. N. S. (2014). Independent component analysis. In *Neural networks and statistical learning* (pp. 419–450). London: Springer.
Duarte, N. C., Becker, S. A., Jamshidi, N., Thiele, I., Mo, M. L., Vo, T. D., et al. (2007). Global reconstruction of the human metabolic network based on genomic and bibliomic data. *Proceedings of the National Academy of Sciences*, *104*(6), 1777–1782.
Dührkop, K., Shen, H., Meusel, M., Rousu, J., & Böcker, S. (2015). Searching molecular structure databases with tandem mass spectra using CSI:FingerID. *Proceedings of the National Academy of Sciences of the United States of America*, *112*, 12580–12585.
Dunn, W. B., Erban, A., Weber, R. J., Creek, D. J., Brown, M., Breitling, R., et al. (2013). Mass appeal: metabolite identification in mass spectrometry-focused untargeted metabolomics. *Metabolomics*, *9*(1), 44–66.
Dunn, W. B., & Ellis, D. I. (2005). Metabolomics: current analytical platforms and methodologies. *TrAC Trends in Analytical Chemistry*, *24*, 285–294.
Dunn, W. B., Wilson, I. D., Nicholls, A. W., & Broadhurst, D. (2012). The importance of experimental design and QC samples in large-scale and MS-driven untargeted metabolomic studies of humans. *Bioanalysis*, *4*, 2249–2264.
Eckert, H., & Bajorath, J. (2007). Molecular similarity analysis in virtual screening: foundations, limitations and novel approaches. *Drug Discovery Today*, *12*(5-6), 225–233.
Eghbalnia, H. R., Romero, P. R., Westler, W. M., Baskaran, K., Ulrich, E. L., & Markley, J. L. (2017). Increasing rigor in NMR-based metabolomics through validated and open source tools. *Current Opinion in Biotechnology*, *43*, 56–61.
Ehlers, B. K., Damgaard, C. F., & Laroche, F. (2016). Intraspecific genetic variation and species coexistence in plant communities. *Biology Letters*, *12*, 20150853.
Eiler, J., Cesar, J., Chimiak, L., Dallas, B., Grice, K., Griep-Raming, J., et al. (2017). Analysis of molecular isotopic structures at high precision and accuracy by Orbitrap mass spectrometry. *International Journal of Mass Spectrometry*, *422*, 126–142.
Eisenhauer, N., Bowker, M. A., Grace, J. B., & Powell, J. R. (2015). From patterns to causal understanding: Structural equation modeling (SEM) in soil ecology. *Pedobiologia*, *58*, 65–72.

Eisenhauer, N., Cesarz, S., Koller, R., Worm, K., & Reich, P. B. (2012). Global change belowground: impacts of elevated CO_2, nitrogen, and summer drought on soil food webs and biodiversity. *Global Change Biology*, *18*, 435–447.

El-Aneed, A., Cohen, A., & Banoub, J. (2009). Mass spectrometry, review of the basics: electrospray, MALDI, and commonly used mass analyzers. *Applied Spectroscopy Reviews*, *44* (3), 210–230.

Emwas, A. H., Luchinat, C., Turano, P., Tenori, L., Roy, R., Salek, R. M., et al. (2015). Standardizing the experimental conditions for using urine in NMR-based metabolomic studies with a particular focus on diagnostic studies: a review. *Metabolomics*, *11*(4), 872–894.

Fan, J., Krautkramer, K. A., Feldman, J. L., & Denu, J. M. (2015). Metabolic regulation of histone post-translational modifications. *ACS Chemical Biology*, *10*(1), 95–108.

Farag, M. A., Porzel, A., & Wessjohann, L. A. (2012). Comparative metabolite profiling and fingerprinting of medicinal licorice roots using a multiplex approach of GC–MS, LC–MS and 1D NMR techniques. *Phytochemistry*, *76*, 60–72.

Feist, A. M., Herrgård, M. J., Thiele, I., Reed, J. L., & Palsson, B. Ø. (2009). Reconstruction of biochemical networks in microorganisms. *Nature Reviews Microbiology*, 7(2), 129.

Fester, T. (2015). Plant metabolite profiles and the buffering capacities of ecosystems. *Phytochemistry*, *110*, 6–12.

Fiehn, O. (2002). Metabolomics—the link between genotypes and phenotypes. *Plant Molecular Biology*, *48*, 155–171.

Fiehn, O., Kind, T., & Barupal, D. K. (2018). Data processing, metabolomic databases and pathway analysis. *Annual Plant Reviews*, 367–406.

Field, K. J., & Lake, J. A. (2011). Environmental metabolomics links genotype to phenotype and predicts genotype abundance in wild plant populations. *Physiologia Plantarum*, *142* (4), 352–360.

Fillbrunn, A., Dietz, C., Pfeuffer, J., Rahn, R., Landrum, G. A., & Berthold, M. R. (2017). KNIME for reproducible cross-domain analysis of life science data. *Journal of Biotechnology*, *261*, 149–156.

Flood, P. J., Kruijer, W., Schnabel, S. K., Schoor, R., Jalink, H., Snel, J. F., et al. (2016). Phenomics for photosynthesis, growth and reflectance in *Arabidopsis thaliana* reveals circadian and long-term fluctuations in heritability. *Plant Methods*, *12*(1), 14.

Föllinger, O., & Konigorski, U. (2013). *Regelungstechnik Einführung in die Methoden und ihre Anwendung*. Berlin: VDE.

Forseth, R. R., & Schroeder, F. C. (2011). NMR-spectroscopic analysis of mixtures: from structure to function. *Current Opinion in Chemical Biology*, *15*(1), 38–47.

Frank, T., Röhlig, R. M., Davies, H. V., Barros, E., & Engel, K. H. (2012). Metabolite profiling of maize kernels—genetic modification versus environmental influence. *Journal of Agricultural and Food Chemistry*, *60*(12), 3005–3012.

Freitag, S., Feldmann, J., Raab, A., Crittenden, P. D., Hogan, E. J., Squier, A. H., et al. (2012). Metabolite profile shifts in the heathland lichen *Cladonia portentosa* in response to N deposition reveal novel biomarkers. *Physiologia Plantarum*, *146*(2), 160–172.

Frolov, A., Bilova, T., Paudel, G., Berger, R., Balcke, G. U., Birkemeyer, C., et al. (2017). Early responses of mature Arabidopsis thaliana plants to reduced water potential in the agar-based polyethylene glycol infusion drought model. *Journal of Plant Physiology*, *208*, 70–83.

Fukushima, A., Kusano, M., Redestig, H., Arita, M., & Saito, K. (2011). Metabolomic correlation-network modules in *Arabidopsis* based on a graph-clustering approach. *BMC Systems Biology*, *5*(1), 1.

Nylund, G. M., Weinberger, F., Rempt, M., & Pohnert, G. (2011). Metabolomic assessment of induced and activated chemical defence in the invasive red alga *Gracilaria vermiculophylla*. *PLoS One*, *6*(12)e29359.

Gallo, V., Intini, N., Mastrorilli, P., Latronico, M., Scapicchio, P., Triggiani, M., et al. (2015). Performance assessment in fingerprinting and multi component quantitative NMR analyses. *Analytical Chemistry*, *87*(13), 6709–6717.
Gao, P., & Xu, G. (2015). Mass-spectrometry-based microbial metabolomics: recent developments and applications. *Analytical and Bioanalytical Chemistry*, *407*, 669–680.
Gargallo-Garriga, A., Sardans, J., Pérez-Trujillo, M., Guenther, A., Llusià, J., Rico, L., et al. (2016). Shifts in plant foliar and floral metabolomes in response to the suppression of the associated microbiota. *BMC Plant Biology*, *16*.
Gargallo-Garriga, A., Sardans, J., Pérez-Trujillo, M., Oravec, M., Urban, O., Jentsch, A., et al. (2015). Warming differentially influences the effects of drought on stoichiometry and metabolomics in shoots and roots. *The New Phytologist*, *207*, 591–603.
Gavard, R., Rossell, D., Spencer, S. E., & Barrow, M. P. (2017). Themis: batch preprocessing for ultrahigh-resolution mass spectra of complex mixtures. *Analytical Chemistry*, *89*(21), 11383–11390.
Gehlenborg, N., O'donoghue, S. I., Baliga, N. S., Goesmann, A., Hibbs, M. A., Kitano, H., et al. (2010). Visualization of omics data for systems biology. *Nature Methods*, 7(3s), S56.
Ghosh, S., Matsuoka, Y., Asai, Y., Hsin, K. -Y., & Kitano, H. (2011). Software for systems biology: from tools to integrated platforms. *Nature Reviews Genetics*, *12*, 821–832. https://doi.org/10.1038/nrg3096.
Giacomoni, F., Le Corguille, G., Monsoor, M., Landi, M., Pericard, P., Petera, M., et al. (2015). Workflow4Metabolomics: a collaborative research infrastructure for computational metabolomics. *Bioinformatics*, *31*, 1493–1495.
Giannarelli, S., Muscatello, B., Bogani, P., Spiriti, M. M., Buiatti, M., & Fuoco, R. (2010). Comparative determination of some phytohormones in wild-type and genetically modified plants by gas chromatography–mass spectrometry and high-performance liquid chromatography–tandem mass spectrometry. *Analytical Biochemistry*, *398*(1), 60–68.
Gika, H. G., Theodoridis, G. A., Plumb, R. S., & Wilson, I. D. (2014). Current practice of liquid chromatography–mass spectrometry in metabolomics and metabonomics. *Journal of Pharmaceutical and Biomedical Analysis*, *87*, 12–25.
Gillard, J., Frenkel, J., Devos, V., Sabbe, K., Paul, C., Rempt, M., et al. (2013). Metabolomics enables the structure elucidation of a diatom sex pheromone. *Angewandte Chemie, International Edition*, *52*, 854–857.
Go, E. P. (2010). Database resources in metabolomics: an overview. *Journal of Neuroimmune Pharmacology*, *5*(1), 18–30.
Godzien, J., Alonso-Herranz, V., Barbas, C., & Armitage, E. G. (2015). Controlling the quality of metabolomics data: new strategies to get the best out of the QC sample. *Metabolomics*, *11*, 518–528.
Goodacre, R., Vaidyanathan, S., Dunn, W. B., Harrigan, G. G., & Kell, D. B. (2004). Metabolomics by numbers: acquiring and understanding global metabolite data. *Trends in Biotechnology*, *22*(5), 245–252.
Gorrochategui, E., Jaumot, J., Lacorte, S., & Tauler, R. (2016). Data analysis strategies for targeted and untargeted LC-MS metabolomic studies: overview and workflow. *TrAC Trends in Analytical Chemistry*, *82*, 425–442.
Gowda, H., Ivanisevic, J., Johnson, C. H., Kurczy, M. E., Paul Benton, H., Rinehart, D., et al. (2014). Interactive XCMS Online: simplifying advanced metabolomic data processing and subsequent statistical analyses. *Analytical Chemistry*, *86*(14), 6931–6939.
Granato, D., Santos, J. S., Escher, G. B., Ferreira, B. L., & Maggio, R. M. (2018). Use of principal component analysis (PCA) and hierarchical cluster analysis (HCA) for multivariate association between bioactive compounds and functional properties in foods: a critical perspective. *Trends in Food Science and Technology*, *72*, 83–90.
Green, P. J., Łatuszyński, K., Pereyra, M., & Robert, C. P. (2015). Bayesian computation: a summary of the current state, and samples backwards and forwards. *Statistics and Computing*, *25*(4), 835–862.

Groessl, M., Graf, S., & Knochenmuss, R. (2015). High resolution ion mobility-mass spectrometry for separation and identification of isomeric lipids. *Analyst, 140*(20), 6904–6911.

Gromski, P. S., Muhamadali, H., Ellis, D. I., Xu, Y., Correa, E., Turner, M. L., et al. (2015). A tutorial review: metabolomics and partial least squares-discriminant analysis—a marriage of convenience or a shotgun wedding. *Analytica Chimica Acta, 879*, 10–23.

Gromski, P. S., Xu, Y., Hollywood, K. A., Turner, M. L., & Goodacre, R. (2015). The influence of scaling metabolomics data on model classification accuracy. *Metabolomics, 11*(3), 684–695.

Grootveld, M. (2014). Recent developments in exploratory data analysis and pattern recognition techniques. In *Metabolic profiling* (pp. 74–116). Royal Society of Chemistry.

Guasch-Ferré, M., Hruby, A., Toledo, E., Clish, C. B., Martínez-González, M. A., Salas-Salvadó, J., et al. (2016). Metabolomics in prediabetes and diabetes: a systematic review and meta-analysis. *Diabetes Care, 39*(5), 833–846.

Guo, X., & Lankmayr, E. (2010). Multidimensional approaches in LC and MS for phospholipid bioanalysis. *Bioanalysis, 2*(6), 1109–1123.

Gutbrodt, B., Dorn, S., Unsicker, S. B., & Mody, K. (2012). Species-specific responses of herbivores to within-plant and environmentally mediated between-plant variability in plant chemistry. *Chemoecology, 22*, 101–111.

Hagel, J. M., Mandal, R., Han, B., Han, J., Dinsmore, D. R., Borchers, C. H., et al. (2015). Metabolome analysis of 20 taxonomically related benzylisoquinoline alkaloid-producing plants. *BMC Plant Biology, 15*(1), 220.

Haug, K., Salek, R. M., Conesa, P., Hastings, J., de Matos, P., Rijnbeek, M., et al. (2013). MetaboLights—an open-access general-purpose repository for metabolomics studies and associated meta-data. *Nucleic Acids Research, 41*, D781–D786.

Hecker, M., Lambeck, S., Toepfer, S., Van Someren, E., & Guthke, R. (2009). Gene regulatory network inference: data integration in dynamic models—a review. *Biosystems, 96* (1), 86–103.

Hendriks, M. M., van Eeuwijk, F. A., Jellema, R. H., Westerhuis, J. A., Reijmers, T. H., Hoefsloot, H. C., et al. (2011). Data-processing strategies for metabolomics studies. *TrAC Trends in Analytical Chemistry, 30*(10), 1685–1698.

Henke, M. T., & Kelleher, N. L. (2016). Modern mass spectrometry for synthetic biology and structure-based discovery of natural products. *Natural Product Reports, 33*(8), 942–950.

Herden, J., Meldau, S., Kim, S. -G., Kunert, G., Joo, Y., Baldwin, I. T., et al. (2016). Shifting *Nicotiana attenuata*'s diurnal rhythm does not alter its resistance to the specialist herbivore *Manduca sexta*: N. attenuata's resistance to M. sexta is diurnally robust. *Journal of Integrative Plant Biology, 58*, 656–668.

Hernández, F., Sancho, J. V., Ibáñez, M., Abad, E., Portolés, T., & Mattioli, L. (2012). Current use of high-resolution mass spectrometry in the environmental sciences. *Analytical and Bioanalytical Chemistry, 403*, 1251–1264.

Hervé, M. R., Nicolè, F., & Lê Cao, K. A. (2018). Multivariate analysis of multiple datasets: a practical guide for chemical ecology. *Journal of Chemical Ecology, 44*(3), 215–234.

Hill, C. B., & Roessner, U. (2013). Metabolic profiling of plants by GC–MS. In *The handbook of plant metabolomics* (pp. 1–23). Wiley.

Hill, K., Porco, S., Lobet, G., Zappala, S., Mooney, S., Draye, X., et al. (2013). Root systems biology: integrative modeling across scales, from gene regulatory networks to the rhizosphere. *Plant Physiology, 163*, 1487–1503. https://doi.org/10.1104/pp.113.227215.

Hjältén, J., Lindau, A., Wennström, A., Blomberg, P., Witzell, J., Hurry, V., et al. (2007). Unintentional changes of defence traits in GM trees can influence plant–herbivore interactions. *Basic and Applied Ecology, 8*(5), 434–443.

Hodgkison, R., Ayasse, M., Kalko, E. K. V., Häberlein, C., Schulz, S., Mustapha, W. A. W., et al. (2007). Chemical ecology of fruit bat foraging behavior in relation to the fruit odors

of two species of paleotropical bat-dispersed figs (Ficus hispida and Ficus scortechinii). *Journal of Chemical Ecology*, *33*, 2097–2110.
Hofmann, J., Hahm, H. S., Seeberger, P. H., & Pagel, K. (2015). Identification of carbohydrate anomers using ion mobility–mass spectrometry. *Nature*, *526*(7572), 241.
Hollert, H., Seiler, T. B., Blaha, L., & Young, A. L. (2007). Multiple stressors for the environment: present and future challenges and perspectives. *Journal of Soils and Sediments*, *7* (4), 272.
Hong, J., Yang, L., Zhang, D., & Shi, J. (2016). Plant metabolomics: an indispensable system biology tool for plant science. *International Journal of Molecular Sciences*, *17*(6), 767.
Hopfgartner, G., Lesur, A., & Varesio, E. (2013). Analysis of biopharmaceutical proteins in biological matrices by LC-MS/MS II. LC-MS/MS analysis. *TrAC Trends in Analytical Chemistry*, *48*, 52–61.
Hopkins, R. J., van Dam, N. M., & van Loon, J. J. A. (2009). Role of glucosinolates in insect-plant relationships and multitrophic interactions. *Annual Review of Entomology*, *54*, 57–83.
Hou, J., Acharya, L., Zhu, D., & Cheng, J. (2015). An overview of bioinformatics methods for modeling biological pathways in yeast. *Briefings in Functional Genomics*, *15*(2), 95–108.
Hufsky, F., Rempt, M., Rasche, F., Pohnert, G., & Böcker, S. (2012). De novo analysis of electron impact mass spectra using fragmentation trees. *Analytica Chimica Acta*, *739*, 67–76.
Huseby, S., Degerlund, M., Zingone, A., & Hansen, E. (2012). Metabolic fingerprinting reveals differences between northern and southern strains of the cryptic diatom Chaetoceros socialis. *European Journal of Phycology*, *47*, 480–489.
Hutschenreuther, A., Birkemeyer, C., Grötzinger, K., Straubinger, R. K., & Rauwald, H. W. (2010). Growth inhibiting activity of volatile oil from Cistus creticus L. against Borrelia burgdorferi ss in vitro. *Die Pharmazie*, *65*, 290–295.
Iacono, F., Cappello, T., Corsaro, C., Branca, C., Maisano, M., Gioffre, G., et al. (2010). Environmental metabolomics and multibiomarker approaches on biomonitoring of aquatic habitats. *Comparative Biochemistry and Physiology Part A: Molecular & Integrative Physiology*, *157*, S50.
Ikeda, S., Abe, T., Nakamura, Y., Kibinge, N., Hirai Morita, A., Nakatani, A., et al. (2013). Systematization of the protein sequence diversity in enzymes related to secondary metabolic pathways in plants, in the context of big data biology inspired by the KNApSAcK Motorcycle database. *Plant and Cell Physiology*, *54*(5), 711–727.
Ioannidis, J. P. A., Allison, D. B., Ball, C. A., Coulibaly, I., Cui, X., Culhane, A. C., et al. (2009). Repeatability of published microarray gene expression analyses. *Nature Genetics*, *41*, 149–155.
Iriti, M., & Vitalini, S. (2018). Plant metabolomics in the global scenario of food security: a systems-biology approach for sustainable crop production. *International Journal of Molecular Sciences*, *19*(7), 2094.
Jamil, T., Ozinga, W. A., Kleyer, M., & ter Braak, C. J. F. (2013). Selecting traits that explain species-environment relationships: a generalized linear mixed model approach. *Journal of Vegetation Science*, *24*, 988–1000.
Jandová, K., Dostál, P., Cajthaml, T., & Kameník, Z. (2015). Intraspecific variability in allelopathy of Heracleum mantegazzianum is linked to the metabolic profile of root exudates. *Annals of Botany*, *115*, 821–831.
Jang, C., Chen, L., & Rabinowitz, J. D. (2018). Metabolomics and isotope tracing. *Cell*, *173* (4), 822–837.
Jansen, J. J., Allwood, J. W., Marsden-Edwards, E., van der Putten, W. H., Goodacre, R., & van Dam, N. M. (2009). Metabolomic analysis of the interaction between plants and herbivores. *Metabolomics*, *5*, 150–161.

Jesch, A., Barry, K. E., Ravenek, J. M., Bachmann, D., Strecker, T., Weigelt, A., et al. (2018). Below-ground resource partitioning alone cannot explain the biodiversity-ecosystem function relationship: a field test using multiple tracers. *Journal of Ecology*.

Johnson, C. H., Ivanisevic, J., & Siuzdak, G. (2016). Metabolomics: beyond biomarkers and towards mechanisms. *Nature Reviews Molecular Cell Biology*, *17*(7), 451.

Jolliffe, I. T., & Cadima, J. (2016). Principal component analysis: a review and recent developments. *Philosophical Transactions of the Royal Society A*, *374*(2065)20150202.

Jones, O. A., Maguire, M. L., Griffin, J. L., Dias, D. A., Spurgeon, D. J., & Svendsen, C. (2013). Metabolomics and its use in ecology: metabolomics in ecology. *Austral Ecology*, *38*(6), 713–720.

Jonsson, P., Johansson, A. I., Gullberg, J., Trygg, J., Grung, B., Marklund, S., et al. (2005). High-throughput data analysis for detecting and identifying differences between samples in GC/MS-based metabolomic analyses. *Analytical Chemistry*, 77(17), 5635–5642.

Jonsson, P., Bruce, S. J., Moritz, T., Trygg, J., Sjöström, M., Plumb, R., et al. (2005). Extraction, interpretation and validation of information for comparing samples in metabolic LC/MS data sets. *Analyst*, *130*, 701–707.

Joyce, A. R., & Palsson, B. Ø. (2006). The model organism as a system: integrating 'omics' data sets. *Nature Reviews Molecular Cell Biology*, 7(3), 198.

Kanani, H., Chrysanthopoulos, P. K., & Klapa, M. I. (2008). Standardizing GC–MS metabolomics. *Journal of Chromatography B*, *871*, 191–201.

Kanehisa, M. (2008). The KEGG database. *'In Silico' Simulation of Biological Processes: Novartis Foundation Symposium*, Vol. 247 (pp. 91–103). John Wiley & Sons, Ltd.

Kanehisa, M. (2016). KEGG bioinformatics resource for plant genomics and metabolomics. In *Plant bioinformatics* (pp. 55–70). New York, NY: Humana Press.

Kanehisa, M., Furumichi, M., Tanabe, M., Sato, Y., & Morishima, K. (2016). KEGG: new perspectives on genomes, pathways, diseases and drugs. *Nucleic Acids Research*, *45*(D1), D353–D361.

Kanehisa, M., Goto, S., Sato, Y., Kawashima, M., Furumichi, M., & Tanabe, M. (2013). Data, information, knowledge and principle: back to metabolism in KEGG. *Nucleic Acids Research*, *42*(D1), D199–D205.

Karp, P. D., Billington, R., Holland, T. A., Kothari, A., Krummenacker, M., Weaver, D., et al. (2015). Computational metabolomics operations at BioCyc.org. *Metabolites*, *5*(2), 291–310.

Karp, P. D., Paley, S. M., Krummenacker, M., Latendresse, M., Dale, J. M., Lee, T. J., et al. (2009). Pathway Tools version 13.0: integrated software for pathway/genome informatics and systems biology. *Briefings in Bioinformatics*, *11*(1), 40–79.

Kattge, J., DíAz, S., Lavorel, S., Prentice, I. C., Leadley, P., BöNisch, G., et al. (2011). TRY—a global database of plant traits. *Global Change Biology*, *17*, 2905–2935.

Kelder, T., van Iersel, M. P., Hanspers, K., Kutmon, M., Conklin, B. R., Evelo, C. T., et al. (2011). WikiPathways: building research communities on biological pathways. *Nucleic Acids Research*, *40*(D1), D1301–D1307.

Kelley, K. (2007). Confidence intervals for standardized effect sizes: theory, application, and implementation. *Journal of Statistical Software*, *20*, 1–24.

Keurentjes, J. J., Molenaar, J., & Zwaan, B. J. (2013). Predictive modelling of complex agronomic and biological systems. *Plant, Cell & Environment*, *36*(9), 1700–1710.

Kikuchi, J., Ito, K., & Date, Y. (2018). Environmental metabolomics with data science for investigating ecosystem homeostasis. *Progress in Nuclear Magnetic Resonance Spectroscopy*, *104*, 56–88.

Kim, H. K., & Verpoorte, R. (2010). Sample preparation for plant metabolomics. *Phytochemical Analysis*, *21*(1), 4–13.

Kim, J., Kim, H. -S., Choi, S. -H., Jang, J. -Y., Jeong, M. -J., & Lee, S. (2017). The importance of the circadian clock in regulating plant metabolism. *International Journal of Molecular Sciences*, *18*, 2680.

Kim, S., Thiessen, P. A., Bolton, E. E., Chen, J., Fu, G., Gindulyte, A., et al. (2015). PubChem substance and compound databases. *Nucleic Acids Research*, *44*(D1), D1202–D1213.

Kim, T., Dreher, K., Nilo-Poyanco, R., Lee, I., Fiehn, O., Lange, B. M., et al. (2015). Patterns of metabolite changes identified from large-scale gene perturbations in Arabidopsis using a genome-scale metabolic network. *Plant Physiology*, *167*(4), 1685–1698.

Kind, T., & Fiehn, O. (2006). Metabolomic database annotations via query of elemental compositions: mass accuracy is insufficient even at less than 1 ppm. *BMC Bioinformatics*, 7(1), 234.

Kind, T., & Fiehn, O. (2010). Advances in structure elucidation of small molecules using mass spectrometry. *Bioanalytical Reviews*, *2*(1-4), 23–60.

Kind, T., Wohlgemuth, G., Lee, D. Y., Lu, Y., Palazoglu, M., Shahbaz, S., et al. (2009). FiehnLib: mass spectral and retention index libraries for metabolomics based on quadrupole and time-of-flight gas chromatography/mass spectrometry. *Analytical Chemistry*, *81* (24), 10038–10048.

Kiontke, A., Oliveira Birkmeier, A., Opitz, A., & Birkemeyer, C. (2016). Electrospray ionization efficiency is dependent on different molecular descriptors with respect to solvent pH and instrumental configuration. *PLoS One*, *11*, e0167502.

Kirwan, J. A., Brennan, L., Broadhurst, D., Fiehn, O., Cascante, M., Dunn, W. B., et al. (2018). Preanalytical processing and biobanking procedures of biological samples for metabolomics research: a white paper, community perspective (for "Precision Medicine and Pharmacometabolomics Task Group"—The Metabolomics Society Initiative). *Clinical Chemistry*, *64*(8), 1158–1182. https://doi.org/10.1373/clinchem.2018.287045.

Klein, E. A., Santiago-Jiménez, M., Yousefi, K., Robbins, B. A., Schaeffer, E. M., Trock, B. J., et al. (2017). Molecular analysis of low grade prostate cancer using a genomic classifier of metastatic potential. *The Journal of Urology*, *197*(1), 122–128.

Koek, M. M., Jellema, R. H., van der Greef, J., Tas, A. C., & Hankemeier, T. (2011). Quantitative metabolomics based on gas chromatography mass spectrometry: status and perspectives. *Metabolomics*, 7(3), 307–328.

Kopka, J., Schauer, N., Krueger, S., Birkemeyer, C., Usadel, B., Bergmuller, E., et al. (2005). GMD@CSB.DB: The Golm Metabolome Database. *Bioinformatics*, *21*, 1635–1638.

Kotera, M., Moriya, Y., Tokimatsu, T., Kanehisa, M., & Goto, S. (2015). KEGG and GenomeNet, new developments, metagenomic analysis. In *Encyclopedia of metagenomics* (pp. 329–339). US: Springer.

Kothari, C. R. (2004). *Research methodology: methods and techniques*. New Age International.

Kouzes, R. T., Anderson, G. A., Elbert, S. T., Gorton, I., & Gracio, D. K. (2009). The changing paradigm of data-intensive computing. *Computer*, *42*(1), 26–34.

Krauss, M., Singer, H., & Hollender, J. (2010). LC–high resolution MS in environmental analysis: from target screening to the identification of unknowns. *Analytical and Bioanalytical Chemistry*, *397*, 943–951.

Kravitz, B., Guenther, A. B., Gu, L., Karl, T., Kaser, L., Pallardy, S. G., et al. (2016). A new paradigm of quantifying ecosystem stress through chemical signatures. *Ecosphere*, 7(11).

Krumsiek, J., Suhre, K., Illig, T., Adamski, J., & Theis, F. J. (2011). Gaussian graphical modeling reconstructs pathway reactions from high-throughput metabolomics data. *BMC Systems Biology*, *5*(1), 21.

Kueger, S., Steinhauser, D., Willmitzer, L., & Giavalisco, P. (2012). High-resolution plant metabolomics: from mass spectral features to metabolites and from whole-cell analysis to subcellular metabolite distributions. *The Plant Journal*, *70*(1), 39–50.

Kuehnbaum, N. L., & Britz-Mckibbin, P. (2013). New advances in separation science for metabolomics: resolving chemical diversity in a post-genomic era. *Chemical Reviews, 113*, 2437–2468. https://doi.org/10.1021/cr300484s.

Kuhlisch, C., & Pohnert, G. (2015). Metabolomics in chemical ecology. *Natural Product Reports, 32*(7), 937–955.

Kuligowski, J., Pérez-Guaita, D., Lliso, I., Escobar, J., León, Z., Gombau, L., et al. (2014). Detection of batch effects in liquid chromatography-mass spectrometry metabolomic data using guided principal component analysis. *Talanta, 130*, 442–448.

Kumar, M., Kuzhiumparambil, U., Pernice, M., Jiang, Z., & Ralph, P. J. (2016). Metabolomics: an emerging frontier of systems biology in marine macrophytes. *Algal Research, 16*, 76–92.

Kursar, T. A., Dexter, K. G., Lokvam, J., Pennington, R. T., Richardson, J. E., Weber, M. G., et al. (2009). The evolution of anti-herbivore defenses and their contribution to species coexistence in the tropical tree genus Inga. *Proceedings of the National Academy of Sciences of the United States of America, 106*, 18073–18078.

Kusano, M., Tohge, T., Fukushima, A., Kobayashi, M., Hayashi, N., Otsuki, H., et al. (2011). Metabolomics reveals comprehensive reprogramming involving two independent metabolic responses of Arabidopsis to UV-B light. *The Plant Journal, 67*(2), 354–369.

Kusari, S., Hertweck, C., & Spiteller, M. (2012). Chemical ecology of endophytic fungi: origins of secondary metabolites. *Chemistry & Biology, 19*(7), 792–798.

Kutyniok, M., & Müller, C. (2012). Crosstalk between above- and belowground herbivores is mediated by minute metabolic responses of the host Arabidopsis thaliana. *Journal of Experimental Botany, 63*, 6199–6210.

Kutyniok, M., & Müller, C. (2013). Plant-mediated interactions between shoot-feeding aphids and root-feeding nematodes depend on nitrate fertilization. *Oecologia, 173*, 1367–1377.

Kuzina, V., Ekstrom, C. T., Andersen, S. B., Nielsen, J. K., Olsen, C. E., & Bak, S. (2009). Identification of defense compounds in Barbarea vulgaris against the herbivore Phyllotreta nemorum by an ecometabolomic approach. *Plant Physiology, 151*, 1977–1990.

Ladero, V., Calles-Enríquez, M., Fernández, M., & Alvarez, M. A. (2010). Toxicological effects of dietary biogenic amines. *Current Nutrition & Food Science, 6*(2), 145–156.

Lamb, E. G. (2008). Direct and indirect control of grassland community structure by litter, resources, and biomass. *Ecology, 89*, 216–225.

Lamb, E. G., Kembel, S. W., & Cahill, J. F. (2009). Shoot, but not root, competition reduces community diversity in experimental mesocosms. *Journal of Ecology, 97*, 155–163.

Landuyt, D., Broekx, S., D'hondt, R., Engelen, G., Aertsens, J., & Goethals, P. L. (2013). A review of Bayesian belief networks in ecosystem service modelling. *Environmental Modelling & Software, 46*, 1–11.

Lankadurai, B. P., Nagato, E. G., & Simpson, M. J. (2013). Environmental metabolomics: an emerging approach to study organism responses to environmental stressors. *Environmental Reviews, 21*(3), 180–205.

Le Cao, K. A., Boitard, S., & Besse, P. (2011). Sparse PLS discriminant analysis: biologically relevant feature selection and graphical displays for multiclass problems. *BMC Bioinformatics, 12*, 253. https://doi.org/10.1186/1471-2105-12-253.

Legendre, P., & Legendre, L. (2012). *Numerical ecology* (3rd. Engl. ed.). *Developments in environmental modelling: (Vol. 25*). Amsterdam, The Netherlands: Elsevier 978-0-444-53868-0.

Leibold, M. A., Holyoak, M., Mouquet, N., Amarasekare, P., Chase, J. M., Hoopes, M. F., et al. (2004). The metacommunity concept: A framework for multi-scale community ecology: the metacommunity concept. *Ecology Letters*, 7, 601–613.

Leipzig, J. (2016). A review of bioinformatic pipeline frameworks. *Briefings in Bioinformatics*, bbw020.

Leiss, K. A., Cristofori, G., van Steenis, R., Verpoorte, R., & Klinkhamer, P. G. L. (2013). An eco-metabolomic study of host plant resistance to Western flower thrips in cultivated, biofortified and wild carrots. *Phytochemistry*, *93*, 63–70.

Leon, Z., Garcia-Canaveras, J. C., Donato, M. T., & Lahoz, A. (2013). Mammalian cell metabolomics: experimental design and sample preparation. *Electrophoresis*, *34*, 2762–2775. https://doi.org/10.1002/elps.201200605.

Lewinsohn, E., & Gijzen, M. (2009). Phytochemical diversity: the sounds of silent metabolism. *Plant Science*, *176*(2), 161–169.

Li, D. X., Gan, L., Bronja, A., & Schmitz, O. J. (2015). Gas chromatography coupled to atmospheric pressure ionization mass spectrometry (GC-API-MS). *Analytica Chimica Acta*, *891*, 43–61.

Li, Q., Zhao, C., Li, Y., Chang, Y., Wu, Z., Pang, T., et al. (2011). Liquid chromatography/mass spectrometry-based metabolic profiling to elucidate chemical differences of tobacco leaves between Zimbabwe and China. *Journal of Separation Science*, *34*(2), 119–126.

Lin, C. Y., Viant, M. R., & Tjeerdema, R. S. (2006). Metabolomics: methodologies and applications in the environmental sciences. *Journal of Pesticide Science*, *31*(3), 245–251.

Lin, C. Y., Wu, H., Tjeerdema, R. S., & Viant, M. R. (2007). Evaluation of metabolite extraction strategies from tissue samples using NMR metabolomics. *Metabolomics*, *3* (1), 55–67.

Liu, H., Han, F., Yuan, M., Lafferty, J., & Wasserman, L. (2012). High-dimensional semiparametric Gaussian copula graphical models. *The Annals of Statistics*, *40*(4), 2293–2326.

Liu, H., Carvalhais, L. C., Crawford, M., Singh, E., Dennis, P. G., Pieterse, C. M. J., et al. (2017). Inner plant values: diversity, colonization and benefits from endophytic bacteria. *Frontiers in Microbiology*, *8*.

Liu, K., Altaf-Ul-Amin, M., Abdullah, A. A., Morita, A. H., Shiraishi, M., & Kanaya, S. (2015, August). A novel plant classification method based on similarities in chemical structures of metabolite contents obtained from the KNApSAcK database. In *II International Symposium on Pyrethrum 1169* (pp. 139–150).

Liu, Y., Ren, P., Zhao, T., Yang, Z., & Gao, J. (2016). Estimation of adjacent substitution rate based on clustering algorithm and its application. In *2016 12th international conference on natural computation, fuzzy systems and knowledge discovery (ICNC-FSKD)* (pp. 794–800). IEEE.

Llaneras, F., & Picó, J. (2008). Stoichiometric modelling of cell metabolism. *Journal of Bioscience and Bioengineering*, *105*(1), 1–11.

Lo, K., Raftery, A. E., Dombek, K. M., Zhu, J., Schadt, E. E., Bumgarner, R. E., et al. (2012). Integrating external biological knowledge in the construction of regulatory networks from time-series expression data. *BMC Systems Biology*, *6*(1), 101.

Loboda, A. V., Krutchinsky, A. N., Bromirski, M., Ens, W., & Standing, K. G. (2000). A tandem quadrupole/time-of-flight mass spectrometer with a matrix-assisted laser desorption/ionization source: design and performance. *Rapid Communications in Mass Spectrometry*, *14*(12), 1047–1057.

Lössl, P., van de Waterbeemd, M., & Heck, A. J. (2016). The diverse and expanding role of mass spectrometry in structural and molecular biology. *The EMBO Journal*, *35*(24), 2634–2657 e201694818.

Luan, H., Liu, L. F., Tang, Z., Zhang, M., Chua, K. K., Song, J. X., et al. (2015). Comprehensive urinary metabolomic profiling and identification of potential noninvasive marker for idiopathic Parkinson's disease. *Scientific Reports*, *5*, 13888.

Lüttge, U. (2007). *Physiological ecology of tropical plants*. Springer Science & Business Media.

Ma, H., Sorokin, A., Mazein, A., Selkov, A., Selkov, E., Demin, O., et al. (2007). The Edinburgh human metabolic network reconstruction and its functional analysis. *Molecular Systems Biology*, *3*(1), 135.

Ma, S., & Chowdhury, S. K. (2013). Data acquisition and data mining techniques for metabolite identification using LC coupled to high-resolution MS. *Bioanalysis, 5*(10), 1285–1297.
MacArthur, R. (1955). Fluctuations of animal populations and a measure of community stability. *Ecology, 36*, 533–536.
Macel, M., Van, D. A. M., Nicole, M., & Keurentjes, J. J. (2010). Metabolomics: the chemistry between ecology and genetics. *Molecular Ecology Resources, 10*(4), 583–593.
Macel, M., de Vos, R. C. H., Jansen, J. J., van der Putten, W. H., & van Dam, N. M. (2014). Novel chemistry of invasive plants: exotic species have more unique metabolomic profiles than native congeners. *Ecology and Evolution, 4*, 2777–2786.
Macintyre, L., Zhang, T., Viegelmann, C., Martinez, I. J., Cheng, C., Dowdells, C., et al. (2014). Metabolomic tools for secondary metabolite discovery from marine microbial symbionts. *Marine Drugs, 12*(6), 3416–3448.
Maier, T. S., Kuhn, J., & Müller, C. (2010). Proposal for field sampling of plants and processing in the lab for environmental metabolic fingerprinting. *Plant Methods, 6*, 6.
Majetic, C. J., Raguso, R. A., & Ashman, T. -L. (2008). The impact of biochemistry vs. population membership on floral scent profiles in colour polymorphic Hesperis matronalis. *Annals of Botany, 102*, 911–922.
Marcoux, J., Champion, T., Colas, O., Wagner-Rousset, E., Corvaïa, N., Van Dorsselaer, A., et al. (2015). Native mass spectrometry and ion mobility characterization of trastuzumab emtansine, a lysine-linked antibody drug conjugate. *Protein Science, 24*(8), 1210–1223.
Mardinoglu, A., Gatto, F., & Nielsen, J. (2013). Genome-scale modeling of human metabolism—a systems biology approach. *Biotechnology Journal, 8*, 985–996. https://doi.org/10.1002/biot.201200275.
Marty, M. T., Baldwin, A. J., Marklund, E. G., Hochberg, G. K., Benesch, J. L., & Robinson, C. V. (2015). Bayesian deconvolution of mass and ion mobility spectra: from binary interactions to polydisperse ensembles. *Analytical Chemistry, 87*(8), 4370–4376.
Masakapalli, S. K., Kruger, N. J., & Ratcliffe, R. G. (2013). The metabolic flux phenotype of heterotrophic *Arabidopsis* cells reveals a complex response to changes in nitrogen supply. *The Plant Journal, 74*, 569–582. https://doi.org/10.1111/tpj.12142.
Masakapalli, S. K., Le Lay, P., Huddleston, J. E., Pollock, N. L., Kruger, N. J., & Ratcliffe, R. G. (2010). Subcellular flux analysis of central metabolism in a heterotrophic Arabidopsis cell suspension using steady-state stable isotope labeling. *Plant Physiology, 152*, 602–619. https://doi.org/10.1104/pp.109.151316.
Matsuda, F., Okazaki, Y., Oikawa, A., Kusano, M., Nakabayashi, R., Kikuchi, J., et al. (2012). Dissection of genotype–phenotype associations in rice grains using metabolome quantitative trait loci analysis. *The Plant Journal, 70*(4), 624–636.
McCartney, P. R. (2015). Big data science. *MCN: The American Journal of Maternal/Child Nursing, 40*(2), 130.
McEachran, A. D., Mansouri, K., Newton, S. R., Beverly, B. E. J., Sobus, J. R., & Williams, A. J. (2018). A comparison of three liquid chromatography (LC) retention time prediction models. *Talanta, 182*, 371–379.
Meier, R., Ruttkies, C., Treutler, H., & Neumann, S. (2017). Bioinformatics can boost metabolomics research. *Journal of Biotechnology, 261*, 137–141.
Meinert, C., & Meierhenrich, U. J. (2012). A new dimension in separation science: comprehensive two-dimensional gas chromatography. *Angewandte Chemie International Edition, 51*(42), 10460–10470.
Melaku, Y. A., Gill, T. K., Taylor, A. W., Adams, R., & Shi, Z. (2017). A comparison of principal component analysis, partial least-squares and reduced-rank regressions in the identification of dietary patterns associated with bone mass in ageing Australians. *European Journal of Nutrition*, 1–15.

Menéndez, R., González-Megías, A., Collingham, Y., Fox, R., Roy, D. B., Ohlemüller, R., et al. (2007). Direct and indirect effects of climate and habitat factors on butterfly diversity. *Ecology*, *88*, 605–611.
Mertler, C. A., & Reinhart, R. V. (2016). *Advanced and multivariate statistical methods: practical application and interpretation*. Routledge.
Meysman, F. J., & Bruers, S. (2010). Ecosystem functioning and maximum entropy production: a quantitative test of hypotheses. *Philosophical Transactions of the Royal Society, B: Biological Sciences*, *365*(1545), 1405–1416.
Miller, G. A., Islam, M. S., Claridge, T. D., Dodgson, T., & Simpson, S. J. (2008). Swarm formation in the desert locust *Schistocerca gregaria*: isolation and NMR analysis of the primary maternal gregarizing agent. *Journal of Experimental Biology*, *211*(3), 370–376.
Miller, M. G. (2007). Environmental metabolomics: a SWOT analysis (strengths, weaknesses, opportunities, and threats). *Journal of Proteome Research*, *6*(2), 540–545.
Mintz-Oron, S., Meir, S., Malitsky, S., Ruppin, E., Aharoni, A., & Shlomi, T. (2012). Reconstruction of Arabidopsis metabolic network models accounting for subcellular compartmentalization and tissue-specificity. *Proceedings of the National Academy of Sciences of the United States of America*, *109*, 339–344. https://doi.org/10.1073/pnas.1100358109.
Misra, B. B. (2018). New tools and resources in metabolomics: 2016–2017. *Electrophoresis*, *39* (7), 909–923.
Misra, B. B., Fahrmann, J. F., & Grapov, D. (2017). Review of emerging metabolomic tools and resources: 2015–2016. *Electrophoresis*, *38*(18), 2257–2274.
Mithöfer, A., & Boland, W. (2012). Plant defense against herbivores: chemical aspects. *Annual Review of Plant Biology*, *63*, 431–450.
Mitra, R., Müller, P., & Ji, Y. (2016). Bayesian graphical models for differential pathways. *Bayesian Analysis*, *11*(1), 99–124.
Mittler, R., & Shulaev, V. (2013). Functional genomics, challenges and perspectives for the future. *Physiologia Plantarum*, *148*(3), 317–321.
Miyagi, A., Takahara, K., Takahashi, H., Kawai-Yamada, M., & Uchimiya, H. (2010). Targeted metabolomics in an intrusive weed, *Rumex obtusifolius* L., grown under different environmental conditions reveals alterations of organ related metabolite pathway. *Metabolomics*, *6*(4), 497–510.
Mochida, K., & Shinozaki, K. (2011). Advances in omics and bioinformatics tools for systems analyses of plant functions. *Plant and Cell Physiology*, *52*(12), 2017–2038.
Moore, B. D., Andrew, R. L., Külheim, C., & Foley, W. J. (2014). Explaining intraspecific diversity in plant secondary metabolites in an ecological context. *New Phytologist*, *201*(3), 733–750.
Moore, M. N., Depledge, M. H., Readman, J. W., & Paul Leonard, D. (2004). An integrated biomarker-based strategy for ecotoxicological evaluation of risk in environmental management. *Mutation Research, Fundamental and Molecular Mechanisms of Mutagenesis*, *552*, 247–268.
Morel, D., & Levy, W. B. (2009). The cost of linearization. *Journal of Computational Neuroscience*, *27*(2), 259–275.
Moseley, H. N. (2013). Error analysis and propagation in metabolomics data analysis. *Computational and Structural Biotechnology Journal*, *4*(5)e201301006.
Moshelion, M., & Altman, A. (2015). Current challenges and future perspectives of plant and agricultural biotechnology. *Trends in Biotechnology*, *33*(6), 337–342.
Muller, E. E., Faust, K., Widder, S., Herold, M., Arbas, S. M., & Wilmes, P. (2018). Using metabolic networks to resolve ecological properties of microbiomes. *Current Opinion in Systems Biology*, *8*, 73–80.
Murren, C. J., Auld, J. R., Callahan, H., Ghalambor, C. K., Handelsman, C. A., Heskel, M. A., et al. (2015). Constraints on the evolution of phenotypic plasticity: limits and costs of phenotype and plasticity. *Heredity*, *115*(4), 293.

Nacoulma, A., Vandeputte, O., De Lorenzi, M., El Jaziri, M., & Duez, P. (2013). Metabolomic-based study of the leafy gall, the ecological niche of the phytopathogen Rhodococcus fascians, as a potential source of bioactive compounds. *International Journal of Molecular Sciences*, *14*, 12533–12549.

Nägele, T. (2014). Linking metabolomics data to underlying metabolic regulation. *Frontiers in Molecular Biosciences*, *1*, 22.

Nägele, T., Fürtauer, L., Nagler, M., Weiszmann, J., & Weckwerth, W. (2016). A strategy for functional interpretation of metabolomic time series data in context of metabolic network information. *Frontiers in Molecular Biosciences*, *3*, 6.

Nägele, T., Mair, A., Sun, X., Fragner, L., Teige, M., & Weckwerth, W. (2014). Solving the differential biochemical Jacobian from metabolomics covariance data. *PLoS One*, *9*(4) e92299.

Nakamura, K., Shimura, N., Otabe, Y., Hirai-Morita, A., Nakamura, Y., Ono, N., et al. (2013). KNApSAcK-3D: a three-dimensional structure database of plant metabolites. *Plant and Cell Physiology*, *54*(2), e4.

Nakamura, Y., Mochamad Afendi, F., Kawsar Parvin, A., Ono, N., Tanaka, K., Hirai Morita, A., et al. (2014). KNApSAcK metabolite activity database for retrieving the relationships between metabolites and biological activities. *Plant and Cell Physiology*, *55*(1), e7.

Nemkov, T., D'Alessandro, A., & Hansen, K. C. (2015). Three-minute method for amino acid analysis by UHPLC and high-resolution quadrupole orbitrap mass spectrometry. *Amino Acids*, *47*(11), 2345–2357.

Nevo, O., Heymann, E. W., Schulz, S., & Ayasse, M. (2016). Fruit odor as a ripeness signal for seed-dispersing primates? A case study on four neotropical plant species. *Journal of Chemical Ecology*, *42*, 323–328.

Nevo, O., OrtsGarri, R., Hernandez Salazar, L. T., Schulz, S., Heymann, E. W., Ayasse, M., et al. (2015). Chemical recognition of fruit ripeness in spider monkeys (Ateles geoffroyi). *Scientific Reports*, *5*, 14895.

Newson, R., & Team, A. S. (2003). Multiple-test procedures and smile plots. *The Stata Journal*, *3*, 109–132.

Nguyen, H. V., Müller, E., Vreeken, J., Efros, P., & Böhm, K. (2014). Multivariate maximal correlation analysis. In *International conference on machine learning* (pp. 775–783).

Nicotra, A. B., Atkin, O. K., Bonser, S. P., Davidson, A. M., Finnegan, E. J., Mathesius, U., et al. (2010). Plant phenotypic plasticity in a changing climate. *Trends in Plant Science*, *15* (12), 684–692.

Noecker, C., Eng, A., Srinivasan, S., Theriot, C. M., Young, V. B., Jansson, J. K., et al. (2016). Metabolic model-based integration of microbiome taxonomic and metabolomic profiles elucidates mechanistic links between ecological and metabolic variation. *mSystems*, *1*(1), e00013–e00015.

Oberhardt, M. A., Palsson, B. Ø., & Papin, J. A. (2009). Applications of genome-scale metabolic reconstructions. *Molecular Systems Biology*, *5*(1), 320.

Okada, T., Afendi, F. M., Katoh, A., Hirai, A., & Kanaya, S. (2013). Multivariate analysis of analytical chemistry data and utility of the KNApSAcK family database to understand metabolic diversity in medicinal plants. In *Biotechnology for medicinal plants* (pp. 413–438). Berlin, Heidelberg: Springer.

Okie, J. G., Van Horn, D. J., Storch, D., Barrett, J. E., Gooseff, M. N., Kopsova, L., et al. (2015). Niche and metabolic principles explain patterns of diversity and distribution: theory and a case study with soil bacterial communities. *Proceedings of the Royal Society B: Biological Sciences*, *282*, 20142630.

Olson, M. T., & Yergey, A. L. (2009). Calculation of the isotope cluster for polypeptides by probability grouping. *Journal of the American Society for Mass Spectrometry*, *20*(2), 295–302.

Orata, F. (2012). Derivatization reactions and reagents for gas chromatography analysis. In *Advanced gas chromatography—progress in agricultural, biomedical and industrial applications*. InTech.

Orgogozo, V., Morizot, B., & Martin, A. (2015). The differential view of genotype–phenotype relationships. *Frontiers in Genetics*, *6*, 179.
Ossipov, V., Ossipova, S., Bykov, V., Oksanen, E., Koricheva, J., & Haukioja, E. (2008). Application of metabolomics to genotype and phenotype discrimination of birch trees grown in a long-term open-field experiment. *Metabolomics*, *4*(1), 39–51.
Ouborg, N. J., & Vriezen, W. H. (2007). An ecologist's guide to ecogenomics. *Journal of Ecology*, *95*(1), 8–16.
Pang, T. Y., & Maslov, S. (2011). A toolbox model of evolution of metabolic pathways on networks of arbitrary topology. *PLoS Computational Biology*, 7(5)e1001137.
Patterson, A. D., Li, H., Eichler, G. S., Krausz, K. W., Weinstein, J. N., Fornace, A. J., et al. (2008). UPLC-ESI-TOFMS-based metabolomics and gene expression dynamics inspector self-organizing metabolomic maps as tools for understanding the cellular response to ionizing radiation. *Analytical Chemistry*, *80*(3), 665–674.
Paules, R. (2003). Phenotypic anchoring: linking cause and effect. *Environmental Health Perspectives*, *111*(6), A338.
Peiris, D., Dunn, W. B., Brown, M., Kell, D. B., Roy, I., & Hedger, J. N. (2008). Metabolite profiles of interacting mycelia fronts differ for pairings of the wood decay basidiomycete fungus, *Stereum hirsutum* with its competitors *Coprinus micaceus* and *Coprinus disseminatus*. *Metabolomics*, *4*, 52–62.
Peng, R. D. (2011). Reproducible research in computational science. *Science*, *334*, 1226–1227.
Peñuelas, J., & Sardans, J. (2009a). Ecological metabolomics. *Chemistry and Ecology*, *25*(4), 305–309.
Peñuelas, J., & Sardans, J. (2009b). Elementary factors. *Nature*, *460*, 803–804.
Peñuelas, J., Sardans, J., Estiarte, M., Ogaya, R., Carnicer, J., Coll, M., et al. (2013). Evidence of current impact of climate change on life: a walk from genes to the biosphere. *Global Change Biology*, *19*, 2303–2338.
Peters, K., Worrich, A., Weinhold, A., Alka, O., Balcke, G., Birkemeyer, C., et al. (2018). Current challenges in plant eco-metabolomics. *International Journal of Molecular Sciences*, *19*(5), 1385.
Peters, K., Breitsameter, L., & Gerowitt, B. (2014). Impact of climate change on weeds in agriculture: a review. *Agronomy for Sustainable Development*, *34*, 707–721.
Petersen, I. L., Tomasi, G., Sørensen, H., Boll, E. S., Hansen, H. C. B., & Christensen, J. H. (2011). The use of environmental metabolomics to determine glyphosate level of exposure in rapeseed (*Brassica napus* L.) seedlings. *Environmental Pollution*, *159*(10), 3071–3077.
Pfeuffer, J., Sachsenberg, T., Alka, O., Walzer, M., Fillbrunn, A., Nilse, L., et al. (2017). OpenMS—a platform for reproducible analysis of mass spectrometry data. *Journal of Biotechnology*, *261*, 142–148.
Pianka, E. R. (2011). *Evolutionary ecology*. New York: Harper and Row.
Pimenta, L. P. S., Kim, H. K., Verpoorte, R., & Choi, Y. H. (2013). NMR-based metabolomics: a probe to utilize biodiversity. U. Roessner & D. A. Dias (Eds.), *Metabolomics tools for natural product discovery* (pp. 117–127). *1055* (pp. 117–127). Totowa, NJ: Humana Press. ISBN 978-1-62703-576-7.
Pitts-Singer, T. L., Hagen, M. M., Helm, B. R., Highland, S., Buckner, J. S., & Kemp, W. P. (2017). Comparison of the chemical compositions of the cuticle and Dufour's gland of two solitary bee species from laboratory and field conditions. *Journal of Chemical Ecology*, *43*, 451–468.
Poisot, T., Bever, J. D., Nemri, A., Thrall, P. H., & Hochberg, M. E. (2011). A conceptual framework for the evolution of ecological specialisation. *Ecology Letters*, *14*(9), 841–851.
Prince, E. K., & Pohnert, G. (2010). Searching for signals in the noise: metabolomics in chemical ecology. *Analytical and Bioanalytical Chemistry*, *396*(1), 193–197.

Prosser, G. A., Larrouy-Maumus, G., & de Carvalho, L. P. S. (2014). Metabolomic strategies for the identification of new enzyme functions and metabolic pathways. *EMBO Reports*, *15*(6), 657–669.

Provost, F., & Fawcett, T. (2013). Data science and its relationship to big data and data-driven decision making. *Big Data*, *1*(1), 51–59.

Qiu, B. -L., Harvey, J. A., Raaijmakers, C. E., Vet, L. E. M., & van Dam, N. M. (2009). Nonlinear effects of plant root and shoot jasmonic acid application on the performance of Pieris brassicae and its parasitoid Cotesia glomerata. *Functional Ecology*, *23*, 496–505.

Raguso, R. A., Agrawal, A. A., Douglas, A. E., Jander, G., Kessler, A., Poveda, K., et al. (2015). The raison d'être of chemical ecology. *Ecology*, *96*, 617–630.

Rahman, M. A. (2018). *Gaussian process in computational biology: covariance functions for transcriptomics*. Doctoral dissertationUniversity of Sheffield.

Rahman, M. M., Bosch, L. V., Baggerman, G., Clynen, E., Hens, K., Hoste, B., et al. (2002). Search for peptidic molecular markers in hemolymph of crowd-(gregarious) and isolated-reared (solitary) desert locusts, *Schistocerca gregaria*. *Peptides*, *23*(11), 1907–1914.

Reed, L. K., Baer, C. F., & Edison, A. S. (2017). Considerations when choosing a genetic model organism for metabolomics studies. *Current Opinion in Chemical Biology*, *36*, 7–14.

Reed, T. E., Schindler, D. E., & Waples, R. S. (2011). Interacting effects of phenotypic plasticity and evolution on population persistence in a changing climate. *Conservation Biology*, *25*(1), 56–63.

Regoli, F., Gorbi, S., Fattorini, D., Tedesco, S., Notti, A., Machella, N., et al. (2006). Use of the land snail *Helix aspersa* as sentinel organism for monitoring ecotoxicologic effects of urban pollution: an integrated approach. *Environmental Health Perspectives*, *114*(1), 63.

Ren, S., Hinzman, A. A., Kang, E. L., Szczesniak, R. D., & Lu, L. J. (2015). Computational and statistical analysis of metabolomics data. *Metabolomics*, *11*(6), 1492–1513.

Reznik, E., & Segre, D. (2010). On the stability of metabolic cycles. *Journal of Theoretical Biology*, *266*, 536–549. https://doi.org/10.1016/j.jtbi.2010.07.023.

Rezzi, S., Axelson, D. E., Héberger, K., Reniero, F., Mariani, C., & Guillou, C. (2005). Classification of olive oils using high throughput flow 1H NMR fingerprinting with principal component analysis, linear discriminant analysis and probabilistic neural networks. *Analytica Chimica Acta*, *552*(1-2), 13–24.

Riedel, T., Biester, H., & Dittmar, T. (2012). Molecular fractionation of dissolved organic matter with metal salts. *Environmental Science & Technology*, *46*(8), 4419–4426.

Riipi, M., Haukioja, E., Lempa, K., Ossipov, V., Ossipova, S., & Pihlaja, K. (2004). Ranking of individual mountain birch trees in terms of leaf chemistry: seasonal and annual variation. *Chemoecology*, *14*(1), 31–43.

Rivas-Ubach, A., Hódar, J. A., Sardans, J., Kyle, J. E., Kim, Y. M., Oravec, M., et al. (2016). Are the metabolomic responses to folivory of closely related plant species linked to macroevolutionary and plant–folivore coevolutionary processes? *Ecology and Evolution*, *6*(13), 4372–4386.

Rivas-Ubach, A., Pérez-Trujillo, M., Sardans, J., Gargallo-Garriga, A., Parella, T., & Penuelas, J. (2013). Ecometabolomics: optimized NMR-based method. *Methods in Ecology and Evolution*, *4*(5), 464–473.

Rivas-Ubach, A., Sardans, J., Pérez-Trujillo, M., Estiarte, M., & Peñuelas, J. (2012). Strong relationship between elemental stoichiometry and metabolome in plants. *Proceedings of the National Academy of Sciences*, *109*(11), 4181–4186.

Rivas-Ubach, A., Sardans, J., Hódar, J. A., Garcia-Porta, J., Guenther, A., Paša-Tolic´, L., et al. (2017). Close and distant: contrasting the metabolism of two closely related subspecies of Scots pine under the effects of folivory and summer drought. *Ecology and Evolution*, 7, 8976–8988.

Rivas-Ubach, A., Sardans, J., Hódar, J. A., Garcia-Porta, J., Guenther, A., Oravec, M., et al. (2016). Similar local, but different systemic, metabolomic responses of closely related pine subspecies to folivory by caterpillars of the processionary moth. *Plant Biology, 18*, 484–494.

Robinson, A. R., Dauwe, R., & Mansfield, S. D. (2018). Assessing the between-background stability of metabolic effects arising from lignin-related transgenic modifications, in two Populus hybrids using non-targeted metabolomics. *Tree Physiology, 38*(3), 378–396.

Rocca-Serra, P., Salek, R. M., Arita, M., Correa, E., Dayalan, S., Gonzalez-Beltran, A., et al. (2016). Data standards can boost metabolomics research, and if there is a will, there is a way. *Metabolomics, 12*.

Rochfort, S. (2005). Metabolomics reviewed: a new "omics" platform technology for systems biology and implications for natural products research. *Journal of Natural Products, 68* (12), 1813–1820.

Rockwood, A. L., & Haimi, P. (2006). Efficient calculation of accurate masses of isotopic peaks. *Journal of the American Society for Mass Spectrometry, 17*(3), 415–419.

Rodríguez, A., Alquézar, B., & Peña, L. (2013). Fruit aromas in mature fleshy fruits as signals of readiness for predation and seed dispersal. *The New Phytologist, 197*, 36–48.

Roessner, U., Luedemann, A., Brust, D., Fiehn, O., Linke, T., Willmitzer, L., et al. (2001). Metabolic profiling allows comprehensive phenotyping of genetically or environmentally modified plant systems. *The Plant Cell, 13*(1), 11–29.

Roessner, U., & Dias, D. A. (Eds.), (2013). *Metabolomics tools for natural product discovery*. In *Methods in molecular biology: Vol. 1055*. Totowa, NJ, USA: Humana Press. ISBN 978-1-62703-576-7.

Rosental, L., Perelman, A., Nevo, N., Toubiana, D., Samani, T., Batushansky, A., et al. (2016). Environmental and genetic effects on tomato seed metabolic balance and its association with germination vigor. *BMC Genomics, 17*(1), 1047.

Rossi, M. T., Kalde, M., Srisakvarakul, C., Kruger, N. J., & Ratcliffe, R. G. (2017). Cell-type specific metabolic flux analysis: a challenge for metabolic phenotyping and a potential solution in plants. *Metabolites*, 7(4), 59.

Röst, H. L., Sachsenberg, T., Aiche, S., Bielow, C., Weisser, H., Aicheler, F., et al. (2016). OpenMS: a flexible open-source software platform for mass spectrometry data analysis. *Nature Methods, 13*, 741–748.

Rowland, S. M., Robbins, W. K., Corilo, Y. E., Marshall, A. G., & Rodgers, R. P. (2014). Solid-phase extraction fractionation to extend the characterization of naphthenic acids in crude oil by electrospray ionization Fourier transform ion cyclotron resonance mass spectrometry. *Energy & Fuels, 28*(8), 5043–5048.

Russell, J. F. (2013). If a job is worth doing, it is worth doing twice: Researchers and funding agencies need to put a premium on ensuring that results are reproducible. *Nature, 496*, 7–8.

Ruttkies, C., Strehmel, N., Scheel, D., & Neumann, S. (2015). Annotation of metabolites from gas chromatography/atmospheric pressure chemical ionization tandem mass spectrometry data using an in silico generated compound database and MetFrag: annotation of metabolites from high-resolution GC/APCI-MS/MS data. *Rapid Communications in Mass Spectrometry, 29*, 1521–1529.

Ryan, R. P., Germaine, K., Franks, A., Ryan, D. J., & Dowling, D. N. (2008). Bacterial endophytes: recent developments and applications. *FEMS Microbiology Letters, 278*, 1–9.

Saccenti, E., Hoefsloot, H. C., Smilde, A. K., Westerhuis, J. A., & Hendriks, M. M. (2014). Reflections on univariate and multivariate analysis of metabolomics data. *Metabolomics, 10*(3), 361–374.

Sali, M., De Maio, F., Tarantino, M., Garofolo, G., Tittarelli, M., Sacchini, L., et al. (2018). Rapid and safe one-step extraction method for the identification of *Brucella* strains at genus and species level by MALDI-TOF mass spectrometry. *PLoS One, 13*(6)e0197864.

Sanchez, B. J., Perez-Correa, J. R., & Agosin, E. (2014). Construction of robust dynamic genome-scale metabolic model structures of *Saccharomyces cerevisiae*through iterative re-parameterization. *Metabolic Engineering*, *25C*, 159–173. https://doi.org/10.1016/j.ymben.2014.07.004.

Sangwan, N. S., Tiwari, P., Mishra, S. K., Yadav, R. K., Tripathi, S., Kushwaha, A. K., et al. (2015). Plant metabolomics: an overview of technology platforms for applications in metabolism. In *PlantOmics: the omics of plant science* (pp. 257–298). New Delhi: Springer.

Santos, C. A. F., Corrêa, L. C., & Costa, S. R. D. (2011). Genetic divergence among *Psidium* accessions based on biochemical and agronomic variables. *Crop Breeding and Applied Biotechnology*, *11*(2), 149–156.

Sardans, J., Penuelas, J., & Rivas-Ubach, A. (2011). Ecological metabolomics: overview of current developments and future challenges. *Chemoecology*, *21*(4), 191–225.

Sardans, J., & Peñuelas, J. (2013). Tree growth changes with climate and forest type are associated to relative allocation of nutrients, especially P, to leaves and wood. *Global Ecology and Biogeography*, *22*, 494–507.

Sardans, J., Peñuelas, J., Coll, M., Vayreda, J., & Rivas-Ubach, A. (2012). Stoichiometryof potassium is largely determined by water availability and growth in Catalonian forests. *Functional Ecology*, *26*, 1077–1089.

Sarpe, V., & Schriemer, D. C. (2017). Supporting metabolomics with adaptable software: design architectures for the end-user. *Current Opinion in Biotechnology*, *43*, 110–117.

Sasidharan, S., Chen, Y., Saravanan, D., Sundram, K. M., & Latha, L. Y. (2011). Extraction, isolation and characterization of bioactive compounds from plants' extracts. *African Journal of Traditional, Complementary, and Alternative Medicines*, *8*(1).

Saunders, M. (2004). Stochastic search for isomers on a quantum mechanical surface. *Journal of Computational Chemistry*, *25*(5), 621–626.

Scheltema, R. A., Decuypere, S., Dujardin, J. C., Watson, D. G., Jansen, R. C., & Breitling, R. (2009). Simple data-reduction method for high-resolution LC–MS data in metabolomics. *Bioanalysis*, *1*(9), 1551–1557.

Scherling, C., Roscher, C., Giavalisco, P., Schulze, E. D., & Weckwerth, W. (2010). Metabolomics unravel contrasting effects of biodiversity on the performance of individual plant species. *PLoS One*, *5*, e12569.

Schrimpe-Rutledge, A. C., Sherrod, S. D., & McLean, J. A. (2018). Improving the discovery of secondary metabolite natural products using ion mobility–mass spectrometry. *Current Opinion in Chemical Biology*, *42*, 160–166.

Schripsema, J. (2010). Application of NMR in plant metabolomics: techniques, problems and prospects. *Phytochemical Analysis*, *21*, 14–21.

Schuhmann, K., Thomas, H., Ackerman, J. M., Nagornov, K. O., Tsybin, Y. O., & Shevchenko, A. (2017). Intensity-independent noise filtering in FT MS and FT MS/MS spectra for shotgun lipidomics. *Analytical Chemistry*, *89*(13), 7046–7052.

Schweiger, R., Baier, M. C., Persicke, M., & Müller, C. (2014). High specificity in plant leaf metabolic responses to arbuscular mycorrhiza. *Nature Communications*, *5*.

Sedio, B. E., Rojas Echeverri, J. C., Boya, P., Cristopher, A., & Wright, S. J. (2017). Sources of variation in foliar secondary chemistry in a tropical forest tree community. *Ecology*, *98*, 616–623.

Seger, C., & Sturm, S. (2007). Analytical aspects of plant metabolite profiling platforms: current standings and future aims. *Journal of Proteome Research*, *6*(2), 480–497.

Sévin, D. C., Fuhrer, T., Zamboni, N., & Sauer, U. (2016). Nontargeted in vitro metabolomics for high-throughput identification of novel enzymes in *Escherichia coli*. *Nature Methods*, *14*(2), 187.

Shahaf, N., Rogachev, I., Heinig, U., Meir, S., Malitsky, S., Battat, M., et al. (2016). The WEIZMASS spectral library for high-confidence metabolite identification. *Nature Communications*, *7*, 12423.

Sheth, B., & Thaker, V. (2014). Plant systems biology: insights, advances and challenges. *Planta*, *240*, 33–54. https://doi.org/10.1007/s00425-014-2059-5.

Shinbo, Y., Nakamura, Y., Altaf-Ul-Amin, M., Asahi, H., Kurokawa, K., Arita, M., et al. (2006). KNApSAcK: a comprehensive species-metabolite relationship database. In *Plant metabolomics* (pp. 165–181). Berlin, Heidelberg: Springer.

Sigurdsson, M. I., Jamshidi, N., Steingrimsson, E., Thiele, I., & Palsson, B. Ø. (2010). A detailed genome-wide reconstruction of mouse metabolism based on human Recon 1. *BMC Systems Biology*, *4*(1), 140.

Silverstein, R. M., Webster, F. X., Kiemle, D. J., & Bryce, D. L. (2014). *Spectrometric identification of organic compounds*. John Wiley.

Simmler, C., Napolitano, J. G., McAlpine, J. B., Chen, S. N., & Pauli, G. F. (2014). Universal quantitative NMR analysis of complex natural samples. *Current Opinion in Biotechnology*, *25*, 51–59.

Simpson, A. J., Simpson, M. J., & Soong, R. (2012). Nuclear magnetic resonance spectroscopy and its key role in environmental research. *Environmental Science and Technology*, *46* (21), 11488–11496.

Skoneczny, D., Weston, P., Zhu, X., Gurr, G., Callaway, R., Barrow, R., et al. (2017). Metabolic profiling and identification of shikonins in root periderm of two invasive Echium spp. weeds in Australia. *Molecules*, *22*, 330.

Smith, C. A., O'Maille, G., Want, E. J., Qin, C., Trauger, S. A., Brandon, T. R., et al. (2005). METLIN: a metabolite mass spectral database. *Therapeutic Drug Monitoring*, *27* (6), 747–751.

Smith, C. A., Want, E. J., O'Maille, G., Abagyan, R., & Siuzdak, G. (2006). XCMS: processing mass spectrometry data for metabolite profiling using nonlinear peak alignment, matching, and identification. *Analytical Chemistry*, *78*, 779–787.

Smith, R., Ventura, D., & Prince, J. T. (2015). LC-MS alignment in theory and practice: a comprehensive algorithmic review. *Briefings in Bioinformatics*, *16*, 104–117.

Smith, Z. D., Sindhu, C., & Meissner, A. (2016). Molecular features of cellular reprogramming and development. *Nature Reviews Molecular Cell Biology*, *17*(3), 139.

Smolinska, A., Blanchet, L., Buydens, L. M., & Wijmenga, S. S. (2012). NMR and pattern recognition methods in metabolomics: from data acquisition to biomarker discovery: a review. *Analytica Chimica Acta*, *750*, 82–97.

Sneddon, T. P., Si Zhe, X., Edmunds, S. C., Li, P., Goodman, L., & Hunter, C. I. (2014). GigaDB: promoting data dissemination and reproducibility. *Database*, *2014*, bau018.

Song, X. W., Dong, Z. Y., Long, X. Y., Li, S. F., Zuo, X. N., Zhu, C. Z., et al. (2011). REST: a toolkit for resting-state functional magnetic resonance imaging data processing. *PLoS One*, *6*(9)e25031.

Song, Y., Westerhuis, J. A., Aben, N., Michaut, M., Wessels, L. F., & Smilde, A. K. (2017). Principal component analysis of binary genomics data. *Briefings in Bioinformatics*. https://doi.org/10.1093/bib/bbx119.

Soranno, P. A., Bissell, E. G., Cheruvelil, K. S., Christel, S. T., Collins, S. M., Fergus, C. E., et al. (2015). Building a multi-scaled geospatial temporal ecology database from disparate data sources: fostering open science and data reuse. *GigaScience*, *4*.

Sousa, S. A. A., Magalhães, A., & Ferreira, M. M. C. (2013). Optimized bucketing for NMR spectra: three case studies. *Chemometrics and Intelligent Laboratory Systems*, *122*, 93–102.

Spicer, J., & Gaston, K. (2009). *Physiological diversity: ecological implications*. John Wiley.

Spicer, R., Salek, R. M., Moreno, P., Cañueto, D., & Steinbeck, C. (2017). Navigating freely-available software tools for metabolomics analysis. *Metabolomics*, *13*.

Stahl, E., Bellwon, P., Huber, S., Schlaeppi, K., Bernsdorff, F., Vallat-Michel, A., et al. (2016). Regulatory and functional aspects of indolic metabolism in plant systemic acquired resistance. *Molecular Plant*, *9*(5), 662–681.

Stein, R. R., Marks, D. S., & Sander, C. (2015). Inferring pairwise interactions from biological data using maximum-entropy probability models. *PLoS Computational Biology*, *11*(7)e1004182.

Steuer, R., Gross, T., Selbig, J., & Blasius, B. (2006). Structural kinetic modeling of metabolic networks. *Proceedings of the National Academy of Sciences of the United States of America*, *103*, 11868–11873. https://doi.org/10.1073/pnas.0600013103.

Stobbe, M. D., Swertz, M. A., Thiele, I., Rengaw, T., Van Kampen, A. H., & Moerland, P. D. (2013). Consensus and conflict cards for metabolic pathway databases. *BMC Systems Biology*, 7(1), 50.

Strogatz, S. H. (1994). *Nonlinear dynamics and chaos with applications to physics, biology, chemistry, and engineering*. Reading, MA: Addison-Wesley.

Sue, T., Obolonkin, V., Griffiths, H., & Villas-Bôas, S. G. (2011). An exometabolomics approach to monitoring microbial contamination in microalgal fermentation processes by using metabolic footprint analysis. *Applied and Environmental Microbiology*, 77, 7605–7610.

Sugimoto, M., Kawakami, M., Robert, M., Soga, T., & Tomita, M. (2012). Bioinformatics tools for mass spectroscopy-based metabolomic data processing and analysis. *Current Bioinformatics*, 7, 96–108. https://doi.org/10.2174/157489312799304431.

Sumner, L. W., Lei, Z., Nikolau, B. J., & Saito, K. (2015). Modern plant metabolomics: advanced natural product gene discoveries, improved technologies, and future prospects. *Natural Product Reports*, *32*(2), 212–229.

Sun, C. X., Gao, X. X., Li, M. Q., Fu, J. Q., & Zhang, Y. L. (2016). Plastic responses in the metabolome and functional traits of maize plants to temperature variations. *Plant Biology*, *18*, 249–261.

Sun, X., & Weckwerth, W. (2012). COVAIN: a toolbox for uni- and multivariate statistics, time-series and correlation network analysis and inverse estimation of the differential Jacobian from metabolomics covariance data. *Metabolomics*, *8*, S81–S93. https://doi.org/10.1007/s11306-012-0399-3.

Swan, M. (2013). The quantified self: fundamental disruption in big data science and biological discovery. *Big Data*, *1*(2), 85–99.

Swoger, B. J., & Helms, E. (2015). An organic chemistry exercise in information literacy using scifinder. *Journal of Chemical Education*, *92*(4), 668–671.

Szecowka, M., Heise, R., Tohge, T., Nunes-Nesi, A., Vosloh, D., Huege, J., et al. (2013). Metabolic fluxes in an illuminated *Arabidopsis rosette*. *The Plant Cell*, *25*(2), 694–714. https://doi.org/10.1105/tpc.112.106989.

Szymańska, E., Saccenti, E., Smilde, A. K., & Westerhuis, J. A. (2012). Double-check: validation of diagnostic statistics for PLS-DA models in metabolomics studies. *Metabolomics*, *8*(1), 3–16.

Tan, K. C., Ipcho, S. V., Trengove, R. D., Oliver, R. P., & Solomon, P. S. (2009). Assessing the impact of transcriptomics, proteomics and metabolomics on fungal phytopathology. *Molecular Plant Pathology*, *10*(5), 703–715.

Tarakhovskaya, E., Lemesheva, V., Bilova, T., & Birkemeyer, C. (2017). Early embryogenesis of brown alga Fucus vesiculosus L. is characterized by significant changes in carbon and energy metabolism. *Molecules*, *22*, 1509.

Tautenhahn, R., Bottcher, C., & Neumann, S. (2008). Highly sensitive feature detection for high resolution LC/MS. *BMC Bioinformatics*, *9*, 504.

Tenenboim, H., & Brotman, Y. (2016). Omic relief for the biotically stressed: metabolomics of plant biotic interactions. *Trends in Plant Science*, *21*(9), 781–791.

Tewes, L. J., Michling, F., Koch, M. A., & Müller, C. (2018). Intracontinental plant invader shows matching genetic and chemical profiles and might benefit from high defence variation within populations. *Journal of Ecology*, *106*, 714–726.

Theodoridis, G. A., Gika, H. G., Want, E. J., & Wilson, I. D. (2012). Liquid chromatography–mass spectrometry based global metabolite profiling: a review. *Analytica Chimica Acta*, *711*, 7–16.

Thiele, I., Swainston, N., Fleming, R. M., Hoppe, A., Sahoo, S., Aurich, M. K., et al. (2013). A community-driven global reconstruction of human metabolism. *Nature Biotechnology*, *31*(5), 419.

Tietenberg, T. H., & Lewis, L. (2016). *Environmental and natural resource economics*. Routledge.
Todeschini, R., & Consonni, V. (2009). *Molecular descriptors for chemoinformatics: volume I: alphabetical listing/volume II: appendices, references*. (Vol. 41). John Wiley.
Toubiana, D., Semel, Y., Tohge, T., Beleggia, R., Cattivelli, L., Rosental, L., et al. (2012). Metabolic profiling of a mapping population exposes new insights in the regulation of seed metabolism and seed, fruit, and plant relations. *PLoS Genetics*, *8*(3)e1002612.
Trogisch, S., Schuldt, A., Bauhus, J., Blum, J. A., Both, S., Buscot, F., et al. (2017). Toward a methodical framework for comprehensively assessing forest multifunctionality. *Ecology and Evolution*, 7, 10652–10674.
Trutschel, D., Schmidt, S., Grosse, I., & Neumann, S. (2015). Experiment design beyond gut feeling: statistical tests and power to detect differential metabolites in mass spectrometry data. *Metabolomics*, *11*, 851–860.
Tufféry, S. (2011). *Data mining and statistics for decision making*. (Vol. 2). Chichester: Wiley.
Tung, C. W. (2014). Public databases of plant natural products for computational drug discovery. *Current Computer-Aided Drug Design*, *10*(3), 191–196.
Valares Masa, C., Alías Gallego, J., Chaves Lobón, N., & Sosa Díaz, T. (2016). Intra-population variation of secondary metabolites in Cistus ladanifer L. *Molecules*, *21*, 945.
Valenta, K., Burke, R. J., Styler, S. A., Jackson, D. A., Melin, A. D., & Lehman, S. M. (2013). Colour and odour drive fruit selection and seed dispersal by mouse lemurs. *Scientific Reports*, *3*, .
Vallino, J. J. (2010). Ecosystem biogeochemistry considered as a distributed metabolic network ordered by maximum entropy production. *Philosophical Transactions of the Royal Society, B: Biological Sciences*, *365*(1545), 1417–1427.
Van Aggelen, G., Ankley, G. T., Baldwin, W. S., Bearden, D. W., Benson, W. H., Chipman, J. K., et al. (2010). Integrating omic technologies into aquatic ecological risk assessment and environmental monitoring: hurdles, achievements, and future outlook. *Environmental Health Perspectives*, *118*(1), 1.
Van Agthoven, M. A., Delsuc, M. A., Bodenhausen, G., & Rolando, C. (2013). Towards analytically useful two-dimensional Fourier transform ion cyclotron resonance mass spectrometry. *Analytical and Bioanalytical Chemistry*, *405*, 51–61. https://doi.org/10.1007/s00216-012-6422-8.
van Dam, N. M., & van der Meijden, E. (2011). A role for metabolomics in plant ecology. In R. D. Hall (Ed.), *Annual Plant Reviews, Biology of Plant Metabolomics* (pp. 87–108). Oxford, UK: Blackwell Publishing Ltd.
van Dam, N. M., & van der Meijden, E. (2018). A role for metabolomics in plant ecology. R. D. Hall (Ed.), *Annual plant reviews* (pp. 87–107). Vol. 43 (pp. 87–107). Oxford, UK: Wiley-Blackwell. ISBN 978-1-4443-3995-6.
Van den Berg, R. A., Hoefsloot, H. C., Westerhuis, J. A., Smilde, A. K., & van der Werf, M. J. (2006). Centering, scaling, and transformations: Improving the biological information content of metabolomics data. *BMC Genomics*, 7, 142.
van der Kooy, F., Maltese, F., Choi, Y. H., Kim, H. K., & Verpoorte, R. (2009). Quality control of herbal material and phytopharmaceuticals with MS and NMR based metabolic fingerprinting. *Planta Medica*, *75*(07), 763–775.
van Iersel, M. P., Kelder, T., Pico, A. R., Hanspers, K., Coort, S., Conklin, B. R., et al. (2008). Presenting and exploring biological pathways with PathVisio. *BMC Bioinformatics*, *9*(1), 399.
Van Meulebroek, L., Hanssens, J., Steppe, K., & Vanhaecke, L. (2016). Metabolic fingerprinting to assess the impact of salinity on carotenoid content in developing tomato fruits. *International Journal of Molecular Sciences*, *17*, 821.
Van Veld, P. A., Sanders, B. M., Fowler, B. A., Lars, F., Di Giulio, R. T., Marius, B., et al. (2018). Molecular responses to environmental contamination: enzyme and protein systems as indicators of chemical exposure and effect. In *Biomarkers* (pp. 235–336). CRC Press.

Vandamme, D., Fitzmaurice, W., Kholodenko, B., & Kolch, W. (2013). Systems medicine: helping us understand the complexity of disease. *QJM*, *106*, 891–895. https://doi.org/10.1093/qjmed/hct163.

Vaniya, A., & Fiehn, O. (2015). Using fragmentation trees and mass spectral trees for identifying unknown compounds in metabolomics. *TrAC Trends in Analytical Chemistry*, *69*, 52–61.

Varmuza, K., & Filzmoser, P. (2016). *Introduction to multivariate statistical analysis in chemometrics*. CRC Press.

Veley, K. M., Berry, J. C., Fentress, S. J., Schachtman, D. P., Baxter, I., & Bart, R. (2017). High-throughput profiling and analysis of plant responses over time to abiotic stress. *Plant Direct*, *1*(4)e00023.

Viant, M. R. (2009). Applications of metabolomics to the environmental sciences. *Metabolomics*, *5*, 1–2. https://doi.org/10.1007/s11306-009-0157-3.

Viant, M. R., & Sommer, U. (2013). Mass spectrometry based environmental metabolomics: a primer and review. *Metabolomics*, *9*, 144–158. https://doi.org/10.1007/s11306-012-0412-x.

Viant, M. R. (2008). Recent developments in environmental metabolomics. *Molecular BioSystems*, *4*(10), 980–986.

Vinaixa, M., Schymanski, E. L., Neumann, S., Navarro, M., Salek, R. M., & Yanes, O. (2016). Mass spectral databases for LC/MS-and GC/MS-based metabolomics: state of the field and future prospects. *TrAC Trends in Analytical Chemistry*, *78*, 23–35.

Vinaixa, M., Samino, S., Saez, I., Duran, J., Guinovart, J. J., & Yanes, O. (2012). A guideline to univariate statistical analysis for LC/MS-based untargeted metabolomics-derived data. *Metabolites*, *2*, 775–795.

Vinciotti, V., Augugliaro, L., Abbruzzo, A., & Wit, E. C. (2016). Model selection for factorial Gaussian graphical models with an application to dynamic regulatory networks. *Statistical Applications in Genetics and Molecular Biology*, *15*(3), 193–212.

Walker, S. H., Budhathoki-Uprety, J., Novak, B. M., & Muddiman, D. C. (2011). Stable-isotope labeled hydrophobic hydrazide reagents for the relative quantification of N-linked glycans by electrospray ionization mass spectrometry. *Analytical Chemistry*, *83*(17), 6738–6745.

Wang, Y., & Bryant, S. H. (2014). *NCBI PubChem BioAssay database*.

Wang, M., Carver, J. J., Phelan, V. V., Sanchez, L. M., Garg, N., Peng, Y., et al. (2016). Sharing and community curation of mass spectrometry data with global natural products social molecular networking. *Nature Biotechnology*, *34*, 828–837.

Wang, Y., Cui, T., Zhang, C., Yang, M., Huang, Y., Li, W., et al. (2010). Global protein—-protein interaction network in the human pathogen *Mycobacterium tuberculosis* H37Rv. *Journal of Proteome Research*, *9*(12), 6665–6677.

Wanichthanarak, K., Fan, S., Grapov, D., Barupal, D. K., & Fiehn, O. (2017). Metabox: a toolbox for metabolomic data analysis, interpretation and integrative exploration. *PLoS One*. *12*(1)e0171046https://doi.org/10.1371/journal. pone.0171046.

Waring, R. H., & Running, S. W. (2010). *Forest ecosystems: analysis at multiple scales*. Elsevier.

Warton, D. I., Wright, S. T., & Wang, Y. (2012). Distance-based multivariate analyses confound location and dispersion effects: mean-variance confounding in multivariate analysis. *Methods in Ecology and Evolution*, *3*, 89–101.

Watson, E., Yilmaz, L. S., & Walhout, A. J. (2015). Understanding metabolic regulation at a systems level: metabolite sensing, mathematical predictions, and model organisms. *Annual Review of Genetics*, *49*, 553–575.

Weber, R., Selander, E., Sommer, U., & Viant, M. (2013). A stable-isotope mass spectrometry-based metabolic footprinting approach to analyze exudates from phytoplankton. *Marine Drugs*, *11*, 4158–4175.

Weckwerth, W. (2008). Integration of metabolomics and proteomics in molecular plant physiology-coping with the complexity by data-dimensionality reduction. *Physiologia Plantarum*, *132*, 176–189. https://doi.org/10.1111/j.1399-3054.2007.01011.x.

Weckwerth, W. (2011a). Unpredictability of metabolism—the key role of metabolomics science in combination with next-generation genome sequencing. *Analytical and Bioanalytical Chemistry*, *400*(7), 1967.

Weckwerth, W. (2011b). Green systems biology—From single genomes, proteomes and metabolomes to ecosystems research and biotechnology. *Journal of Proteomics*, *75*, 284–305.

Wehrens, R., Hageman, J. A., van Eeuwijk, F., Kooke, R., Flood, P. J., Wijnker, E., et al. (2016). Improved batch correction in untargeted MS-based metabolomics. *Metabolomics*, *12*(5), 88.

Wei, Z., Yang, T., Friman, V. -P., Xu, Y., Shen, Q., & Jousset, A. (2015). Trophic network architecture of root-associated bacterial communities determines pathogen invasion and plant health. *Nature Communications*, *6*.

Weisser, H., Nahnsen, S., Grossmann, J., Nilse, L., Quandt, A., Brauer, H., et al. (2013). An automated pipeline for high-throughput label-free quantitative proteomics. *Journal of Proteome Research*, *12*, 1628–1644.

Welzenbach, J., Neuhoff, C., Looft, C., Schellander, K., Tholen, E., & Große-Brinkhaus, C. (2016). Different statistical approaches to investigate porcine muscle metabolome profiles to highlight new biomarkers for pork quality assessment. *PLoS One*, *11*(2)e0149758.

Widarto, H. T., Van Der Meijden, E., Lefeber, A. W., Erkelens, C., Kim, H. K., Choi, Y. H., et al. (2006). Metabolomic differentiation of *Brassica rapa* following herbivory by different insect instars using two-dimensional nuclear magnetic resonance spectroscopy. *Journal of Chemical Ecology*, *32*(11), 2417–2428.

Wiggins, N. L., Forrister, D. L., Endara, M. -J., Coley, P. D., & Kursar, T. A. (2016). Quantitative and qualitative shifts in defensive metabolites define chemical defense investment during leaf development in Inga, a genus of tropical trees. *Ecology and Evolution*, *6*, 478–492.

Wijaya, S. H., Husnawati, H., Afendi, F. M., Batubara, I., Darusman, L. K., Altaf-Ul-Amin, M., et al. (2014). Supervised clustering based on DPClusO: prediction of plant-disease relations using Jamu formulas of KNApSAcK database. *BioMed Research International*. *2014*, 831751https://doi.org/10.1155/2014/831751.

Wilkinson, M. D., Dumontier, M., Aalbersberg, I. J., Appleton, G., Axton, M., Baak, A., et al. (2016). The FAIR guiding principles for scientific data management and stewardship. *Scientific Data*, *3*, 160018.

Williams, T. D., Davies, I. M., Wu, H., Diab, A. M., Webster, L., Viant, M. R., et al. (2014). Molecular responses of European flounder (*Platichthys flesus*) chronically exposed to contaminated estuarine sediments. *Chemosphere*, *108*, 152–158.

Wishart, D. S. (2009). Computational strategies for metabolite identification in metabolomics. *Bioanalysis*, *1*, 1579–1596.

Wolf, S., Schmidt, S., Müller-Hannemann, M., & Neumann, S. (2010). In silico fragmentation for computer assisted identification of metabolite mass spectra. *BMC Bioinformatics*, *11*(1), 148.

Wolfender, J. L., Marti, G., Thomas, A., & Bertrand, S. (2015). Current approaches and challenges for the metabolite profiling of complex natural extracts. *Journal of Chromatography A*, *1382*, 136–164.

Woodward, G., Bonada, N., Brown, L. E., Death, R. G., Durance, I., Gray, C., et al. (2016). The effects of climatic fluctuations and extreme events on running water ecosystems. *Philosophical Transactions of the Royal Society of London B: Biological Sciences*, *371*(1694) 20150274.

Wu, H., Li, X., Yan, X., An, L., Luo, K., Shao, M., et al. (2015). An untargeted metabolomics-driven approach based on LC–TOF/MS and LC–MS/MS for the screening of xenobiotics and metabolites of Zhi-Zi-Da-Huang decoction in rat plasma. *Journal of Pharmaceutical and Biomedical Analysis, 115*, 315–322.

Xanthopoulos, P., Pardalos, P. M., & Trafalis, T. B. (2013). Principal component analysis. In *Robust data mining* (pp. 21–26). New York, NY: Springer.

Xia, J., & Wishart, D. S. (2011). Metabolomic data processing, analysis, and interpretation using MetaboAnalyst. *Current Protocols in Bioinformatics, 34*(1), 14.10.1–14.10.48.

Xia, J., Psychogios, N., Young, N., & Wishart, D. S. (2009). MetaboAnalyst: a web server for metabolomic data analysis and interpretation. *Nucleic Acids Research, 37*(suppl_2), W652–W660.

Xiao, J. F., Zhou, B., & Ressom, H. W. (2012). Metabolite identification and quantitation in LC-MS/MS-based metabolomics. *TrAC Trends in Analytical Chemistry, 32*, 1–14.

Yao, F., Coquery, J., & Lê Cao, K. A. (2012). Independent principal component analysis for biologically meaningful dimension reduction of large biological data sets. *BMC Bioinformatics, 13*(1), 24.

Yi, L., Dong, N., Yun, Y., Deng, B., Ren, D., Liu, S., et al. (2016). Chemometric methods in data processing of mass spectrometry-based metabolomics: a review. *Analytica Chimica Acta, 914*, 17–34.

Yonekura-Sakakibara, K., Fukushima, A., Nakabayashi, R., Hanada, K., Matsuda, F., Sugawara, S., et al. (2012). Two glycosyltransferases involved in anthocyanin modification delineated by transcriptome independent component analysis in *Arabidopsis thaliana*. *The Plant Journal, 69*(1), 154–167.

Zampieri, M., Sekar, K., Zamboni, N., & Sauer, U. (2017). Frontiers of high-throughput metabolomics. *Current Opinion in Chemical Biology, 36*, 15–23.

Zeng, X., Zhang, P., He, W., Qin, C., Chen, S., Tao, L., et al. (2018). NPASS: natural product activity and species source database for natural product research, discovery and tool development. *Nucleic Acids Research, 46*, D1217–D1222.

Zhang, A., Sun, H., Wang, P., Han, Y., & Wang, X. (2012). Modern analytical techniques in metabolomics analysis. *Analyst, 137*(2), 293–300.

Zhang, L., Wang, X., Guo, J., Xia, Q., Zhao, G., Zhou, H., et al. (2013). Metabolic profiling of Chinese tobacco leaf of different geographical origins by GC-MS. *Journal of Agricultural and Food Chemistry, 61*(11), 2597–2605.

Zhang, P., Dreher, K., Karthikeyan, A., Chi, A., Pujar, A., Caspi, R., et al. (2010). Creation of a genome-wide metabolic pathway database for *Populus trichocarpa* using a new approach for reconstruction and curation of metabolic pathways for plants. *Plant Physiology, 153*(4), 1479–1491.

Zhou, J., Liu, H., Liu, Y., Liu, J., Zhao, X., & Yin, Y. (2016). Development and evaluation of a parallel reaction monitoring strategy for large-scale targeted metabolomics quantification. *Analytical Chemistry, 88*(8), 4478–4486.

Zhu, J., Sova, P., Xu, Q., Dombek, K. M., Xu, E. Y., Vu, H., et al. (2012). Stitching together multiple data dimensions reveals interacting metabolomic and transcriptomic networks that modulate cell regulation. *PLoS Biology, 10*(4)e1001301.

Zuppinger-Dingley, D., Flynn, D. F. B., Brandl, H., & Schmid, B. (2015). Selection in monoculture vs. mixture alters plant metabolic fingerprints. *Journal of Plant Ecology, 8*, 549–557.

FURTHER READING

Abd Rahman, S. (2015). *A systems biology approach towards child obesity and obesity-related diseases: integration of clinical, genetic, hormonal and NMR metabonomic factor*. PhD ThesisInstitute for Molecular Bioscience, The University of Queensland.

Cakır, T., Hendriks, M. M., Westerhuis, J. A., & Smilde, A. K. (2009). Metabolic network discovery through reverse engineering of metabolome data. *Metabolomics*, *5*(3), 318–329.

Dziekonski, E. T., Johnson, J. T., Lee, K. W., & McLuckey, S. A. (2018). Determination of collision cross sections using a Fourier transform electrostatic linear ion trap mass spectrometer. *Journal of the American Society for Mass Spectrometry*, *29*(2), 242–250.

Foygel, R., & Drton, M. (2010). Extended Bayesian information criteria for Gaussian graphical models. In *Advances in neural information processing systems* (pp. 604–612).

Fuhrer, T., & Zamboni, N. (2015). High-throughput discovery metabolomics. *Current Opinion in Biotechnology*, *31*, 73–78.

Guo, W., & Romano, J. (2007). A generalized Sidak-Holm procedure and control of generalized error rates under independence. *Statistical Applications in Genetics and Molecular Biology*, *6*, .

Helyer, N., Cattlin, N. D., & Brown, K. C. (2014). *Biological control in plant protection: a colour handbook*. CRC Press.

Højsgaard, S., Edwards, D., & Lauritzen, S. (2012). Gaussian graphical models. In *Graphical models with R* (pp. 77–116). Boston, MA: Springer.

Kind, T., Scholz, M., & Fiehn, O. (2009). How large is the metabolome? A critical analysis of data exchange practices in chemistry. *PLoS One*, *4*(5)e5440.

Kroymann, J. (2011). Natural diversity and adaptation in plant secondary metabolism. *Current Opinion in Plant Biology*, *14*(3), 246–251.

Mohamed, A., Nguyen, C. H., & Mamitsuka, H. (2015). Current status and prospects of computational resources for natural product dereplication: a review. *Briefings in Bioinformatics*, *17*(2), 309–321.

Petrarca, M. H., Ccanccapa-Cartagena, A., Masiá, A., Godoy, H. T., & Picó, Y. (2017). Comparison of green sample preparation techniques in the analysis of pyrethrins and pyrethroids in baby food by liquid chromatography–tandem mass spectrometry. *Journal of Chromatography A*, *1497*, 28–37.

Pleil, J. D., & Isaacs, K. K. (2016). High-resolution mass spectrometry: basic principles for using exact mass and mass defect for discovery analysis of organic molecules in blood, breath, urine and environmental media. *Journal of Breath Research*, *10*(1)012001.

Schymanski, S. J., Kleidon, A., Stieglitz, M., & Narula, J. (2010). Maximum entropy production allows a simple representation of heterogeneity in semiarid ecosystems. *Philosophical Transactions of the Royal Society of London B: Biological Sciences*, *365*(1545), 1449–1455.

Sedio, B. E. (2017). Recent breakthroughs in metabolomics promise to reveal the cryptic chemical traits that mediate plant community composition, character evolution and lineage diversification. *The New Phytologist*, *214*, 952–958.

Wang, X., Dalkic, E., Wu, M., & Chan, C. (2008). Gene module level analysis: identification to networks and dynamics. *Current Opinion in Biotechnology*, *19*(5), 482–491.

INDEX

Note: Page numbers followed by *t* indicate tables.

FSC
www.fsc.org
MIX
Papier | Fördert
gute Waldnutzung
FSC® C083411

Kolibri 360 GmbH
Ferdinand-Jühlke-Straße 7
99095 Erfurt, Deutschland
produktsicherheit@kolibri360.de